Advances in Geographical and Environmental Sciences

Advances in Geographical and Environmental Sciences synthesizes series diagnostigation and prognostication of earth environment, incorporating challenging interactive areas within ecological envelope of geosphere, biosphere, hydrosphere, atmosphere and cryosphere. It deals with land use land cover change (LUCC), urbanization, energy flux, land-ocean fluxes, climate, food security, ecohydrology, biodiversity, natural hazards and disasters, human health and their mutual interaction and feedback mechanism in order to contribute towards sustainable future. The geosciences methods range from traditional field techniques and conventional data collection, use of remote sensing and geographical information system, computer aided technique to advance geostatistical and dynamic modeling.

The series integrate past, present and future of geospheric attributes incorporating biophysical and human dimensions in spatio-temporal perspectives. The geosciences, encompassing land-ocean-atmosphere interaction is considered as a vital component in the context of environmental issues, especially in observation and prediction of air and water pollution, global warming and urban heat islands. It is important to communicate the advances in geosciences to increase resilience of society through capacity building for mitigating the impact of natural hazards and disasters. Sustainability of human society depends strongly on the earth environment, and thus the development of geosciences is critical for a better understanding of our living environment, and its sustainable development.

Geoscience also has the responsibility to not confine itself to addressing current problems but it is also developing a framework to address future issues. In order to build a 'Future Earth Model' for understanding and predicting the functioning of the whole climatic system, collaboration of experts in the traditional earth disciplines as well as in ecology, information technology, instrumentation and complex system is essential, through initiatives from human geoscientists. Thus human geoscience is emerging as key policy science for contributing towards sustainability/survivality science together with future earth initiative.

Advances in Geographical and Environmental Sciences series publishes books that contain novel approaches in tackling issues of human geoscience in its broadest sense-books in the series should focus on true progress in a particular area or region. The series includes monographs and edited volumes without any limitations in the page numbers.

All submitted proposals are carefully reviewed by the series editor.

Publication Ethics and Publication Malpractice Statement: https://www.springernature.com/gp/policies/book-publishing-policies

Ethical, Conflict-of-Interest, and Editorial Policy Statement: https://www.springer.com/us/authors-editors/journal-author/journal-author-helpdesk/before-you-start/before-you-start/1330#c14214

Rayees Ahmed · Shahid Saleem · Pervez Ahmed
Editors

Climate Change, Water Resources, and Natural Hazards in the Himalayas

Springer

Editors
Rayees Ahmed
Divecha Centre for Climate Change
Indian Institute of Science Bangalore
Bengaluru, Karnataka, India

Shahid Saleem
Department of Geography and Disaster Management
University of Kashmir
Srinagar, Jammu and Kashmir, India

Pervez Ahmed
Department of Geography and Disaster Management
University of Kashmir
Srinagar, Jammu and Kashmir, India

ISSN 2198-3542 ISSN 2198-3550 (electronic)
Advances in Geographical and Environmental Sciences
ISBN 978-981-92-0834-0 ISBN 978-981-92-0835-7 (eBook)
https://doi.org/10.1007/978-981-92-0835-7

This Springer imprint is published by the registered company Springer Nature Singapore Pte Ltd.
The registered company address is: 152 Beach Road, #21-01/04 Gateway East, Singapore 189721, Singapore

This book is dedicated to the fragile yet resilient Himalayas and to the millions of people whose lives, livelihoods, cultures, and identities are inseparably intertwined with these mountains. It is also dedicated to the mountain communities—pastoralists, farmers, forest dwellers, pilgrims, and indigenous peoples—who stand on the frontlines of climate change, water scarcity, and natural hazards, often contributing least to the crisis yet bearing its greatest burdens. We further dedicate this volume to researchers, practitioners, and policymakers who work tirelessly to understand, protect, and sustain Himalayan ecosystems and to the next generation of scholars and students who will carry forward the responsibility of safeguarding mountain environments in an era of unprecedented environmental change. May this work contribute, in however modest a way, to fostering knowledge, resilience, and stewardship for the Himalayas and other high-mountain regions of the world.

Preface

The genesis of this volume lies in a growing recognition that the Himalayas—often referred to as the "Third Pole"—are undergoing profound and accelerating transformations under a changing climate. Over the past two decades, advances in remote sensing, geospatial modelling, field-based observations, and socio-environmental research have dramatically improved our understanding of Himalayan cryospheric dynamics, hydrological systems, and natural hazards. Yet, despite this progress, scientific knowledge remains fragmented across disciplinary silos, spatial scales, and thematic domains. This book emerges from the need to bring these diverse strands together into a coherent, integrated framework that captures the coupled nature of climate change, water resources, natural hazards, and human systems in the Himalayan region. The Himalayas are not only a global climate regulator and a critical freshwater reservoir for nearly two billion people downstream, but also one of the most hazard-prone mountain systems on Earth. Rapid glacier retreat, the formation and expansion of glacial lakes, destabilization of debris-covered glaciers, changing precipitation regimes, groundwater depletion, increasing landslide frequency, and heightened seismic risks are collectively reshaping both landscapes and livelihoods. These processes do not operate in isolation. Rather, they interact across scales—from cryospheric processes at high elevations to river basin hydrology, ecosystems, infrastructure, and socio-cultural systems in downstream valleys and plains.

This edited volume was conceived to address this complexity by adopting an explicitly interdisciplinary perspective. The chapters bring together physical geography, geomorphology, hydrology, cryospheric science, geoinformatics, ecology, disaster risk science, social sciences, and governance studies. By doing so, the book seeks to move beyond single-hazard or single-discipline narratives and instead emphasize systemic linkages between environmental change and societal vulnerability in the Himalayas. The early chapters focus on the cryosphere and its evolving dynamics. Detailed syntheses of glacial lake formation, distribution, and criticality across the Himalayan arc highlight the growing risk of glacial lake outburst floods (GLOFs) under continued warming. Reviews of debris-covered glacier mapping techniques demonstrate both the progress made through remote sensing and the persistent challenges posed by surface heterogeneity. These cryospheric assessments are complemented by analyses of long-term glacier–climate interactions and their implications for water availability and hazard cascades.

Subsequent chapters examine geomorphic and tectonic controls on landscape evolution, particularly through river response to differential uplift and neotectonics. These studies underscore how tectonics, climate, and surface processes jointly shape Himalayan topography and influence hazard susceptibility. Basin-scale syntheses, such as those focused on the Jhelum and Indus River systems, further integrate cryosphere–hydrology interactions with land use change, groundwater dynamics, and demographic pressures, offering a comprehensive picture of water security challenges in a transboundary context. A central strength of this volume lies in its engagement with natural hazards and risk. Chapters addressing landslides, floods, seismic risk, tunnel safety, and GLOFs demonstrate how advances in geospatial technologies—such as GRACE satellite gravimetry, Google Earth Engine, and high-resolution DEM-based modelling—are transforming hazard assessment and monitoring. At the same time, these contributions emphasize that technical sophistication alone is insufficient without meaningful integration into planning, governance, and community-based risk reduction.

Equally important are the chapters that foreground ecological systems, cultural landscapes, and human dimensions. Analyses of Himalayan wetlands, forests, sacred landscapes, pastoral livelihoods, and human–wildlife conflict reveal how climate change and development pressures are reconfiguring socio-ecological relationships. These chapters challenge purely technocratic approaches to sustainability by highlighting indigenous knowledge, cultural resilience, environmental education, and participatory governance as essential components of adaptation and risk management. The final sections of the book explicitly engage with policy, governance, and sustainability pathways. By examining environmental education, institutional frameworks, and community-based governance models, the volume bridges science and practice, emphasizing that resilience in the Himalayas depends as much on social learning, equity, and inclusive decision-making as on scientific innovation.

Collectively, the chapters in this book reflect a shared conviction: that the Himalayan crisis is not a future possibility but a present reality. Climate change is already altering water resources, intensifying hazards, and reshaping human–environment interactions across the region. Addressing these challenges requires integrative science, cross-border cooperation, and sustained dialogue between researchers, policymakers, practitioners, and mountain communities.

This volume is intended for researchers, graduate students, planners, disaster risk professionals, and policymakers seeking a comprehensive and nuanced understanding of Himalayan environmental change. By synthesizing cutting-edge research with grounded case studies and policy-relevant insights, *Climate Change, Water Resources, and Natural Hazards in the Himalayas* aims to contribute to more informed, equitable, and resilient futures for one of the world's most critical mountain regions.

Bengaluru, India	Rayees Ahmed
Srinagar, India	Shahid Saleem
Srinagar, India	Pervez Ahmed

Acknowledgments

This edited volume, *Climate Change, Water Resources, and Natural Hazards in the Himalayas*, is the result of sustained collective effort, intellectual exchange, and institutional support spanning multiple disciplines, regions, and years. Bringing together such a wide range of perspectives—encompassing cryospheric science, hydrology, geomorphology, ecology, geoinformatics, disaster risk, governance, and human dimensions—would not have been possible without the generous contributions of many individuals and organizations.

First and foremost, we extend our sincere gratitude to all contributing authors for their commitment, patience, and scholarly rigor. Each chapter reflects not only deep disciplinary expertise but also a shared willingness to engage beyond conventional boundaries, which has been central to the interdisciplinary spirit of this volume. The authors' readiness to revise, refine, and align their contributions with the broader vision of the book has greatly strengthened its coherence and impact. We are deeply grateful to the reviewers—both formal and informal—whose constructive comments, critical insights, and careful reading significantly improved the quality, clarity, and scientific robustness of the chapters. Their engagement ensured that the volume meets high academic standards while remaining accessible to a broad readership that includes researchers, practitioners, and policymakers.

We acknowledge with appreciation the support of academic and research institutions that provided the intellectual environment, infrastructure, and encouragement necessary for this work. Universities, research centers, and laboratories across India and abroad have played a vital role in facilitating fieldwork, data access, computational resources, and interdisciplinary dialogue. In particular, we recognize institutions working in the Himalayan region whose long-term engagement with mountain environments has laid the foundation for much of the research synthesized in this book. Several chapters benefited from access to satellite data, geospatial platforms, and open-source datasets. We gratefully acknowledge data providers and agencies such as space and earth observation programs, hydrological and meteorological organizations, and open geospatial initiatives whose commitment to data accessibility has enabled cutting-edge research in data-scarce mountain regions. The use of platforms such as Google Earth Engine and GRACE-derived products highlights the importance of open science in advancing hazard assessment and climate research.

We also acknowledge the support of funding agencies and research programs that indirectly contributed to the studies presented in this volume. While individual chapters draw on diverse funding sources, the collective emphasis on climate change adaptation, disaster risk reduction, sustainability, and resilience reflects the broader priorities of national and international research frameworks aligned with the Sustainablc Development Goals and the Sendai Framework for Disaster Risk Reduction.

Special thanks are due to colleagues, mentors, and collaborators who provided intellectual guidance, encouragement, and critical feedback during the conceptualization of this volume. Informal discussions at conferences, workshops, field campaigns, and academic exchanges—often across disciplinary and national boundaries—played a crucial role in shaping the themes and structure of the book. Importantly, we acknowledge the mountain communities of the Himalayas, whose lived experiences, indigenous knowledge, and resilience are reflected—directly and indirectly—throughout this volume. Pastoralists, farmers, forest-dependent communities, and residents of hazard-prone valleys continue to bear the frontline impacts of climate change and environmental transformation. Their observations, narratives, and adaptive practices provide essential context for scientific inquiry and remind us that research in the Himalayas carries both ethical responsibility and societal relevance.

We are grateful to the editorial and production teams of the publisher for their professional support, guidance, and patience throughout the preparation of this volume. Their assistance during manuscript coordination, copyediting, and production has been invaluable in bringing this book to fruition. Finally, we acknowledge the support and understanding of our families and friends, who provided encouragement and balance during the long and often demanding process of compiling this volume. Their patience made it possible to dedicate the time and focus required for this collective scholarly effort.

It is our hope that this book will serve as a useful resource for advancing scientific understanding, informing policy, and strengthening resilience in the Himalayan region. Any remaining errors or omissions are solely the responsibility of the editors.

About This Book

The series editor(s) remains neutral with regard to jurisdictional claims in published maps published in this book.

Introduction

The Himalayan region occupies a critical position in the global Earth system. As the highest and one of the most geodynamically active mountain systems on the planet, the Himalayas exert a profound influence on regional climate, hydrology, ecosystems, and human societies across South and Central Asia. Often referred to as the "Third Pole" due to the vast extent of snow and ice stored in its cryosphere, the Himalayas sustain some of the world's largest river systems and provide freshwater to hundreds of millions of people downstream (Adger et al., 2005; Immerzeel et al., 2010). At the same time, the region is characterized by steep environmental gradients, fragile geomorphology, and complex socio-ecological systems that are highly sensitive to climatic and anthropogenic change. In recent decades, climate change has emerged as a dominant force reshaping Himalayan environments (Gorelick et al., 2017; Bolch et al., 2019). Observational and modeling studies consistently indicate that the Himalayas are warming at rates exceeding the global average, with pronounced elevation-dependent warming. These changes are altering cryospheric processes, accelerating glacier retreat, modifying snow accumulation and melt regimes, and increasing the formation and expansion of glacial lakes. Such transformations have cascading impacts on surface and subsurface hydrology, ecosystem functioning, and the frequency and magnitude of natural hazards. Floods, landslides, glacial lake outburst floods (GLOFs), and terrain instability are increasingly reported across the region, often with severe consequences for lives, livelihoods, and infrastructure (Xu et al., 2009). The challenge of addressing climate change impacts in the Himalayas is inherently interdisciplinary (Schwanghart et al., 2016). Changes in temperature and precipitation regimes interact with tectonic activity, land-use change, and socio-economic transitions, producing outcomes that cannot be adequately understood through single-discipline approaches (Portner et al., 2022). While many of the stresses manifest locally—such as drying springs, destabilized slopes, or shifting vegetation zones—their drivers often operate at regional or global scales, including atmospheric circulation changes, upstream–downstream hydrological linkages, and broader development trajectories. Conversely, local responses to climate stress, including land management practices, infrastructure development, and resource governance, can generate far-reaching consequences beyond the mountain region itself (Seto et al., 2012).

This edited volume, *Climate Change, Water Resources, and Natural Hazards in the Himalayas*, is motivated by the need to better understand these interconnected processes and to bridge persistent gaps between physical science, geospatial analysis, and the human dimensions of environmental change. The chapters assembled here collectively examine the Himalayas as a coupled human–earth system, emphasizing interactions between climate dynamics, cryospheric change, hydrology, geomorphology, ecosystems, land systems, and society. Rather than treating hazards, water resources, or livelihoods as isolated themes, the volume highlights their interdependence and the feedbacks that shape vulnerability and resilience in mountain environments.

Himalayan Cryosphere and Earth System Processes

The Himalayan cryosphere forms the foundation of regional water security and plays a central role in regulating seasonal and long-term hydrological flows. Glaciers, snowfields, permafrost, and high-altitude lakes collectively govern runoff dynamics in major river basins such as the Indus, Ganga, and Brahmaputra. However, these cryospheric components are undergoing rapid and uneven transformation. Glacier retreat, the expansion of debris-covered ice, and the proliferation of glacial lakes have introduced new uncertainties into water availability and hazard risk assessments. Understanding these changes requires the integration of cryospheric science with geomorphology and tectonics. River incision, differential uplift, and long-term landscape evolution influence sediment transport, valley morphology, and flood behaviour, shaping how climate signals propagate through Himalayan river systems. The chapters in the first part address these issues by examining cryospheric processes responsible for lake formation, advances in remote sensing of debris-covered glaciers, and geomorphic responses to tectonic and climatic forcing, situating contemporary climate change within longer-term Earth system dynamics.

Water Resources, Wetlands, and Hydroclimatic Variability

Water availability in the Himalayas is increasingly characterized by variability and uncertainty. Shifts in monsoon behaviour, changes in snowfall and snowmelt timing, and declining groundwater reserves are altering both surface and subsurface hydrological systems. These changes have significant implications for agriculture, domestic water supply, ecosystems, and transboundary water governance. The second part focuses on hydroclimatic variability and water resources, highlighting both biophysical trends and human responses. Quantitative assessments of groundwater storage change using remotely sensed data provide insights into otherwise poorly monitored aquifers, while analyses of rainfall variability reveal emerging patterns of extremes. Importantly, this section also examines Himalayan wetlands from community perspectives, recognizing them as dynamic socio-ecological systems that function simultaneously as livelihood resources, ecological buffers, and hazard-prone landscapes under climate stress.

Natural Hazards, Terrain Instability, and Risk Reduction

The Himalayan region is inherently prone to natural hazards due to its steep topography, active tectonics, and intense climatic gradients. Climate change has amplified these risks by increasing hydrological extremes, destabilizing slopes, and intensifying cryosphere-related hazards. Rapid infrastructure expansion into fragile mountain terrain has further heightened exposure. The third part adopts an integrated risk perspective, examining natural hazards through geomorphological, ecological, engineering, and socio-economic lenses. The chapters address glacial lake outburst floods, terrain instability, and seismic risks associated with tunnelling and infrastructure development, emphasizing that effective risk reduction requires an understanding of hazard processes alongside vulnerability, exposure, and institutional capacity.

Ecosystems, Biodiversity, and Land Systems Under Climate Change

Himalayan ecosystems are experiencing rapid shifts in response to warming temperatures, altered precipitation regimes, and land-use pressures. Forest composition, alpine pastures, wildlife habitats, and soil systems are changing in ways that affect biodiversity, carbon storage, ecosystem services, and agricultural sustainability. These changes also intersect with human–wildlife conflict and forest- and land-dependent livelihoods. The fourth part explores climate impacts on forests, ecosystems, biodiversity, and land management systems. Through assessments of forest sector vulnerability, habitat suitability analysis, and soil quality under different tillage systems, this section highlights the role of ecological and land systems in shaping climate resilience and sustainable resource management in mountain environments.

Geospatial Technologies and Earth Observation

Advances in remote sensing and cloud-based geospatial platforms have transformed climate and hazard research in the Himalayas. Satellite data enable monitoring of floods, land cover change, cryospheric dynamics, and atmospheric processes in regions where ground observations remain sparse. The fifth part highlights methodological developments in flood mapping and satellite image processing, emphasizing reproducible and scalable geospatial workflows. These tools are increasingly central to scientific research as well as operational decision-making for disaster management and climate adaptation.

Human Dimensions, Culture, Education, and Governance

Climate change in the Himalayas is deeply entwined with culture, livelihoods, and governance. Indigenous knowledge systems, pastoral mobility, sacred landscapes, and educational practices are all being reshaped by environmental change and development pressures. The sixth part foregrounds these human dimensions, presenting lived narratives of pastoral communities, analyses of pilgrimage landscapes under climate stress, and assessments of environmental education and governance frameworks. This section underscores the importance of inclusive, culturally grounded, and participatory approaches to sustainability and adaptation in the Himalayan region.

Scope and Contribution of the Volume

Taken together, the chapters in this volume provide a comprehensive and interdisciplinary assessment of climate change, water resources, and natural hazards in the Himalayan region. By integrating insights from climate science, cryosphere studies, hydrology, geomorphology, geospatial technologies, and the social sciences, the book advances a systems-level understanding of Himalayan vulnerability and resilience. The contributions highlight how large-scale climatic drivers interact with local environmental processes and socio-economic conditions to shape risk, resource availability, and adaptation outcomes in mountain landscapes. A key contribution of the volume lies in its emphasis on linking physical processes with human dimensions. Several chapters demonstrate how advances in Earth observation and spatial analysis enhance the monitoring of cryospheric change, water resources, and hazards, while others foreground community perspectives, governance challenges, and cultural contexts. This integrative approach helps bridge the gap between scientific assessment and policy-relevant knowledge, particularly in data-scarce and ecologically fragile regions. In an era of accelerating climate uncertainty, the Himalayas demand not only improved observation and modeling but also inclusive governance, knowledge co-production, and long-term commitment to resilience. This volume contributes to that collective effort by documenting emerging challenges, methodological innovations, and context-specific pathways toward sustainable and resilient futures in the Himalayan region.

Rayees Ahmad
Shahid Saleem

References

Adger, W. N., Arnell, N. W., & Tompkins, E. L. (2005). Successful adaptation to climate change across scales. *Global Environmental Change*, 15(2), 77–86. https://doi.org/10.1016/j.gloenvcha.2004.12.005

Bolch, T., Shea, J. M., Liu, S., et al. (2019). Status and change of the cryosphere in the extended Hindu Kush Himalaya region. *In: The Hindu Kush Himalaya Assessment*. Springer, Cham. https://doi.org/10.1007/978-3-319-92288-1_7

Gorelick, N., Hancher, M., Dixon, M., et al. (2017). Google Earth Engine: Planetary-scale geospatial analysis for everyone. *Remote Sensing of Environment*, 202, 18–27. https://doi.org/10.1016/j.rse.2017.06.031

Immerzeel, W. W., van Beek, L. P. H., & Bierkens, M. F. P. (2010). Climate change will affect the Asian water towers. *Science*, 328(5984), 1382–1385. https://doi.org/10.1126/science.1183188

Pörtner, H.-O., Roberts, D. C., et al. (2022). Climate Change 2022: Impacts, Adaptation and Vulnerability. *IPCC Sixth Assessment Report, Working Group II*. Cambridge University Press. https://doi.org/10.1017/9781009325844

Schwanghart, W., Wulf, H., et al. (2016). How long will it take for the Himalaya to erode? *Geology*, 44(9), 723–726. https://doi.org/10.1130/G38126.1

Seto, K. C., Güneralp, B., & Hutyra, L. R. (2012). Global forecasts of urban expansion to 2030 and direct impacts on biodiversity and carbon pools. *Proceedings of the National Academy of Sciences*, 109(40), 16083–16088. https://doi.org/10.1073/pnas.1211658109

Xu, J., Grumbine, R. E., Shrestha, A., et al. (2009). The melting Himalayas: Cascading effects of climate change on water, biodiversity, and livelihoods. *Conservation Biology*, 23(3), 520–530. https://doi.org/10.1111/j.1523-1739.2009.01237.x

Contents

Climate, Cryosphere, and Himalayan Earth System Processes

Cryospheric Processes Responsible for Lake Formation: A Review on Lake Types, Threats, Status, and Their Distribution within the Himalaya

Litan Kumar Mohanty, Parmanand Sharma, Rayees Ahmed, and Ankur Dixit

Abstract Identification of critical lakes in the Himalaya is a crucial job for glacial lake hazard assessment, risk mitigation and disaster management planning. This study investigates the role of climate driven cryospheric processes controlling lake formation and evolution, along with the identification and distribution of critical lakes with their hazard potential in the Himalaya. Datasets were gathered from a few related literatures and compared with research generated datasets, confirming that majority of the lakes are found to be concentrated in the Langtang, Everest, Sikkim, and Bhutan regions. New lakes are continuously forming higher elevation along with up glacial expansion of existing lakes. Additionally, most of these lakes are gathered at an elevation between 4500–5500 m. A total of 10,101 lakes with size >0.01 km^2 are located in the Himalayas which have been extracted from all the used inventories in this study. Besides, the lake count is maximum in Koshi basin (sub basin Arun = 1125) followed by Yarlung Zangbo (773), Manas (sub basin Dangme Chu = 715) and Upper Indus (619). In terms of total lake size distribution, the largest contributions come from the Qinghai-Tibetan, Yarlung Zangbo, Panjnad, Kosi, Manas, and Upper Indus basins. Moreover, the basin wise average lake size change rate is higher in Upper Indus, Karnali, Tista, Yamuna and Manas, while maximum lake size change rate is seen in Kosi, Yarlung Zangbo, Manas, Karnali, Upper Indus and Tista basin, indicating future GLOF prone basins among the Himalaya. Here, a total of 221 critical lakes were found in the Himalaya. Critical lake density is higher in Kosi River basin followed by (47), Yarlung Zangbo (22), Tista (22), Manas (21), Qinghai-Tibetan (15), Panchnad (12) and Phelku (8). Different research groups are

L. K. Mohanty (✉) · P. Sharma
National Centre for Polar and Ocean Research, Vasco Da Gama, Goa, India

L. K. Mohanty
Wadia Institute of Himalayan Geology, Dehradun, India

R. Ahmed
Indian Institute of Science, Bengaluru, India

A. Dixit
Cornell University, Ithaca, NY, USA

R. Ahmed et al. (eds.), *Climate Change, Water Resources, and Natural Hazards in the Himalayas*, Advances in Geographical and Environmental Sciences,
https://doi.org/10.1007/978-981-92-0835-7_1

working on ways to minimize the impact of GLOF in many different ways, such as by using remote sensing identification of critical lakes, modelling and field surveys to find potential impacted zone and hazard mapping, reducing lake volumes artificially, building concrete dams and making artificial drainage systems, etc. For the management of the GLOF, governing bodies have also set up early warning systems and information dissemination to the local people for community awareness.

Keywords Glacial lake · Himalaya · GLOF · Hazard

1 Introduction

The word cryosphere is derived from the Greek word 'kryos', which means the part of the surface of Earth containing a solid form of water (Barry & Gan, 2022; Ding et al., 2021). The component of cryosphere are glaciers (Rignot et al., 2013; Zemp et al., 2019), permafrost (Obu et al., 2019), snow cover (Xu et al., 2016; Bormann et al., 2018), ice sheets (Barry & Gan, 2011; Ding et al., 2021) and sea ice (Parkinson, 2014; Parkinson & DiGirolamo, 2016). The present time ice and glaciers are a result of Little Ice age extending from the 16th to nineteenth century (Quaternary period) (Kargel et al., 2014; Harrison et al., 2018). Besides, cryosphere in the Earth (i.e., high-latitude and high-altitude) have suffered 2–3 times temperature rise than the global average which helps the glacier to retreat at a faster rate, subsequently promoting the formation and expansion of glacial lakes (Taylor et al., 2023; Sattar et al., 2022, 2025; Evans & Clauge, 2000). Glacial lakes are any water body on or near or at a distance to the glacier in a periglacial environment. Change in temperature (rising) and precipitation (declining) pattern affects glacial lake formation as well as expansion by glacier retreats (Furian et al., 2021; Fujita et al., 2013; Bolch et al., 2008). When the natural dam of a glacial lake fails, it triggers a Glacial Lake Outburst Flood (GLOF) — a sudden and large discharge of water that rushes through narrow downstream river channels as a result of the upstream dam breach. Due to the current trend in climatic parameters, many cryosphere hazards are prominent in the glaciated terrain. Cryosphere threats can be broadly classified into three parts based on a component of the Earth, i.e., land, ocean and atmosphere (Ding et al., 2021). Besides, the land Cryosphere hazards include glacier hazards, permafrost hazards and snow hazards (Kattelmann, 2003; Mool et al., 2008; Bolch et al., 2008). There are four types of glacial hazards in the periglacial environment mostly: GLOF, debris flow, landslide and avalanches. GLOF could be in most cases even more devastating than other glacial hazards such as debris flow or avalanches due to its far-reaching limits, speed movement and compounding behaviour (Sattar et al., 2022). The Himalayas are no exception and witnessed a high rate of retreat of many glaciers providing lake formation conditions. Accelerated melting of glaciers potentially can expand these lakes and under the favorable geomorphic conditions can lead to catastrophic consequences such as GLOF (Evan & Clauge, 2000; Fujita et al., 2009; Maskey et al., 2020). The proposed study will focus on land hazards only more specifically on GLOF.

1.1 *Cryospheric Processes and Lake Formation*

Glacial landforms are primarily shaped by processes of erosion during glacial advance and deposition during retreat (Staiger et al., 2005; Leith, 2012). The geomorphic features observed in contemporary glaciated landscapes result from the combined influence of both present-day glaciers and past glaciation episodes. Accordingly, many modern glacial lakes have also originated through these geomorphic processes. Subglacial and glacial erosion processes are predominantly governed by factors such as lithology, bedrock shear strength, fracture resistance, penetration toughness, moisture content, structural configuration, and the geometry of fractures and joints, alongside basal thermal conditions (Staiger et al., 2005; Leith, 2012, 2014; Koppes et al., 2015). These erosional mechanisms frequently generate depressions within the glacial bed, which subsequently become sites for meltwater accumulation following glacier retreat, leading to the formation of glacial lakes (Farinotti et al. 2009; Gantayat et al., 2014). In certain cases, meltwater may be naturally impounded by resistant bedrock barriers or, alternatively, by landslide deposits occurring within or adjacent to glacier-fed channels, giving rise to landslide-dammed lakes.

Debris accumulation on glaciers is primarily governed by three key processes: subglacial erosion, landslides or rockfall, and avalanches. The volume of snow input directly influences glacier size and, consequently, the associated debris cover (Bajracharya et al., 2008; Racoviteanu et al., 2015). It is important to note that not all large glaciers exhibit extensive supraglacial debris cover. For instance, **Samudra Tapu** and **Sutridhaka** glaciers, despite their considerable sizes, have less than 5% of their surfaces mantled by debris. In contrast, **Batal Glacier**, a relatively small valley glacier within the Chandra basin, is characterized by a debris cover exceeding 80%. Interestingly, all three glaciers are valley-type glaciers, yet display markedly different debris cover extents, highlighting the influence of local topography, debris supply, and catchment characteristics on debris distribution patterns in the region. Besides, among glacier types, valley and cirque glaciers typically exhibit higher debris content due to their geomorphological settings. Additionally, the nature of the underlying bedrock plays a significant role in controlling glacier dimensions — glaciers flowing over sedimentary bedrock tend to attain larger sizes compared to those underlain by metamorphic or igneous formations. Besides, Seismic events also indirectly contribute to debris loading on glaciers by triggering landslides and avalanches, occasionally enhancing glacier flow velocities (Nagai et al., 2013). Furthermore, the rate of debris delivery from surrounding slopes, alongside topography, bedrock lithology, and snow accumulation, collectively modulate glacier dynamics, particularly surface velocity and erosional capacity (Scherler et al., 2008, 2011). An increase in supraglacial debris content typically results in a decline in surface velocity and a reduction in surface slope ($< 2°$), creating favourable conditions for the formation

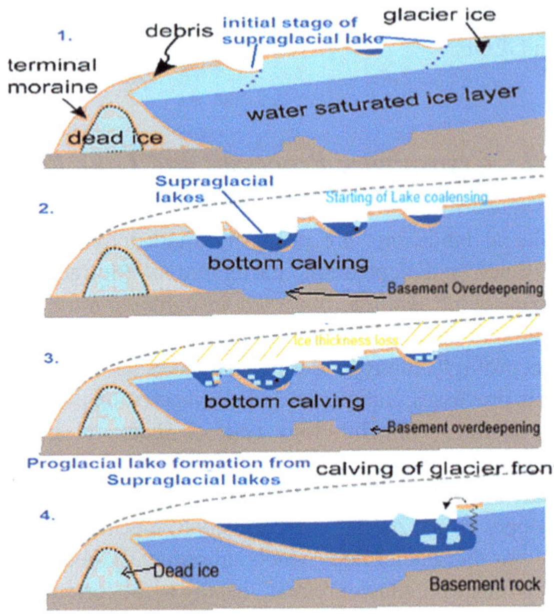

Fig. 1 Schematic illustration of the lake coalescing process, depicted in four sequential stages (Aggarwal et al., 2017; Ahmed et al., 2021; Allen et al., 2011a). The mechanism of supraglacial lake formation is also explained in this sequence (Modified after Sakai, 2012)

of supraglacial lakes. These water bodies tend to coalesce in areas characterized by low gradients and quasi-stagnant ice zones (Quincey et al., 2007; Watson et al., 2018). Once established, glacial lakes absorb significant solar radiation, enhancing localized ice melt and facilitating lake expansion. In proglacial and supraglacial lakes, processes such as glacier calving, subaqueous melting, and waterline ablation predominantly govern lake growth and evolution (Fig. 1).

With the increasing rate of glacier retreat, notable shifts such as glacier fragmentation, upward migration of the snowline, and a rise in the equilibrium line altitude (ELA) are being observed (Bhambri et al., 2012; Racoviteanu et al., 2015). Globally, sustained glacier recession has already led—and will likely continue to lead—to fragmentation, wherein tributary glaciers detach from their main trunks and formerly contiguous glacier bodies break into smaller, disconnected units. This process of fragmentation has, in certain cases, particularly among larger glaciers, accelerated retreat rates when compared to neighbouring, still-connected glacier systems (Paul et al., 2004; Jiskoot et al., 2009). As glaciers fragment, proglacial lakes commonly form at the junctions where tributary glaciers once merged with their parent glaciers (Mohanty & Maiti, 2022, 2024). In these scenarios, the lateral moraine of the main glacier, combined with the terminal moraine of the detached tributary, often acts as a natural barrier, impounding meltwater and facilitating the formation of moraine-dammed lakes (Fig. 2a). However, the occurrence and stability of such lakes depend greatly on local geomorphic conditions and glacier morphodynamics, and their development is not universally observed in all instances of glacier fragmentation (Fig. 2b).

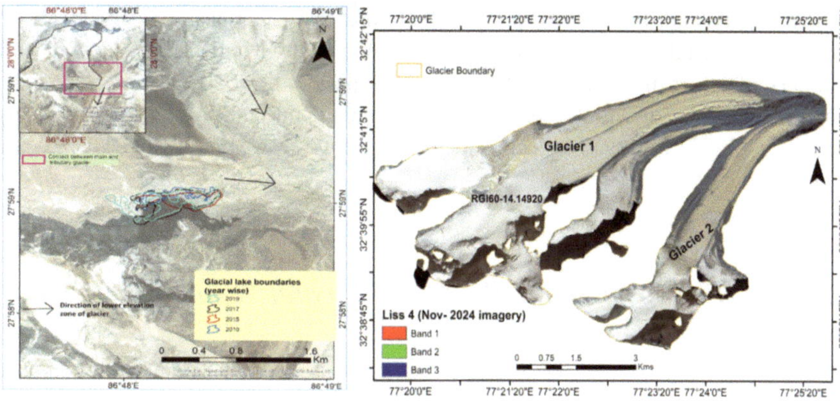

Fig. 2 **a** Illustration showing the fragmentation of valley glaciers under mass loss conditions, along with the identification of probable zones for future lake formation. **b** Example where glacier fragmentation occurred after 2015, but no visible signs of lake formation have been observed

1.2 Glacial Lake Inventory

Glacial lake inventories have been created by various scientists and organisations across various glaciated regions to know the glacial lake's status and characteristics in a certain region. These datasets play a crucial role in analysing the spatio-temporal dynamics of glacial lakes and understanding how they are responding to climatic variations in specific regions. The earliest inventories of glacial lakes were available in 1967 by concentrating on their morphological and morphometric characteristics and bathymetric surveys in the central Himalaya, and this dataset was updated in 1980 (Nikitin, 1987; Staviskiy & Jukov, 1968; Nikitin & Gorelkin, 1979). In that inventory, more than 300 glacial lakes were observed in High Mountain Asia between 1966–1975 (Nikitin, 1987; Staviskiy & Jukov, 1968). Later, much works has been performed by researchers on lake inventory of the mountain glaciers based on multisource remote sensing images on a regional basis in the whole Hindu Kush Himalaya (HKH); viz., Wang et al. (2020a, b), Nie et al. (2017), Maharjan et al. (2018), Gardelle et al. (2011), and ICIMOD (2010). The latest regional coverage of glacial lakes shows a total of 13,000 lakes in the Himalayas (Nie et al., 2017). Additionally, Wang et al. (2020a, b) showed 27,205 and 30,121 glacial lakes (size 0.0054–6.46 km^2) in High Mountain Asia in 1990 and 2018, respectively. In the Hindu Kush Himalaya region, 25,614 glacial lakes with a total area of 14440 km^2 were recognized with an elevation range between 2200 and 6200 m by Maharjan et al. (2018). However, a study jointly carried out in-between 1999 to 2003 by the International Centre for Integrated Mountain Development (ICIMOD), United Nations Environment Programme (UNEP), and Asia Pacific Network (APN) reported a total of 9000 glacial lakes for 15,000 glaciers in Bhutan, Nepal, Karakoram, China, and India. Moreover, Zhang et al. (2015) showed that the lake counts are distributed in different basins: the Amu Darya, 594, Indus, 1607, Ganges, 364 and Brahmaputra, 2247, based on Landsat image 2010. Mohanty and Maiti (2021) showed 4091 numbers of lakes distributed widely in the Himalaya in 2020 with a size > 0.01 km^2. Lakes are most widely distributed in Himalaya and these are present densely in eastern (3282) followed by central (2946) and western Himalaya (1848) (Wang et al., 2020a, b).

1.3 GLOF Distribution in the Himalaya and Their Impacts

Almost 800 million people who live near catchment areas of the Ganga, Brahmaputra, and Indus basins will be exposed to the risk of severe flooding and future freshwater availability (Bolch et al., 2012). In the context of Asia, flood is the predominant natural hazard, causing almost 90% of the destruction (Mirza, 2011). Moreover, the tropical rivers in Southeast Asia are experiencing regular floods (Bandyopadhyay et al., 2016). Some of these floods are caused by GLOF. GLOF is mostly dangerous in a short time among all types of glaciated hazards, and it has been noticed in different glacierized parts of the world: the Himalayas, Tien Shan, the Karakoram, the Andes, the Rockies, etc. (Evan & Clague, 2000; Komori, 2008; Bolch et al., 2012; Worni et al., 2013; Emmer et al., 2016, 2022; Huggel et al., 2006; Aggarwal et al., 2017). Many researchers predicted flood occurrences will upsurge in the twenty-first century in the lower Indus, Ganges and Brahmaputra rivers (Hirabayashi et al., 2013; Wijngaard et al., 2017), while this theory was contradicted by Veh et al. (2019) and Harrison et al. (2018). GLOFs are even precarious to aquatic life, like the recent Chamoli flash flood in Uttarakhand, which caused a huge loss in aquatic animals like fish in the river basin, creating breathing problems by adding vast amounts of debris to the water (Meena et al., 2021). Among the various existing global GLOF inventories, the most recent and comprehensive database was compiled by Lützow et al. (2023), whereas for the High Mountain Asia region, the latest and extensive inventory was developed by Shrestha (2023). A total of 3151 GLOF was recorded globally between 850 and 2022 CE (Lützow et al., 2023). Among all the glaciated parts of the world, the Himalaya has the third highest GLOF occurrences after North America and Iceland (Lützow et al., 2023; Carrivick & Tweed, 2016). In the Himalaya alone, Lützow et al. (2023) have listed 422 GLOF events since 850 CE, while Shrestha et al. (2023) have documented 703 GLOF events since 1533. Analysis of these GLOF databases revealed two regional peak distributions of previous GLOFs in the Himalayas: one near the western Himalayas and the other in the Everest, Bhutan, and Sikkim regions (Eastern Himalaya). Global GLOF assessments show that among all continents, Central Asia is the most vulnerable region to GLOFs, causing societal impact, followed by South America (Carrivick & Tweed, 2016). Among all the global mountain regions, the High Mountain Asia (HMA) region exhibits the highest exposure to Glacial Lake Outburst Flood (GLOF) hazards, with the Himalaya constituting a significant and highly vulnerable sub-region within it. Additionally, more than half of the global exposed population are found in just four countries:

India, Pakistan China (HMA), and Peru (Andes) (Taylor et al., 2023). A few most common triggers are landslide, avalanche, cloudburst, earthquake, and accelerated melting rate of glaciers (e.g., Harrison et al., 2018; Veh et al., 2019). Out of 703 Himalayan GLOF events recorded since 1833, 28 were triggered by heavy precipitation, followed by temperature rise (Carrivick & Tweed, 2016), ice avalanches (Hirabayashi et al., 2013), rock falls (Ahmed et al., 2021), ice calving (Aggarwal et al., 2017), upstream floods (Aggarwal et al., 2017), and landslides (Aggarwal et al., 2017; Shrestha et al., 2023). From the GLOF inventory datasets, it is clear that the GLOF occurrence has been increasing year-wise in the Himalaya from 1533 to 1990 after that this trend got less slope (Harrison et al., 2018; Carrivick & Tweed, 2016; Nie et al., 2020). Additionally, the GLOF frequency is seen to have increased after 2023 with three GLOF events including South Lhonak lake (04 Oct 2023) (Zhang et al., 2024) in the Sikkim Himalaya, Birendra Lake GLOF in the Gorkha district of Nepal (21 April 2024) (Khadak et al., 2024), and the GLOF in the Everest region in Nepal (16 August 2024) (Khadka et al., 2024). Most of the GLOFs have created havoc on the community, livestock, agricultural land, as well as infrastructure downstream in the Himalaya (Nie et al., 2020). This highlights the urgent need to assess risks from various triggering mechanisms, particularly in the Himalayas, where 68% of hydropower plants are situated along potential GLOF pathways (Schwanghart et al., 2016). The Luggey Tso flood totally damaged a hydropower plant estimated to be ~$500 million in 1994 (Shrestha et al. 2010). Recently, South Lhonak GLOF destroyed the Chungthang hydropower project, which cost around 14,000 crores Indian rupees (16 billion dollars) (Yu et al., 2024). The destruction of the Namche Small Hydro-plant by the Dig Tsho GLOF in 1985 was valued at $ US$three million. The Chamoli flash flood swept away the unfinished Tapovan Vishnugad Hydropower Project and inflicted substantial damage on the Rishi Ganga Hydropower Project (Shugar et al., 2021). Moreover, the most significant impact of Dig Tsho in 1985 GLOF was the complete destruction of a new hydropower station at Thamo and about 30 houses, and the loss of 14 bridges between Mingbo and Jubing (Bajracharya et al., 2007). In 1981, a GLOF event in Nepal damaged the Friendship Bridge of the Nepal-China Highway and the Koshi power station with heavy economic losses (Sattar et al., 2022). A GLOF event (Ciremacho Lake/Zhangzangbo Lake GLOF) that occurred during August 2000 damaged 98 bridges in the Tibetan Plateau and destroyed around 10,000 houses, accounting for about 75 million US dollars in terms of financial losses (Nie et al., 2018). Thus, measures should be taken to minimize the devastating nature of GLOF towards costly infrastructure. Moreover, documentation of critical glacial lakes, downstream impact analysis, and application of remedial measures can prevent various upcoming calamitous GLOFs.

1.4 Controls of Lakes on Glacier Mass Balance in the Himalaya

The presence of glacial lakes, either on the glacier surface or at its terminus, significantly influences glacier mass loss processes by accelerating melting through calving and waterline melting (Watson et al., 2016). In earlier studies, the contribution of calving to overall mass loss was often overlooked in glacier mass balance estimations; however, this component has received increasing attention in recent years (Zhang et al., 2022, 2023). It has been well established that glaciers terminating in lakes experience higher rates of calving, ice thickness loss, and increased ice velocity compared to land-terminating glaciers (King et al., 2018; Song et al., 2014). When in contact with a proglacial lake, glacier mass loss is primarily driven by contact melting at the ice-water interface and active calving near the lake margin (Mool et al., 2008; Komori et al., 2012; Song et al., 2017). Furthermore, debris-covered glaciers with supraglacial lakes or those in contact with proglacial lakes can retreat more rapidly than clean-ice glaciers, especially when the debris thickness remains below a critical threshold (Basnett et al., 2013). Frontal mass loss, particularly through calving, is a major factor contributing to the rapid expansion of proglacial lakes, making these lakes particularly hazardous, as demonstrated by the recent Glacial Lake Outburst Flood (GLOF) event at South Lhonak Lake in Sikkim, India. Additionally, processes such as waterline and contact melting around supraglacial lakes, along with the back-wasting of ice cliffs, are considered principal causes of high ice thickness loss in the stagnant, debris-covered glacier zones. This phenomenon has been notably observed in glaciers like Bara Shigri, Khumbu, Ngozumpa, and Imja, which host multiple supraglacial lakes (Garg et al., 2025; Mohanty & Maiti, 2021, 2022).

Glacier mass balance is a key parameter that quantifies the amount of water stored as ice or released as meltwater from a glacier over a defined time period (Cogley et al., 2011). Ice loss from glaciers typically occurs in three primary forms: area loss, thickness loss, and length (terminus) loss (Bhambri et al., 2012; Racoviteanu et al., 2015). These different modes of ice loss contribute to the development of various types of glacial lakes (Jain & Mir, 2017; Komori et al., 2008). Among these, thickness loss is often the earliest detectable sign of glacier depletion, which is subsequently followed by reductions in area and length (Berthier et al., 2014; Kääb et al., 2015; Garg et al., 2023). In the case of larger glaciers, which are predominantly debris-covered, ice loss through thickness reduction is typically more pronounced than through area or length reduction. The lateral or areal shrinkage in these highly debris-covered glaciers tends to be a delayed response to ongoing thickness loss (Garg et al., 2023; Maurer et al., 2019). Conversely, clean-ice glaciers exhibit relatively higher rates of area and length loss compared to thickness loss (Racoviteanu et al., 2015; Maurer et al., 2016). Thus, in this book chapter, we will be discussing four aspects of glacial lakes, i., latest lake distribution in the Himalaya from various inventories, ii. Lakes change and their dynamics, III. critical lake distribution in the Himalaya and iv. Various lake types and their distribution in the Himalaya.

2 Study Area

2.1 Physiography and Topography

The Himalaya constitutes an approximately 2500-km-long mountain belt formed by the ongoing convergence and collision between the Indian and Eurasian plates, a process that initiated around 55 Ma (Rubatto et al., 2013). In this study, the Himalayan region is considered within the geographical extent of 69°48′36″ E to 98°22′22″ E longitude and 27°12′16″ N to 39°31′26″ N latitude. Across its length, the mountain chain exhibits considerable variation in width, ranging from nearly 400 km in some western segments to about 150 km toward the eastern margin. The study domain encompasses four major drainage systems: the Indus, Ganges, Brahmaputra, and the internally drained Mansarovar basin. The Himalaya hosts the world's highest peak, Mount Everest (8848 m), and represents one of the most elevated regions on Earth. Elevation patterns vary longitudinally, with the central Himalaya attaining the highest mean altitude (~ 5600 m), followed by the eastern (~ 5395 m) and western sectors (~ 5155 m). Within the Indian Himalayan sector alone, glaciers occupy roughly 23,300 km^2 and store an estimated ice volume of about 1071 km^3 (Bolch et al., 2012; Worni et al., 2013). As a core component of High Mountain Asia (HMA), the Himalaya plays a critical hydrological role and is widely recognized as the "Water Tower of Asia." The region supplies nearly 8.6 million units of freshwater annually to densely populated downstream areas, supporting domestic water needs, irrigated agriculture, and hydropower production (Singh & Bengtsson, 2004; Immerzeel et al., 2010; Bolch et al., 2012; Pratap et al., 2016; Ahmed et al., 2026). In recent decades, the number and areal extent of glacial lakes in the Himalaya have expanded markedly. Between 1990 and 2015, the total count of glacial lakes increased by approximately 8.8% (from 4549 to 4950), while their cumulative area expanded by about 14.1% (from 398.9 to 455.3 km^2) (Nie et al., 2017). This growth reflects ongoing cryospheric changes and underscores the evolving geomorphic and hydrological dynamics of the region.

2.2 Climatic Pattern

Climate exerts a fundamental control on Himalayan cryospheric and hydrological processes, primarily through variations in atmospheric circulation, precipitation regimes, and temperature patterns (Fig. 1). The Himalayan arc is influenced by three dominant circulation systems that deliver snowfall: (i) northwesterly disturbances (westerlies), (ii) the southwest (SW) monsoon, and (iii) northeasterly (NE) wind flow (Bolch et al., 2012; Bookhagen & Burbank, 2006, 2010). In the eastern Himalaya, snowfall is largely associated with the SW monsoon and NE wind systems. Moisture transported from the Indian Ocean during the summer months produces substantial precipitation, leading to a predominance of summer-accumulation-type glaciers in this sector. In contrast, the western Himalaya—including the Karakoram and Pamir ranges—is mainly influenced by winter precipitation delivered by mid-latitude westerlies originating from the Mediterranean region. As snowfall in this region occurs primarily during winter, glaciers here are typically classified as winter-accumulation-type (Bolch et al., 2012; Bookhagen & Burbank, 2010). A pronounced longitudinal gradient characterizes monsoonal precipitation across the Himalaya, with rainfall intensity generally declining from east to west. The western limit of significant monsoonal influence lies near 78° E longitude, around the Sutlej Valley (Bookhagen & Burbank, 2010). Orographic effects further modulate spatial precipitation patterns: the Lesser Himalaya receives comparatively higher rainfall, followed by the central and Greater Himalaya, reflecting the strong influence of relief and topographic barriers on moisture transport and condensation (Bookhagen & Burbank, 2010).

3 Results and Discussion

3.1 *Glacial Lake Types and Classification Schemes*

There are various kinds of glacial lakes which can be classified based on various parameters: dam type (ice, moraine and rock), proximal distance (connected and non-connected), mother glacier characteristics (valley, cirque, rock), position of lake with respect to mother glacier (proglacial, supraglacial, subglacial), lake size (big, small), source of meltwater (glacier-fed, non-glacier fed), position of glacial lake with respect to Himalaya slope (northern slope or southern slope) and position of moraine (lateral moraine, end moraine, moraine thaw) by various scientist (Zhang et al., 2015; Yao et al., 2018; Nie et al., 2018). Furthermore, a complete classification system of glacial lakes was proposed by (Yao et al., 2018) based on its formation mechanism, topographic feature and geographical position. In this study, glacial lakes were systematically categorized into six primary classes and eight subordinate subclasses to ensure a comprehensive typological framework. The six main classes comprise: (Aggarwal et al., 2017) glacial erosion lakes, (Ahmed et al., 2021) moraine-dammed lakes, (Ahmed et al., 2021) ice-blocked lakes, (Allen et al., 2011a) supraglacial lakes, (Allen et al., 2016) subglacial lakes, and (Allen et al., 2019) other glacial lakes. Glacial erosion lakes were further divided into cirque lakes, glacial valley lakes, and other erosion-related lakes formed by glacial scouring processes. Moraine-dammed lakes were subdivided into end moraine-dammed lakes, lateral moraine-dammed lakes, and moraine thaw lakes, reflecting differences in dam structure and formation mechanisms. Ice-blocked lakes were classified into advancing glacier-blocked lakes and other glacier-blocked lakes, depending on the dynamics and position of the blocking ice mass. To enhance consistency and applicability across regions, the study also identified distinct geomorphological characteristics observable in remote sensing imagery for each lake type. In addition, quantitative indicators were proposed to support objective identification and differentiation among classes. These criteria collectively contribute to the development of a standardized, transferable, and operational classification system for glacial lakes. Further classification tells that there are primarily 3 types of glacial lakes based on retreat pattern and presence of lake with respect to mother glaciers like pro-glacial lakes (Figs. 3 and 4), supra-glacial lake and other types (e.g. landslide dam lake, permafrost lake) (Mohanty & Maiti, 2024). Pro-glacial lakes are formed by lateral glacier loss (Fujita et al., 2008). However, supra-glacial lakes are formed due to thickness loss of debris covered glaciers, while other types of lakes formed by accumulation of meltwater at a distance (<1 km) to the main glacier (Mool et al., 2008; Komori et al., 2012; Nie et al., 2017, 2019) (Figs. 3 and 4). Pro-glacial lakes from where glacial ice or a lateral moraine blocked a side valley or between the end moraine and glacier snout (Kettelmann, 2000; McKillop & Clague, 2007). Sometimes conversions of supra-glacial lakes to pro-glacial lakes have also been observed in the Himalayas (Imja lake in Nepal, Rongbuk lake in Everest and Thorthormi in Bhutan). Most supra-glacial lakes merging have been detected in the southern slope of Himalaya, i.e., Thorthormi lake in Bhutan and Imja lake in Everest; whereas a very few examples are seen for the northern slope of Himalaya, i.e., in the Rongbuk glacier (Bolch et al., 2008; Fujita et al., 2008). Those types of lakes have a higher growth rate (Komori et al., 2008; Mool et al., 2008; Bajracharya et al., 2008; Bolch et al., 2012). Supra-glacial lakes are always associated with ice cliffs, where the mass loss is normally high (Chand & Watanabe, 2019; Taylor & Carr, 2019). On the moraine dam, sometimes numerous smaller lakes are present called moraine thaw lakes that form when the moraine is ice cored (Yao et al., 2018), which is also a type of supra-glacial lake. Moreover, other types of lakes, valley-blocked and landslide dam lakes are formed by melt-water gathered and accumulated by converging valleys/blocked valleys. Other than these two; the melt water blocked by the landslide derived material from nearby mountains in case of the landslide dam lake. This type of lake can grow by the accumulation of glacial melt-water or liquid precipitation. Besides, permafrost lakes are formed by the sinking of ground ice by increasing in temperature (Figs. 3 and 4).

3.2 *Lake Type and Its Distribution*

3.2.1 Northern Side and Southern Side Lakes

Northern side and southern side lakes have been classified based on similar topography and climatic patterns, as discussed in studies by Nagai et al. (2013) and King et al. (2018). These are the lakes connected to northern and southern flowing glaciers, respectively (Fig. 5). The similarities in climatic and topographic patterns on either side of the Himalaya contribute to the overall homogeneity in glacial characteristics. Additionally, the rock types differ between the two sides, with the northern side featuring hard crystalline rocks and the southern side having less hard rocks.

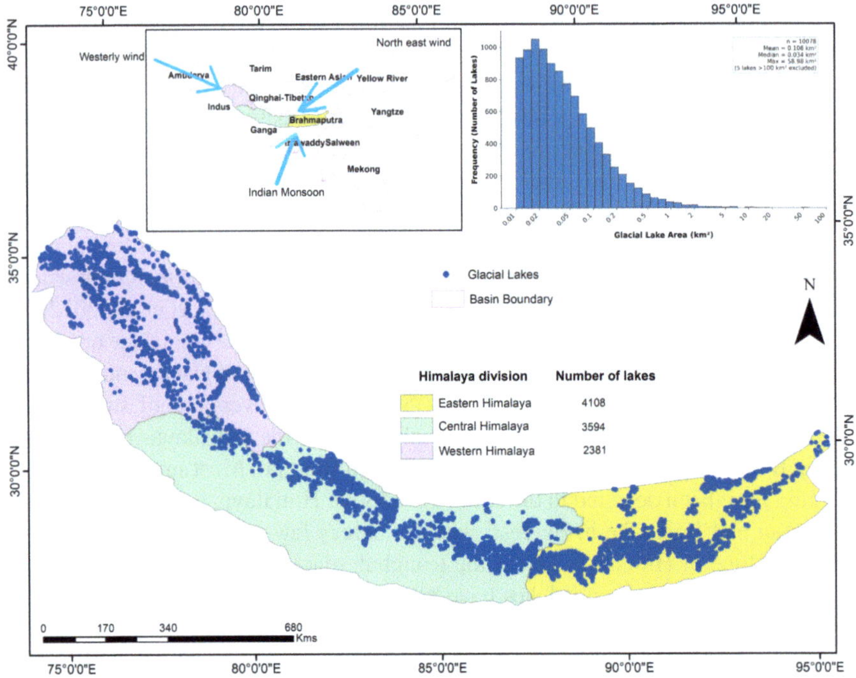

Fig. 3 Map of the study area displaying the various sub-regions of the Himalaya, along with the spatial distribution of glacial lakes and the regional precipitation pattern. Arrows indicate the prevailing directions of major wind flows. Additionally, a histogram plot illustrates the size distribution and frequency of glacial lakes within the study area

A total of 2021 lakes were identified on the northern side, while 1103 lakes were observed on the southern side. The average size of lakes is larger on the northern side (0.22 km^2) compared to the southern side (0.11 km^2). Northern side glaciers are described as cleaner and less dendritic, making them more susceptible to retreat (Nagai et al., 2013; King et al., 2018). Consequently, the growth rate of lakes is higher on the northern side than on the southern side (Debnath et al., 2018). Furthermore, northern side lakes are more prone to glacial lake outburst floods (GLOFs) due to narrower dams compared to the southern side lakes (Fig. 5). This difference is attributed to the availability of more and larger debris for dam construction on southern side glaciers (Nagai et al., 2013; King et al., 2018).

3.2.2 Glacier-Fed and Non-glacier-Fed Lake

Glacier-fed lakes get their water from glacier meltwater, whereas non-glacier-fed lakes get water from precipitation only (Zhang et al., 2015). Glacier-fed lakes are always present downstream of glacier flow, unlike a non-glacier-fed lake. The non-glacier-fed lake may have a higher proximal distance (Fig. 5). Glacier-fed lakes have

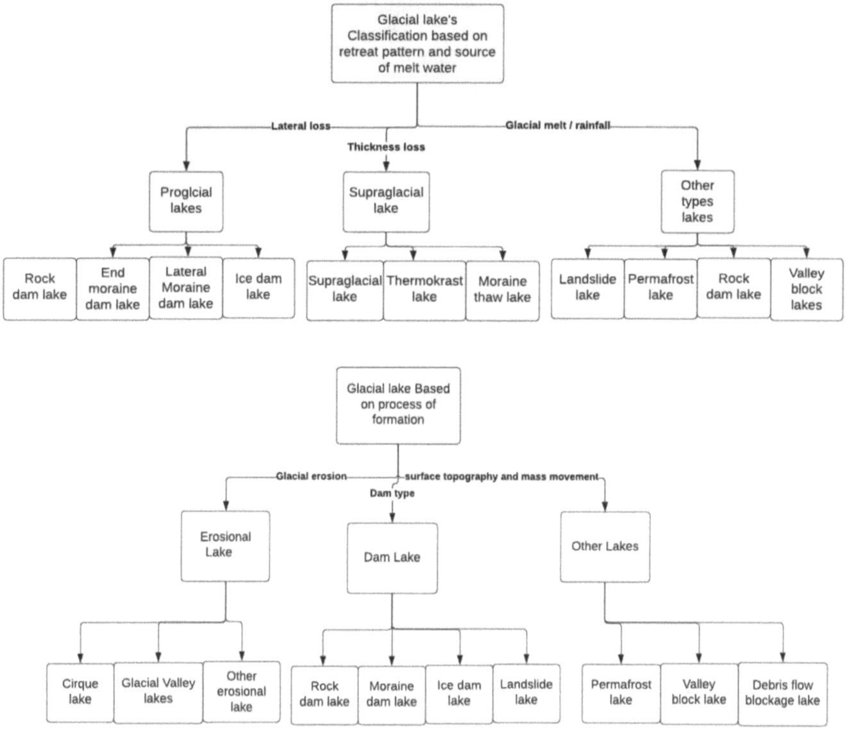

Fig. 4 Flow diagram illustrating the glacial lake classification scheme for the Himalayan region, highlighting **a** glacier retreat patterns and associated sources of meltwater, and **b** the processes involved in glacial lake formation

a higher increase rate, hence dangerous (Zhang et al., 2015; Song et al., 2017). An increase in lake size of glacier-fed lakes denotes climate change or glacier retreat. In the Himalaya, we noticed 326 and 3872 non-glacier-fed and glacier-fed lakes, respectively. The non-glacier-fed lakes were gathered at the mountain extremity of the Himalayas. Moreover, glacier-fed (0.14 km^2) lakes have a lesser average lake size than the non-glacier-fed lake (0.17 km^2).

3.2.3 Proximal Distance-Based Lakes

Lakes can be classified based on proximal distance like connected (0 m), moderately connected (0–500 m) and non-connected lakes (> 500 m) (Fig. 5). Connected lakes have a higher growth rate, hence dangerous as the waterline melting and calving process is prominent. With the continual increase in glacier retreat, some lakes will become unconnected, thus decreasing in expansion rate (Mohanty & Maiti, 2024). Erosional lakes were connected earlier, but with time, they can be converted to unconnected lakes with continual frontal loss of the glacier. Moreover, unconnected lakes

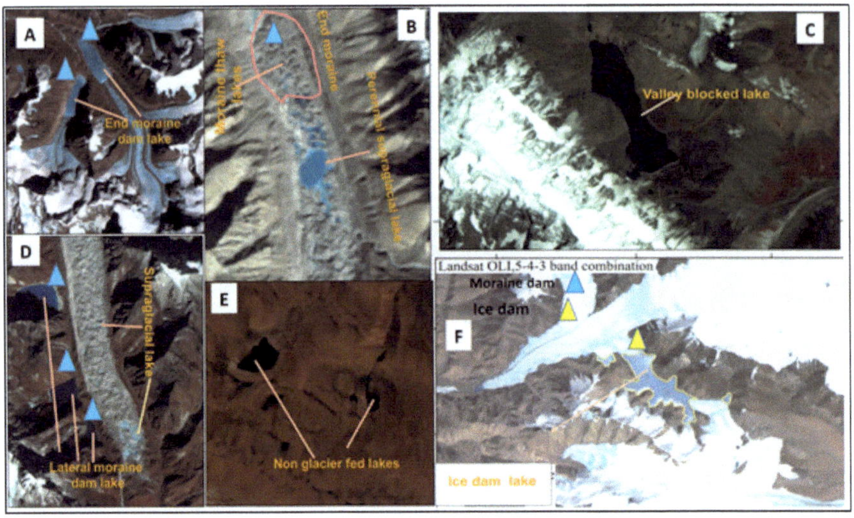

Fig. 5 Google Earth imagery displaying various types of glacial lakes in the Himalayas. **a** Moraine-dammed Lake (end moraine), **b** perennial supraglacial lake and moraine thaw lake, **c** valley-blocked lake, **d** supraglacial lakes and lateral moraine-dammed lakes, **e** non-glacier-fed lake, and **f** ice-dammed lake

(2105) are prominent in the Himalaya followed by connected (1200) and moderately connected lakes (893) (Table 1). In the Himalaya, the average lake size is higher for moderately connected lakes, followed by connected and non-connected lakes. Higher proximal lakes have a lower average lake size (Table 1). Most of the higher proximal distance lakes are permafrost or landslide dam lakes, or the mother glacier of those lakes has wholly retreated. Most of the moderately connected lakes are erosional in type.

3.2.4 Meltwater Origin-Based Lake

Glacial lakes can be divided into supra-glacial and other types of lakes (Rock Dam Lake, Landslide Lake, erosional lake, Pro-glacial Lake) based on meltwater origin (i.e., dominated retreat pattern) (Fig. 5). Supra-glacial lakes are gathered within the glacier surface, whereas other types of lakes form close to the glacier tongue or at a distance from the glacier tongue. Other types of lakes have a higher increase rate, large volume, and are unstable, hence dangerous. In the Himalayas, we noticed 26,111 and 5040 numbers of supra-glacial and other lakes, respectively. The average lake size of supra-glacial and other types of lakes is 0.003 km^2 and 0.14 km^2, respectively. Other types of lakes are more prone to GLOF as these are dammed by moraine or ice and have a higher average lake size than a supraglacial lake. Besides, GLOF has also occurred from the supraglacial lake in the Everest region (Watson et al., 2018; Mohanty & Maiti, 2021).

Table 1 Lake classification scheme with counts and average size of each group (size > 0.01 km^2) for the Himalayan

Controlling factors	Basis of classification	Lake type	Average lake size (m^2/yr)	Lake count
Climate	Topography	Northern side	0.22	1174
		Southern side	0.11	2039
Rock type	Dam type	Ice dam lake	0.16	6
		Moraine dam lake	0.18	3088
		Rock dam lake	0.09	994
		Other dam lake	0.67	110
Lake glacier relationship	Proximal distance	Connected (< 100)	0.15	1200
		Moderately connected (100–500 m)	0.16	893
		Non-connected (> 500 m)	0.13	2105
Glacier characteristics	Source of meltwater	Glacier-fed lake	0.14	3872
		Non-glacier-fed lake	0.17	326
	Discharge volume	Supra-glacial lake	0.003	26,111
		Other lakes	0.14	5040

3.2.5 Dam Type Lake Distribution

Dam type can be used as a lake classification parameter: Ice dam, moraine dam, rock dam, and others (valley blocked, landslide Dam, Lake) (Fig. 5). A glacial lake dammed by moraine is called a moraine dam lake, as well as an ice dam lake and a rock dam lake. In the Himalayas, the ice dam lake is very minimal (6), which is the most dangerous, followed by the moraine dam lake (Carrivick et al., 2016). Smaller ice dam lakes are available in the Himalaya, which are also called thermokarst lakes. However, ice dam lakes with a size >0.01 km^2 are absent in the eastern and central Himalaya. Most moraine dam lakes are present in the central and eastern Himalaya. Besides, rock dam lakes are gathered in the eastern Himalayas. Other types of lakes, i.e., the landslide dam lakes, are lower in number and present mostly in the southern part of the Himalayas. The average lake size of other types (0.67 km^2) of lakes is higher than that of the moraine dam (0.18 km^2) and the rock dam lake (0.09 km^2) (Table 1).

3.3 Glacial Lake Distribution and Dynamics in the Himalaya

In this study, we analysed the spatial distribution of glacial lakes across the Himalaya using five published datasets: Zhang et al. (2015), Zhang et al. (2023), Shugar et al. (2020), Mohanty et al. (2023a, b), and Wang et al. (2020a, b) (Fig. 3). For a comprehensive, basin-scale statistical assessment of lake area change rates and spatial patterns, we primarily utilized the Zhang et al. (2023) dataset, which represents the most extensive and up-to-date glacial lake inventory among the five sources. These glacial lakes are predominantly distributed in regions experiencing pronounced glacial retreat, which is governed by a combination of glaciological, topographic, and climatic factors (Bhambri et al., 2023; Bolch et al., 2012; Pandey et al., 2024). The histogram analysis (Fig. 3) reveals a concentration of glacial lakes within the Langtang, Bhutan, Sikkim, and Everest regions, corresponding to the longitudinal extent of 83° E to 94° E. In total, 31,115 glacial lakes have been identified in the Himalaya, comprising approximately 26,111 supraglacial lakes, with the remainder categorized as proglacial, permafrost-related, erosional, rock-dammed, and valley-blocked lakes. Furthermore, the majority of these lakes are situated at elevations between 4500 m and 5500 m, with the highest lake expansion rates observed within the 5000–5500 m elevation band (Fig. 4). Across the inventories analyzed, 10,101 glacial lakes exceeding 0.01 km^2 in area have been documented within the Himalaya. The persistent frontal retreat of glaciers has progressively created new accommodation space for lake expansion at higher elevations, a trend clearly visible in Fig. 4. In addition to the growth of existing lakes, numerous new pristine lakes are continuously forming in these higher elevation zones. Both glacial lake counts, and cumulative lake area have shown a consistent year-on-year increase across the Himalayan region (Tables 2 and 3).

3.4 Distribution of Lake Area, Number, and their Dynamics

We compiled and analysed the distribution of glacial lake counts, total lake area, and average lake area across different sub-basins of the Himalaya (Fig. 5). The highest number of lakes was recorded in the Koshi basin, particularly within its Arun sub-basin (1125 lakes), followed by the Yarlung Zangbo (773), Manas (Dangme Chu sub-basin with 715), and the Upper Indus (619). In terms of total lake area, the largest contributions were observed from the Qinghai-Tibetan, Yarlung Zangbo, Panjnad, Koshi, Manas, and Upper Indus basins. Regarding average lake area, the Qinghai-Tibetan, Yarlung Zangbo, Panjnad, Phelku, Upper Indus, and Koshi basins exhibited the highest values (Fig. 5). Through an overlay analysis integrating total lake counts, cumulative lake area, and mean lake size, we identified four critical basins—Koshi, Yarlung Zangbo, Manas, and Upper Indus—as particularly susceptible to glacial lake-related hazards.

Table 2 Lake size and number distribution in different years by different researchers

	Lake size (km^2)						Lake count					
	1990	2000	2005	2010	2015	2018/2020	1990	2000	2005	2010	2015	2018/2020
Mohanty et al. (2023a, b)	664.75	660.22	*	800.3	492.01	2075.35	5194	5222	*	6432	6481	8357
Shugar et al. (2020)	*	139.17	217.02	225.73	223.69	224.95	*	588	876	888	861	950
Wang et al. (2020a. b)	525.19	*	*	*	*	615.94	7111	*	*	*	*	8070
Zhang et al. (2015)	253.38	272.95	*	329.77	*	*	2089	2251	2585	*	*	*
Zhang et al. (2023)	666.41	676.61		724.17	749.61	779.01	7334	7361		7833	8143	9208

Table 3 Geomorphological based classification of glacial lakes and their characteristics

Basis of classification	Lake types	Change rate (m^2/yr)	Count
Glacier and lake relation	Connected	492.45	2707
	Non connected	201.58	1491
Proximal distance	< 500 m	800.56	945
	500–1000 m	268.23	378
	> 1000 m	287.08	2875
Dam type	Moraine dam	740.56	1826
	Rock dam	250.31	994
	Others	504.96	111
Geographical position	Northern side	772.61	1192
	Southern side	825.32	2038
Water supply agent	Glacier fed	414.95	2891
	Non-glacier fed	331.98	1307
Mixed classification	Connected (moraine dam & connected lake)	964.07	711
	Non connected	276.08	3488

In the Himalaya, lake-area growth has been quantified at 26%, 17%, 16%, 20%, and 30% based on the datasets of Mohanty et al. (2023a, b), Wang et al. (2020a, b), Shugar et al. (2020), Zhang et al. (2023), and Zhang et al. (2015), respectively, over varying study periods, predominantly spanning 1990–2020. Notably, all datasets consistently reported the highest rates of lake-area expansion during the 2010–2020 decade, while the lowest growth was observed between 1990 and 2000. These decadal trends clearly indicate a progressive increase in glacial lake area over time, closely associated with the accelerating retreat of glaciers. Elevational analysis revealed that the most significant lake-area expansion occurred within the 4000–5000 m elevation band (Mohanty et al., 2023a, b). Furthermore, studies have shown that lake-area growth rates are markedly higher for glacier-connected moraine-dammed lakes compared to unconnected or non-glacier-fed lakes (Wang et al., 2013; Mohanty & Maiti, 2021; Song et al., 2014, 2017; Zhang et al., 2019).

Basin-wise analysis of average glacial lake area changes rates revealed the highest values in the Upper Indus, Karnali, Tista, Yamuna, and Manas basins. Similarly, the maximum recorded lake area change rates were observed in the Kosi, Yarlung Zangbo, Manas, Karnali, Upper Indus, and Tista basins, indicating these regions as highly susceptible to potential future GLOF events (Fig. 6). Furthermore, the basins exhibiting the lowest minimum lake size change rates—notably in the Kosi, Tista, Upper Indus, Manas, and Yarlung Zangbo—suggest these areas may have previously experienced GLOF occurrences, resulting in reduced lake sizes following past outburst events. This inference is supported by the GLOF event inventory, which highlights several basins with a significant number of recorded outbursts: Hunza (23 events), Yarkant (Bolch et al., 2012), Chitral (Armstrong et al., 2016), Gilgit (Allen

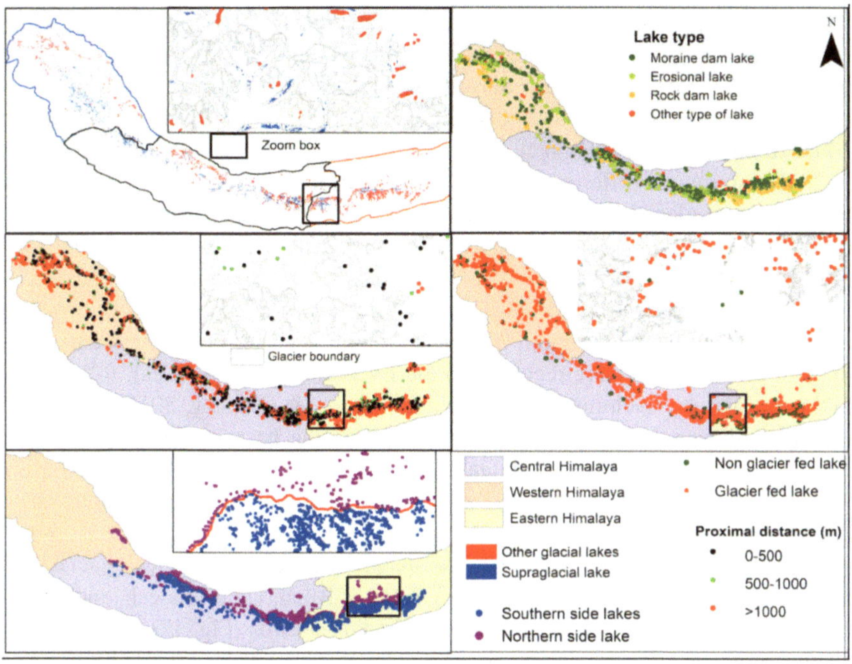

Fig. 6 Distribution of different types of glacial lakes across various parts of the Himalaya. The figure is divided into five segments, each representing a distinct classification scheme: **a** based on genesis, **b** by proximal distance from the glacier terminus, **c** according to topographic setting, **d** by dam type, and **e** based on the source of meltwater

et al., 2011b), Humla (Allen et al., 2019), Dudh Kosi (Allen et al., 2016), and Poique (Ahmed et al., 2021).

3.5 *Critical Lake Distribution in the Himalaya*

Several researchers have identified critical glacial lakes in the Himalayan using diverse data integration approaches and multi-parameter assessment techniques (Washakh et al., 2019; Mohanty & Maiti, 2021; Maurer et al., 2020). These methods include analysing the rate of glacial lake-area change through multi-temporal satellite imagery, assessing the risk of landslides or avalanches that could potentially trigger GLOF events, estimating lake volume using either in-situ bathymetric surveys or empirical relationships between lake area and volume, and evaluating moraine dam stability by detecting the presence of dead ice through field-based observations. Additionally, calving-prone lakes have been identified via satellite image analysis, while probable flood volumes and intensities have been estimated using hydrodynamic modelling techniques (Mohanty & Maiti, 2021; Aggarwal et al., 2017; Wang et al.,

2013). Other approaches involve GIS-based statistical analysis, overlay techniques, matrix-based evaluations, and the fusion of vulnerability and exposure indices—methodologies that have been widely applied by researchers to systematically classify and prioritize critical lakes (McKillop & Clague, 2007; Kääb & Reichmuth, 2005; Allen et al., 2016; Huggel, 2004).

In this study, six different datasets were integrated to map the spatial distribution of critical glacial lakes across the Himalaya (Fig. 7). An overlay analysis was conducted to identify regions of overlap and concentration, resulting in the identification of 221 critical lakes. These critical lakes are predominantly concentrated in the eastern and central sectors of the Himalaya, while comparatively fewer are observed in the western region. At the sub-regional scale, the highest concentrations of critical lakes were recorded in the Everest region (38 lakes), followed by Sikkim (Kääb & Reichmuth, 2005), Bhutan (Cook & Quincey, 2015), and Langtang (Bookhagen & Burbank, 2006) (Fig. 7). At the basin scale, the Kosi basin hosts the highest number of critical lakes (Maurer et al., 2020), followed by Yarlung Zangbo (Dubey & Goyal, 2020), Tista (Dubey & Goyal, 2020), (Donnelly et al., 2004), Qinghai-Tibetan (Carrivick & Tweed, 2016), Panjnad (Bolch et al., 2008), and Phelku (Armstrong et al., 2016). The spatial distribution of critical glacial lakes in the Himalaya indicates that the Indian states of Bihar, Uttar Pradesh, and Sikkim are among the region's most vulnerable to GLOF hazards (Fig. 12). With ongoing glacial retreat and lake expansion driven by climate change, the risk of GLOF events impacting these areas is expected to increase in the future. These findings indicate that settlements, ecosystems, and critical infrastructure located in proximity to the Everest, Langtang, Bhutan, and Sikkim regions are under elevated threat from potential GLOF events. Nevertheless, the precise hazard impact zones and downstream risk profiles require further assessment through hydrodynamic modelling and scenario-based simulations across these priority basins (Figs. 8, 9, 10, and 11).

3.6 Management Plans

As per the Disaster Management (DM) Act 2005 and the National Disaster Management Policy 2009, the National Disaster Management Authority (NDMA) is the apex body for disasters, and the prime minister is the head of the authority. The main objective of these departments is to provide guidelines regarding awareness, preparedness. Moreover, research work has been going on for the future prediction of natural hazards and their intensity, the creation of vulnerable areas, etc. Many research organizations (CSIR, MOES, DST, and IITs) are creating various maps and installing early warning systems to reduce losses from various natural hazards like glacial hazards. A prominent example of glacial lake hazard assessment and mitigated work was done at the South Lhonak glacier lake in Sikkim, where major work was carried out by the Geological Survey of India (GSI), DST Sikkim and Defence Research & Development Organization (DRDO), including remote sensing and field observations to assess Geotechnical and geomorphic factors. Many researchers are focusing

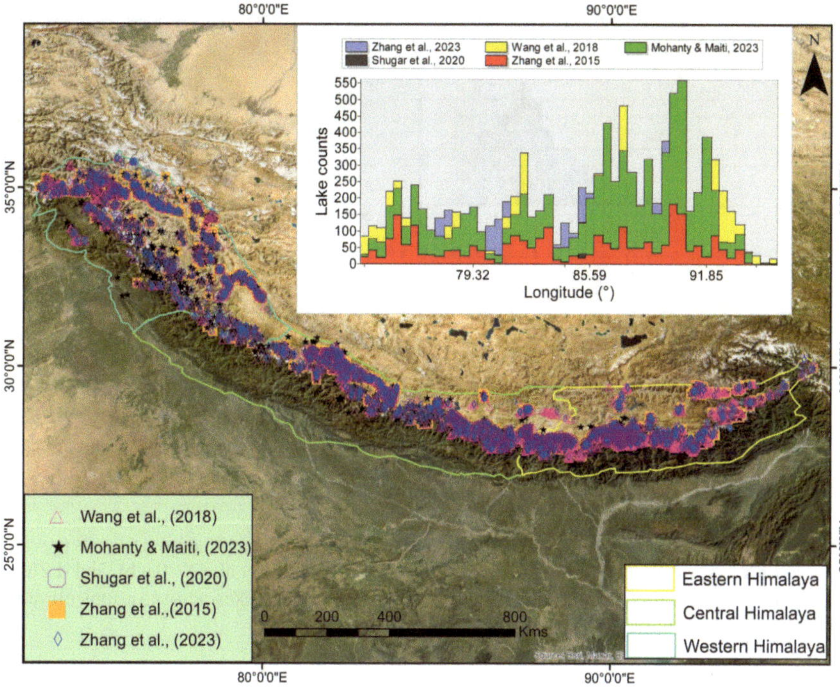

Fig. 7 Latest distribution of glacial lakes across the Himalaya, compiled from various previously conducted studies. The symbols represent data sources as follows: black star (Mohanty et al., 2023a, b), hollow diamond (Zhang et al., 2023), red hollow circle (Shugar et al., 2020), green hollow triangle (Wang et al., 2018), and brown square (Zhang et al., 2015)

on creating GLOF modelling in many particular critical glacial lakes, which are identified as critical by various researchers. Various hazard maps should be distributed to the local authority for the eradication of the GLOF hazard. Similar to the artificial lowering of water levels implemented at South Lhonak Lake by DST Sikkim to reduce hydrostatic pressure and mitigate the risk of a GLOF event, such intervention techniques can be strategically applied to other critically hazardous glacial lakes across the Himalaya to minimize their potential for catastrophic outburst events.

3.7 *Implications for Stakeholders for Establishing a Strong Early Warning System*

It is imperative that a dedicated central authority focusing exclusively on high mountain hazards be established by the Government of India. Alternatively, all activities related to GLOF risk assessment, monitoring, and management could be systematically undertaken by the National Institute of Disaster Management (NIDM), with

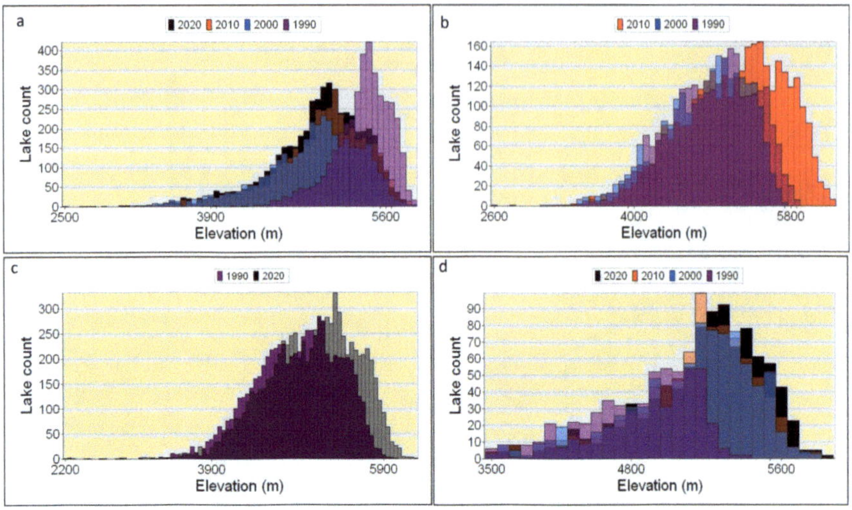

Fig. 8 Elevation-wise distribution of glacial lakes in the Himalaya, compiled from various previous studies: **a** Mohanty et al. (2023a, b), **b** Shugar et al. (2020), **c** Wang et al. (2020a, b), and **d** Zhang et al. (2015). Different colors represent the distribution of glacial lakes across various elevation ranges for different study years

a specific mandate to oversee environmental, glaciological, and geological changes across the Himalayan region. This body should involve a dedicated team of scientists responsible for the continuous observation of critical glacial lakes and the assessment of associated risks in every Himalayan basin across India. Comprehensive field data collection is crucial for developing a robust and standardized model for GLOF risk assessment in downstream regions of the Himalayas.

Glacial lakes, especially those connected to glaciers, undergo routine calving monitoring and moraine subsidence to minimise the GLOF effect. Such events should be tracked using a time series of satellite imagery, such as Sentinel-1 and Sentinel-2. Moreover, these **320** critical lakes identified here should be monitored regularly. Additionally, monitoring should extend to potential mass movement triggers, such as extreme climatic events and earthquakes, to ensure a proactive approach to hazard detection and risk mitigation. Regular and systematic surveillance of these factors will be essential to enhance early warning systems and strengthen disaster preparedness in vulnerable regions.

We recommend the installation of an Automatic Weather Station (AWS), a Broadband Seismometer (BS), and an Automatic Water Level Recorder (AWLR), along with a new device called the water pressure recorder, which has not yet been included in the GLOF early warning system, in every basin across the Himalayas. These instruments should be equipped with V-sat connectivity to enable real-time monitoring and to enhance the effectiveness of the early warning system. The water pressure recorder has been an essential component of tsunami early warning systems for many years. Multiple pressure recorders should be placed at the bottom of each critical lake and

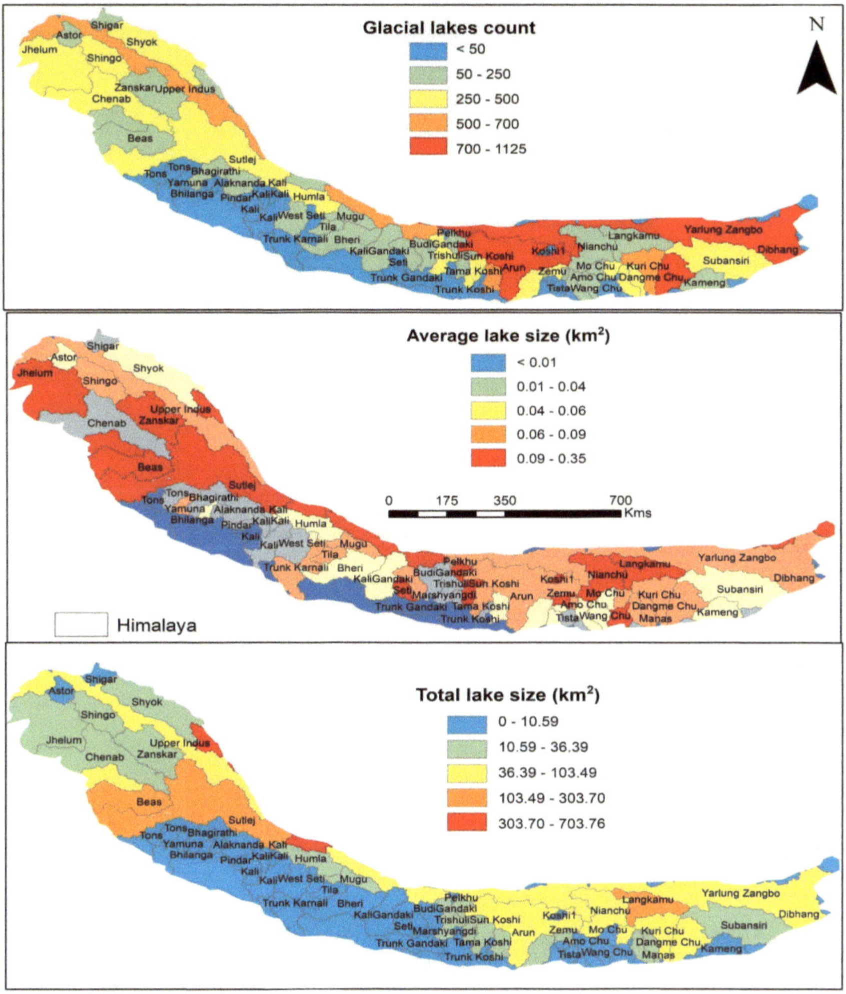

Fig. 9 Basin-wise distribution of **a** glacial lake counts, **b** average lake size, and **c** total lake area across the Himalaya. Color coding represents value ranges, with red indicating high values, yellow indicating medium values, and green indicating low values

along their downstream river channels, similar to their use in oceans. In the event of mass movement into the lake, seiche waves will form, and the subsequent changes in water pressure will be captured and monitored in real-time. This data will facilitate timely early warnings to communities downstream.

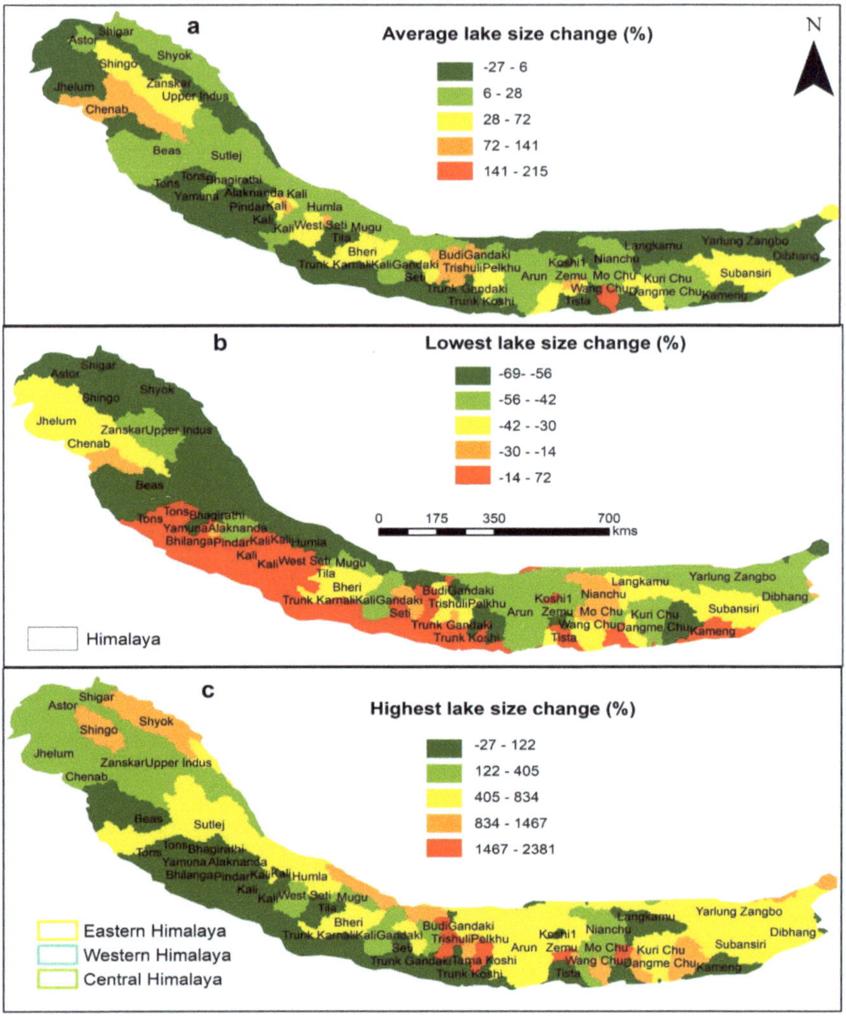

Fig. 10 Percentage rates of **a** average, **b** minimum, and **c** maximum glacial lake size changes across various basins in the Himalaya. Color shading represents value categories: red for higher rates, light yellow for medium rates, and blue for lower rates

4 Summary

- This chapter shows the distribution of lake numbers, sizes, lake types, GLOF events, and critical lakes in the Himalayas. Here, we found most of the lake numbers, sizes, and critical lakes observed in Everest, Sikkim, Langtang, and Bhutan regions. Moreover, these critical lakes are connected to the southern

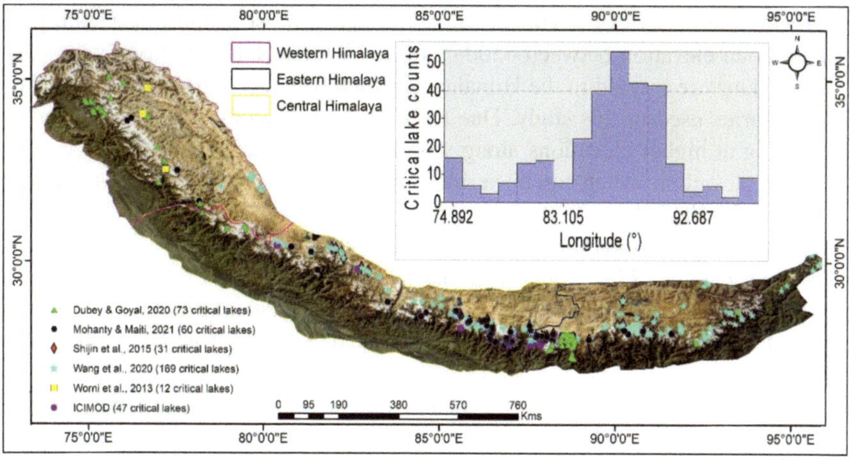

Fig. 11 Distribution of GLOF-prone lakes across the Himalaya as identified by different studies: violet circle—Bajracharya et al. (2020, ICIMOD), green triangle—Dubey and Goyal (2020), blue circle—Mohanty et al. (2023a, b), yellow square—Wang et al. (2013), red diamond—Shijin et al. (2015), and green plus—Wang et al. (2020a, b). The lower two graphs display the frequency of critical lakes in relation to latitude and longitude across the Himalayan region

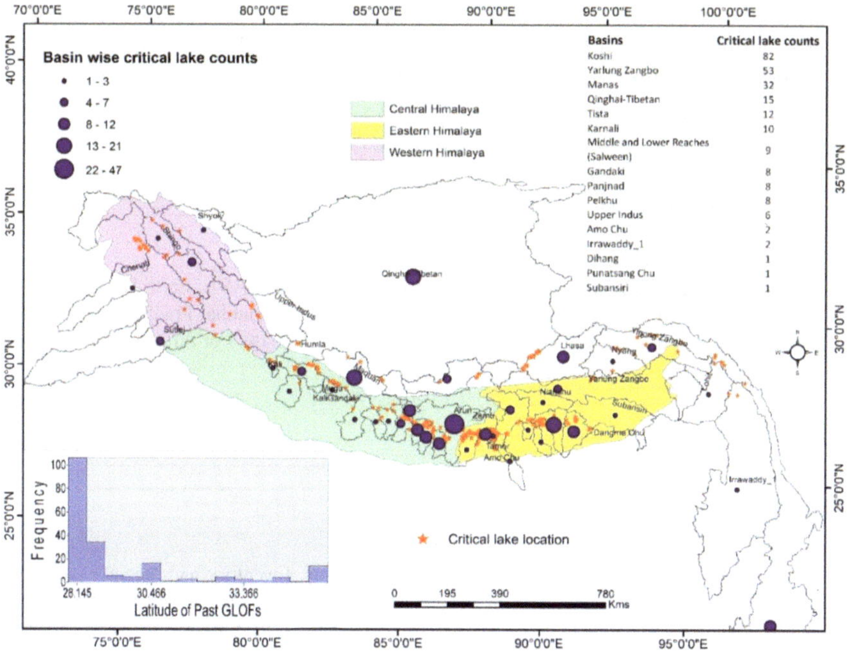

Fig. 12 Basin-wise distribution of critical lakes across various Himalayan basins. The histogram also shows the latitude-wise distribution of past GLOF events in the Himalaya

flowing drainages of the Himalaya. Additionally, most of these lakes are gathered at an elevation between 4500–5500 m. A total of 10,101 lakes with a size >0.01 km^2 are located in the Himalayas, which have been extracted from all the inventories used in this study. Due to continuous glacier retreat, new lakes are forming at higher elevations along with a continual increase in lake size to the higher elevation, which can be verified from 4 datasets from various researchers used in this study.

- The lake count is maximum in Koshi basin (sub basin Arun = 1125), followed by Yarlung Zangbo (773), Manas (sub basin Dangme Chu = 715), and Upper Indus (619). In terms of total lake size distribution, the largest contributions come from the Qinghai-Tibetan, Yarlung Zangbo, Panjnad, Kosi, Manas, and Upper Indus basins.
- The basin-wise average lake size change rate is higher in Upper Indus, Karnali, Tista, Yamuna, and Manas, while the maximum lake size change rate is seen in Kosi, Yarlung Zangbo, Manas, Karnali, Upper Indus, and Tista basins, indicating future GLOF-prone basins among the Himalaya.
- A total of 221 critical lakes were found in the Himalaya. Critical lake density is higher in the Kosi River basin, followed by Yarlung Zangbo, Tista, Qinghai-Tibetan, Panchnad, and Phelku.
- The spatial distribution of critical glacial lakes in the Himalaya indicates that the Indian states of Bihar, Uttar Pradesh, and Sikkim are among the region's most vulnerable to GLOF hazards. With ongoing glacial retreat and lake expansion driven by climate change, the risk of GLOF events impacting these areas is expected to increase in the future."
- It is also inferred from the study that the eastern and central Himalaya are more prone to GLOF than the western Himalaya based on lake type, average lake size, and distribution of critical lakes.

Acknowledgements Satellite images were collected from USGS, NASA, Earth Data and glacier databases from GLIMPS. I want to give my special thanks to Mr. Benudhar Pradha, Dr. Manik Das Adhikari and Mr. Anup Bera for helping me to create this geodatabase. For further data access, SM can be contacted.

References

Aggarwal, S., Rai, S. C., Thakur, P. K., & Emmer, A. (2017). Inventory and recently increasing GLOF susceptibility of glacial lakes in Sikkim, eastern Himalaya. *Geomorphology, 295*, 39–54.

Ahmed, R., Wani, G. F., Ahmad, S. T., Sahana, M., Singh, H., & Ahmed, P. (2021). A review of glacial lake expansion and associated glacial lake outburst floods in the Himalayan region. *Earth Systems and Environment, 5*(3), 695–708.

Ahmed, R., Sam, L., Bhardwaj, A., Kulkarni, A. V., & Yamasaki, S. (2026). Increasing global threat of outburst floods from overlooked small alpine lakes. *Nature Sustainability*, 1–10.

Allen, D. E., Singh, B. P., & Dalal, R. C. (2011a). Soil health indicators under climate change: A review of current knowledge. In B. P. Singh, A. L. Cowie, & K. Y. Chan (Eds.), *Soil health and climate change* (pp. 25–45). Springer.
Allen, D. E., Singh, B. P., & Dalal, R. C. (2011b). Soil health indicators under climate change: A review of current knowledge. In *Soil health and climate change* (pp. 25–45).
Allen, S. K., Linsbauer, A., Randhawa, S. S., Huggel, C., Rana, P., & Kumari, A. (2016). Glacial lake outburst flood risk in Himachal Pradesh, India: An integrative and anticipatory approach considering current and future threats. *Natural Hazards, 84*, 1741–1763.
Allen, S. K., Zhang, G., Wang, W., Yao, T., & Bolch, T. (2019). Potentially dangerous glacial lakes across the Tibetan plateau revealed using a large-scale automated assessment approach. *Science Bulletin, 64*(7), 435–445.
Armstrong, W. H., Anderson, R. S., Allen, J., & Rajaram, H. (2016). Modeling the WorldView-derived seasonal velocity evolution of Kennicott glacier, Alaska. *Journal of Glaciology, 62*(234), 763–777.
Bajracharya, S. R. (2009). *Glacial lake outburst floods risk reduction activities in Nepal. In: Asia-Pacific symposium on new Technologies for Prediction and Mitigation of sediment disasters.* Japan Society of Erosion Control Engineering (JSECE). 2009 Nov 18–19.
Bajracharya, S. R., Maharjan, S. B., Shrestha, F., Sherpa, T. C., Wagle, N., & Shrestha, A. B. (2020). *Inventory of glacial lakes and identification of potentially dangerous glacial lakes in the Koshi, Gandaki, and Karnali River basins of Nepal, the Tibet autonomous region of China.* International Centre for Integrated Mountain Development (ICIMOD).
Bajracharya, B., Shrestha, A. B., & Rajbhandari, L. (2007). Glacial lake outburst floods in the Sagarmatha region. *Mountain Research and Development, 27*(4), 336-344.
Barry, R. G., & Gan, T. Y. (2022). *The global cryosphere: Past, present, and future.* Cambridge University Press.
Bandyopadhyay, J., Rodda, J. C., Kattelmann, R., Kundzewicz, Z. W., & Kraemer, D. (2016). *High mountain waters: Risks and adaptation in a changing climate.* Cambridge University Press.
Basnett, S., Kulkarni, A. V., & Bolch, T. (2013). The influence of debris cover and glacial lakes on the recession of glaciers in Sikkim Himalaya, India. *Journal of Glaciology, 59*(218), 1035–1046.
Berthier, E., Vincent, C., Magnússon, E., Gunnlaugsson, Á. Þ., Pitte, P., Le Meur, E., ... & Wagnon, P. (2014). Glacier topography and elevation changes derived from Pléiades sub-meter stereo images. *The Cryosphere, 8*(6), 2275–2291.
Bhambri, R., Schmidt, S., Chand, P., Nüsser, M., Haritashya, U., Sain, K., & Yadav, J. S. (2023). Heterogeneity in glacier thinning and slowdown of ice movement in the Garhwal Himalaya, India. *Science of the Total Environment, 875*, 162625.
Bolch, T., Buchroithner, M. F., Peters, J., Baessler, M., & Bajracharya, S. (2008). Identification of glacier motion and potentially dangerous glacial lakes in the Mt. Everest region/Nepal using spaceborne imagery. *Natural Hazards and Earth System Sciences, 8*(6), 1329–1340.
Bolch, T., Kulkarni, A., Kääb, A., Huggel, C., Paul, F., Cogley, J. G., & Stoffel, M. (2012). The state and fate of Himalayan glaciers. *Science, 336*(6079), 310–314.
Bookhagen, B., & Burbank, D. W. (2006). Topography, relief, and TRMM-derived rainfall variations along the Himalaya. *Geophysical Research Letters, 33*(8), L08405.
Bormann, K. J., Brown, R. D., Derksen, C., & Painter, T. H. (2018). Estimating snow-cover trends from space. *Nature Climate Change, 8*(11), 924–928.
Carrivick, J. L., & Tweed, F. S. (2016). A global assessment of the societal impacts of glacier outburst floods. *Glob Planet Change., 144*, 1–16.
Chand, M. B., & Watanabe, T. (2019). Development of supraglacial ponds in the Everest region, Nepal, between 1989 and 2018. *Remote Sensing, 11*(9), 1058.
Cogley, J. G., Hock, R., Rasmussen, L. A., Arendt, A. A., Bauder, A., Braithwaite, R. J., ... & Zemp, M. (2011). Glossary of glacier mass balance and related terms.
Cook, S. J., & Quincey, D. J. (2015). Estimating the volume of alpine glacial lakes. *Earth Surf Dynam., 3*(4), 559–575.

Ding, Y., Mu, C., Wu, T., Hu, G., Zou, D., Wang, D., & Wu, X. (2021). Increasing cryospheric hazards in a warming climate. *Earth-Science Reviews, 213*, 103500.
Donnelly, A., Jones, M. B., & Sweeney, J. (2004). A review of indicators of climate change for use in Ireland. *International Journal of Biometeorology, 49*(1), 1–12.
Dubey, S., & Goyal, M. K. (2020). Glacial lake outburst flood hazard, downstream impact, and risk over the Indian Himalayas. *Water Resources Research, 56*(4), e2019WR026533.
Emmer, A., Allen, S. K., Carey, M., Frey, H., Huggel, C., Korup, O., & Yde, J. C. (2022). Progress and challenges in glacial lake outburst flood research (2017–2021): A research community perspective. *Natural Hazards and Earth System Sciences Discussions, 2022*, 1–34.
Emmer, A., Klimeš, J., Mergili, M., Vilímek, V., & Cochachin, A. (2016). 882 lakes of the Cordillera Blanca: An inventory, classification, evolution and assessment of susceptibility to outburst floods. *Catena, 147*, 269–279.
Eyzvikh, V. F., Vilesov, E. N., and colleagues. (1971). Catalogue and morphometric characteristics of mountain glacial lakes in Central Asia (in Russian). Institute of Geography, Academy of Sciences of the USSR, Moscow.
Fujita, K., Sakai, A., Nuimura, T., Yamaguchi, S., & Sharma, R. R. (2009). Recent changes in Imja Glacial Lake and its damming moraine in the Nepal Himalaya revealed by in situ surveys and multi-temporal ASTER imagery. *Environmental Research Letters, 4*(4), 045205.
Fujita, K., Sakai, A., Takenaka, S., Nuimura, T., Surazakov, A. B., Sawagaki, T., & Yamanokuchi, T. (2013). Potential flood volume of Himalayan glacial lakes. *Natural Hazards and Earth System Sciences, 13*(7), 1827–1839.
Fujita, K., Suzuki, R., Nuimura, T., & Sakai, A. (2008). Performance of ASTER and SRTM DEMs, and their potential for assessing glacial lakes in the Lunana region, Bhutan Himalaya. *Journal of Glaciology, 54*(185), 220–228.
Furian, W., Loibl, D., & Schneider, C. (2021). Future glacial lakes in High Mountain Asia: An inventory and assessment of hazard potential from surrounding slopes. *Journal of Glaciology, 67*(264), 653–670.
Gantayat, P., Kulkarni, A. V., & Srinivasan, J. (2014). Estimation of ice thickness using surface velocities and slope: Case study at Gangotri Glacier, India. *Journal of Glaciology, 60*(220), 277–282.
Gardelle, J., Arnaud, Y., & Berthier, E. (2011). Contrasted evolution of glacial lakes along the Hindu Kush Himalaya mountain range between 1990 and 2009. *Global and Planetary Change, 75*(1–2), 47–55.
Garg, S., Agarwal, V., Haritashya, U. K., Kargel, J. S., & Shukla, A. (2025). Evolving glacier patterns in the Chenab River Basin (1993–2021): Drivers and environmental implications. *Science of The Total Environment, 962*, 178430.
Garg, P. K., Malviya, A., Shukla, A., Garg, S., & Singh, N. (2023). On periodic growth and shrinkage of glaciers in the Warwan sub-basin, western Himalaya, between 1990 and 2020. *Environmental Monitoring and Assessment, 195*(3), 390.
Harrison, S., Kargel, J. S., Huggel, C., Reynolds, J., Shugar, D. H., Betts, R. A., & Vilímek, V. (2018). Climate change and the global pattern of moraine-dammed glacial lake outburst floods. *The Cryosphere, 12*(4), 1195–1209.
Hirabayashi, Y., Mahendran, R., Koirala, S., Konoshima, L., Yamazaki, D., Watanabe, S., & Kanae, S. (2013). Global flood risk under climate change. *Nature Climate Change, 3*(9), 816–821.
Huggel, C. (2004). Assessment of glacial hazards based on remote sensing and GIS modeling. *Physical geography Series, 44.*
Huggel, C., Kääb, A., & Salzmann, N. (2006). Evaluation of QuickBird and IKONOS imagery for assessment of high-mountain hazards. *EARSeL eProceedings, 5*(1), 51–62.
Immerzeel, W. W., Droogers, P., de Jong, S. M., & Bierkens, M. F. P. (2010). Satellite derived snow and runoff dynamics in the Upper Indus River basin. *Grazer Schriften der Geographie und Raumforschung, 45*, 303–312.

Ives, J. D., Shrestha, R. B., & Mool, P. K. (2010). Formation of glacial lakes in the Hindu Kush-Himalayas and GLOF risk assessment. Kathmandu: ICIMOD—ISBN 978 92 9115 137 0 (printed) 978 92 9115 138 7 (electronic).

Kääb, A., & Reichmuth, T. (2005). Advance mechanisms of rock glaciers. *Permafrost and Periglacial Processes, 16*(2), 187–193.

Kääb, A., Treichler, D., Nuth, C., & Berthier, E. J. T. C. (2015). Brief communication: Contending estimates of 2003–2008 glacier mass balance over the Pamir–Karakoram–Himalaya. *The Cryosphere, 9*(2), 557–564.

Khadka, N., Zheng, G., Chen, X., Zhong, Y., Allen, S. K., & Gouli, M. R. (2024). An ice-snow avalanche triggered small glacial Lake outburst flood in Birendra Lake, Nepal Himalaya. *Natural Hazards*, 1–9.

Komori, J. (2008). Recent expansions of glacial lakes in the Bhutan Himalayas. *Quaternary International, 184*(1), 177–186.

Komori, J., Koike, T., Yamanokuchi, T., & Tshering, P. (2012). Glacial lake outburst events in the Bhutan Himalayas. *Global Environmental Research, 16*(1), 59–70.

Leith, K. J. (2012). Stress development and geomechanical controls on the geomorphic evolution of alpine valleys (Doctoral dissertation, ETH Zurich).

Leith, K., Moore, J. R., Amann, F., & Loew, S. (2014). Subglacial extensional fracture development and implications for Alpine Valley evolution. *Journal of Geophysical Research: Earth Surface, 119*(1), 62–81.

Lützow, N., Veh, G., & Korup, O. (2023). A global database of historic glacier lake outburst floods. *Earth System Science Data Discussions, 2023*, 1–27.

Maharjan, S. B., Mool, P. K., Lizong, W., Xiao, G., Shrestha, F., Shrestha, R. B., et al. (2018). *The status of Glacial Lakes in the Hindu Kush Himalaya-ICIMOD Research Report 2018/1.* International Centre for Integrated Mountain Development (ICIMOD).

Maskey, S., Kayastha, R. B., & Kayastha, R. (2020). Glacial lakes outburst floods (GLOFs) modelling of Thulagi and lower Barun glacial lakes of Nepalese Himalaya. *Progress in Disaster Science, 7*, 100106.

Maurer, J. M., Rupper, S. B., & Schaefer, J. M. (2016). Quantifying ice loss in the eastern Himalayas since 1974 using declassified spy satellite imagery. *The Cryosphere, 10*(5), 2203–2215.

Maurer, J. M., Schaefer, J. M., Russell, J. B., Rupper, S., Wangdi, N., Putnam, A. E., & Young, N. (2020). Seismic observations, numerical modeling, and geomorphic analysis of a glacier lake outburst flood in the Himalayas. *Science Advances, 6*(38), eaba3645.

Maurer, J. M., Schaefer, J. M., Rupper, S., & Corley, A. J. S. A. (2019). Acceleration of ice loss across the Himalayas over the past 40 years. *Science Advances, 5*(6), eaav7266.

McKillop, R. J., & Clague, J. J. (2007). Statistical, remote sensing-based approach for estimating the probability of catastrophic drainage from moraine-dammed lakes in southwestern British Columbia. *Global and Planetary Change, 56*(1–2), 153–171.

Meena, S. R., Chauhan, A., Bhuyan, K., & Singh, R. P. (2021). Chamoli disaster: Pronounced changes in water quality and flood plains using Sentinel data. *Environmental Earth Sciences, 80*(17), 601.

Mirza, M. M. Q. (2011). Climate change, flooding in South Asia and implications. *Regional Environmental Change, 11*(Suppl 1), 95–107.

Mohanty, L., & Maiti, S. (2021). Probability of glacial lake outburst flooding in the Himalaya. *Resources, Environment and Sustainability, 5*, 100031.

Mohanty, L., & Maiti, S. (2024). A GIS and remote sensing-based approach on supraglacial lake dynamics study in the Himalayan glaciers. *Journal of the Geological Society of India, 100*(10), 1457–1465.

Mohanty, L., Maiti, S., & Dixit, A. (2023a). Spatio-temporal assessment of regional scale evolution and distribution of glacial lakes in Himalaya. *Frontiers in Earth Science, 10*, 1038777.

Mohanty, L., Pateswary, V., Maiti, S., & Mohanty, D. D. (2023b). Future prediction of existing glacial lake's size in the Himalaya by Markov model and glacial surface topography. *All Earth, 35*(1), 329–343.

Mohanty, L. K., & Maiti, S. (2022). Glacial lake formation probability mapping in the Himalayan glacier: A probabilistic approach. *Journal of Earth System Science, 131*(1), 54.

Nagai, H., Fujita, K., Nuimura, T., & Sakai, A. (2013). Southwest-facing slopes control the formation of debris-covered glaciers in the Bhutan Himalaya. *The Cryosphere, 7*(4), 1303–1314.

Nie, Y., Liu, Q., Wang, J., Zhang, Y., Sheng, Y., & Liu, S. (2018). An inventory of historical glacial lake outburst floods in the Himalayas based on remote sensing observations and geomorphological analysis. *Geomorphology, 308*, 91–106.

Nie, Y., Liu, W., Liu, Q., Hu, X., & Westoby, M. J. (2020). Reconstructing the Chongbaxia Tsho glacial lake outburst flood in the eastern Himalaya: Evolution, process and impacts. *Geomorphology, 370*, 107393.

Nie, Y., Sheng, Y., Liu, Q., Liu, L., Liu, S., Zhang, Y., & Song, C. (2017). A regional-scale assessment of Himalayan glacial lake changes using satellite observations from 1990 to 2015. *Remote Sensing of Environment, 189*, 1–13.

Nikitin, A. M. (1987). *Lakes of Central Asia* (p. 106). Gidrometeoizdat.

Nikitin, S. A., & Gorelkin, N. I. (1979). Catalogue of mountain lakes of central Asia and assessment of their Outburst Hazard (in Russian). State Hydrometeorological Service (Hydromet), USSR.

Obu, J., Westermann, S., Bartsch, A., Berdnikov, N., Christiansen, H. H., Dashtseren, A., ... & Zou, D. (2019). Northern Hemisphere permafrost map based on TTOP modelling for 2000–2016 at 1 km^2 scale. *Earth-Science Reviews, 193*, 299–316.

Pandey, A., Parashar, D., Palni, S., Sarkar, M. S., Mishra, A. P., Singh, A. P., & MYA, K. (2024). Spatiotemporal snowline status and climate variability impact assessment: A case study of Pindari River basin, Kumaun Himalaya, India. *Environmental Sciences Europe, 36*(1), 104.

Parkinson, C. L. (2014). Spatially mapped reductions in the length of the Arctic sea ice season. *Geophysical Research Letters, 41*(12), 4316–4322.

Parkinson, C. L., & DiGirolamo, N. E. (2016). New visualizations highlight new information on the contrasting Arctic and Antarctic sea-ice trends since the late 1970s. *Remote Sensing of Environment, 183*, 198–204.

Paul, F., Kääb, A., Maisch, M., Kellenberger, T., & Haeberli, W. (2004). Rapid disintegration of Alpine glaciers observed with satellite data. *Geophysical Research Letters, 31*(21).

Quincey, D. J., Richardson, S. D., Luckman, A., Lucas, R. M., Reynolds, J. M., Hambrey, M. J., & Glasser, N. F. (2007). Early recognition of glacial lake hazards in the Himalaya using remote sensing datasets. *Global and Planetary Change, 56*(1–2), 137–152.

Racoviteanu, A. E., Arnaud, Y., Williams, M. W., & Manley, W. F. (2015). Spatial patterns in glacier characteristics and area changes from 1962 to 2006 in the Kanchenjunga–Sikkim area, eastern Himalaya. *The Cryosphere, 9*(2), 505–523.

Rignot, E., Jacobs, S., Mouginot, J., & Scheuchl, B. (2013). Ice-shelf melting around Antarctica. *Science, 341*(6143), 266–270.

Rubatto, D., Chakraborty, S., & Dasgupta, S. (2013). Timescales of crustal melting in the higher Himalayan Crystallines (Sikkim, eastern Himalaya) inferred from trace element-constrained monazite and zircon chronology. *Contributions to Mineralogy and Petrology, 165*, 349–372.

Sattar, A., Cook, K. L., Rai, S. K., Berthier, E., Allen, S., Rinzin, S., & Bhat, S. Y. (2025). The Sikkim flood of October 2023: Drivers, causes, and impacts of a multihazard cascade. *Science, 387*(6740), eads2659.

Sattar, A., Haritashya, U. K., Kargel, J. S., & Karki, A. (2022). Transition of a small Himalayan glacier lake outburst flood to a giant transborder flood and debris flow. *Scientific Reports, 12*(1), 12421.

Scherler, D., Bookhagen, B., & Strecker, M. R. (2011). Spatially variable response of Himalayan glaciers to climate change affected by debris cover. *Nature Geoscience, 4*(3), 156–159.

Scherler, D., Leprince, S., & Strecker, M. R. (2008). Glacier-surface velocities in alpine terrain from optical satellite imagery—Accuracy improvement and quality assessment. *Remote Sensing of Environment, 112*(10), 3806–3819.

Schwanghart, W., Worni, R., Huggel, C., Stoffel, M., & Korup, O. (2016). Uncertainty in the Himalayan energy–water nexus: Estimating regional exposure to glacial lake outburst floods. *Environmental Research Letters, 11*(7), 074005.

Shijin, W., Dahe, Q., & Cunde, X. (2015). Moraine-dammed lake distribution and outburst flood risk in the Chinese Himalaya. *Journal of Glaciology, 61*(225), 115–126.

Shrestha, A. B., Eriksson, M., Mool, P., Ghimire, P., Mishra, B., & Khanal, N. R. (2010). Glacial lake outburst flood risk assessment of sun Koshi basin, Nepal. *Geomatics, Natural Hazards & Risk, 1*(2), 157–169.

Shrestha, F., Steiner, J. F., Shrestha, R., Dhungel, Y., Joshi, S. P., Inglis, S., & Zhang, T. (2023). A comprehensive and version-controlled database of glacial lake outburst floods in High Mountain Asia. *Earth System Science Data, 15*(9), 3941–3961.

Shugar, D. H., Burr, A., Haritashya, U. K., Kargel, J. S., Watson, C. S., Kennedy, M. C., & Strattman, K. (2020). Rapid worldwide growth of glacial lakes since 1990. *Nature Climate Change, 10*(10), 939–945.

Shugar, D. H., Jacquemart, M., Shean, D., Bhushan, S., Upadhyay, K., Sattar, A., & Westoby, M. J. (2021). A massive rock and ice avalanche caused the 2021 disaster at Chamoli, Indian Himalaya. *Science, 373*(6552), 300–306.

Singh, P., & Bengtsson, L. (2004). Hydrological sensitivity of a large Himalayan basin to climate change. *Hydrological Processes, 18*(13), 2363–2385.

Staviskiy, Y. N., & Jukov, V. M. (1968). Morphometric investigations and inventory of glacial lakes in Central Asia (in Russian). *Soviet Glaciological Survey Reports.*

Taylor, C., Robinson, T. R., Dunning, S., Carr, J. R., & Westoby, M. J. (2023). Glacial lake outburst floods threaten millions globally. *Nature Communications, 14*(1), 487.

Taylor, C. J., & Carr, J. R. (2019). Supraglacial pond evolution in the Everest region, central Himalaya, 2015–2018. *The Cryosphere Discussions, 2019*, 1–33.

Veh, G., Korup, O., von Specht, S., Roessner, S., & Walz, A. (2019). Unchanged frequency of moraine-dammed glacial lake outburst floods in the Himalaya. *Nature Climate Change, 9*(5), 379–383.

Wang, S., Che, Y., & Xinggang, M. (2020a). Integrated risk assessment of glacier lake outburst flood (GLOF) disaster over the Qinghai–Tibetan Plateau (QTP). *Landslides, 17*, 2849–2863.

Wang, X., Ding, Y., Liu, S., Jiang, L., Wu, K., Jiang, Z., & Guo, W. (2013). Changes of glacial lakes and implications in Tian Shan, Central Asia, based on remote sensing data from 1990 to 2010. *Environmental Research Letters, 8*(4), 044052.

Wang, X., Guo, X., Yang, C., Liu, Q., Wei, J., Zhang, Y., & Tang, Z. (2020b). Glacial lake inventory of high-mountain Asia in 1990 and 2018 derived from Landsat images. *Earth System Science Data, 12*(3), 2169–2182.

Wang, X., Liu, S., Guo, W., & Xu, J. (2010). Glacial lake inventory and its recent changes in the Chinese Himalaya. *Chinese Science Bulletin, 55*(36), 4177–4187

Washakh, R. M. A., Chen, N., Wang, T., Almas, S., Ahmad, S. R., & Rahman, M. (2019). GLOF risk assessment model in the Himalayas: A case study of a hydropower project in the upper Arun River. *Water., 11*(9), 1839.

Watson, C. S., Quincey, D. J., Carrivick, J. L., & Smith, M. W. (2016). The dynamics of supraglacial ponds in the Everest region, central Himalaya. *Global and Planetary Change, 142*, 14–27.

Watson, C. S., Quincey, D. J., Carrivick, J. L., Smith, M. W., Rowan, A. V., & Richardson, R. (2018). Heterogeneous water storage and thermal regime of supraglacial ponds on debris-covered glaciers. *Earth Surface Processes and Landforms, 43*(1), 229–241.

Wijngaard, R. R., Lutz, A. F., Nepal, S., Khanal, S., Pradhananga, S., Shrestha, A. B., & Immerzeel, W. W. (2017). Future changes in hydro-climatic extremes in the upper Indus, Ganges, and Brahmaputra river basins. *PLoS One, 12*(12), e0190224.

Worni, R., Huggel, C., & Stoffel, M. (2013). Glacial lakes in the Indian Himalayas—From an area-wide glacial lake inventory to on-site and modeling-based risk assessment of critical glacial lakes. *Science of the Total Environment, 468*, S71–S84.

Yu, Y., Li, B., Li, Y., & Jiang, W. (2024). Retrospective analysis of glacial Lake outburst flood (GLOF) using AI earth InSAR and optical images: A case study of south Lhonak Lake, Sikkim. *Remote Sensing, 16*(13), 2307.

Zemp, M., Huss, M., Thibert, E., Eckert, N., McNabb, R., Huber, J., ... & Cogley, J. G. (2019). Global glacier mass changes and their contributions to sea-level rise from 1961 to 2016. *Nature, 568*(7752), 382–386.

Zhang, G., Bolch, T., Yao, T., Rounce, D. R., Chen, W., Veh, G., & Wang, W. (2023). Underestimated mass loss from lake-terminating glaciers in the greater Himalaya. *Nature Geoscience, 16*(4), 333–338.

Zhang, G., Yao, T., Xie, H., Wang, W., & Yang, W. (2015). An inventory of glacial lakes in the third pole region and their changes in response to global warming. *Glob Planet Change., 131*, 148–157.

Zhang, T., Wang, W., & An, B. (2024). A massive lateral moraine collapse triggered the 2023 south Lhonak Lake outburst flood, Sikkim Himalayas. *Landslides*, 1–13.

Zhang, T., Wang, W., Gao, T., An, B., & Yao, T. (2022). An integrative method for identifying potentially dangerous glacial lakes in the Himalayas. *Sci Total Environ., 806*, 150442.

Mapping of Debris-Covered Glaciers Using Remote Sensing Techniques—A Review

Iram Ali, Aparna Shukla, F. A. Lone, Purushottam K. Garg, and Bisma Yousuf

Abstract Understanding the health and dynamics of Himalayan glaciers has become a global scientific priority due to their critical role in sustaining freshwater supplies for drinking water, agriculture, industry, and hydropower across densely populated downstream regions. Accelerated glacier change under a warming climate has intensified the need for systematic monitoring and improved understanding of glacier responses to climatic variability. Consequently, global-scale glacier mapping, monitoring, and inventory development have emerged as essential components of cryospheric research and water resource assessment. Given the vast extent, complex terrain, and inaccessibility of mountain glaciers, remote sensing remains the most effective and economically viable approach for repeated, large-scale glacier mapping. A wide range of techniques has been developed, including spectral band ratios, index-based classifications, morphometric analysis using topographic attributes, multi-source and texture-based approaches, and supervised classification methods. While these techniques have proven effective for clean-ice and snow-covered glaciers, their applicability is significantly constrained in regions dominated by debris-covered glaciers. Debris-covered glaciers present a major challenge for satellite-based mapping due to the spectral similarity between supraglacial and periglacial debris and the surrounding valley rock, which severely limits discrimination using conventional multispectral imagery. This complexity often leads to underestimation of glacier extent and uncertainty in glacier inventories, with implications for mass balance assessment, hydrological modelling, and hazard evaluation. Over the past decade, substantial methodological advances have been made to address these challenges, incorporating thermal data, surface morphology, texture measures, multi-temporal analysis, and data fusion techniques. This chapter critically reviews the evolution of remote sensing–based methods for mapping debris-covered glaciers,

I. Ali (✉) · F. A. Lone · B. Yousuf
Sher-e-Kashmir Kashmir University of Agricultural Sciences, Shalimar Srinagar, India

A. Shukla
Ministry of Earth Sciences, New Delhi, India

P. K. Garg
G.B. Pant National Institute of Himalayan Environment, Ladakh Regional Centre, Leh, India

R. Ahmed et al. (eds.), *Climate Change, Water Resources, and Natural Hazards in the Himalayas*, Advances in Geographical and Environmental Sciences,
https://doi.org/10.1007/978-981-92-0835-7_2

evaluating their strengths, limitations, and suitability across different physiographic and climatic settings. By synthesizing methodological progress and remaining challenges, the chapter highlights key research gaps and outlines future directions for improving the accuracy, automation, and reliability of debris-covered glacier mapping in high-mountain regions.

Keywords Glacier · Himalaya · Techniques · Debris-cover · Remote sensing

1 Introduction

1.1 Debris-Cover Origin and Types

Glacial debris represents the loose fragments of rock, dust, sand, silt and gravel that are deposited by a melting glacier. Most of the glaciers worldwide are extensively debris-mantled that masks their ablation zones, and hence the name 'debris-covered glaciers (DCGs)'. Several definitions of DCGs can be found in the literature (Racoviteanu et al., 2009; Kirkbride, 2011). The debris entrainment in DCGs occurs from three main sources: supraglacial (at the glacier surface), englacial (through the glacier interior), and sub-glacial (at the glacier bed) sources (Schomacker & Benediktsson, 2018). Supraglacial debris (SGD) may accumulate from different processes and sediment sources, such as thrusting of sub-glacial material, melt-out of englacial debris bands, channel fill material, rockfall from mountain sides and nunataks, snow and ice avalanching or other mass movements, meltwater bursts through the crevasse and conduit system or aeolian deposition directly on the glacier surface (Schomacker, 2008; Kirkbride, 2011; Schomacker & Benediktsson, 2018). As the SGD gets detached from the glacial ice due to downwasting and retreat, it forms the periglacial debris (PGD). Englacial debris is disseminated throughout the glacier interior. It originates either as surface debris buried in the accumulation area or falls into crevasses, or in basal debris emerged as a result of thrusting or folding. Subglacial debris originates from the glacial valley bottom or from the material penetrating the cracks, crevasses and moulins and is generated by plucking, abraded bedrock and over-ridden sediments (Kirkbride, 2011). (Fig. 1) shows the different appearance of different types of debris cover on famous Kashmir Himalayan glaciers.

The evolution of a clean glacier to a DCG is characterized by the reduction in the proportion of internal ice in relation to its SGD and englacial debris. Although the satellite technology enables us to quantify the SGD estimates of a DCG, but there is no standard percent value of SGD that makes a glacier to be characterized as DCG. Some studies have categorized the glaciers with 25–50% and > 50% of SGD as sparsely DCGs and DCGs respectively (Kirkbride, 2011; Banerjee & Shankar, 2013; Ali & Shukla, 2015; Ali et al., 2017). Others assume that DCGs have partially or fully debris-mantled ablation zones (Mattson, 1993; Nakawo, 1993; Singh et al., 2000). The formation conditions of DCGs are generally associated with: (a) their

Fig. 1 An example of spatial distribution of debris-cover types on **a** Kolahoi glacier, largest glacier in west Lidder, Kashmir Himalayas **b** Hoksar glacier, excessively debris-covered glacier in west Lidder, Kasmir Himalayas. Here SI = snow-ice, IMD = ice-mixed debris, SGD = supraglacial debris, PGD = periglacial debris, VR = valley rock, and VEG = vegetation

topographic settings (steeper accumulation and gentler ablation areas, and slope-orientation), (b) diurnal fluctuations in surface temperatures, (c) sustained negative mass balance, (d) permafrost degradation and slope failure, and (e) deglaciation of rock glaciers (Nagai et al., 2013; Ojha et al., 2017; Gibson et al., 2017). Also, the glacier dynamics of DCGs responds differently to a warming climate than the clean glaciers.

1.2 Distribution of Debris-Cover

DCGs occur globally, particularly in high-relief mountain ranges with significant debris supplies including the Himalayas, Karakoram, Caucasus, Alaska, New Zealand, and parts of the Andes, as well as in Iceland, Greenland, Svalbard and the west-coast of the continental United States, and also in the east-Antarctic Dry Valleys (Koutnik et al., 2016). Landsat 8 and Sentinel-2 optical data-derived global datasets indicate ~ 4.4% of all glacier surfaces (excluding the Greenland ice sheet and Antarctica) to be debris-covered, exhibiting heterogeneous distribution. The global assessment of DCGs reveal an inverse trend between the percentage of debris-cover and glacier size in most of the regions (Scherler et al., 2018). Debris usually accumulates in planar geometric shapes (debris septa) or into discrete bodies of variable geometry on the DCG surface. The mass of debris originating from valley sides that gets distributed along the sides of the glacier are called lateral moraines. The remnant of lateral moraines along the valley sides can prove the former existence of past glaciers. If the debris piles up longitudinally downstream from junctions between confluent glaciers, it is termed as medial moraine. It includes debris derived from the lateral moraines of each tributary glacier. Generally, medial moraines are in continuous form only where debris is supplied almost continuously from a point source. Otherwise, they show a beaded form where inputs of debris are pulsed such as infrequent rockfalls from the same source (Menzies, 2002; Goudie, 2004).

The spatial pattern of supraglacial debris (SGD) is governed by the complex interactions between tributary glaciers and the main trunk glacier. Its distribution is primarily influenced by the supply of rock material from surrounding valley walls and tributaries, along with its subsequent transport by glacier flow. As glaciers undergo mass loss and ice velocity decreases, the redistribution of debris continuously evolves, leading to dynamic changes in SGD coverage (Bolch et al., 2008; Rowan et al., 2015; Gibson et al., 2017). Supraglacial debris exhibits considerable variability in both grain size and surface extent, ranging from fine particles to large boulders, and from sparse, patchy cover to complete burial of glacier tongues. Fine materials such as dust and soot are typically dispersed across the ablation zone, where they progressively reduce surface albedo. In contrast, coarser debris components—including pebbles and boulders—tend to accumulate along well-defined medial moraines (Paul et al., 2004).

Investigating the spatio-temporal characteristics and distribution of SGD is crucial because debris cover exerts a strong control on glacier behaviour and mass balance processes (Wang et al., 2011; Zhang et al., 2011; Banerjee & Shankar, 2013; Pratap et al., 2015; Racoviteanu et al., 2015; Gibson et al., 2017; Sharma et al., 2016; Shukla & Qadir, 2016). Variations in debris lithology and thickness over space and time significantly modify the surface energy balance, thereby influencing melt rates and patterns of glacier ablation. Despite its importance, accurately quantifying spatial and temporal changes in debris thickness and distribution—either through remote sensing techniques or field-based measurements—remains challenging (Nicholson & Benn, 2006; Reid et al., 2012; Pellicciotti et al., 2015; Rowan et al., 2015). These limitations hinder a comprehensive understanding of how debris-covered glaciers (DCGs) respond to ongoing climatic changes. Identifying debris sources and analysing their interaction with glacier dynamics are therefore essential for explaining observed variations in SGD distribution (Gibson et al., 2017).

1.3 Characteristics of a DCG

Some of the major characteristics of DCGs inferred from a number of pioneering studies are summarized below (Hambrey et al., 2008; Schomacker, 2008; Scherler et al., 2011; Benn and Evans, 2014; Bolch et al., 2012; Banerjee & Shanker, 2013; Juen et al., 2014; Rowan et al., 2015; Anderson & Anderson, 2016; Scherler et al., 2018; Ali et al., 2024).

(i) DCGs are either extensively or sparsely debris-mantled with stagnant or advancing/retreating fronts.
(ii) DCGs are mostly flanked by large lateral and terminal moraines, formed by reworking of SGD by gravitational and glacifluvial processes.
(iii) DCGs may be characterized by the presence of moraine-impounded proglacial lakes and melt hot spots such as ice-cliffs and supraglacial lakes (SGLs).
(iv) DCGs have thinner and longer ablation zones.
(v) The SGD present in the ablation zone perturbs the ice flow velocity (IFV) pattern. The velocities are maximum up-glacier and decrease in the downstream direction.
(vi) Debris thickness generally increases down-glacier with the accumulation of melt-out debris, debris from rockfall on the valley sides and gradual decrease in IFV down-glacier.
(vii) The ablation and surge of DCGs are highly dependent on their debris thickness and slope.
(viii) DCGs have larger climate sensitivities than clean glaciers with the same equilibrium line altitude (ELA).
(ix) DCGs can persist at lower elevations and attain longer lengths than clean glaciers, resulting in much longer timescales of mass loss and melt-water production.
(x) The terminus positions of DCGs tend to be stable for longer periods. Mass loss in DCGs occurs mostly by surface lowering rather than terminus recession. The accumulation area ratio (AAR) of DCGs tends to be smaller than clean glaciers. Benn and Evans (1998) reported that DCGs in equilibrium have AARs of 0.2–0.4, compared with values of 0.6–0.7 for clean glaciers. The specific mass-balance profile of DCGs decreases to a minimum at certain altitude in the ablation zone and then either flattens out or increases in the downstream direction. While for clean glaciers, it decreases with decreasing altitude and is minimum at the glacier front.

1.4 Impact of Debris-Cover on Glacier Parameters

Debris-cover exerts a strong impact on different glacier parameters such as ablation, IFV, retreat/deglaciation, surface elevation changes (SECs), and mass balance (Kayastha et al., 2000; Takeuchi et al., 2001; Berthier et al., 2007; Brock et al., 2010; Scherler et al., 2011; Dhobal et al., 2013; Collier et al., 2015; Sharma et al., 2016; Garg et al., 2017). The impact of debris on glacial ablation rates is significantly affected by both geographic location and altitude. The effect of debris cover is magnified at high altitude and in lower latitudes (Shukla et al., 2010). The shallow debris layers lower the surface albedo, causing it to absorb more solar radiation and thus enhance the glacial ablation rate. Whereas, thicker layers inhibit ablation rates by reducing the amount of surface heat conducted to the ice. At a critical debris thickness, the ablation rates of the debris-covered ice and the clean ice are equal. Above this critical thickness, the ablation rate of underlying ice decreases exponentially with increasing debris thickness (Østrem, 1959; Kayastha et al., 2000; Nicholson & Benn, 2006). The ablation rate of debris-covered ice depends on the physical characteristics of debris layer such as lithology, porosity, surface albedo, debris thickness, thermal conductivity, and heat capacity as well as their meteorological conditions (Gibson et al., 2017). However, it is often difficult to quantify the degree to which debris alters the glacial ablation rate and the critical thickness at which ablation is retarded (Scherler et al., 2011).

SGD alters the glacier surface-energy and mass fluxes at the atmosphere–glacier interface. It allows the surface temperature to exceed the melting point and modification of the surface roughness and saturation conditions, which impacts the surface-energy fluxes and the atmospheric boundary layer (Collier et al., 2015). Change in SGD thickness also changes the morphology of a DCG. The smaller ablation rates due to the increased SGD thickness cause ablation zones to enlarge to offset mass gains in the accumulation zone, and also the glacier length profile becomes concave over time. It has been suggested that for Himalayan DCGs, the surface topographic profile may indicate an increasing or decreasing mass, with thickening/advancing glaciers indicated by a convex surface profile, and thinning/receding glaciers depicting a concave surface profile (Singh et al., 2011). The spatial variability in SGD also promotes the supraglacial lake (SGL) formation in the ablation zone of the glacier through differential ablation pattern, thermokarst processes and surface slope reduction. SGLs are known to accelerate the ablation rate of DCGs by producing internal ablation in the englacial channels which transfer warm water to the interior of the DCGs. Following the collapse of englacial channels, new ice cliffs and ponds are created (Thompson, 2017).

DCG length is highly sensitive to the surface debris flux and the site of SGD deposition. High surface debris fluxes can greatly increase glacier lengths relative to the clean glaciers under given climatic conditions. SGD deposition near the ELA, where ice discharge is high, leads to longer glaciers with greater fractional debris-cover. Whereas, SGD deposition near the headwall or terminus, where ice discharge is low, leads to shorter glaciers with smaller fractional debris-cover (Scherler et al.,

2011; Anderson & Anderson, 2016). Various studies have reported higher recession rates for clean glaciers than DCGs (Scherler, 2011; Dhobal et al., 2013; Shukla & Qadir, 2016; Shukla & Ali, 2016; Ali et al., 2017; Ali et al., 2024). Debris-cover on glaciers are found to be controlled by slope, higher deglaciation rates, higher IFVs in the upper ablation zone coupled with lower IFVs in the lower ablation zone and surface lowering (Scherler et al., 2011; Garg et al., 2017). The variations in different glacier parameters are intimately interlinked with each other through active positive and negative feedback mechanisms. An increase in debris-cover leads to deglaciation, which in turn affects the glacier retreat and thinning. Reduced IFVs cause stagnation, favor increased accumulation of debris-cover, and enhance surface lowering. Surface lowering is again linked with slope gradient decrease, IFV dropdown, and debris-cover increase (Garg et al., 2017).

2 Significance of Mapping DCGs

At present glaciologists are actively concerned over DCGs particularly with respect to their anomalous behavior towards ongoing climate change (Østrem, 1959; Scherler et al., 2011; Racoviteanu & Williams, 2012; Ali et al., 2017; Ali et al., 2024). The DCG mapping has been addressed in many studies (Shukla et al., 2010; Scherler et al., 2011; Kirkbride & Deline, 2013; Banerjee & Shanker, 2013; Racoviteanu et al., 2015; Shukla & Qadir, 2016; Ali et al., 2017) emphasizing the requirement for the inclusion and increased monitoring of DCGs within current global glacier monitoring programs. The significance of mapping DCGs is outlined below:

(i) Since the sub-debris melt rates of DCGs is strongly influenced by the extent and thickness of debris layer, therefore a comprehensive and detailed mapping of these glaciers will give the clear status of the total water storage capacity within the glacierized basin which subsequently is of utmost importance in quantification of glacier runoff and in management of downstream water system.
(ii) The characteristic association of SGLs and ice-cored moraines with the DCGs necessitates their large-scale mapping for realistic glacier hazard assessments. This is because, overall melt from SGLs, englacial drainage, pro-glacial lake calving and/or ice-cored moraine disintegration can cause catastrophic glacial lake outburst floods (GLOFs). In this context, large scale DCG mapping is critical
(iii) The change in the spatial patterns of mass balance of DCGs is a function of climate change as well as the extent and thickness of debris cover. In addition, the degree day factor of such glaciers is considerably different from clean glaciers. Therefore, the assessment of DCGs at regular intervals would greatly determine the precision of hydrological modeling and improve mass-balance studies.

(iv) Considering their reverse ablation gradient, precise mapping of DCGs and understanding their subsequent response to climate change need a special consideration.
(v) The increased understanding of DCGs and inclusion of debris impact upon glacial surface melt and their retreat will serve as the baseline data for modeling the past and future contribution of glacier melt to global sea level rise.
(vi) Downwasting being the common source of mass loss in DCGs, monitoring their IFVs will provide key information on glacier dynamics and its stability. Moreover, it provides valuable information regarding icefall, glacier hazard and surge events.

3 Constraints/Challenges in Satellite-Based Delineation of DCGs

The availability of low-cost remote sensing data along with its global coverage provides the opportunity for regular glacier monitoring at regional scales. The characteristic higher reflectance of snow in the visible region allows its mapping through the semi-automated approaches. However, debris-mantled snouts of DCGs critically impede their accurate satellite-based mapping due to their spectral similarity with the surrounding rocky moraines and glacier forefields. Thus, spectral information alone has proved to be inadequate for mapping of DCGs (Paul et al., 2004; Raup et al., 2007; Shukla et al., 2010; Bhambri et al., 2011; Frey et al., 2012; Burns & Nolin, 2014; Shukla & Ali, 2016; Shukla & Yousuf, 2016). Besides, manual delineation of debris-covered boundaries over an extensive area is limited due to its laborious and time-consuming nature (Bayr et al., 1994; Paul et al., 2002; Stokes et al., 2007). Therefore, studies have long been using the geomorphometric parameters (Bolch & Kamp, 2005) or a combination of geomorphometric parameters and band ratios (Paul et al., 2004), along with the spectral data for semi-automated mapping of DCGs. Although these multisource approaches have yielded improved results but the mapping accuracy is usually subjected to the quality of DEM. Further, their usage in differentiating SGD and PGD remains unproven.

Another technique for DCG mapping is based on the thermal difference between the SGD (expected to be relatively colder), and the surrounding moraines (Taschner & Ranzi, 2002). However, the applicability of thermal approaches is limited by the fact that the temperature of thick debris overlying the ice is similar to that of the surrounding moraines, thus confounding the signal. To overcome this problem, several studies have combined morphometric and thermal parameters (Bolch et al., 2008; Bhambri et al., 2011) and optical/thermal and geomorphometric parameters (Shukla et al., 2010; Karimi et al., 2012). Although these synergistic methods satisfactorily mapped the debris-covered areas but these methods were majorly optimized for a single glacier and their applicability over an extended area is complicated by the heterogeneous distribution of the debris thickness and the thermal properties along the glacier margins. Alternately, the DCGs were delineated by employing optical and texture data analysis which increased their mapping accuracy to certain degree (Racoviteanu & Williams, 2012). In this case the efficiency of the results is challenged by the precision of thresholds applied for mapping debris-cover which may vary from one glacier to another. Recently knowledge-based classification technique was employed for mapping DCGs (Shukla & Ali, 2016) which though showed promise in mapping a single DCG but still needed manual input when applied for large-scale mapping of DCGs (Ali et al., 2017).

Several researchers (Robson et al., 2015; Bajracharya et al., 2015) have used object-based image analysis which differentiates the debris-cover not only on the basis of spectral characteristics but also incorporates shape, texture and relationships between the objects. This approach has shown better performance in DCG mapping and still higher accuracy could be achieved by post-processing possibilities such as neighborhood analysis.

It should be noted that all of the above mentioned approaches require cloud free optical imagery that are not always available for many glaciers due to forecast conditions. This gap was complimented by applying passive microwave systems. With this advantage many recent studies have utilized coherence pattern between two SAR images for their differentiation (Frey et al., 2012; Saraswat et al., 2013). However, the main challenge in applying this technique over the mountainous regions is complicated by the complex topography where layover and foreshortening can cause no signal return to the sensor over sizable areas (Frey et al., 2012). In general, there is no single best technique for DCG mapping as these advanced techniques have not so far been compared and a superior method has thus not yet emerged. Thus, the development in DCG mapping is still in progress.

4 Remote Sensing Based Mapping of DCGs

Considering the aforesaid challenges in RS based mapping of DCGs several methods have been used for their accurate mapping using various types of satellite RS datasets. A review of the image processing methods applied till date for delineating the DCGs has been discussed here.

4.1 Manual Delineation

Glacier boundaries are often mapped through on-screen digitization of satellite imagery and derived products, including false colour composites (FCCs) and image enhancement techniques such as spectral ratios, principal component analysis, and contrast stretching (Fig. 2). Delineation is based on visual interpretation, where key image elements—tone, texture, shape, size, pattern, association, and shadow—are evaluated either individually or in combination to distinguish glacier ice from surrounding terrain.

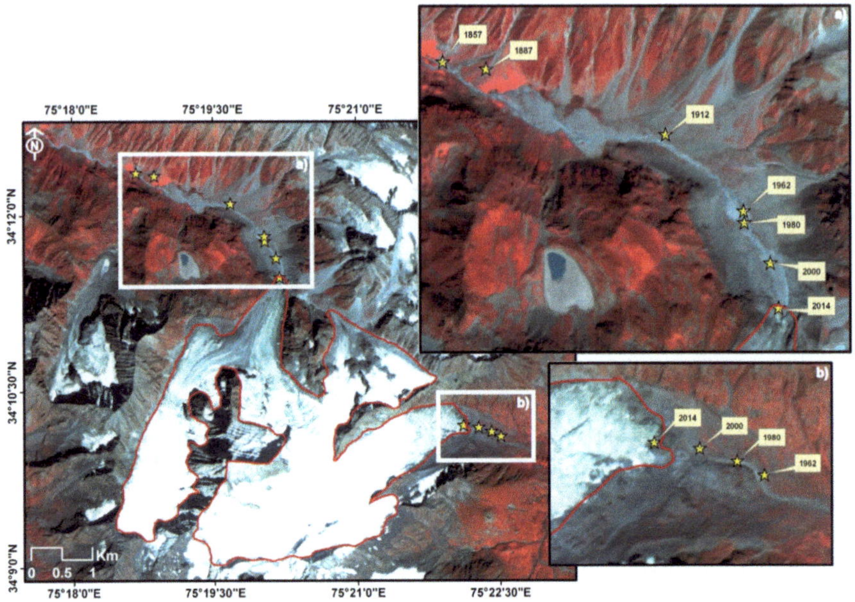

Fig. 2 An example of manual delineation (shown by red boundary) using a false colour composite (321) of 19th Sept 2013 IRS P6-LISS IV image of Kolahoi glacier, Kashmir Himalaya. The yellow-coloured stars mark the position of both snouts at different points in time

Mapping debris-covered glaciers (DCGs) through manual interpretation is particularly challenging. The spectral similarity between supraglacial debris and adjacent morainic or valley materials makes boundary identification complex, demanding both time and strong expertise in glacial geomorphology. Several ancillary cues can improve interpretation accuracy, including shaded relief generated from digital elevation models (DEMs), valley morphology, visible ice-flow features, and the coarse surface texture characteristic of debris-covered ablation zones. Additional indicators such as shadows cast by steep peaks, vegetation-free high-altitude terrain, and geomorphic contrasts further assist glacier mapping. Accurate identification of DCG termini can be supported by observing proglacial streams, ice-cliff shadows, supraglacial lakes (SGLs), and distinctive proglacial landforms, which serve as diagnostic features (Shukla & Qadir, 2016). The use of optimized band combinations and image transformations can further enhance feature discrimination during manual interpretation.

Manual digitization has been widely applied to document glacier extent changes and assess retreat and deglaciation trends across different mountain regions (Hall et al., 1995; Stokes et al., 2007; Burns & Nolin, 2014; Shukla & Qadir, 2016; Garg et al., 2017; Ahmed et al., 2026). Despite its reliability, this approach is labor-intensive and time-consuming, limiting its suitability for large-scale or operational glacier inventories. Furthermore, because the method depends heavily on analyst judgment, discrepancies may arise between different studies due to variations in expertise, interpretation criteria, and subjective bias (Rai et al., 2016).

4.2 Semiautomatic and Automatic Methods

Considering the limitations of manual technique for large-scale mapping and monitoring of DCGs, research led to the development of various semiautomatic and automatic methods. These may be further sub-divided into methods using single and hybrid approaches.

(i) Using single approach

(a) *Spectral ratios and indices*

Spectral ratios and indices usually aim at enhancing the spectral information in the images and reduce the variable effects of solar illumination and topography (Gupta, 2003). Spectral ratios and indices were primarily devised to differentiate the snow-ice covered regions from the non-snow-ice covered regions. These work on the basic premise that snow-ice regions have higher reflectance in the visible part of the electromagnetic spectrum and exhibit strong absorption in the SWIR region. Spectral ratios such as NIR/SWIR, Red/SWIR and Red/NIR have been used extensively in glaciology (Table 1). Out of these NIR/SWIR ratios could also be found suitable for mapping of clean glaciers, however, they fail to map the ice covered by debris (Burns & Nolin, 2014).

A detailed review of spectral ratios and indices used for mapping of snow-ice classes has been given by Arora et al. (2011). Table 2 shows the spectral indices used in glaciological research. Amongst these, the Normalized Difference Snow Index (NDSI) has been the most widely used and robust methods for segregation of snow-ice covered areas from the non-snow ice covered regions (Hall et al., 1995; Burns & Nolin, 2014). NDSI has also been utilized along with a DEM for mapping dry- and

Table 1 Basic formulations of spectral ratios used for mapping snow cover and clean glaciers (after Arora et al., 2011).

S. No.	Popular Name	Formulation	Reference(s)
1	TM4/ TM5	$\frac{NearInfrared(NIR)}{ShortwaveInfrared(SWIR)}$	Hall et al. (1987, 1988); Bayr et al. (1994); Rott and Markl (1989); Jacobs et al. (1997); Williams et al. (1991); Bronge and Bronge (1999); Paul (2000, 2003); Paul et al. (2004)
2	TM3/ TM5	$\frac{Red}{ShortwaveInfrared(SWIR)}$	Rott and Markl (1989); Rott (1994); Jacobs et al. (1997); Bronge and Bronge (1999)
3	TM3/ TM4	$\frac{Red}{NearInfrared(NIR)}$	Bronge and Bronge (1999)

wet-snow in using IRS LISS-III data (Gupta et al., 2005). Haq et al. (2014) modified the NDSI to make a new index Normalised Difference Snow Thermal Index (NDSTI) using blue and thermal infrared bands for mapping of snow-ice regions and excluding the water bodies.

Recently, attention has been paid towards devising spectral indices for discriminating the snow-ice facies (Keshri et al., 2009; Bhardwaj et al., 2015). Keshri et al. (2009) proposed two new indices (Normalized Difference Glacier Index (NDGI) and Normalized Difference Snow and Ice Index (NDSII-2)) for detailed mapping of supraglacial terrain using ASTER data of a Himalayan glacier. Based on the spectral

Table 2 Formulations and utility of various spectral indices for mapping of snow and glacier covered areas (modified after Arora et al., 2011)

Name of the index	Reference(s)	Formulation	Utility
Normalized difference snow index (NDSI)	Dozier (1989), Hall et al. (1995)	$NDSI = \frac{Green-SWIR}{Green+SWIR}$	For mapping and differentiation of snow-ice covered areas from non-snow and ice areas
Normalized difference snow and ice index (NDSII-1)	Xiao et al. (2001)	$NDSII = \frac{Red-SWIR}{Red+SWIR}$	For mapping and differentiation of the snow-ice covered areas from non-snow and ice areas
S3 Index	Saito and Yamazaki (1999), Shimamura et al. (2006)	$S3 = \frac{NIR(red-SWIR)}{(NIR+red)(NIR+SWIR}$	For mapping snow and ice cover under forest covered areas
Normalized difference glacier index (NDGI)	Keshri et al. (2009)	$NDGI = \frac{Green-Red}{Green+Red}$	For mapping and differentiating between snow-ice and mixed ice and debris class
Normalized difference snow and ice index (NDSII-2)		$NDSII = \frac{Green-NIR}{Green+NIR}$	For mapping and differentiating between snow and ice class
Normalised difference snow thermal index (NDSTI)	Haq et al. (2014)	$NDSI = \frac{Blue-TIR}{Blue+TIR}$	For mapping and differentiation of snow-ice covered areas from non-snow and ice areas without mapping water bodies as snow-ice
A Index	Bhardwaj et al. (2015)	$A = \frac{T^2}{(OB1/OB9)\times(OB5/OB6)}$ where T is at-satellite brightness temperature and OB is Landsat 8 OLI bands	For mapping various supraglacial features

difference between classes, the indices facilitate the sequential 3-step discrimination. First, the snow and ice are separated from rest of the terrain using NDSI, then snow and ice from ice-mixed-debris class using NDGI and finally snow from ice using NDSII-2 (Table 2). Bhardwaj et al. (2015) devised a Landsat8 Operational Land Imager (OLI) sensor based 'A' index for automated mapping of various supraglacial features (crevasses, ice, dirty ice, slush zone, snow and supraglacial debris) by thresholding the index values. Though robust because they are mostly spectrally-based, a major issue with the ratios and indices is the subjectivity involved in deciding the threshold values (Gjermundsen et al., 2011; Burns & Nolin, 2014). The threshold values for ratios and indices tend to vary from scene to scene probably owing to differences in illumination and topography. However, Burns and Nolin (2014) have shown that with effective atmospheric and topographic corrections, a single threshold value can be used for mapping snow-ice cover from different scenes staying within the permissible error limits.

Literature suggests that spectral ratio and indices proposed till date have not been exclusively employed for mapping of DCGs. Though, indices have been used for the mapping of SGD (Keshri et al., 2009; Bhardwaj et al., 2015) but within a pre-defined manually delineated glacier boundary. However, the potential of the ratios and indices has been successfully utilized by combining them with other data sources for effective mapping of DCGs (Paul et al., 2004; Bhambri et al., 2011; Racoviteanu & Williams, 2012; Alifu et al., 2015b; Bharadwaj et al., 2015; Shukla & Ali, 2016; Ali et al., 2017).

(b) *Digital image classification (DIC)*

DIC is an image information extraction technique which aims at obtaining spatial distribution and quantitative data for classes (land cover type). DIC can be per-pixel or sub-pixel depending upon the type of input image and may be performed in supervised (Parallelepiped Classifier, Minimum Distance to Mean Classifier, Maximum Likelihood Classifier, Artificial Neural Network (ANN) classifier, Decision Tree Classifier, Support Vector Machines) or unsupervised manner (K-means clustering, Iterative Self Organising Data Analysis Techniques (ISODATA, a terminal 'A' added for aesthetic reasons) depending on the prior knowledge of the analyst about the study area. DIC has been widely used in both supervised and unsupervised modes for snow and ice mapping (e.g., Paul, 2000, 2003; Paul et al., 2004). Classification techniques have also been applied for specific problems like estimation of debris-cover over Himalayan glaciers (Raikot, Siachen and Shaigiri) (Bishop et al., 1995). Bishop et al. (1995) applied an ANN to estimate the SGD loads of Himalayan glaciers using pre-defined glacier outlines and also compared the performance of ANNs to the unsupervised ISODATA clustering algorithm. Results indicated that minimal training of ANN is sufficient to produce accurate information regarding supraglacial characteristics and ANN performs better than ISODATA clustering algorithm. Contrary to this, when Paul et al. (2004) tried to map five glacier terrain classes (debris, bare ice, rock, vegetation, and shadowed areas) using ANN on spectral data, they did not achieve satisfactory results as compared to the semi-automatic approach which utilized data from varied sources.

However, with the advancements in glaciology, the needs to segregate the spectrally similar glacier facies and correctly mapping the glacier boundaries (with special reference to ice-extensions beneath SGD cover) emerged. DIC on spectral data alone suffered severe limitations in this regard (Gjermundsen et al., 2011). Therefore, classification was performed on data from multiple sources (image transformations, geomorphometric data, surface temperature and textural information etc.) which could aid better characterization of complex features like DCGs (Paul et al., 2004; Shukla et al., 2010; Karimi et al., 2012; Khan et al., 2015). Such classification approaches are referred as 'multisource classification'. These techniques have been discussed under the forthcoming sections.

(c) *Geomorphometric Parameters*

Terrain-based geomorphological parameters derived from digital elevation models (DEMs)—such as elevation, slope, aspect, profile curvature, and plan curvature—are closely linked to landscape-forming processes and therefore provide valuable information for mapping debris-covered glaciers (DCGs). Among these, slope gradient and curvature metrics are particularly effective in characterizing glacier surfaces and identifying transitions between glacier ice, lateral moraines, adjacent mountain slopes, and proglacial areas.

Profile curvature, which describes surface curvature along the direction of maximum slope, is useful for highlighting linear features such as mountain crests and lateral moraines. In contrast, plan (or plane) curvature, representing curvature perpendicular to the direction of maximum slope, is well suited for detecting flow convergence and divergence patterns and is often applied in hydrological analyses (Buchroithner & Bolch, 2007). Variations in curvature values have been shown to mark the contact zone between glacier ice and adjoining moraines. For instance, Kieffer et al. (2000) demonstrated that noticeable curvature changes occur at the ice–moraine boundary and combined this information with central flow-line analysis to delineate glacier tongues.

Hydrologically derived flow paths have also been used to define glacier margins. Bishop et al. (2000) employed modeled flow paths initiated from seed points together with a V-shaped operator along the outer margins of lateral moraines to identify glacier perimeters and termini. In a subsequent study, Bishop et al. (2001) applied a hierarchical object-oriented classification approach at Nanga Parbat (northern Pakistan), integrating DEM-derived variables such as elevation, slope, aspect, and curvature. Their findings indicated improved performance when aspect was excluded from the model. Although DCGs could be mapped, uncertainties persisted in areas where glacier boundaries transitioned gradually into surrounding terrain, particularly where lateral moraines were indistinct or poorly resolved in the DEM (Bolch et al., 2008). Similarly, Bolch and Kamp (2006) introduced a morphometry-based glacier mapping (MGM) approach in the Alps, relying on clustering of curvature parameters (plan and profile curvature) to delineate debris-covered ice. Despite these advances, methods that depend solely on geomorphometric indicators require substantial expertise and face limitations in regions where glacier–terrain contrasts are subtle or inadequately captured by elevation data. Furthermore, the limited availability of accurate, high-resolution topographic datasets in high mountain regions such as the Himalaya remains a major constraint for DEM-based glacier mapping (Shukla et al., 2010).

(d) *Thermal Data*

Earlier field investigations and remote sensing (RS) analyses have demonstrated consistent thermal contrasts between supraglacial debris (SGD) resting on glacier ice and periglacial debris (PGD) situated beyond glacier margins (Lougeay, 1974; Taschner & Ranzi, 2002; Ranzi et al., 2004). These temperature differences provide a useful basis for distinguishing debris-covered glacier surfaces from surrounding non-glaciated terrain. The underlying mechanism is straightforward: SGD overlies glacier ice, which acts as a persistent cooling source, whereas PGD lacks subsurface ice and therefore tends to attain comparatively higher surface temperatures (Fig. 3). Lougeay (1974) first suggested that ice-cored moraines could be identified by their relatively lower surface temperatures. Building on this concept, Taschner & Ranzi (2002) applied thermal data from Landsat TM and ASTER sensors to delineate the debris-covered Belvedere Glacier in the Italian Alps. Their methodology relied on a threshold-based classification of thermal imagery, subsequently improved through NDVI integration and contextual filtering. They highlighted the advantage of ASTER's finer thermal spatial resolution (90 m) over Landsat TM (120 m) for such applications.

Further field-based observations by Ranzi et al. (2004) revealed that when debris thickness is relatively shallow (less than ~ 30–40 cm), surface temperatures are significantly influenced by the underlying ice, which remains near 0 °C during the melt season. Consequently, SGD surfaces are measurably cooler than adjacent PGD at similar elevations. Field measurements conducted in summer 2003 showed that debris layers 5–30 cm thick were, on average, 4.5 °C (± 3 °C) cooler than nearby PGD. Using combined VNIR and TIR datasets and a methodology similar to Taschner and Ranzi (2002), they mapped the Belvedere and Miage glaciers. However, they noted that thermal discrimination becomes less reliable when debris thickness exceeds approximately 40–50 cm, as the insulating effect reduces the cooling influence of underlying ice. More recently, Veettil (2012) employed Landsat TM thermal data to delineate debris-covered ice on the Baltoro Glacier in the Karakoram. A threshold-based segmentation of thermal digital numbers (DNs) was applied, with

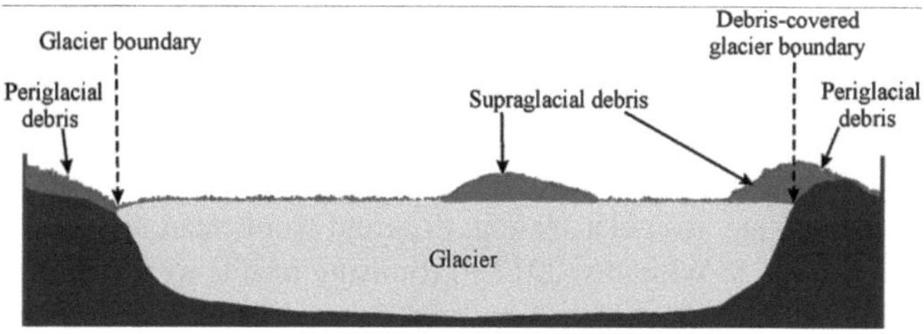

Fig. 3 Schematic cross-section of a typical debris-covered glacier showing DCGs boundary and occurrence of supraglacial and periglacial debris

values between 105 and 115 identified as optimal for separating debris-covered ice in the ablation zone. The resulting outlines showed reasonable agreement with the South Asia glacier inventory. Nonetheless, it was acknowledged that threshold values are site-specific and may require adjustment when applied to different climatic or geomorphological settings.

Though the surface temperature-based delineation of DCGs has considerable potential, but the magnitude of temperature difference between the DCG and adjacent terrain is likely to be affected by many factors, such as, thickness of the debris-cover, moisture content, differential illumination, lithology of the debris-cover (its thermal conductivity), sensor specifications and time of over-pass etc. However, giving due consideration to most of the above stated factors, the thermal data has been widely used in conjunction with the other datasets for mapping DCGs with varying success (Shukla et al., 2010; Karimi et al., 2012; Racoviteanu & Williams, 2012; Bharadwaj et al., 2015; Shukla & Ali, 2016; Nijhawan et al., 2016; Ali et al., 2017).

(e) *Textural Parameters*

Characterizing various glacier surfaces on the basis of their textural parameters is a domain which is yet to be fully explored with regard to the DCG mapping. As observed till now, a major limitation in the large-scale mapping of DCGs using spectral or morphometric or thermal approaches is the heterogeneous distribution of the debris-cover thickness and thermal properties over the glaciers. In this regard, texture analysis may provide an alternate method for capturing the spatial distribution of debris-cover properties (Racoviteanu & Williams, 2012). Texture analysis has the potential to distinguish land cover types on the basis of surface characteristics such as coarseness/smoothness, roughness, and symmetry.

Racoviteanu and Williams (2012) carried out a detailed textural analysis of snow/ice, debris, bare rock, and cloud surfaces on the Zemu glacier in the Kangchenjunga area using ASTER data to explore its potential for the DCG mapping. They derived the standard deviation, variance and gray-level co-occurrence (mean, entropy, homogeneity and energy) measures from the first principal component (PC1) of the ASTER VNIR and SWIR bands. They also explored the utility of the geo-statistics measures (the semi-variogram) and filtering (in spatial and frequency domain). The texture images resulting from the co-occurrence measures revealed that the variance image showed maximum textural information. It specially enhanced the SGLs and bordering ice-cliffs. Besides, textural statistics could characterize different surface types to great extents as clean ice and clouds had low variance, low entropy and high homogeneity, which in general are characteristic of a smooth surface. Debris-cover, on the other hand, had the highest variance of all classes, highest entropy and lowest homogeneity compared to the other classes. This suggests that the debris-covered ice surfaces have a unique, irregular texture, different from clean ice and surrounding terrain (Racoviteanu & Williams, 2012). Promising results were obtained from the frequency domain filtering which were further used as input to an unsupervised clustering algorithm to map the clean ice and debris-covered ice. Results from geo-statistics measures also showed differences in texture between debris-covered ice and clean ice or lateral moraine. However, authors find that further work is needed for an in-depth evaluation of the use of geo-statistics for mapping of these features. Racoviteanu and Williams (2012) finally concluded that though textural information of the glacier surface showed considerable potential but their use for development of an automatic approach for mapping of DCGs requires further in-depth analysis and study. Recently, Nijhawan et al. (2016) have used textural measures (variance, skewness, entropy, homogeneity, mean and dissimilarity) as one of the inputs to a multisource classification for mapping DCGs of Alaknanda basin, Himalaya.

(ii) Using hybrid approach

Certain issues (outlined in Sect. 4.2 (i)) arise while applying single approaches to the mapping of DCGs. Therefore, use of hybrid or combined approaches assimilating different mapping techniques and multisource data were considered. These have been discussed here and recapitulated in Table 3.

(a) *Spectral ratio and indices + Geomorphometric Parameters*

These studies have shown the utility of geomorphometric parameters (slope, plan and profile curvature; used mainly for differentiating between SGD and surrounding terrain) together with the spectral ratios (used mainly for mapping the snow-ice part of the glaciers) (Paul et al., 2004; Veettil, 2012; Veettil et al., 2014; Ghosh et al., 2014; Alifu et al., 2015a, 2015b).

Paul et al. (2004) devised a semi-automatic approach which used NIR/SWIR ratio image, Intensity Hue Saturation (HIS) transformation, and a slope map (with slope $< 24°$) to map a debris-covered glacier in Swiss Alps. Further improvements in the glacier map were achieved by integrating information from neighborhood analyses (to exclude the debris-cover areas not associated with ice) and multi-temporal change detection (to detach flat and vegetation-free glacier forefields). Though this approach had drawbacks related to derivation of multiple thresholds at several stages and transferability issues, nonetheless, it had a robust basis and clearly demonstrated the importance of slope in DCG mapping. Moreover, this study also focused on the issue of quality of DEM which ascertains the quality of its derivatives (slope etc.) and thus DCG mapping. Later, Veettil (2012) implemented the Paul et al.'s (2004) approach for mapping the changes in debris-covered tongues of Baltoro and Bilafond glaciers, Karakoram Himalaya. Landsat data, ASTER and SRTM GDEMs were used by Veettil et al. (2014) with a modified slope threshold value of 12°. A combination of manual and semi-automatic method (Paul et al., 2004) was advocated as a better approach. Ghosh et al. (2014) also followed the same methodizing the slope threshold of $< 15°$ to map the DCGs in parts of Ladakh region, western Himalaya, India. Lot of misclassifications in the distal parts of the glaciers was obvious in their results. Alifu et al. (2015a) devised a new band ratio (TIR/(NIR/SWIR)) and used it in combination with slope ($< 12°$) for mapping of two DCGs in China. They report that the newly proposed index enhanced the difference between the DCGs and surrounding non-glaciated terrain facilitating their extraction. They also compared their results with the Randolph Glacier Inventory (RGI), Glacier Inventory GIS data of China, and manually delineated glacier boundaries on Google Earth (GE) images. Later improving upon their previous work, Alifu et al. (2015a, 2015b) used the newly devised band

Table 3 Overview of studies on DCG mapping using hybrid approaches

Hybrid method	References	Techniques used	Inferences
Spectral ratio, indices and image transformations + geomorphometric parameters	Alifu et al. (2015a)	Thresholding of (a) new band ratio image TM6/(TM4/TM5); and, (b) slope image to delineate DCGs	• Combination of multispectral and slope data is well suited for glacier mapping • Integration of slope data considerably improves the glacier mapping. However, there is a requirement for high precision DEMs • Slope information accounts for misclassification between SGD and adjacent shaded and/or elevated terrain. Curvatures can separate SGDPGD in the glacier margin side • Threshold selection for any parameter is a crucial step which needs to be done carefully and is often scene-specific, and/or area-specific • This approach is faster than manual delineation, particularly for larger number of glaciers. However, manual corrections are still required
	Alifu et al. (2015b)	Thresholded TM6/(TM4/TM5) band ratio image along with mean slope, plan and profile curvatures were used to improve their previous work (Alifu et al., 2015a) on a larger glacierized region	
Multispectral image classification; + geomorphometric analysis	Paul et al. (2004)	Thresholding of (a) TM4/TM5 band ratio image; (b) H-component of IHS (Intensity Hue Saturation) color space transformations from TM bands 3, 4 and 5, and; (c) slope image was done for the delineation of a DCG Neighborhood analysis was done to exclude the PGD, wrongly mapped as part of glacier Multi-temporal change detection was carried out to detach flat and vegetation-free glacier forfield, and to update glacier outlines	

(continued)

Table 3 (continued)

Hybrid method	References	Techniques used	Inferences
	Veettil (2012)	Mapping of DCG using (a) manual delineation; (b) semi-automatic method (Paul et al., 2004) and modified semi-automatic method (wherein a slope range of 0° to 24° was used to identify the glacier terminus); and, (c) thresholding of Landsat thermal band (TM6)	• Manual approach produces more accurate results but is time consuming • The accuracy depends highly on the researcher's skill and the spectral similarity of various surrounding objects makes difficulty in decision making • Though, Paul et al. (2004) method applied here could distinguish between SGD, debris connected to glaciers and valley rock. But the modified semi-automatic method was found to be more accurate, especially near the snout position and gave a similar result compared to manual and automatic approaches • If the ice-cored debris is covered by snow, this method would not work properly because the thermal radiation can be similar to that of debris-free glacier/snow/ice • The choice of threshold value may vary with local weather conditions, cloud cover and snow cover and such applications are limited to images having very similar surface and seasonal characteristics • Use of high spatial resolution images and DEM can improve delineation of DCGs. High temporal imageries and DEMs can be effective in change detection analysis. Satellite based studies should be combined with field-based studies to reveal the expansion or ablation characteristics of DCGs

(continued)

Table 3 (continued)

Hybrid method	References	Techniques used	Inferences
	Ghosh et al. (2014)	Landsat TM4/TM5 band ratio image and slope derived from Cartosat-1 DEM were used for the delineation of DCGs. Paul et al. (2004) method was followed which involved thresholding of band ratio image, hue component of IHS transform and slope image Principal component analysis was also performed on snow-covered and debris-covered areas. Different combinations of PCs were used to further categorize different types of snow/ice and supraglacial debris on the glacier surface The glacier boundaries were also delineated manually using high-resolution Cartosat-1 satellite data	The classification technique used here for DCG mapping corresponded well with manually delineated glacier outlines and field observations This approach was faster than the manual delineation Inclusion of different combinations of PCs provided accurate identification of different components of glaciers. Manual corrections are needed to differentiate SGD from PGD

(continued)

Table 3 (continued)

Hybrid method	References	Techniques used	Inferences
	Khan et al. (2015)	Maximum Likelihood Classification (MLC) based approach was used to map distinct glacier zones. Landsat TM bands 1, 2, 5 and slope mages were utilized Thresholding of band ratios TM3/TM5, TM4/TM5 and NDSI were also tested	Combination of Landsat TM bands with surface slope information showed promising results in distinguishing debris covered areas from surrounding terrain Manually digitized glacier outlines showed good agreement with the derived glacier outlines for most glaciers Addition of the shadowed areas to the perennial snow-covered ice resulted in an over-estimation of the snow-covered glacier surfaces compared to the manually delineated outlines Manual corrections are required to separate out the spectrally similar surrounding terrain mapped as supraglacial debris Among all the ratio images used, TM3/TM5 provided better classification of snow/ice areas on glaciers, particularly in shadowed regions. But, overall, the thresholding methods were not successful in differentiating perennial snow-cover from exposed ice and/or debris cover from surrounding terrain
Geomorphometric analysis + surface temperature	Buchroithner and Bolch (2007) Bolch et al. (2008)	Glacier delineation using ASTER thermal band and topographic parameters (elevation, slope, plan and profile curvatures) derived from ASTER DEM	This approach could delineate the debris-covered glaciers with satisfying accuracy. Problems occurred mainly at the distal parts of the glaciers where stagnant ice exists and where it is even in the field difficult to identify the glacier boundary Ancillary data and higher-resolution DEMs are required to describe the complex topography of the DCGs and their surroundings

(continued)

Table 3 (continued)

Hybrid method	References	Techniques used	Inferences
	Veettil et al. (2014)	Delineation of DCG boundaries using (a) manual delineation; (b) Paul et al. (2004) method; (c) segmentation of land surface temperature images (derived using Landsat TM6 and NDVI images) which are then overlaid with segmented slope images.	Manual approach is simple but is time consuming. The accuracy is dependent on expertise of the user Thermal and semi-automatic approaches are faster than manual method All the three techniques applied showed a change in the glacier terminus position. The study glacier had been retreated during the last 30 years which is in contrast to the 'Karakoram anomaly' A combination of manual and semi-automatic method has been found to be better from this study. In such a case, semi-automatic method can be applied first and then edited manually based on the user expertise
Multispectral image classification + geomorphometric analysis + Surface temperature based	Shukla et al. (2010)	(a) Multisensor classification using optical (AWiFS) and thermal (ASTER bands 10-14) data; (b) thresholding of geomorphometric (slope and elevation) data; and, (c) manual delineation on ASTER data	The optical-thermal approach, compared to geomorphometric-based techniques, has a good potential to map the glacier boundary more precisely Results from former approach show similarity to manual delineation, even though manual editing is required The optical-thermal approach resolves the ambiguities due to the inability of optical data in deciphering the ice-extensions underneath the debris cover and enable mapping of the glacier boundary in the debris-covered regions
	Karimi et al. (2012)	Glacier outlines were mapped by employing (a) neural network classification technique using combination of optical (WorldView-2 image) and thermal data (Landsat TM band 6); (b) threshold method using WorldView-2 image and slope; and (c) manual delineation using high resolution panchromatic imagery	Even though high-quality DEM (1m, derived from LiDAR data) was used, morphometric-based approach could not distinguish between actual glacier area from its surrounding near the ablation area due to the significant recession of glacier and accumulation area due to the existence of large amount of ice in steep slopes High spatial resolution thermal data with additional field data on thermal properties (e.g., thermal inertia) can improve the mapping of DCGs, however, it gets affected by the thickness of debris cover

(continued)

Table 3 (continued)

Hybrid method	References	Techniques used	Inferences
	Tiwari et al. (2014)	Maximum Likelihood classification (MLC and Hybrid classification (MLC and knowledge-based classification (KBC)) using transformed and original ASTER VNIR bands, thermal band and geomorphometric parameters (slope, aspect, elevation) Manual delineation of glacier boundary using ASTER data	Transformed bands integrated with thermal data maps SGD and PGD more accurately than the original bands using MLC. Temperature information is not always sufficient enough for delineating supraglacial part from the non-supraglacial part of glaciers using KBC It is possible that PGD has lower temperature than SGD e.g., in shaded areas. Knowledge based classifier combining the temperature and geomorphometric parameters (slope, aspect and elevation) has shown the ability to separate SGD and PGD in such cases and map the glacier boundary more accurately Less manual editing is needed, since the user's knowledge about the classes is inherently taken care of, during KBC
Spectral ratio, indices and image transformations + Geomorphometric analysis + Surface temperature based	Bhambri et al. (2011)	Thresholding on ASTER band ratio 3/4 image, ASTER thermal band 12, Landsat ETM+ thermal band, cluster analysis of geomorphometric parameters (slope, plan and profile curvatures) and manual delineation on Cartosat-1 and IRS IC-PAN data were conducted for DCG mapping	This approach provided promising results for larger DCGs while it could not accurately map the smaller DCGs The post-depositional sedimentation by debris flow/mass movement obscures the semi-automated mapping of DCGs Shaded regions strongly affect the single thermal band thresholding, thus hampering the automated mapping of debris-covered areas Thresholds and number of cluster groups need to be carefully selected because the regional settings of relief, quantity of debris load, orientation and climatic conditions vary considerably throughout the Himalayas and other mountainous regions and even within one basin

(continued)

Table 3 (continued)

Hybrid method	References	Techniques used	Inferences
	Racoviteanu and Williams, (2012)	Two approaches were used a) Decision Tree based classification using optical (ASTER VNIR-SWIR bands), thermal bands 10 and topographic (slope, elevation and curvatures) data; and, b) Texture analysis. Spectral indices (NDSI, NDVI) and band ratios (ASTER 3/4 and 3/5) were also employed using appropriate thresholds for the delineation of various glacier features	Results obtained from both the approaches were almost comparable, however, when compared with high resolution data, both the approaches overestimated the debris-covered area of glaciers Texture analysis helped detecting debris covered-ice in areas where the decision tree algorithm failed to do so, due to either the patchy supraglacial debris in the upper ablation areas of the glaciers, or the thick debris at the glacier termini Both these approaches present some limitations which make it difficult to apply these at large scales. The sophisticated and laborious processing of the decision tree prevents a completely automated process for delineating the debris-covered glacier tongues. The texture approach is limited by the fact that while debris-cover training areas may have distinctive texture values, these values or patterns may not be characteristic for all the debris cover across the region The profile and plan curvatures based on the ASTER DEM proved inconclusive in distinguishing debris-covered tongues from the surrounding terrain NDSI produced more satisfactory results than other band ratios for delineating snow and clean-ice Manual corrections are still required due to the ambiguity caused by shadows and similar spectra of SGD and surrounding terrain

(continued)

Table 3 (continued)

Hybrid method	References	Techniques used	Inferences
	Bharadwaj et al. (2015)	The glacier boundaries were delineated using (a) cluster analysis of slope, plan and profile curvatures; (b) thresholded surface temperature images obtained from Landsat TM/ETM+ thermal band for mapping SGD; (c) thresholded TM/ETM+ 4/5 ratio image for snow/clean-ice mapping; and, d) manual delineation using pan-sharpened ETM+ image	Glacier boundaries derived from cluster analysis of geomorphometric parameters were almost comparable with the manually delineated ones Refining the glacier boundary (derived from cluster analysis) using Landsat thermal band (as reported in Bhambri et al. 2011) did not work out well because of poorer spectral as well as spatial resolution. Thus, this study used a combination of two different clustering processes, Iterative Minimum Distance and Hill-Climbing to refine their results. Low spatial resolution of ASTER DEM and low spectral (only 1 band) and spatial resolution of Landsat TM/ETM+ thermal bands limited the automated mapping of small DCGs. Manual editing is still a requirement in areas where these semi-automatic approaches fail

(continued)

Table 3 (continued)

Hybrid method	References	Techniques used	Inferences
	Shukla and Ali (2016)	Hierarchical knowledge-based classification (HKBC) and MLC approaches incorporating ASTER derived NIR/SWIR band ratio image, spectral indices (normalized-difference glacier index (NDGI), normalized-difference water index (NDWI), normalized-difference debris index (NDDI), image transformations (IHS image), geomorphometric parameter (slope) and the thermal glacier mask (TGM) were used to map the glacier boundary Onscreen digitization using ASTER VNIR image	Compared to MLC, HKBC maps the glacier boundary most accurately where it is concealed by debris cover The boundary delineated by means of visual interpretation relies mostly on the optical properties of the debris cover while HKBC utilizes information from SWIR and thermal data to map the extent of debris-covered ice Appropriate selection of thresholds is must for accurate mapping of glacier boundaries and can vary from glacier to glacier
	Ali et al. (2017)	Glacier boundaries were obtained from multiple data sets (Landsat TM/ETM+/OLI/ASTER) and validated against the glacier boundaries derived from Cartosat-1 PAN and IRS P6-LISS IV images An HKBC approach was implemented for the differentiation of various glacier features. Appropriate thresholds on several indices like NDGI, NDWI, NDVI, NDDI, band ratio (NIR/SWIR), thermal mask and slope layer were defined for mapping different glacier classes The glacier boundaries were obtained by merging the glacier components as one class and the non-glacier components classes as another class	Spatiotemporal changes in glacier terrain classes can be linked with variations in glacier dimensions An increase in the number of DCGs has been found over the study period DCGs have been found to have lower recession rates compared to sparsely debris-covered and clean-ice glaciers The size of the glaciers greatly influenced the rate of conversion of glaciers, from clean to sparse to heavily debris-covered glaciers Fluctuations in temperature and precipitation patterns appeared to be the primary factor controlling the changes in glacier dimensions

ratio image along with mean slope, plan and profile curvature image for mapping the Yengisogat glacier, Shaksgam valley, China. They compared the results with glacier boundaries derived from previously mentioned reference sources and also with their earlier study and reported improvements specifically in the differentiation of SGD and PGD on glacier boundary.

(b) *Geomorphometric Parameters + Thermal Data*

Buchroithner and Bolch (2007) used the ASTER DEM and ASTER imagery for mapping of DCGs in the Mt. Everest area. They used several geomorphometric parameters (altitude, slope, plan and profile curvature) together with thermal data for delineating the debris-covered glacier tongues. Later, Bolch et al. (2008) followed a similar approach combining the thermal information from ASTER data with various shape parameters (slope gradient and curvature) for mapping of Khumbu glacier in Nepal Himalayas. These approaches encountered problems in correctly mapping the glacier boundary in distal parts of the glaciers where the transition between the glacier and the adjacent terrain was gradual and the lateral moraines were too small to be represented on the ASTER DEM. They concluded that the use of ancillary data and higher-resolution DEMs are required to describe the complex topography of the DCGs and their surroundings. Veettil et al. (2014) attempted the delineation of debris-covered boundaries of the Bilafond glacier, Karakoarm Himalaya by segmentation of land surface temperature images (derived using Landsat TM6 and NDVI images) in combination with binary slope map ($< 12°$). They conclude that owing to the heterogeneity in the thickness of debris-cover and consequent variations in the thermal properties, pose severe constraints in application of a combined thermal and topographic approach for mapping of DCGs and found a combination of semi-automatic (Paul et al., 2004) and manual method most appropriate for their area.

(c) *Multisource Classification (DIC+ Spectral ratio and indices/Geomorphometric Parameters/Thermal Data/Textural Parameters)*

As discussed earlier (Sect. 4.2 (ii), multisource classification involves the usage of input data from different sources (transformed spectral data, geomorphometric parameters, thermal data, textural information etc.). This technique probably evolved because image classification algorithms are good assimilator of information from multiple sources (Arora et al., 2011). It has been extensively employed for mapping of DCGs (Sidjak & Wheate, 1999; Shukla et al., 2010; Karimi et al., 2012; Ghosh et al., 2014; Khan et al., 2015).

Sidjak and Wheate (1999) applied the supervised Maximum Likelihood Classification (MLC) using different input bands (PCs, TM4/TM5 ratio, NDSI, TM543 composite image) for mapping the glacier extent and discriminating glacier zones. The most accurate classifications were obtained using the combination of PCs 2, 3, and 4 of the masked glacier area, the ratio TM4/TM5, and the NDSI. This method also avoided the problems of sensor saturation and shadowed areas, and was able to discriminate debris-mantled ice and ice marginal water bodies. Shukla et al. (2010) tried to differentiate between the SGD and PGD based on classification of integrated optical (IRS P6 AWiFS) and thermal (Terra ASTER) data for Samudratapu glacier,

Himachal Himalaya. Later, they extended the same approach for the delineation of DCG boundaries together with geomorphometric (slope (< 15°) and elevation) data and some manual editing. Results were appraised against the manually derived boundary. This method seemed to have the potential to identify the ice-extensions underneath the debris-cover. Later, Karimi et al. (2012) applied the same approach for mapping of Alamkouh glacier, Iran. They performed a neural network classification on combination of optical (WorldView-2 image) and thermal data (Landsat TM band 6) together with the geomorphometric parameters derived from a high-resolution LiDAR DEM. They reported that even though high-quality DEM was used, morphometric-based approach alone could not distinguish between actual glacier areas from its surroundings due to the significant recession of glacier.

Supervised classification (MLC) was performed in PCs 236 and 432 of Landsat TM image for mapping of various supraglacial types (thick debris, shallow debris, shallow debris with moisture) and snow-ice types (very fresh snow, compact snow, snow mixed with fine sediments, clear ice, ice mixed with fine sediments, and ice mixed with debris) on five glaciers located in Ladakh region, western Himalaya by Ghosh et al. (2014). However, their method does not qualify as a DCG mapping technique because they used manually delineated glacier boundaries for glacier delineation and further classification. Moreover, they did not define the basis of characterization of the SGD or snow-ice classes which they attempted to map.

Khan et al. (2014) used a hybrid approach for mapping of various glacier terrain classes, namely, perennial snow, exposed ice, debris-covered ice, water, shadow and other terrain in Upper Indus Basin. Firstly, they mapped and masked out the perennial snow by applying unsupervised clustering algorithm on TM bands 3, 4 and 5. Subsequently, they performed supervised classification (MLC) on TM band 1, 2, 5 and terrain slope layer for mapping rest of the classes. Lot of post-processing in form of filtering, removing small polygons and applying elevation threshold were done to get the final glacier map. An artificial neural network (ANN)-based information extraction approach was applied by Shukla and Yousuf (2016) for mapping the Kolahoi glacier and adjoining areas, using Landsat TM (Thematic Mapper) data and several ancillary layers such as image transformations and topographic attributes. They tried to differentiate and map the SGD and PGD using the multisource data and report that the incorporation of ancillary data into the classification process significantly enhances the mapping accuracy (> 9%), particularly the addition of NIR/SWIR data to the spectral data. Previously, Nijhawan et al. (2016) performed multisource classification on a combined dataset composed of geomorphometric (slope, surface curvature), thermal and textural (variance, skewness, entropy, homogeneity, mean and dissimilarity) data using Support vector machine (SVM) and ANN algorithms for mapping DCGs of the Alaknanda basin. While both ANN and SVM produced good results, accuracy of SVM was reported to be slightly higher than that of ANN.

Multisource classification for mapping of DCGs presents an excellent opportunity for integrating data from different sources which characterize different glacier surfaces and features. However, one needs to be careful while choosing the appropriate multisource data and suitable classification algorithm which is able to handle a multidimensional dataset efficiently.

(d) *Spectral ratio and indices + Geomorphometric Parameters + Thermal Data*

These techniques tried to use the combination of spectral ratios (for mapping snow-ice), various geomorphometric parameters and thermal information (for identifying the debris-covered glacier body) (Bhambri et al., 2011; Racoviteanu & Williams, 2012; Bharadwaj et al., 2015; Shukla & Ali, 2016; Ali et al., 2017) as shown by an example in (Fig. 4) Bhambri et al. (2011) carried out the thresholding of NIR/SWIR ratio, thermal bands, cluster analysis of geomorphometric parameters (slope, plan and profile curvatures) and overlay operation for mapping the debris-covered Gangotri group of glaciers, central Himalaya. They validated the derived glacier boundary with the one obtained by manual delineation on high resolution satellite data. Though, this approach provided optimal results for larger DCGs, it could not accurately map the smaller ones. It was realized that while the presence of debris-fans deposited as a result of mass movements pose problems in geomorphometric-based boundary delineation, the shaded regions preclude the single thermal band thresholding, thus hampering the automated mapping of DCGs. Moreover, the thresholds and number of cluster groups may be area specific as the regional settings of relief, quantity of debris load, orientation and climatic conditions vary considerably throughout the Himalayas and other mountainous regions and even within one basin. Assimilation of information from varied sources such as VNIR-SWIR bands (in original form and as ratios and transformations), thermal data, and topographic attributes (slope, elevation and curvatures) using a decision tree classification has also been attempted for mapping of DCGs in Sikkim Himalaya (Racoviteanu & Williams, 2012). As compared to high resolution data this approach overestimated the area of DCGs. The sophisticated and expertise-intensive processing of the decision tree prevents a completely automated process for delineating the debris-covered glacier tongues. With regard to the morphometric parameters, the profile and plan curvatures did not of much use in distinguishing debris-covered tongues from the surrounding terrain. Manual corrections were required to rectify the derived boundaries.

An approach similar to that applied previously by Bhambri et al. (2011) was used for mapping of Hamtah and Patsio glaciers, Himachal Himalaya (Bharadwaj et al., 2015). Besides mapping the glacier boundary, they mapped temporal variations (for 1989, 2000 and 2010) in three glacier classes namely, clean ice, SGD and ice mixed with debris. Poor spectral and spatial resolution of the thermal data, insensitivity of the thermal data to SGD depth in case of thick debris-cover, presence of mixed pixels in optical data, low spatial and temporal resolution of the available DEMs, influence of differential shadowing were identified as some of the factors influencing the DCG mapping approach. Recently, a hierarchical knowledge-based classification (HKBC) approach was devised for incorporating spectral ratio (NIR/SWIR) and indices (normalized-difference glacier index (NDGI), normalized-difference water index (NDWI), normalized-difference debris index (NDDI)), image transformations (IHS image), geomorphometric information (slope) and thermal data for DCG mapping and tested at the Kolahoi glacier, western Himalaya, Jammu and Kashmir, India (Shukla & Ali, 2016). A new normalized-difference debris index (NDDI) using SWIR and TIR bands was also proposed in this study for mapping

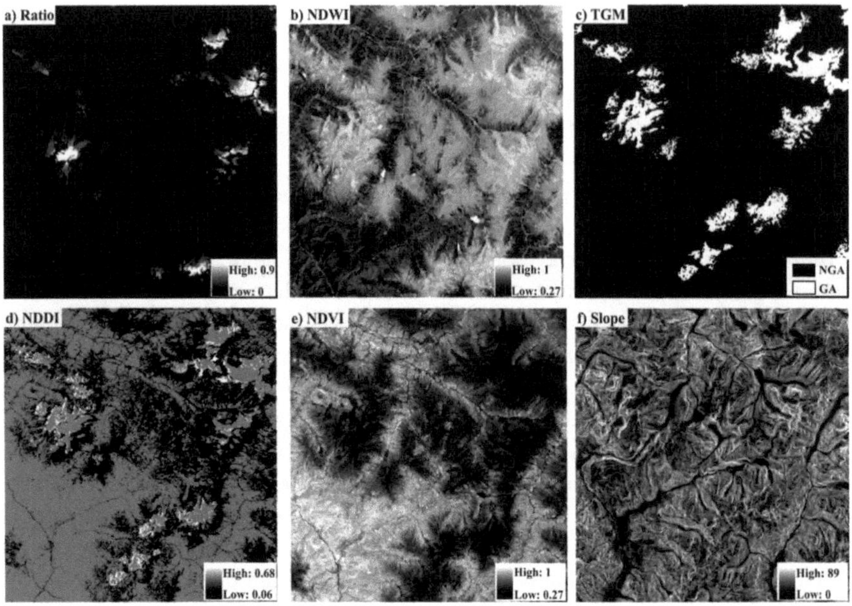

Fig. 4 An example of different indices and ratios that are used for differentiation of different glacial and non-glacial parameters

of debris-cover. These layers were then used in several discrete tests in HKBC, to map various glacier terrain classes. The results obtained were far better than those produced from MLC and matched closely with glacier boundary derived from high resolution image of the area. Though having a robust basis, the major issue with this approach is the derivation of several thresholds, the accuracy of which depends greatly on the expertise of the analyst. However, the same approach has later been applied over an extended area (parts of Lidder and Sindh River basins in J and K) for mapping the temporal variations in glacier facies and dimensions proving its utility and wider applicability (Ali et al., 2017).

4.3 Synthetic Aperture Radar (SAR) Based Techniques

In general SAR data has the capability to separate snow and ice and its surrounding areas due to the different backscatter signal. The main limitation is that the spaceborne satellite SAR data is usually limited to one frequency which hampers the correct classification of DCGs (Rott, 1994). This is especially true if glaciers have different surface roughness e.g., due to icefalls, crevasses and debris cover. Experiments have proven that C- and X-band single polarization SAR data have the ability to discriminate wet snow from ice (Rott & Mätzler, 1987; Shi & Dozier, 1993). However, SAR data have only limited application and usage in many parts of the world particularly due to severe geometric (layover, foreshortening, and radar shadows) and radiometric distortions and speckle (noise), that require intensive preprocessing and accurate digital elevation models (DEMs) (which, unfortunately are unavailable for most high mountainous snow-covered areas). Other limitations associated to SAR data include the difficulties in interpretation of the recorded backscatter (Könnig et al., 2001), inability to map dry-snow cover using single polarization and single frequency SAR data (Rott & Nagler, 1994; Koskinen & Hallikainen, 1997), and inability to distinguish between wet-snow and other smooth surfaces (Shi & Dozier, 1997). To date SAR data were used extensively for glacier facies identification (Huang et al., 2014, Kundu & Chakraborty, 2015) while as InSAR coherence (Atwood et al., 2010; Strozzi et al., 2010; Zongli et al., 2012; Frey et al., 2012) and polarmetric SAR (Huang et al., 2014; Kundu & Chakraborty, 2015) data have recently been used for mapping of debris-covered ice and glaciers.

(i) Use of InSAR Coherence

Synthetic aperture radar (SAR) coherence is highly sensitive to surface motion and structural changes, making it a valuable tool for identifying glacier extent and distinguishing dynamically active ice bodies, including those mantled by debris. Areas exhibiting low coherence are typically associated with glacier movement and deformation, and therefore can be used as a key indicator of glacierized terrain (Atwood et al., 2010). This capability is particularly useful for debris-covered glaciers, where optical contrast with surrounding terrain is often limited. However, coherence measurements are influenced by sensor characteristics and imaging geometry. These system- and geometry-related effects can be minimized through careful interferometric processing and calibration. In addition, coherence loss (decorrelation) may result from volume scattering within snow and ice, as well as from temporal surface changes between acquisitions. While such decorrelation can complicate interpretation, it also provides insights into surface properties and dynamic processes, thereby contributing to improved characterization of glacier surfaces and their activity state.

Surface-dependent decorrelation generally increases with increasing temporal separation of the interferometric partners. Whereas, loss of coherence is generally seen as alimiting factor in interferometric techniques which can act as a tool for detecting glaciers irrespective of debris-cover. Taking advantage of this, Atwood et al. (2010) used the coherence images generated from the Advanced Land Observing Satellite (ALOS) Phased Array L-band SAR (PALSAR) together with the interferometric phase slope, and patch size in the glacier candidate mask, for mapping two glaciers (one clean and one DCG) in Alaska. Comparison of results with manually delineated boundaries suggested high accuracy for the coherence derived outlines. Similarly, Strozzi et al. (2010) demonstrated the use of the degree of coherence over ALOS PALSAR interferograms in summer period over parts of western Himalaya including both clean as well as heavily debris covered. Zongli et al. (2012) also used the ALOS PALSAR derived SAR coherence data with MLC to map glaciers in Kongur Mountains of China. They verified their results from Chinese glacier inventory/field points and reported inaccuracies in low coherence areas arising from surface water bodies and topography. For preparing the glacier inventory of the western Himalaya, Frey et al. (2012) used the ALOS PALSAR coherence images for manually delineating the debris-covered parts of the glaciers and merged them with the automatically extracted snow-clean ice areas (using TM3/TM5 ratio on ETM+ images). They reported considerable amount of post-processing which included elimination of erroneously classified features like turbid water surfaces (lakes and wide rivers), clouds and remaining snow avalanche deposits and delineation of DCG parts. Further, Saraswat et al. (2013) used high value coherence images derived from the Envisat ASAR images for mapping the terminus changes of the Gangotri glacier and suggested better coherence near the terminus compared to the pixels within the glacier.

Summer-acquired InSAR coherence imagery has proven particularly effective for delineating debris-covered glacier tongues, especially for large valley glaciers where surface motion induces measurable decorrelation (Frey et al., 2012). During the melt season, glacier dynamics and surface changes enhance coherence contrasts between actively flowing ice and surrounding stable terrain, thereby improving boundary detection in debris-mantled areas that are otherwise difficult to map using optical data alone.

Despite these advantages, there remains ongoing debate regarding whether stagnant or inactive glacier ice should be included within glacier inventories (Robson et al., 2015, 2016). Several widely accepted definitions emphasize that glaciers are bodies of ice exhibiting active flow under their own weight (Benn and Evans, 2010; Cuffey & Paterson, 2010). From this perspective, stagnant ice masses lacking measurable motion may not strictly qualify as glaciers. However, the inclusion of stagnant ice becomes critical when the focus shifts to hazard assessment, particularly glacial lake outburst flood (GLOF) risk. Down-wasting and degradation of stagnant, ice-cored moraine complexes can significantly influence lake formation, expansion, and dam stability (Richardson & Reynolds, 2000; Bolch et al., 2008). Therefore, while the classification of stagnant ice may be conceptually debated in glaciological terms, its identification remains essential in applied studies addressing glacier-related hazards and landscape evolution.

(ii) Use of Polarimetric SAR

Polarimetric SAR has promising potential for glacial facies identification and area mapping (Huang et al., 2016). However, use of this technique for DCG mapping is yet to be explored. Huang et al. (2014) studied the physical properties of the debris-cover on Koxkar glacier, Tianshan Mountains, using L-band and C-band SAR images and reported that the contrast between bare ice and debris is obvious in cross polarization (HV) in both L and C-bands. They further notified that bare ice is smoother and its backscatter is lower than the debris. Further, they applied two debris extraction methods for glacier mapping. The first method used HV backscatter coefficient threshold, complimented with layover and shadow maps. The second method used multi-polarimetric features along with the SVM classifier. Though more complicated, this approach yielded higher classification accuracy in L-band compared to C-band.

Besides this, a very important and interesting observation was made regarding the sensitivity of the L- and C-band responses to debris thickness.

Similarly, Kundu and Chakraborty (2015) used multi-temporal RISAT-I MRS data of the Gangotri and Mana glaciers, central Himalaya for mapping the glacier facies, namely, snow-cover not fully wet, wet-snow zone, frozen percolation zone, superimposed ice zone, bare-ice zone and debris-covered ice zone. They did not attempt segregating the debris-covered glacier from the surrounding terrain using Pol-SAR, for this they carried out manual delineation aided by the slope and curvature data derived from the SRTM DEM. This way the study doesn't exactly come under the ambit of techniques for DCG mapping, however, it was considered since it maps the debris-covered ice zone as one of the classes. The said glacier facies were separated and mapped based on the thresholds derived from the two-dimensional (in HV-HH space) signatures of the facies. However, the basis of identifying the glacier facies spatially, differences in their responses in the co- and cross-polarization data and reasons behind such response were not discussed thoroughly. Besides, delineation of DCG boundary on the basis of just slope and curvature was also less convincing and doubtful. Thus, Pol SAR data plays a minor role for DCG mapping.

This brief review confirms that the use of SAR data is an appealing proposition as it provides a way to distinguish DCG areas that are optically similar to the surrounding non-glacier terrain. It is specifically of use for the identification of glaciers that are under persistent orographic cloud cover (Frey et al., 2012). However, application of SAR data for DCG mapping in different parts of the world has suggested that despite its proven utility in identification of DCGs, it is not possible to classify debris-covered ice based solely on SAR data (Atwood et al., 2010, Frey et al., 2012; Robson et al., 2015, 2016). Moreover, as already discussed earlier, there are some areas where SAR data is not received at the sensor due to the effects of steep topography on the radar image, namely layover and shadow, as well as problems in orthorectification in the absence of a high-quality DEM. In order to overcome these problems, the integration of SAR data with optical images and topographic data can provide valuable information for classification.

4.4 *Object Based Image Analysis (OBIA)*

The mapping techniques of DCGs discussed earlier operate at pixel-level i.e., they use the spectral and ancillary (Sect. 4.3) information of the pixel while trying to assign a particular class to it (e.g., 'glacier' or 'non-glacier', 'snow' or 'ice', 'SGD' or 'PGD'). Often in implementing most of these techniques to the best of our capabilities, varying amounts of manual corrections are required to reduce the misclassifications and achieve the desirable accuracy. This reduces the robustness of pixel-based methods to adequately depict the complex natural phenomenon such as glaciers (Robson et al. 2015). These misclassifications usually occur where the surface types are spectrally mapped correctly but are semantically wrong or vice-versa. For example, debris-cover lying outside the glacier boundary (periglacial debris) is mapped as supraglacial debris (spectrally correct), but it is not a part of the glacier (semantically wrong). In addition, pixels in the transition zone (at or near the glacier boundary) are often mixed (i.e., contain mixture of various land covers and reflectance from multiple classes) and thus, not assignable to a particular class. So, while the pixel-based image analysis (PBIA) is restricted to the spectral dimension, the object-oriented approach adds the semantic dimension to the classification. Thus, the object-based image analysis (OBIA) is a technique where near-homogenous objects are the basis of classification instead of pixels. Working on the object-level not only facilitates the combined use of spectral, spatial, textural, hierarchical and contextual information but also allows for better control of the analyst at the time of object formation and classification. The OBIA approach usually involves four main steps i.e. the segmentation process (where homogeneous 'objects' are identified on the basis of certain parameters such as scale, shape, color, compactness and smoothness), classification (where the defined objects are designated to one of the classes based on some well-defined rule-set or any other algorithm), post-processing (where in neighborhood operations can be applied and contextual information can be used to automatically reduce the semantic errors, if any). A more detailed account of OBIA terminology and working can be found in Blaschke (2010) and Baatz and Schäpe (2000).

The major advantage of the OBIA compared to the PBIA, lies majorly in the handling of the image information in object-space and in the potential for automatic post-processing possibilities. Various studies have proved the superiority of OBIA over PBIA for many applications such as land use mapping, landslides and glacier delineation (Rastner et al., 2014). As OBIA is found more apt for high resolution RS data analysis (Blaschke, 2010), thus, with increased availability of data at finer resolutions, it has been extensively used for several applications (Salehi et al., 2012; Jawak et al., 2015). Consequently, the potential of OBIA for glacier facies mapping (Höfle et al., 2007) and delineation of DCGs (Bajracharya & Shreshtha, 2011; Bajracharya et al., 2014a, 2014b, 2015; Rastner et al., 2014; Robson et al., 2015, 2016; Biddle, 2015) is also being explored using both fine and medium resolution RS data.

The International Centre for Integrated Mountain Development (ICIMOD) used OBIA to assess the decadal changes in Nepal (Bajracharya et al., 2014a, 2014b) and Bhutan (Bajracharya et al., 2014a, 2014b), as well as to map glaciers over the entire Himalayas (Bajracharya & Shreshtha, 2011; Bajracharya et al., 2015). These studies have used the Landsat series of data (MSS and ETM+) together with the SRTM DEM for glacier mapping. After identifying the objects via segmentation, they adopted separate rule-set to map out the clean and debris-covered glacier areas. For clean glaciers, they employed NDSI images with appropriate thresholds and later removed vegetation (using NDVI), water (using Land Water Mask (LWM)) and shadowed (using Hue image) areas. Slope and elevation thresholds were used additionally to remove further misclassifications. After mapping the clean glacier areas, the debris-covered areas were mapped by applying slope, elevation thresholds and subsequently eliminating the misclassified vegetation, snow and water pixels using NDVI, NDSI and LWM, respectively. Finally, the clean ice and debris-covered ice areas were merged and manually edited with respect to GE images, however, the amount of manual editing required is not known.

Rastner et al. (2014) used a combination of spectral (Red/SWIR), topographic (slope) and thermal information for mapping glaciers in three different challenging conditions (debris-cover (Everest region, Himalaya), adjacent water body (Watkins range, Greenland) and ice in cast shadows (Everest region, Himalaya and Coast Mountains, Canada). They selected appropriate parameters values for the input data (Red, SWIR, TIR bands and slope) used for the segmentation process. High scale (20) and weights were given to layers for correct mapping of debris-covered areas. For the classification step, Red/SWIR ratio (> 2), slope (< 12) and temperature (> 273 K) data layers were suitably thresholded. These thresholds were iteratively adjusted. Where 'gentle-slope (< 12)' overlapped with 'positive temperature (> 273 K) regions' those areas were considered as debris-cover on ice. Finally, automatic post-processing was done to reduce misclassifications.

Robson et al. (2015) used optical (Landsat OLI), topographic (void filled SRTM DEM) and InSAR coherence (ALOS PALSAR) data in OBIA to separately map the snow-clean ice and debris-covered parts of glaciers in the Manaslu region of Nepal Himalaya. They did not include thermal data because they didn't find it useful for this study area. They performed the classification with (OBIA_OTS) and without (OBIA_OT) incorporation of SAR coherence data to evaluate its utility in DCG mapping. A multi-resolution and multistage (3 stages; two for snow-ice mapping and third before initiating mapping of debris covered ice) segmentation was done using blue, green, NIR, PAN, red, SWIR I & II, slope and SAR data with varying size, scale, compactness and weight parameters. High weights were given to the slope and SWIR bands. Thereafter, first the water bodies and clean ice areas were mapped out using NDWI and NIR/SWIR ratios, respectively. Then for mapping the debris-covered ice, NDVI, NDSI and slope (OBIA_OT) and NDVI, NDSI, slope and InSAR coherence (OBIA_OTS) were used. The rule-based classification was followed by post-processing and manual correction of the outlines with respect to the GE image, Rapid eye image and field photos. Results showed that while both OBIA_OT and OBIA_OTS did considerably well, both of them failed in some cases like in identifying the steep tributary glaciers. Incorporation of SAR data enhanced the classification accuracies; however, it had problems of layover and for shortening (effects of steep topography), loss of coherence over water bodies etc. Besides, in this study the stagnant parts of the DCGs were not considered as part of the glacier body, which is quite debatable. Also, though it is found that SAR data was able to increase the mapping accuracy but it could not be used alone for mapping of DCGs because of inability of InSAR coherence due to the vegetation cover, water bodies, steeper topography with unstable slopes and inactive debris-covered ice.

Recently, Robson et al. (2016) used optical (Landsat TM and OLI), topographic (Lidar DEM) and thermal (Landsat TM and OLI) data in OBIA for monitoring the decadal changes in glacier extents and transient snowline (TSL) in the Austrian Alps from 1985 to 2013. SAR coherence data was not used in this study because it was found to be unsuitable for small sized glaciers of this region. They also used multi-resolution and multilevel (4) segmentation process for creating the objects. After running the first two segmentations, the snow-clean ice areas were mapped out using NIR/SWIR ratio (≥ 1.4–2.0), NDSI (> 0.05–0.10), NDWI (< 0.15–0.4), altitude ($\geq$ 2000 m) and slope (40°–60°). Then, a segmentation (using only NIR band) was run only on the objects classified as snow-ice. Thereafter the snow-ice objects were categorized into snow and ice using the intensity image obtained from SWIR, NIR and Red bands. This resulted in the mapping of the TSL. Third segmentation was then performed using the NIR, Red, slope, NDVI, Canny edge detection layers (derived from slope). The debris-covered ice was classified using NDVI, Red band, slope, thermal data and NDWI. This was followed by classification expansion, post-processing and manual correction of the derived boundaries. SAR coherence data, profile curvature, surface elevation changes (between 2000 and 2006) and a hillshade model were used for manual correction.

Differing from the norm, Biddle (2015) integrated two machine learning algorithms (Random Forests (RF) and Conditional Inference Tree (CIT)) with OBIA for mapping of DCGs in Cordillera Blanca using optical (Landsat OLI), textural and topographic (SRTM) data. First, the multi-resolution segmentation was performed using Pan-sharpened OLI image and slope with appropriate (iteratively chosen) shape (0.3), scale (12) and compactness (0.2) parameters. Thereafter, in an initial classification the snow-clean ice and vegetation areas were mapped out using NIR/SWIR (> 2.0) and NDVI (> 0.1) layers, respectively. On the remaining image two machine learning algorithms were applied with Landsat OLI bands, terrain variable (slope, elevation, aspect and curvature) and textural information as input. These algorithms were first trained with the existing/manually delineated DCG boundaries (70% were used for training and 30% for accuracy assessment). This was followed by post-processing and accuracy assessment. Results revealed that the RF algorithm performed much better than the CIT and showed that amongst the input variable, the terrain information was of greatest use in mapping the DCGs. The study demonstrates the potential of integrating the machine learning algorithm for classification with OBIA, however, the data required for training the algorithms is one of the cons.

4.5 *Use of Unmanned Aerial Vehicles (UAVs) in Monitoring of DCGs*

Mapping of DCGs is often impeded by the limitation of spatial and temporal resolution of conventional satellite remote sensing (Immerzeel et al., 2017). Few satellite sensors (such as Worldview, GeoEye), though, provide data in high resolution with sub-meter pixel sizes but the cost of these images is too high (Bhardwaj et al., 2016). These limitations of conventional satellite remote sensing may easily be overcome by use of Unmanned Aerial Vehicle (UAV). UAVs are remotely controlled air vehicles with no pilot on-board. These are reusable devices that can carry various types of payloads (such as camera, GPS, sensors etc.) designed for specific purposes. The use of UAVs is advantageous as one can acquire the data of appropriate type (optical, thermal, hyperspectral, microwave, LiDAR) and on appropriate date and time with substantially lower cast.

During the past decade, the UAVs have been successfully deployed in various environmental (Nex & Remondino, 2013; Whitehead & Hugenholtz, 2014) as well as in glaciological studies (Hodson et al., 2007; Bhardwaj et al., 2016). The use of UAV in glaciological research has accelerated since 2007 (Whitehead & Hugenholtz, 2014; Bhardwaj et al., 2016), however, their uses are mainly concentrated in polar or sub-polar regions (90% of total; Bhardwaj et al., 2016) where the glacier are usually debris free. Hodson et al. (2007), for first time, demonstrated a spatial gradient in the ratio of supraglacial cryoconite using UAV-borne images on Midtre Lovenbreen, Svalbard. In the Polar region, the UAV-borne images were mainly used for studying supraglacial melt-ponds (Hodson et al., 2007; Tschudi et al., 2008), ice sheet topography Crocker et al. (2011), supraglacial drainage (Rippin et al., 2015), sub-glacial elevation profile and landforms (Solboa & Storvolda, 2013; Clayton, 2012), calving (Ryan et al., 2015) and terminus dynamics (Whitehead et al., 2013). Further, Tilg et al. (2011) used UAV acquired data to detect photosynthetic pigments in glacial ecosystem. This shows that UAVs have been successfully used in polar and sub-polar regions to monitor various glacier parameters. However, UAVs are beginning to be used in monitoring of mountain glaciers.

Walter et al. (2009), for the very first time, performed UAV based study in the French Alps to study, not a glacier strictly but, a mudslide event in a glaciated area. The study by Friedli (2013) is considered to be the first work in mountain glaciology (Bhardwaj et al., 2016) who compared the UAV, aircraft-based areal photogrammetric acquisition with respect to the terrestrial photogrammetric acquisition on Rhone glacier (Swiss Alps). In Forni glacier, in the Italian Alp, Fugazza et al. (2018) applied semi-automatic segmentation approach on very high-resolution visible RGB images to map large-scale morphological features i.e., bare ice and medial moraines. They suggested that this approach has ability to map these features with better accuracy than methods relying on medium-resolution satellite imagery, with only slight misclassification at the margins.

Despite of obvious challenges in rugged Himalayan terrain (e.g., high altitude, rugged terrain, low air pressure, poor GPS reception), Immerzeel et al., 2014, for the first time successfully deployed a UAV on a heavily debris-covered Lirung glacier located in Nepal Himalaya. They acquired ortho-mosaics and high resolution DEMs over the debris-covered tongue of the Lirung glacier, before and after the melt and monsoon season (May and October 2013) for a detailed study of surface lowering, frontal dynamics, ice cliffs and their impact on glacier down-wasting. Later, Miles et al. (2016) used the high-resolution DEM from Immerzeel et al. (2014) as one of data sources for energy-balance modelling of a supraglacial pond on Lirung glacier. Kraaijenbrink et al. (2016) also used the ortho-mosaic data from Immerzeel et al. (2014) in combination with 2014 UAV-borne data for estimating the seasonal surface velocity of Lirung glacier using frequency cross-correlation. Object-based image analysis (OBIA) of UAV imagery (May 2015) was carried out by Kraaijenbrink et al. (2016) to delineate ice cliffs and supraglacial ponds on the debris-covered Langtang Glacier, Nepal. They found that with a combination of high-resolution UAV imagery and digital elevation models, OBIA is able to objectively derive these features that makes it a valuable tool for accurate delineation of surface features on debris-covered glaciers.

In manual delineation of debris-covered glacier boundary on satellite images, the identification of glacier termini is, generally, hampered by extensive debris cover. However, the same can be done with a great ease using UAV-borne images. Debris-covered glacier surface is generally characterized by uneven and heterogeneous morphology. Presence of glacier features such as supraglacial ponds, ice cliffs, small ice mounds and other flow patterns are easily identifiable on UAV-borne orthomosacis and digital surface models (DSMs). These features may facilitate accurate demarcation of glacier boundaries.

Efficiency of automatic and semi-automatic techniques heavily depend on the spatial, temporal and radiometric resolution of the remote sensing data being used (Sect. 5). Previous studies clearly highlighted that the poor-quality DEM and coarse resolution thermal data are amongst the major challenges in fully automation of existing approaches. The deployment of UAV can bridge this gap as it provides bundle of information and has manifold advantage over conventional satellite remote sensing. UAVs can provide data with sub-centimeter spatial resolution and with high positional accuracy which can facilitate the efficient and accurate mapping of debris-covered glaciers. The high-resolution digital elevation model (DEM) can be easily developed using UAV-borne stereo images (Whitehead et al., 2013; Whitehead & Hugenholtz, 2014; Rippin et al. 2015). Further, Whitehead and Hugenholtz (2014) emphasized that thermal imaging is one area that is particularly suited to the use of small low-flying aerial platforms which can help to downscale the coarse resolution satellite thermal remote sensing data. Terrain parameters such as slope, aspect and curvature derived from high accuracy UAV DEMs along with high resolution optical, thermal and elevation information which UAVs now provide together with state of art information extraction techniques could easily segregate active debris-covered ice from dead ice and thus locate the actual glacier boundary. The high-resolution imageries obtained from UAVs can also aid developing the object-based classifiers for the automation of the debris-covered glacier mapping (Kraaijenbrink et al., 2016) which can save a lot of time and efforts and can provide results with high classification accuracy (Bhardwaj et al., 2016).

This brief review suggests that despite of great applicability, the use of UAVs in glaciology is in its infancy. However, future prospects of this technology are very high due to its immense potential. Technological improvements will indeed play an important part in the development more sophisticated UAV instruments. Therefore, it is likely that the next few years will see such systems being used to provide data for an ever-increasing range of glaciological applications.

5 Sources of Uncertainties in Debris-Covered Glacier Mapping

This section aims at evaluating the performance of various approaches (manual delineation, unsupervised and supervised classification) using optical, thermal and other ancillary data for the mapping of Gangotri glacier, central Himalaya, India. Continuous efforts have been made to map the extensively debris-covered Gangotri glacier (Philip & Ravindran, 1998; Naithani et al., 2001; Ahmad et al., 2004; Bahuguna et al., 2004, 2007; Gupta et al. 2005; Bhambri et al. 2011; Haq et al., 2011; Rai et al., 2016). However, most of these studies are either restricted to the snow classification or identification of various glacial landforms and/or manual delineation of the glacier outline using visual image interpretation to assess the glacier changes. For instance, Philip and Ravindran (1998) attempted to delineate the glacial landforms in parts of the Gangotri glacier using Landsat Thematic Mapper (TM) bands 4, 5 and 7 in the NIR and SWIR regions. Edge enhancement was performed on bands 7 and 5 which were then combined with linearly stretched band 4 to generate a color composite of bands 457 (RGB). This digitally processed image helped in identifying various glacial features in the Gangotri glacier. Naithani et al. (2001) studied various geomorphological features and morphometric parameters of Gangotri glacier to access the retreat of the glacier. Ahmad et al. (2004) performed principal component analysis on ASTER data for mapping of the debris-covered area of Gangotri glacier. Bahuguna et al. (2004) performed the Brovey transformation technique for identification of the snout and debris-covered ice of Gangotri Glacier using IRS LISS III and Indian Remote Sensing Satellite (IRS)-1C-PAN data. However, these techniques are appropriate for manual delineation of the debris-covered tongue based on visual interpretation. Gupta et al. (2005) investigated the snow cover pattern of Gangotri glacier using repetitive IRS-1C/-1D LISS-III multispectral data. NIR band thresholding was done over the snow-covered area computed from NDSI for the mapping of dry- and wet-snow zones on a temporal scale. The elevation information was also incorporated in order to reclassify shadowed pixels (wrongly classified as wet snow) with elevation above the dry-wet snow boundary as dry snow pixels using Boolean logic. Bahuguna et al. (2007) identified the snout position of Gangotri glacier using the merged product of IRS-1C PAN and LISS III data and measured the retreat of this glacier. The first attempt at semi-automated mapping of the debris-covered Gangotri Glacier was made by Bhambri et al. (2011). The authors developed a multi-criteria technique for Gangotri glacier mapping using cluster analysis of ASTER DEM derived morphometric parameters such as slope, plan curvature, profile curvature, thresholding of NIR/SWIR ratio image and ASTER thermal band 12. Vector maps were drawn up from the output of the cluster analysis, thermal mask and band ratio mask for the preparation of the final outlines of the Gangotri glacier using general overlay operations. A similar methodology was applied for validation, while replacing the ASTER thermal band (90 m) with Landsat thermal band (30 m). Glacier outlines were also delineated manually on high-resolution Cartosat-1 and IRS IC-PAN images based on visual interpretation for evaluation of the developed approach. Bhambri et al. (2012) monitored the frontal recession of the Gangotri group of glaciers from 1965–2006 using high resolution remote sensing data. The glacier extents were manually delineated from panchromatic data of Corona, Hexagon, IRS PAN and Cartosat-1 imageries. Haq et al. (2011) used repeat satellite optical imagery to deduce the Gangotri glacier extent change for the year 1972–2010. Negi et al. (2012) followed the Gupta et al.'s (2005) approach for snow classification and mapping of wet-snow line using multi-temporal IRS-1C/-1D/-P5 PAN/LISS III/Cartosat-1 data. The glacier boundary was also delineated using false color composite (FCC) of VNIR images due to significant differences in reflectance from glacial and non-glacial features. Chakraborty et al. (2014) employed a semi-automated approach to extract the boundary of Gangotri glacier using the glacio-morphological characteristics (slope, aspect, plan and profile curvatures) derived from SRTM DEM (90 m). The slope-based boundary was checked for consistency with the boundary from profile curvature and combined manually to generate the final glacier boundary. Kundu and Chakraborty (2015) delineated the zones of Gangotri glacier using the 2D signatures from dual-polarization C-band SAR data. The use of elevation information along with SAR signatures helped in classifying zones of similar backscattering signature situated at different altitude ranges. The authors studied the identified snowlines and other boundaries of glacial facies on a temporal scale. Bhattacharya et al. (2016) used NIR/SWIR ratio, thermal data, slope and curvature information to derive the boundary of Gangotri and its tributary glaciers using Terra-ASTER image of the year 2014, This outline was then manually adjusted on Landsat series and declassified imageries (KH-4A Corona, KH-9 Hexagon) to extract the glacier boundary for the period 1965–2015, and thus assess the multi-decadal changes in the Gangotri glacier. Rai et al. (2016) also used the manual approach to determine the multi-decadal changes in Gangotri glacier during 1972–2013 using Terra-ASTER and Landsat series data. Amongst these, only few studies are dedicated to the debris-cover mapping of Gangotri glacier. Also, there is a lack of comparison studies for evaluating the performances outlined DCG mapping approaches.

6 Summary and Conclusions

This chapter reviewed the multiple techniques used so far for the accurate mapping of debris covered glaciers. As mentioned in previous sections, accurate assessment of DCGs holds vital importance for proper planning and management of the water resources. However, the presence of debris serves as major bottleneck in precise mapping of the DCGs (Racoviteanu & Williams, 2012; Shukla et al., 2010). To overcome this problem several techniques have emerged throughout the world for precise mapping and monitoring of debris covered glaciers. All the techniques presented above are limited by one or the other reason. Therefore, future work should focus on eliminating the limitations of the techniques and enhancing their robustness, particularly by sensitivity analysis to explore the possibility of standardizing the various thresholds applied in the approach. Transferability of the methodologies presented here to some other area, in its absolute form, would necessitate thorough understanding of the datasets, study area and basis behind each condition applied for retrieval of debris cover. Also, the mapping accuracy can be greatly enhanced by improving the spatial resolution of the thermal bands and DEMs. Besides, future studies should focus on improving the spatial and temporal resolution of remote sensing datasets, particularly through the integration of high-resolution satellite imagery and UAV-based surveys. The application of artificial intelligence and machine learning algorithms can significantly improve debris–ice classification accuracy by leveraging multi-sensor and multi-temporal data. Advances in radar technology, such as multi-frequency and polarimetric SAR, along with the development of low-cost UAV-mounted LiDAR systems, will further enhance the detection of buried ice and debris thickness. Additionally, the use of cloud-based geospatial platforms and open-access archives can facilitate large-scale, automated mapping and temporal analysis. Overall, future progress in debris-covered glacier mapping will depend on synergistic integration of technologies, data fusion frameworks, and algorithmic innovations aimed at reducing uncertainty and improving the monitoring of glacier dynamics under changing climatic conditions.

Acknowledgements Iram Ali gratefully acknowledges funding from the Department of Science and Technology (DST), New Delhi, India, under WOS-A scheme (DST/WOS-A/EA-52/2021).

References

Ahmad, S., Hasnain, S. I., & Selvan, M. T. (2004). Morpho-metric characteristics of glaciers in the Indian Himalayas. *Asian Journal of Water, Environment and Pollution, 1*(1–2), 109–118.

Ahmed, R., Sam, L., Bhardwaj, A., Kulkarni, A. V., & Yamasaki, S. (2026). Increasing global threat of outburst floods from overlooked small alpine lakes. *Nature Sustainability*, 1–10.

Ali, I., & Shukla, A. (2015, July). A knowledge-based approach for assessing debris cover dynamics and its linkages to glacier recession. In *2015 IEEE International Geoscience and Remote Sensing Symposium (IGARSS) (2072–2075)*. IEEE.

Ali, I., Shukla, A., & Romshoo, S. A. (2017). Assessing linkages between spatial facies changes and dimensional variations of glaciers in the upper Indus Basin, western Himalaya. *Geomorphology, 284*, 115–129.

Ali, I., Shukla, A., Romshoo, S. A., Lone, F. A., Garg, P. K., & Yousuf, B. (2024). Empirical and thermal resistance approaches for debris thickness estimation on the Hoksar Glacier, Kashmir Himalaya. *Frontiers in Water, 6*, 1480585.

Alifu, H., Johnson, B. A., & Tateishi, R. (2015a). Delineation of debris-covered glaciers based on a combination of geomorphometric parameters and a TIR/NIR/SWIR band ratio. *IEEE Journal of Selected Topics in Applied Earth Observations and Remote Sensing, 9*(2), 781–792.

Alifu, H., Tateishi, R., & Johnson, B. (2015b). A new band ratio technique for mapping debris-covered glaciers using Landsat imagery and digital elevation model. *International Journal of Remote Sensing, 36*, 2063–2075.

Anderson, L. S., & Anderson, R. S. (2016). Modeling debris-covered glaciers: response to steady debris deposition. *The Cryosphere, 10*(3), 1105–1124.

Atwood, D. K., Meyer, F., & Arendt, A. (2010). Using L-band SAR coherence to delineate glacier extent. *Canadian Journal of Remote Sensing, 36*(S2), S186–S195.

Arora, M. K., Shukla, A., & Gupta, R. P. (2011). Digital image information extraction techniques for snow cover mapping from remote sensing data. In *Encyclopedia of Snow, Ice and Glaciers* (pp. 213–232). Springer.
Baatz, M., & Schäpe, A. (2000). Multiresolution segmentation: An optimization approach for high quality multi-scale image segmentation. In J. Strobl, T. Blaschke, & G. Griesebner (Eds.), *Angewandte Geographische Informationsverarbeitung XII* (pp. 12–23).
Bahuguna, I. M., Kulkarni, A. V., & Nayak, S. (2004). DEM from IRS-1C PAN stereo coverages over Himalayan glaciated region—accuracy and its utility. *International Journal of Remote Sensing, 25*(19), 4029–4041.
Bahuguna, I. M., Kulkarni, A. V., Nayak, S., Rathore, B. P., Negi, H. S., & Mathur, P. (2007). Himalayan glacier retreat using IRS 1C PAN stereo data. *International Journal of Remote Sensing, 28*(2), 437–442.
Bharadwaj, A., Joshi, P. K., Sam, L., Singh, M. K., Singh, S., & Kumar, R. (2015). Applicability of Landsat 8 data for characterizing glacier facies and supraglacial debris. *International Journal of Applied Earth Observation and Geoinformation, 38*, 51–64.
Bhambri, R., Bolch, T., & Chaujar, R. K. (2012). Frontal Recession of Gangotri Glacier, Garhwal Himalayas, from 1965 to 2006, Measured Through High-Resolution Remote Sensing Data. *Current Science, 102*, 489–494.
Buchroithner, M. F., & Bolch, T. (2007). An automated method to delineate the ice extension of the debris-covered glaciers at Mt. Everest based on ASTER Imagery. Grazer Schriften für Geographie und Raumforschung 43. In *Proceedings of the 9th International Symposium on High Mountain Remote Sensing Cartography* (pp. 71–78). 14–22 Sept 2006 – Graz, Austria.
Bajracharya, S. R., & Shrestha, B. (Eds.). (2011). *The status of glaciers in the Hindu Kush–Himalayan region*. International Centre for Integrated Mountain Development (ICIMOD).
Bajracharya, S. R., Maharjan, S. B., & Shrestha, F. (2014a). The status and decadal change of glaciers in Bhutan from the 1980s to 2010 based on satellite data. *Annals of Glaciology, 55*(66), 159–166.
Bajracharya, S. R., Maharjan, S. B., & Shrestha, F. (2014b). *Glacier status in Nepal and decadal change from 1980 to 2010 based on Landsat data*. ICIMOD.
Bajracharya, S. R., Maharjan, S. B., Shrestha, F., Guo, W., Liu, S., Immerzeel, W., & Shrestha, B. (2015). The glaciers of the Hindu Kush Himalayas: current status and observed changes from the 1980s to 2010. *International Journal of Water Resources Development, 31*(2), 161–173.
Banerjee, A., & Shankar, R. (2013). On the response of Himalayan glaciers to climate change. *Journal of Glaciology, 59*(215), 480–490.
Burns, P., & Nolin, A. (2014). Using atmospherically-corrected Landsat imagery to measure glacier area change in the Cordillera Blanca, Peru from 1987 to 2010. *Remote Sensing of Environment, 140*, 165–178.
Bayr, K. J., Hall, D. K., & Kovalick, W. M. (1994). Observations on glaciers in the eastern Austrian Alps using satellite data. *International Journal of Remote Sensing, 15*(9), 1733–1742.
Benn, D. I., & Evans, D. J. A. (1998). *Glaciers and glaciation* (p. 734). Arnold Publishers.
Benn, D. I., & Evans, D. J. A. (2010). *Glaciers and Glaciation* (p. 802). Hodder Education.
Benn, D., & Evans, D. J. (2014). *Glaciers and glaciation*. Routledge.
Berthier, E., Arnaud, Y., Kumar, R., Ahmad, S., Wagnon, P., & Chevallier, P. (2007). Remote sensing estimates of glacier mass balances in the Himachal Pradesh (Western Himalaya, India). *Remote Sensing of Environment, 108*(3), 327–338.
Bhardwaj, A., Joshi, P. K., Sam, L., Singh, M. K., Singh, S., & Kumar, R. (2015). Applicability of Landsat 8 data for characterizing glacier facies and supraglacial debris. *International Journal of Applied Earth Observation and Geoinformation, 38*, 51–64.
Bhardwaj, A., Sam, L., Akanksha, M.-T., & F. J., & Kumar, R. (2016). UAVs as remote sensing platform in glaciology: Present applications and future prospects. *Remote Sensing of Environment, 175*, 196–204.

Bhambri, R., Bolch, T., Chaujar, R. K., & Kulshreshtha, S. C. (2011). Glacier changes in the Garhwal Himalaya, India, from 1968 to 2006 based on remote sensing. *Journal of Glaciology, 57*(203), 543–556.

Bhattacharya, A., Bolch, T., Mukherjee, K., Pieczonka, T., Kropáček, J., & Buchroithner, M. F. (2016). Overall recession and mass budget of Gangotri Glacier, Garhwal Himalayas, from 1965 to 2015 using remote sensing data. *Journal of Glaciology, 62*(236), 1115–1133.

Biddle, D. J. (2015). Mapping debris-covered glaciers in the Cordillera Blanca, Peru: An object-based image analysis approach.

Bishop, M. P., Bonk, R., Kamp, U., Jr., & Shroder, J. F., Jr. (2001). Terrain analysis and data modeling for alpine glacier mapping. *Polar Geography, 25*(3), 182–201.

Bishop, M. P., Kargel, J. S., Kieffer, H. H., MacKinnon, D. J., Raup, B. H., & Shroder Jr, J. F. (2000). Remote-sensing science and technology for studying glacier processes in high Asia. *Annals of glaciology, 31*, 164–170.

Bishop, M. P., Shroder Jr., J. F., & Ward, J. L. (1995). SPOT multispectral analysis for producing supraglacial debris-load estimates for Batura glacier,.

Blaschke, T. (2010). Object-based image analysis for remote sensing. *ISPRS Journal of Photogrammetry and Remote Sensing, 65*(1), 2–16.

Bolch, T., & Kamp, U. (2005). Glacier mapping in high mountains using DEMs, Landsat and ASTER data.

Bolch, T., Buchroithner, M. F., Kunert, A., & Kamp, U. (2007). Automated delineation of debris-covered glaciers based on ASTER data. In M. A. Gomarasca (Ed.), *GeoInformation in Europe. Proceedings of the 27th EARSeL Symposium, 4–6 June 2007, Bolzano, Italy* (pp. 403–410). Millpress.

Bolch, T., Buchroithner, M., Pieczonka, T., & Kunert, A. (2008). Planimetric and volumetric glacier changes in the Khumbu Himal, Nepal, since 1962 using Corona, Landsat TM and ASTER data. *Journal of Glaciology, 54*(187), 592–600.

Bolch, T., & Kamp, U. (2006). Glacier mapping in high mountains using DEMs, Landsat and ASTER data. *Grazer Schriften der Geographie und Raumforschung, 41*, 13–24.

Bolch, T., Kulkarni, A., Kääb, A., Huggel, C., Paul, F., Cogley, J. G., Frey, H., Kargel, J. S., Fujita, K., Scheel, M., & Stoffel, M. (2012). The state and fate of Himalayan glaciers. *Science, 336*(6079), 310–314.

Brock, B. W., Mihalcea, C., Kirkbride, M. P., Diolaiuti, G., Cutler, M. E., & Smiraglia, C. (2010). Meteorology and surface energy fluxes in the 2005–2007 ablation seasons at the Miage debris-covered glacier, Mont Blanc Massif, Italian Alps. *Journal of Geophysical Research: Atmospheres, 115*(D9).

Bronge, L., & Bronge, C. (1999). Glacier changes in the Tarfala Valley, Swedish Lapland, monitored by remote sensing. *Geografiska Annaler: Series A, Physical Geography, 81*(4), 683–692.

Chakraborty, M., Panigrahy, S., & Kundu, S. (2014). Semi-automated technique to extract boundary of valley/mountain glaciers using glacio-morphological information from digital elevation model. *The International Archives of the Photogrammetry, Remote Sensing and Spatial Information Sciences, 40*, 511–516.

Clayton, A. I. (2012). *Remote sensing of subglacial bedforms from the British Ice Sheet using an unmanned aerial system (UAS): Problems and potential* (MScR thesis). Durham University.

Collier, E., Maussion, F., Nicholson, L. I., Mölg, T., Immerzeel, W. W., & Bush, A. B. G. (2015). Impact of debris cover on glacier ablation and atmosphere–glacier feedbacks in the Karakoram. *The Cryosphere, 9*(4), 1617–1632.

Crocker, R. I., Maslanik, J. A., Adler, J. J., Palo, S. E., Herzfeld, U. C., & Emery, W. J. (2011). A sensor package for ice surface observations using small unmanned aircraft systems. *IEEE Transactions on Geoscience and Remote Sensing, 50*(4), 1033–1047.

Cuffey, K. M., & Paterson, W. S. B. (2010). *The physics of glaciers* (4th ed.). Academic Press.

Dobhal, D. P., Mehta, M., & Srivastava, D. (2013). Influence of debris cover on terminus retreat and mass changes of Chorabari Glacier, Garhwal region, central Himalaya, India. *Journal of Glaciology, 59*(217), 961–971.

Dozier, J. (1989). Spectral signature of alpine snow cover from the Landsat Thematic Mapper. *Remote Sensing of Environment, 28*, 9–22.
Frey, H., Paul, F., & Strozzi, T. (2012). Compilation of a glacier inventory for the western Himalayas from satellite data: Methods, challenges, and results. *Remote Sensing of Environment, 124*, 832–843.
Friedli, E. (2013). *Photogrammetric methods for the reconstruction and monitoring of glaciers*. Master's Thesis, ETH Zürich.
Fugazza, D., Scaioni, M., Corti, M., D'Agata, C., Azzoni, R. S., Cernuschi, M., et al. (2018). Combination of UAV and terrestrial photogrammetry to assess rapid glacier evolution and map glacier hazards. *Natural Hazards and Earth System Sciences, 18*(4), 1055–1071.
Garg, P. K., Shukla, A., Tiwari, R. K., & Jasrotia, A. S. (2017). Assessing the status of glaciers in part of the Chandra basin, Himachal Himalaya: a multiparametric approach. *Geomorphology, 284*, 99–114.
Ghosh, S., Pandey, A. C., & Nathawat, M. S. (2014). Mapping of debris-covered glaciers in parts of the Greater Himalaya Range, Ladakh, western Himalaya, using remote sensing and GIS. *Journal of Applied Remote Sensing, 8*(1), 083579–083579.
Gibson, M. J., Glasser, N. F., Quincey, D. J., Mayer, C., Rowan, A. V., & Irvine-Fynn, T. D. (2017). Temporal variations in supraglacial debris distribution on Baltoro Glacier, Karakoram between 2001 and 2012. *Geomorphology, 295*, 572–585.
Gjermundsen, E. F., Mathieu, R. S. A., Kaab, A., Chinn, T., Fitzharris, B., & Hagen, J. O. (2011). Assessment of multispectral glacier-mapping methods and derivation of glacier area changes, 1978–2002, in the central Southern Alps, New Zealand, from ASTER satellite data, field survey and existing inventory data. *Journal of Glaciology, 57*(204), 667–683.
Goudie, A. (Ed.). (2004). *Encyclopedia of geomorphology* (Vol. 2). Psychology Press.
Gupta, R. P. (2003). *Remote sensing geology* (p. 656). Springer.
Gupta, R. P., Haritashya, U. K., & Singh, P. (2005). Mapping dry/wet snow cover in the Indian Himalayas using IRS multispectral imagery. *Remote Sensing of Environment, 97*(4), 458–469.
Hall, D. K., Bayr, K. J., Schöner, W., Bindschadler, R. A., & Chien, J. Y. L. (1987). *Satellite-derived glacier and snow cover data for climate monitoring*. Glaciological Data Report GD-25. World Data Center for Glaciology.
Hall, D. K., Bayr, K. J., Schöner, W., Bindschadler, R. A., & Chien, J. Y. L. (1988). Glacier monitoring with Landsat imagery: Observations from Western North America. *Annals of Glaciology, 11*, 46–52.
Hall, D. K., Riggs, G. A., & Salomonson, V. V. (1995). Development of methods for mapping global snow cover using Moderate Resolution Imaging Spectroradiometer (MODIS) data. *Remote Sensing of Environment, 54*(2), 127–140.
Hambrey, M. J., Quincey, D. J., Glasser, N. F., Reynolds, J. M., Richardson, S. J., & Clemmens, S. (2008). Sedimentological, geomorphological and dynamic context of debris-mantled glaciers, Mount Everest (Sagarmatha) region, Nepal. *Quaternary Science Reviews, 27*(25–26), 2361–2389.
Haq, M. A., Jain, K., & Menon, K. P. R. (2011, December). Change monitoring of gangotri glacier using satellite imagery. In *12th ESRI India User Conference* (pp. 6–8).
Haq, M. A., Jain, K., & Menon, K. P. R. (2014). Modelling of Gangotri glacier thickness and volume using an artificial neural network. *International Journal of Remote Sensing, 35*(16), 6035–6042.
Hodson, A., Anesio, A. M., Ng, F., Watson, R., Quirk, J., Irvine-Fynn, T., Dye, A., Clark, C., McCloy, P., Kohler, J., & Sattler, B. (2007). A glacier respires: Quantifying the distribution and respiration CO_2 flux of cryoconite across an entire Arctic supraglacial ecosystem. *Journal of Geophysical Research. Biogeosciences, 112*(G4), G04S36.
Höfle, B., Pfeifer, N. (2007). Correction of laser scanning intensity data: Data and model-driven approaches. *ISPRS Journal of Photogrammetry and Remote Sensing*. https://doi.org/10.1016/j.isprsjprs.2007.05.008

Huang, X., Deng, J., Ma, X., Wang, Y., Feng, Q., Hao, X., & Liang, T. (2016). Spatiotemporal dynamics of snow cover based on multi-source remote sensing data in China. *The Cryosphere, 10*, 2453–2463.

Huang, X.-D., Hao, X.-H., Feng, Q.-S., Wang, W., & Liang, T. G. (2014). A new MODIS daily cloud-free snow cover mapping algorithm on the Tibetan Plateau. *Sciences in Cold and Arid Regions, 6*, 116–123.

Immerzeel, W., Kraaijenbrink, P., & Andreassen, L. (2017). Use of an Unmanned Aerial Vehicle to assess recent surface elevation change of Storbreen in Norway. *The Cryosphere Discussions, 2017*, 1–31.

Immerzeel, W. W., Kraaijenbrink, P. D. A., Shea, J. M., Shrestha, A. B., Pellicciotti, F., Bierkens, M. F. P., & de Jong, S. M. (2014). High-resolution monitoring of Himalayan glacier dynamics using unmanned aerial vehicles. *Remote Sensing of Environment, 150*, 93–103.

Jacobs, J. D., Simms, A., & Simms, A. E. (1997). Recession of the White Glacier, Axel Heiberg Island, N.W.T., Canada, observed by satellite imagery. *Annals of Glaciology, 24*, 423–429.

Jawak, S. D., Panditrao, S. N., Luis, A. J., & Naik, S. S. (2015). A comprehensive review on pixel-based, object-based, and deep learning classification approaches for mapping of glacial and periglacial geomorphology. *Advances in Remote Sensing, 4*(3), 177–195.

Juen, M., Mayer, C., Lambrecht, A., Han, H., & Liu, S. (2014). Impact of varying debris cover thickness on ablation: a case study for Koxkar Glacier in the Tien Shan. *The Cryosphere, 8*(2), 377–386.

Karimi, N., Farokhnia, A., Karimi, L., Eftekhari, M., & Ghalkhani, H. (2012). Combining optical and thermal remote sensing data for mapping debris-covered glaciers (Alamkouh Glaciers, Iran). *Cold Regions Science and Technology, 71*, 73–83.

Kayastha, R. B., Takeuchi, Y., Nakawo, M., & Ageta, Y. (2000). Practical prediction of ice melting beneath various thickness of debris cover on Khumbu Glacier. Nepal, using a positive degree-day factor. *IAHS-AISH P, 264*, 71–81.

Keshri, A. K., Shukla, A., & Gupta, R. P. (2009). ASTER ratio indices for supraglacial terrain mapping. *International Journal of Remote Sensing, 30*(2), 519–524.

Khan, A., Naz, B. S., & Bowling, L. C. (2015). Separating snow, clean and debris covered ice in the Upper Indus Basin, Hindukush-Karakoram-Himalayas, using Landsat images between 1998 and 2002. *Journal of Hydrology, 521*, 46–64.

Khan, S. A., Kjeldsen, K. K., Kjær, K. H., Bevan, S., Luckman, A., Aschwanden, A., ... & Fitzner, A. (2014). Glacier dynamics at Helheim and Kangerdlugssuaq glaciers, southeast Greenland, since the little ice age. *The Cryosphere, 8*(4), 1497–1507.

Kieffer, H., Kargel, J. S., Barry, R., Bindschadler, R., Bishop, M., MacKinnon, D., et al. (2000). New eyes in the sky measure glaciers and ice sheets. *EOS, Transactions American Geophysical Union, 81*(24), 265–271.

Kirkbride, M. P. (2011). Debris-covered glaciers. In *Encyclopedia of snow, ice and glaciers* (pp. 180–182). Springer.

Kirkbride, M. P., & Deline, P. (2013). The formation of supraglacial debris covers by primary dispersal from transverse englacial debris bands. *Earth Surface Processes and Landforms, 38*(15), 1779–1792.

Könnig, M., Winther, J.-G., & Isaksson, E. (2001). Measuring snow and glacier ice properties from satellite. *Reviews of Geophysics, 39*(1), 1–27.

Koskinen, J., & Hallikainen, M. T. (1997). The use of ERS-1 SAR data in snow melt monitoring. *Physics and Chemistry of the Earth, 22*(3–4), 285–289.

Koutnik, M. R., Fudge, T. J., Conway, H., Waddington, E. D., Neumann, T. A., Cuffey, K. M., Buizert, C., & Taylor, K. C. (2016). Holocene accumulation and ice flow near the West Antarctic Ice Sheet Divide ice core site. *Journal of Geophysical Research: Earth Surface, 121*(5), 907–924.

Kraaijenbrink, P. D. A., Meijer, S. W., Shea, J. M., Pellicciotti, F., de Jong, S. M., & Immerzeel, W. W. (2016). Seasonal surface velocities of a Himalayan glacier derived by automated correlation of unmanned aerial vehicle imagery. *Annals of Glaciology, 57*(71), 103–113.

Kundu, S., & Chakraborty, M. (2015). Delineation of glacial zones of Gangotri and other glaciers of Central Himalaya using RISAT-1 C-band dual-pol SAR. *International Journal of Remote Sensing, 36*(6), 1529–1550.

Lougeay, R. (1974). Detection of buried glacial and ground ice with thermal infrared remote sensing. In *Advanced Concepts and Techniques in the Study of Snow and Ice Resources*. National Academy of Sciences.

Mattson, L. E. (1993). Ablation on debris covered glaciers: an example from the Rakhiot Glacier, Punjab, Himalaya. Snow and glacier hydrology.

Menzies, J. (2002). *Modern and past glacial environments: Revised student edition*. Elsevier.

Miles, E. S., Pellicciotti, F., Willis, I. C., Steiner, J. F., Buri, P., & Arnold, N. S. (2016). Refined energy-balance modelling of a supraglacial pond, Langtang Khola, Nepal, Annals of Glaciology. *Annals of Glaciology, 57*, 29–40. https://doi.org/10.3189/2016AoG71A421

Nagai, H., Fujita, K., Nuimura, T., & Sakai, A. (2013). Southwest-facing slopes control the formation of debris-covered glaciers in the Bhutan Himalaya. *The Cryosphere, 7*(4), 1303–1314.

Naithani, A. K., Nainwal, H. C., Sati, K. K., & Prasad, C. (2001). Geomorphological evidences of retreat of Gangotri Glacier and its characteristics. *Current Science, 80*(1), 87–94.

Nakawo, M. (1993). Satellite data utilization for estimating ablation of debris covered glaciers. In *Snow and Glacier Hydrology (Proceedings of the Kathmandu Symposium, November 1992)* (Vol. 218, pp. 75–83). IAHS Publication.

Negi, H. S., Thakur, N. K., Ganju, A., & SNEHMANI. (2012). Monitoring of Gangotri glacier using remote sensing and ground observations. *Journal of earth system science, 121*(4), 855–866.

Nex, F., & Remondino, F. (2013). UAV for 3D mapping applications: A review. *Applied Geomatics, 6*(1), 1–15.

Nicholson, L., & Benn, D. I. (2006). Calculating ice melt beneath a debris layer using meteorological data. *Journal of Glaciology, 52*(178), 463–470.

Nijhawan, R., Garg, P. K., & Thakur, P. K. (2016). Monitoring of glacier in Alaknanda basin using remote sensing data. *Perspectives in Science, 8*, 381–383.

Ojha, S., Fujita, K., Sakai, A., Nagai, H., & Lamsal, D. (2017). Topographic controls on the debris-cover extent of glaciers in the Eastern Himalayas: Regional analysis using a novel high-resolution glacier inventory. *Quaternary International, 455*, 82–92.

Østrem, G. (1959). Ice melting under a thin layer of moraine, and the existence of ice cores in moraine ridges. *Geografiska Annaler, 41*(4), 228–230.

Paul, F. (2000). Evaluation of different methods for glacier mapping using Landsat TM. *EARSeL eProceedings, 1*(1), 239–245.

Paul, F. (2003). *The new Swiss glacier inventory 2000: Application of remote sensing and GIS*. Doctoral dissertation, University of Zürich.

Paul, F., Huggel, C., & Kääb, A. (2004). Combining satellite multispectral image data and a digital elevation model for mapping debris-covered glaciers. *Remote sensing of Environment, 89*(4), 510–518.

Pellicciotti, F., Stephan, C., Miles, E., Herreid, S., Immerzeel, W. W., & Bolch, T. (2015). Mass-balance changes of the debris-covered glaciers in the Langtang Himal, Nepal, from 1974 to 1999. *Journal of Glaciology, 61*(226), 373–386.

Philip, G., & Ravindran, K. V. (1998). Glacial mapping using Landsat Thematic Mapper data: A case study in parts of Gangotri Glacier, NW Himalaya. *Journal of the Indian Society of Remote Sensing, 26*(1), 29–34.

Pratap, B., Dobhal, D. P., Mehta, M., & Bhambri, R. (2015). Influence of debris cover and altitude on glacier surface melting: a case study on Dokriani Glacier, central Himalaya, India. *Annals of Glaciology, 56*(70), 9–16.

Racoviteanu, A. E., Paul, F., Raup, B., Khalsa, S. J. S., & Armstrong, R. (2009). Challenges and recommendations in mapping of glacier parameters from space: results of the 2008 Global Land Ice Measurements from Space (GLIMS) workshop, Boulder, Colorado, USA. *Annals of Glaciology, 50*(53), 53–69.

Racoviteanu, A., & Williams, M. W. (2012). Decision tree and texture analysis for mapping debris-covered glaciers in the Kangchenjunga area. *Eastern Himalaya. Remote sensing, 4*(10), 3078–3109.

Racoviteanu, A. E., Arnaud, Y., Williams, M. W., & Manley, W. F. (2015). Spatial patterns in glacier characteristics and area changes from 1962 to 2006 in the Kanchenjunga–Sikkim area, Eastern Himalaya. *The Cryosphere, 9*(2), 505–523.

Rai, P. K., Mishra, V. N., & Singh, S. (2016). Geospatial approach for quantitative drainage morphometric analysis of Varuna River Basin. *India. Journal of Landscape Ecology, 9*(2), 67–86.

Ranzi, R., Grossi, G., Iacovelli, L., & Taschner, S. (2004). Use of multispectral ASTER images for mapping debris-covered glaciers within the GLIMS project. In *Proceedings of the 2004 IEEE International Geoscience & Remote Sensing Symposium (IGARSS 2004)*, Anchorage, AK, USA (Vol. 2, pp. 1144–1147). IEEE.

Rastner, P., Bolch, T., Mölg, N., Machguth, H., Le Bris, R., & Paul, F. (2014). The first complete inventory of the local glaciers and ice caps on Greenland. *The Cryosphere, 8*, 1297–1315.

Raup, B., Racoviteanu, A., Khalsa, S. J. S., Helm, C., Armstrong, R., & Arnaud, Y. (2007). The GLIMS geospatial glacier database: A new tool for studying glacier change. *Global and Planetary Change, 56*(1–2), 101–110.

Reid, T. D., Carenzo, M., Pellicciotti, F., & Brock, B. W. (2012). Including debris cover effects in a distributed model of glacier ablation. *Journal of Geophysical Research: Atmospheres, 117*(D18).

Richardson, S. D., & Reynolds, J. M. (2000). An overview of glacial hazards in the Himalayas. *Quaternary International, 65/66*, 31–47.

Rippin, D. M., Pomfret, A., & King, N. (2015). High resolution mapping of supra-glacial drainage pathways reveals link between micro-channel drainage density, surface roughness and surface reflectance. *Earth Surface Processes and Landforms, 40*(10), 1279–1290.

Robson, B. A., Hölbling, D., Nuth, C., Strozzi, T., & Dahl, S. O. (2016). Decadal scale changes in glacier area in the Hohe Tauern National Park (Austria) determined by object-based image analysis. *Remote Sensing, 8*(1), 67.

Robson, B. A., Nuth, C., Nielsen, P. R., Girod, L., Hendrickx, M., & Dahl, S. O. (2015). Spatial variability in patterns of glacier change across the Manaslu Range, Central Himalaya. *Frontiers. Earth Science, 6, Article 12.*

Rott, H. (1994). Thematic Mapper and ERS-1 SAR measurements for glacier monitoring. *Annals of Glaciology, 20*, 255–261.

Rott, H., & Markl, G. (1989). Seasonal and annual variations of snow cover in the Alps derived from satellite data. *Annals of Glaciology, 13*, 157–163.

Rott, H., & Mätzler, C. (1987). Possibilities and limits of synthetic aperture radar for snow and glacier surveying. *Annals of Glaciology, 9*, 195–199.

Rott, H., & Nagler, T. (1994). Capabilities of ERS-1 SAR for snow and glacier monitoring in alpine areas. In Proceedings of the Second ERS-1 Symposium: Space at the Service of Our Environment (Vol. 2, pp. 965–970). European Space Agency.

Rowan, A. V., Egholm, D. L., Quincey, D. J., & Glasser, N. F. (2015). Modelling the feedbacks between mass balance, ice flow and debris transport to predict the response to climate change of debris-covered glaciers in the Himalaya. *Earth and Planetary Science Letters, 430*, 427–438.

Saito, A., & Yamazaki, T. (1999). Characteristics of spectral reflectance for vegetation ground surfaces with snow-cover: Vegetation indices and snow indices. *Journal of Japan Society of Hydrology and Water Resources, 12*, 28.

Salehi, B., Zhang, Y., Zhong, M., & Dey, V. (2012). Object-based classification of urban areas using VHR imagery and height point data. *International Journal of Remote Sensing, 33*(15), 4606–4629.

Saraswat, P., Syed, T. H., Famiglietti, J. S., Fielding, E. J., Crippen, R., & Gupta, N. (2013). Recent changes in the snout position and surface velocity of Gangotri glacier observed from space. *International Journal of Remote Sensing, 34*(24), 8653–8668.

Scherler, D., Bookhagen, B., & Strecker, M. R. (2011). Spatially variable response of Himalayan glaciers to climate change affected by debris cover. *Nature geoscience, 4*(3), 156–159.

Scherler, D., Wulf, H., & Gorelick, N. (2018). Global assessment of supraglacial debris-cover extents. *Geophysical Research Letters, 45*(21), 11–798.

Schomacker, A. (2008). What controls dead-ice melting under different climate conditions? *A discussion. Earth-Science Reviews, 90*(3–4), 103–113.

Schomacker, A., & Benediktsson, Í. Ö. (2018). Supraglacial environments. In *Past glacial environments* (pp. 159–179). Elsevier.

Sharma, P., Patel, L. K., Ravindra, R., Singh, A., & K, M., & Thamban, M. (2016). Role of debris cover to control specific ablation of adjoining Batal and Sutri Dhaka glaciers in Chandra Basin (Himachal Pradesh) during peak ablation season. *Journal of Earth System Science, 125*(3), 459–473.

Shi, J., & Dozier, J. (1993). Measurements of snow- and glacier-covered areas with single-polarization SAR. *Annals of Glaciology, 17*, 72–76.

Shi, J., & Dozier, J. (1997). Mapping seasonal snow cover with SIR-C/X-SAR in mountainous areas. *Remote Sensing of Environment, 59*(2), 294–307.

Shimamura, Y., Izumi, T., & Matsuyama, H. (2006). Evaluation of a useful method to identify snow-covered areas under vegetation — comparisons among a newly proposed snow index, normalized difference snow index, and visible reflectance. *International Journal of Remote Sensing, 27*(21–22), 4867–4884.

Sidjak, R. W., & Wheate, R. D. (1999). Glacier mapping of the Illecillewaet icefield, British Columbia, Canada, using Landsat TM and digital elevation data. *International Journal of Remote Sensing, 20*, 273–284.

Shukla, A., & Ali, I. (2016). A hierarchical knowledge-based classification for glacier terrain mapping: a case study from Kolahoi Glacier. *Kashmir Himalaya. Annals of Glaciology, 57*(71), 1–10.

Shukla, A., & Qadir, J. (2016). Differential response of glaciers with varying debris cover extent: evidence from changing glacier parameters. *International Journal of Remote Sensing, 37*(11), 2453–2479.

Shukla, A., Gupta, R. P., & Arora, M. K. (2010). Delineation of debris-covered glacier boundaries using optical and thermal remote sensing data. *Remote Sensing Letters, 1*(1), 11–17.

Singh, P., Kumar, A., & Kishore, N. (2011). Meltwater storage and delaying characteristics of Gangotri Glacier (Indian Himalayas) during ablation season. *Hydrological Processes, 25*(2), 159–166.

Singh, P., Kumar, N., & Arora, M. (2000). Degree–day factors for snow and ice for Dokriani Glacier, Garhwal Himalayas. *Journal of Hydrology, 235*(1–2), 1–11.

Solbø, S., & Storvold, R. (2013). Mapping Svalbard glaciers with the CryoWing UAS. *International Archives of the Photogrammetry, Remote Sensing and Spatial Information Sciences, XL-1/W2*, 373–377.

Stokes, C. R., Popovnin, V., Aleynikov, A., Gurney, S. D., & Shahgedanova, M.: Recent glacier retreat in the caucasus mountains, Russia, and associated increase in supraglacial debris cover and supra-/proglacial lake development. *Annals of Glaciology, 46*, 195–203 (2007).

Strozzi, T., Paul, F., & Kääb, A. (2010). Glacier mapping with ALOS PALSAR within the ESA GlobGlacier project. In *Proceedings of the ESA Living Planet Symposium* (Vol. 686 of ESA Special Publication).

Takeuchi, N., Kohshima, S., & Seko, K. (2001). Structure, formation, and darkening process of albedo-reducing material (cryoconite) on a Himalayan glacier: a granular algal mat growing on the glacier. *Arctic, Antarctic, and Alpine Research, 33*(2), 115–122.

Taschner, S., & Ranzi, R. (2002). Comparing the opportunities of Landsat-TM and ASTER data for monitoring a debris-covered glacier in the Italian Alps within the GLIMS project. In *Proceedings of the 22nd International Geoscience & Remote Sensing Symposium (IGARSS 2002), Toronto, Canada* (Vol. 2, pp. 1044–1046). IEEE.

Thompson, L. G. (2017). Past, present, and future of glacier archives from the world's highest mountains. *Proceedings of the American Philosophical Society, 161*(3), 226–243.

Tilg, M., Storrie-Lombardi, M., Kohstall, C., Trenkwalder, A., Psenner, R., & Sattler, B. (2011, September). LIFE: laser induced fluorescence emission, a non-invasive tool to detect photosynthetic pigments in glacial ecosystems. In *Instruments, Methods, and Missions for Astrobiology XIV* (Vol. 8152, pp. 130–137). SPIE.

Tiwari, R. K., Gupta, R. P., & Arora, M. K. (2014). Estimation of surface ice velocity of Chhota Shigri glacier using sub-pixel ASTER image correlation. *Current Sciences, 106*(6), 853–859.

Tschudi, M. A., Maslanik, J. A., & Perovich, D. K. (2008). Derivation of melt pond coverage on Arctic Sea ice using MODIS observations. *Remote Sensing of Environment, 112*(5), 2605–2614.

Veettil, B. K. (2012). A remote sensing approach for monitoring debris-covered glaciers in the high altitude Karakoram Himalayas. *International Journal of Geomatics and Geosciences, 2*(3), 833–841.

Veettil, K., Bremer, B. F. U., Grondona, E. B., & Florencio De Souza, S. A. (2014). Recent changes occurred in the terminus of the debris covered Bilafond glacier in the karakoram Himalayas using remotely sensed images and digital elevation models (1978–2011). *Journal of Mountain Science, 11*(2), 398–406.

Walter, F., Clinton, J. F., Deichmann, N., Dreger, D. S., Minson, S. E., & Funk, M. (2009). Moment tensor inversions of icequakes on Gornergletscher, Switzerland. *Bulletin of the Seismological Society of America 99*(2A), https://doi.org/10.1785/0120080110.

Wang, S., Zhang, M., Li, Z., Wang, F., Li, H., Li, Y., & Huang, X. (2011). Glacier area variation and climate change in the Chinese Tianshan Mountains since 1960. *Journal of Geographical Sciences, 21*(2), 263–273.

Whitehead, K., & Hugenholtz, C. H. (2014). Remote sensing of the environment with small unmanned aircraft systems (UASs), Part 1: A review of progress and challenges. *Journal of Unmanned Vehicle Systems, 2*(3), 69–85.

Whitehead, K., Moorman, B. J., & Hugenholtz, C. H. (2013). Brief communication: Low-cost, on-demand aerial photogrammetry for glaciological measurement. *The Cryosphere, 7*, 1879–1884.

Williams, R. S., Jr., Hall, D. K., & Benson, C. S. (1991). Analysis of glacier facies using satellite techniques. *Journal of Glaciology, 37*(125), 120–128.

Xiao, X., Zhang, Q., Boles, S., Rawlins, M., & Moore, B., III. (2001). A normalized difference snow and ice index (NDSII) for mapping snow/ice cover from Landsat TM data. *International Journal of Remote Sensing, 22*(13), 2479–2487.

Zhang, Y., Fujita, K., Liu, S., Liu, Q., & Nuimura, T. (2011). Distribution of debris thickness and its effect on ice melt at Hailuogou glacier, southeastern Tibetan Plateau, using in situ surveys and ASTER imagery. *Journal of Glaciology, 57*(206), 1147–1157.

Zongli, J., Shiyin, L., Sichun, L., & Xin, W. (2012). Estimate Yengisogat glacier surface flow velocities using ALOS PALSAR Data feature-tracking, Karakoram, China. *Procedia Environmental Sciences, 12*, 646–652.

Quaternary River Response to Differential Uplift, Inferences from Steepness Index and Concavity

Ahsan Afzal Wani, Bikram Singh Bali, Mohammad Irfan, and Naseer Ahmad

Abstract The thrust-bound Kashmir valley experiences differential uplift as a result of the presence of two thrust fault systems: the Zanskar Thrust and the Panjal Thrust in the Great Himalayan Range (GHR) and the Pir Panjal Range (PPR), respectively. These sub basins have been greatly affected by these thrust faults, with episodic uplift from the past five million years. The rivers from both the GHR and PPR respond to this differential uplift by adopting changes in their concavity and steepness. The influence of neotectonics on the topography of river longitudinal profiles was performed using semi-automated river profile analysis. Profiles of nine longitudinal rivers of the Jhelum basin, extracted from 30 m DEM, were modelled using area-slope analysis and stream power models in Arc GIS and Matlab software. Two sets of knick points, vertical-step and slope-break knick points, have been recognized, which were later compared with the geological settings and field observations to determine their genesis. We also used the stream-power incision model to calculate the channel steepness index ratio (k_{sn}) and concavity (θ) of separated parts of the river system to compare differential uplift of sub basins. The present study shows that the transient states represented by the occurrence of slope break knick points and vertical-step knick points, related to inhomogeneous substrates and the presence of faults in these segments, are the findings of the present study. However, the relative differential uplift of the two mountain ranges was a key finding in the present work. Based on the steepness index ratio, the rivers from the PPR show higher uplift while the sub basins from the GHR show lower uplift, except for the Madhumati River and the Erin River, which show the higher uplift. However, the concavity index values clearly infer that the sub basins from the PPR show relatively higher uplift than the sub basins from the GHR. The higher uplift of Erin and Madhumati clearly indicates that, in addition to the regional geological structures, local geological structures also play an important role in the differential uplift of a particular area.

Keywords Kashmir valley · Differential uplift · Longitudinal profiles · Steepness index ratio · Concavity · Faults

A. A. Wani (✉) · B. S. Bali · M. Irfan · N. Ahmad
Department of Earth Sciences, University of Kashmir, Srinagar, India

R. Ahmed et al. (eds.), *Climate Change, Water Resources, and Natural Hazards in the Himalayas*, Advances in Geographical and Environmental Sciences,
https://doi.org/10.1007/978-981-92-0835-7_3

1 Introduction

The striking features that incise into the earth's surface are river systems, which are in turn shaped by uplift and erosion under the joint influence of climate and tectonics. The concavity changes from one basin to the next as the river descends downwards, reflecting the interplay between uplift and erosional processes. A revolution has occurred over the past decades in quantitative landscape development with the introduction of high-quality global Digital Elevation Models (DEMs) (Pipaud et al., 2015; Harel et al., 2016; Irfan et al., 2024a, c; Bali et al., 2025a). The data from these DEMs has also helped in the development of sophisticated computer models of landscape evolution and advances in geochronology (Whipple & Tucker, 2002; Di Biase et al., 2010; Balco et al., 2008; Wani et al., 2025b). Time, tectonics, and climate influences are responsible for dynamics in landforms (Whipple & Tucker; 1999; Whipple, 2004; Kirby & Whipple, 2003, 2001). The combined influence of tectonics and climate on the underlying rock type causes landforms to change over time. In this way, landforms can notice tectonic signals, whereas climatic signals can be ruled out. Anoop et al. (2012) say that the history of uplift is an important factor for understanding the tectonic mechanisms that cause different uplift in tectonically active places. The geomorphic indicators are the best to relate the tectonics, climate, and geology of the area. The most sensitive geomorphic markers which quickly response to regional tectonic uplift are river longitudinal profiles (Holbrook & Schumm, 1999, Burbank & Anderson, 2001; Wani et al., 2019; Bali & Wani, 2020, 2021; Bali et al., 2025b; Wani et al., 2025a). The identification and interpretation of over-steepened or convex upstream reaches of river profiles can indicate evolutionary aspects of the sub basins. The construction of mathematical correlations between channel parameters (usually employing power functions) is frequently predicted by stream profile analysis, which is ideal at steady-state equilibrium (SSE). The longitudinal profile of river systems and mature rivers at SSE will be smooth and concave up (uplift and erosion will be equal), and the channel bed elevation will be steady (uplift and erosion will be equal) (Hack, 1973). This implies that there are precise correlations between local stream gradient and stream length (Hack, 1957) or between local channel steepness and upstream drainage area (Hack, 1957; Flint, 1974). A stream's local channel gradient will fall as a power law function at SSE due to the larger catchment. As a result, local disequilibrium may be linked to an unpredictable relationship between channel gradient and upstream drainage area. Our main goal was to use a longitudinal profile form analysis of DEMs to identify the current differential uplift scenario of the sub basins of the two mountain ranges (GHR and PPR) which rampart the Kashmir Valley based on the steepness index ratio and concavity.

2 Geology

The Kashmir valley is one of the seismically and tectonically active thrust-bound drainage basins in the northwest part of the Himalayas. The valley is bounded by the union territory of Ladakh from the east, west by Muzafarabad, Pakistan; north by Gilgat; and from the south by the Jammu Region. The whole valley is situated between two major geological domains, namely the Pir Panjal Range and the Zanskar Range. The valley has diverse lithology with the younger deposits of alluvium followed by the deposits of karewas of quarternary age. The oldest formations are granites, which are exposed at Kangan, Hant, and Kazinag.

Figure 1 shows the lithological distribution in the Kashmir basin (Middlemiss, 1910, 1911; Ahmad et al., 2018). The Panjal Traps and the Agglomeratic Slate of Upper Carboniferous-Permian age (Middlemiss, 1910) cover the northeast part of the Valley. The mountains and slopes towards the great Himalayan side are composed of Panjal traps, Dogra slates, Triassic-Jurassic limestones, agglomeratic slates, slates with some patches of granite (Kangan, Kazinag, and Hant) and the valleys are chiefly composed of recent alluvium and, in some places, Karewas (Irfan et al., 2025). The Pir Panjal Range is composed largely of Panjal traps and limestones with small patches of Muth-quartizites and agglomeratic slates. The foothills of the Pir Panjal range are mostly largely covered by the deposits of plio-pleistocene age, generally called Karewa deposits. However, the valley floor is composed of recent alluvium of the Jhelum River and its tributaries (Fig. 1).

3 Methodology

We solely used DEM-based analysis to conduct this study since DEMs are the simplest and most reliable technique of obtaining catchment-level drainage area statistics. The channel profiles and drainage network were extracted using the 30 m SRTM DEM. Before extracting these characteristics, DEM defects such as sinks and peaks were corrected using conventional Geographic Information System (GIS) procedures to avoid drainage network discontinuities. The rasterization approach used to create DEMs involves pixel-scale errors, resulting in longitudinal profiles that are not as smooth as those created by other techniques. As a result, DEM-derived stream profiles necessitate the use of a smoothing method prior to calculating local slopes (Jenson & Domingue, 1988). Calculating flow directions at all sites yielded the drainage network of the study region. Because flow direction is strongly reliant on upslope area and particular watershed area, the D-8 method was used to effectively compute both. To determine elevation, distance from division, contributing area, and other metrics, MATLAB-based smoothing methods are used to chosen individual streams (Shahzad et al., 2009). The slope area profile of all chosen streams is sketched using the stream power law. The breaks in the slope area data are used to choose segments (trends) from each profile (Wobus et al., 2006). Flint's empirical

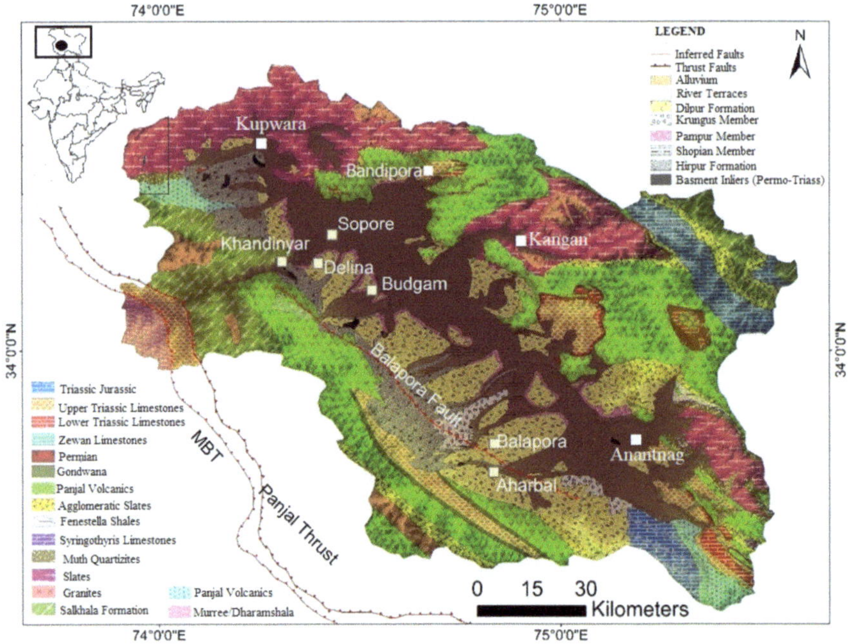

Fig. 1 Lithological distribution in the Kashmir basin after Middlemiss (1910) Thakur and Rawat (1992), Wani et al. (2019); Bali et al. (2025b)

power law equation, which connects the upstream contributing drainage basin area (A) to the local slope (S), is used to analyse the stream profile:

$$S = KSA - \theta \quad (1)$$

Where is the stream concavity and Ks is the steepness index, which is quantified directly from slope and area data analysis by regression (Montgomery et al., 1996; Homke et al., 2004; Whipple, 2004; Wobus et al., 2006). Differential uplift rates or erosion reduction can be linked to abnormally high steepness (Ks) values (Snyder et al., 2000; Kirby & Whipple, 2001; Vanlaningham et al., 2006; Wenliang et al., 2016). The gradient of a stream experiencing pulsating uplift is higher than the gradient of a stream experiencing a low uplift rate for the same upstream location (Cristea, 2015). Concavity indicates the distribution of the heights along a stream profile, and it is empirically considered to range between 0.35 and 0.6 (Wobus et al., 2006). The previous equation illustrates the Lague 2014 stream power incision model, which is excellent for steady-state bedrock streams with homogenous geology. Sharp changes in gradient and slope will occur in the transient phases, as will spatiotemporal fluctuations in rock uplift rate or varying rock strength along the stream profile. A two-stage approach was developed to identify and analyse the selected stream segments. In the first phase, using Eq. (Ahmed et al., 2018) as a regression model, concavity (θ) and steepness (Ks) indices are calculated for area-slope values of the selected stream segments. In the second step, a reference mean concavity (ref = 0.45) is used to measure the normalised steepness index (K_{sn}) for each stream segment.

$$K_{sn} = K_s A_{cent} \theta_{ref} - \theta \quad (2)$$

K_{sn} is the normalised steepness index, and Ks is derived using regression for each individual stream segment. Acent is the drainage area upstream in the stream segment's middle, and ref is the provided reference concavity (Cristea, 2015). The choice of ref, on the other hand, has no effect on relative disparities in K_{sn} (Duvall et al., 2004; Wobus et al., 2006). The steepness index (regression intercept), concavity (regression slope), and upstream drainage regions have a strong relationship with one another (Wobus et al., 2006). According to the methodologies and tools described by Snyder et al. (2000) and Kirby and Whipple (2003), slope-area regression analysis was carried out using MATLAB and ArcGIS software programmes (Shahzad & Gloaguen, 2011). These values were placed on maps to detect their spatial distribution and fluctuations after computing their concavity (θ) and steepness (K_s). Relative uplift rates were estimated using a normalised steepness index, as recommended by Wobus et al. (2006), and displayed to observe how they differed across the whole investigating region.

4 Results and Discussion

Kashmir valley is bounded by two mountain ranges, GHR and PPR, on its northern and southern sides, respectively. These two mountain ranges are the result and response of impulsive tectonic uplift episodes along two major thrust faults that were active in the geological past. Three uplift episodes were observed in the Pir Panjal Range along the Panjal Thrust while the Zanskar mountain range was the output of two episodic uplift phases along the Zanskar Thrust. The three episodes on the Panjal Thrust occurred at approximately 4.8 Ma, 1.6–1.7 Ma, and 0.4–0.5 Ma (Burbank & Renoylds, 1984; Dar et al., 2024). However, only two uplift episodes have been encountered from the Zanskar Range side at 4.5 and 2.8 Ma. With these pulsating uplift phases, the basin comes into existence. These two mountain ranges act as major obstacles and are responsible for temperate climate conditions in the valley. These tectonic and climatic disturbances have greatly affected these sub basins from time to time. Keeping these tectonic disturbances and climatic inhomogeneities in view, the steepness index ratio and concavity index were calculated for nine sub basins of the Kashmir valley, four from the Great Himalayan side and five from the Pir Panjal side, to quantify and correlate the dynamic uplift history of the area (Fig. 2).

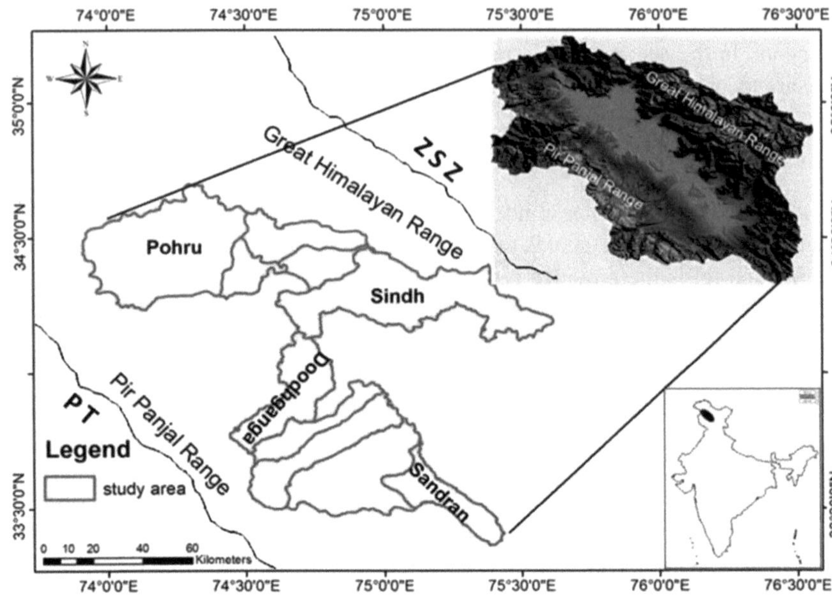

Fig. 2 Location and major sub basins of Kashmir valley

Table 1 summarises the steepness index (K_{sn}) and concavity (θ) values for the main channels of each of the nine drainage basins (Fig. 3). The results show pronounced variations in the steepness and concavity of rivers depending on the location and geology of the area. The steepness index (K_{sn}) ranges from 93 to 310, and concavity values (θ) vary from 0.03 to 2.2. All the rivers show three knick points and two segments except the Sindh river, which shows three knick points with three segments, Madhumati and Erin show one knick point with one segment (Fig. 3). The steepness index ratio was calculated from the longitudinal profiles of the rivers. The segments were differentiated on the basis of breaks in the longitudinal profile. On the basis of the generated results, we found two types of breaks in the longitudinal profiles, viz. vertical step knick points and slope break knick points.

On the basis of field observations and the mutual combination of a lithological map and the drainage of the area, we found that the vertical step knick points were linked with the structural activity and the slope break knick points were associated either with the lithological imbalance or with the climatic factor (Fig. 3). The high values of the steepness index ratio show high uplift, and the lower values indicate lower uplift. The high uplift rate was associated with Erin and Madhumati, followed by Romushi and Doodhganga (Table 1). The lower uplift rate was associated with the Pohru and Sindh sub basins, followed by Veshaw and Sandran sub basins. The high steepness index ratio corresponding to high uplift found in the Madhumati and Erin sub basins could be the effect of local faults (Banakoot and Erin) identified in the respective sub basins (Bali et al., 2016; Ahmad et al., 2018). The lower values of the steepness index ratio and lower uplift rate of the Pohru and Sindh sub basins may

Table 1 Summaries of channel details and indices of fitted channel segments

No.	River	Length (Km)	Area (Km^2)	Segment no.	K_{sn}	θ
R_1	Pohru	94	1775	**N1**	131	0.03 ± 0.07
				N2	115	1.7 ± 0.55
R_2	Madhumati	40	453	**N1**	310	0.86 ± 0.14
R_3	Erin	28	215	**N1**	310	0.86 ± 0.14
R_4	Sindh	111	1585	**N1**	93.7	1.7 ± 0.38
				N2	169	1.3 ± 0.3
				N3	121	2.2 ± 0.68
R_5	Sandran	46	482	**N1**	103	0.84 ± 0.06
				N2	223	0.14 ± 0.07
R_6	Veshaw	51	1016	**N1**	91.1	0.3 ± 0.11
				N2	205	0.69 ± 0.11
R_7	Rambaria	66	660	**N1**	138	1.4 ± 0.33
				N2	185	0.96 ± 0.07
R_8	Romushi	55	365	**N1**	122	0.21 ± 0.05
				N2	184	0.7 ± 0.05
R_9	Doodhganga	60	664	**N1**	219	0.35 ± 0.09
				N2	151	1.7 ± 0.41

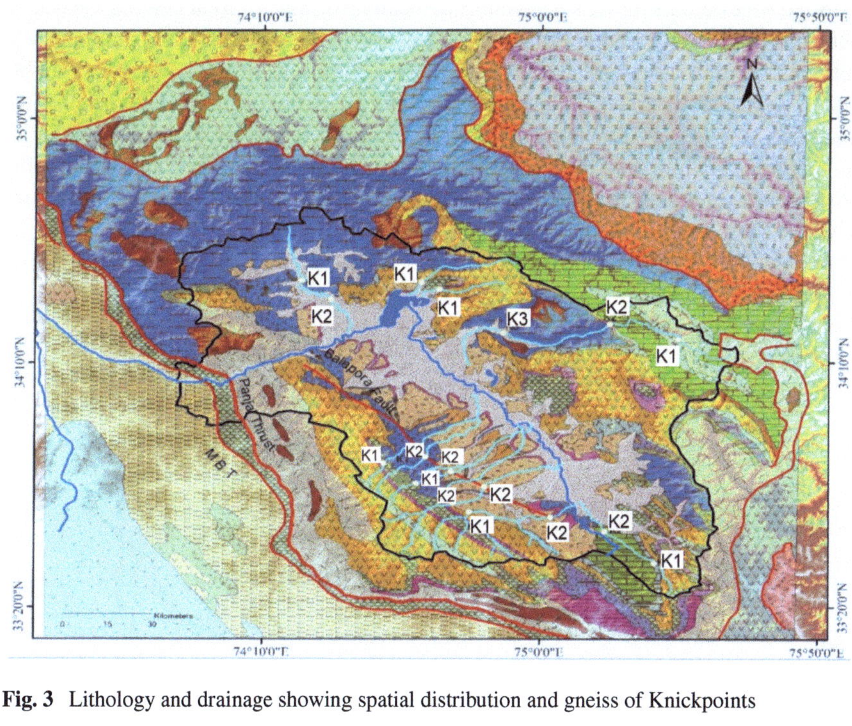

Fig. 3 Lithology and drainage showing spatial distribution and gneiss of Knickpoints

possibly be the effect of response to the local base level (Kirby & Whipple, 2012; Bhat et al., 2016; Bali et al., 2016; Wani and Bali, 2017; Wani et al., 2019; Wani et al., 2020). The intermediate values of the steepness index ratio from the Pir Panjal side infer intermediate uplift of the sub basins and are a response to the Balapora Fault and Panjal Thrust crossing through these river basins.

The concavity index (θ) was calculated for each segment in the sub-basins of the Pir Panjal and Great Himalayan Ranges to analyze the relationship between river profiles and tectonic activity, particularly in terms of uplift. The concavity index, which measures the shape of the longitudinal river profile, serves as an important indicator of tectonic processes. In the sub-basins of the Great Himalayan Range, higher θ values, ranging from 0.03 to 2.2, were found downstream, which suggests less active tectonic uplift and results in more gradual, less concave river profiles. This indicates that these regions are undergoing relatively stable tectonic conditions, where the rivers are less affected by recent or active tectonic deformation. Conversely, in the Pir Panjal Range, lower θ values (ranging from 0.14 to 1.7) were observed towards the source and downstream, reflecting higher rates of uplift and steeper river profiles. These low values suggest that the Pir Panjal sub-basins are undergoing significant tectonic deformation, with active uplift contributing to more pronounced topographic changes and steeper river gradients. Overall, the sub-basins of the Great Himalayan Range are experiencing less tectonic activity and more gradual erosion processes, while the Pir Panjal Range shows more active tectonics, particularly near the river sources, resulting in more rugged landscapes. However, exceptions were noted in the Erin and Madhumati sub-basins, which are located within the Great Himalayan Range but exhibit relatively higher uplift rates, suggesting localized tectonic anomalies or active faulting in otherwise stable regions. These findings provide valuable insights into the tectonic dynamics of the region, revealing a distinct contrast between the two mountain ranges: the Great Himalayan Range displays relatively stable, low-tectonic-activity zones, while the Pir Panjal Range is characterized by more dynamic tectonic processes, with variations in concavity index helping to capture the differential rates of uplift across the region (Fig. 4a–i).

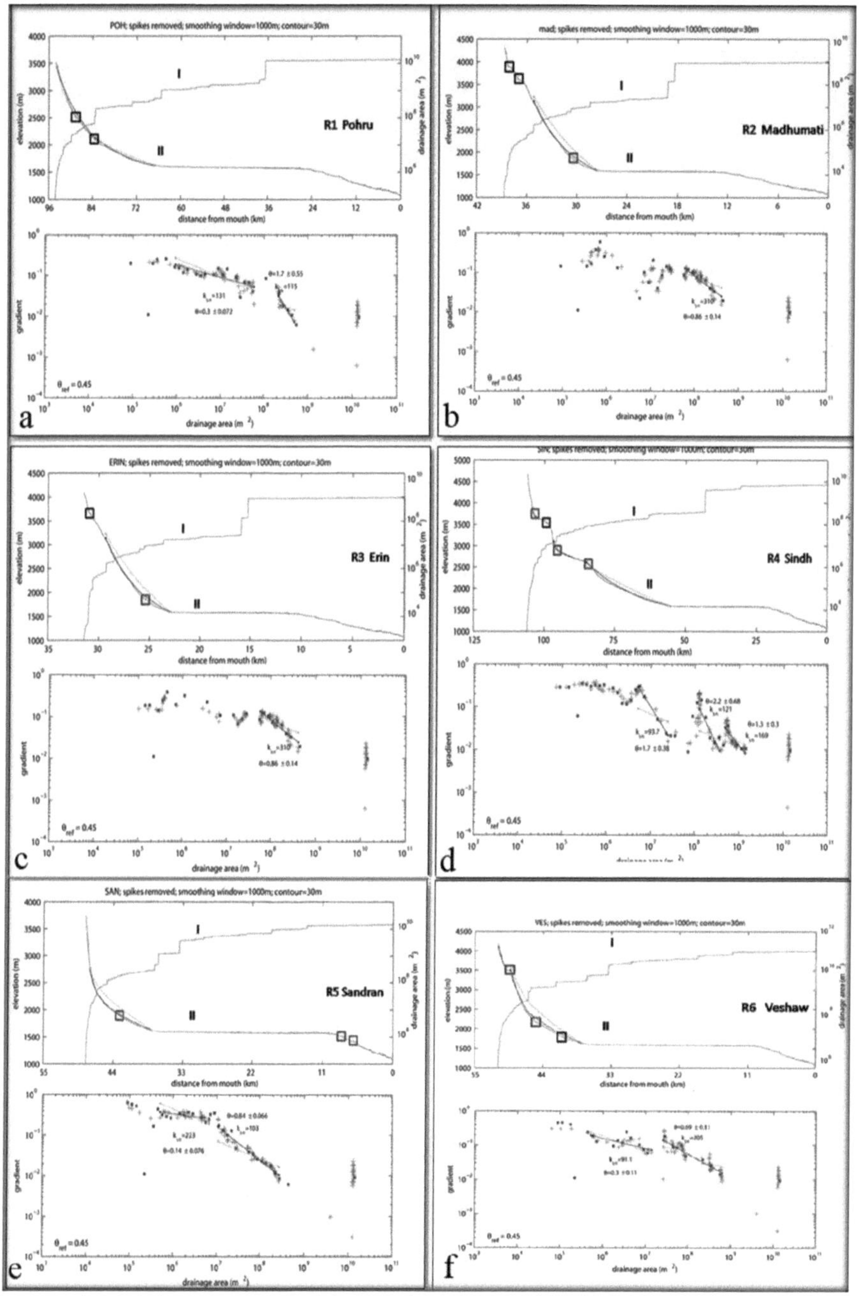

Fig. 4 Longitudinal profiles of nine tributaries of Jhelum river. Profile I and II indicate stream profile and drainage area as a function of downstream distance. The blue line is the fitted line for each segment. The second part of each river shows log-log regression of channel slope and drainage area. Black and red boxes shows vertical step and slope break knickpoints

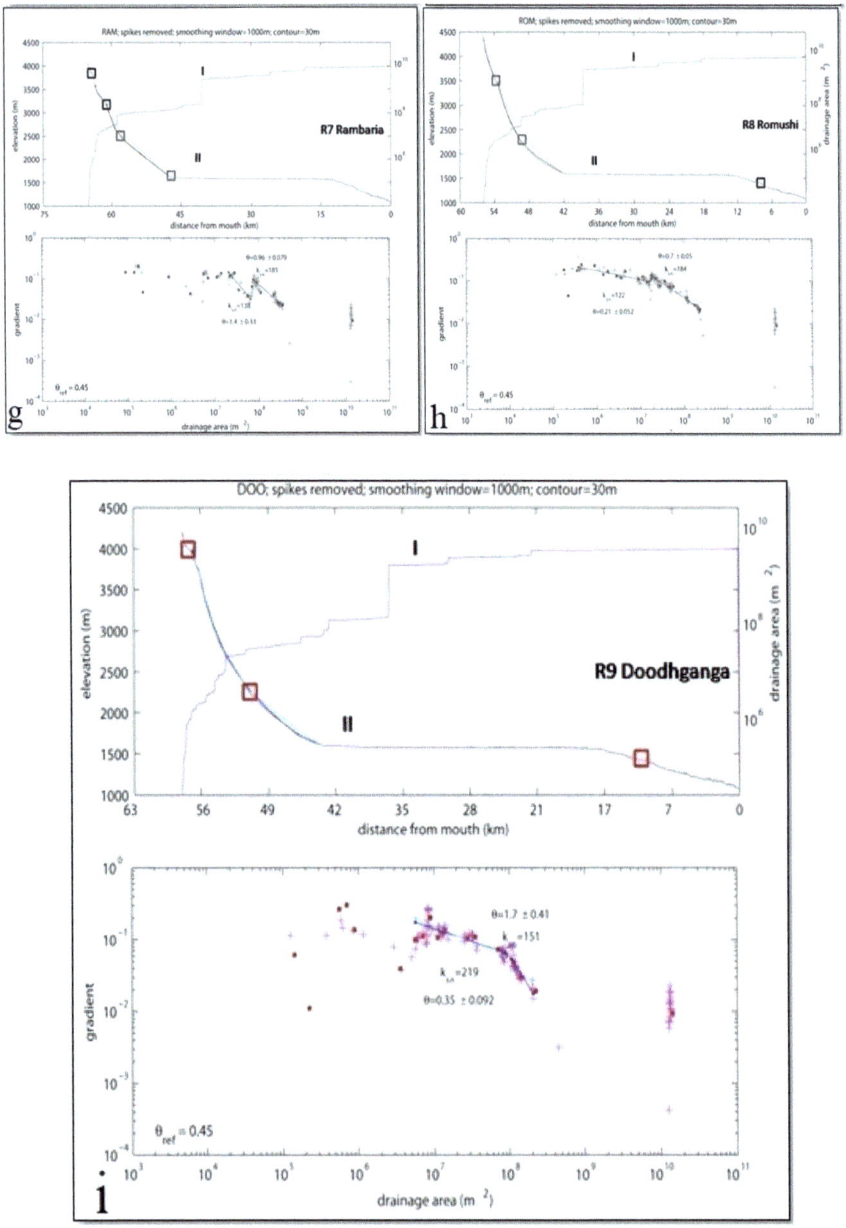

Fig. 4 (continued)

5 Conclusion

The basins falling towards the Great Himalayan Range (GHR) have higher concavity (θ) and lower steepness index values (K_{sn}). However, the lower values of θ and higher values of K_{sn}were found in the sub basins falling towards the Pir Panjal range. The lower values of θ and higher K_{sn} values strongly indicate the sub basins of the Pir Panjal range have a higher uplift rate than the Great Himalayan sub basins. The higher uplift is possibly because these basins are traversed by the active Balapora fault and Panjal Thrust. Except for the Madhumati and Erin subbasins, all GHR subbasins show lower uplift. The higher uplift found in the Madhumati and Erin sub basins could be because of movement along the local faults (Banakoot and Erin faults) present in these basins. Relatively higher θ values in the Sindh basin may be because the river traverses through hard rock terrain and is traversed by a normal fault. Thus, the quantified results infer that the Great Himalayan subbasins have lower uplift and the Pir Panjal subbasins have relatively higher uplift.

Acknowledgments The author expresses gratitude to all individuals and institutions especially Department of Earthsciences, University of Kashmir that provided academic and technical support during the preparation of this manuscript.

Funding statement The author did not receive any financial support from any organization or institution for the research, authorship, and/or publication of this manuscript.

Data availability statement The data supporting the findings of this study are available from the corresponding author upon reasonable request.

Declaration of conflict of interest The author declares that there are no potential conflicts of interest with respect to the research, authorship, and/or publication of this manuscript.

Ethical Approval Not applicable. This study did not involve human participants, animals, or any ethical issues requiring approval.

References

Ahmed, S., Alam, A., Ahmad, B., Wani, A. A., Bhat, M. I., Bhat, M. S., & Farooq, H. (2018). Tectono geomorphic indices of Erin Basin, Northeast Kashmir valley, India. *Asian Earth Sciences, 151*, 16–30.

Anoop, A., Prasad, S., Basavaiah, N., Brauer, A., Shahzad, F., & Deenadayalan, K. (2012). Tectonic versus climate influence on landscape evolution: A case study from the upper Spiti valley, NW Himalaya. *Geomorphology, 145*, 32–44.

Balco, G., Stone, J. O., Lifton, N. A., & Dunai, T. J. (2008). A complete and easily accessible means of calculating surface exposure ages or erosion rates from 10Be and 26Al measurements. *Quaternary Geochronology, 3*, 174–195. https://doi.org/10.1016/j.quageo.2007.12.001

Bali, B. S., Bhat, N. A., Wani, A. A., Mir, S. A., & Bhat, N. A. (2025a). Assessment of seismicity using local broadband seismic network: A case study from Kashmir valley, NW, Himalaya, India. *Geoenvironmental Disasters, 12*(1), 32.

Bali, B. S., & Wani, A. A. (2020). Earthquake hazard potential along the Panjal thrust and the Zanskar thrust, NW Himalaya. *Research Square*. https://doi.org/10.21203/rs.3.rs-62862/v1

Bali, B. S., & Wani, A. A. (2021). Analysis of neotectonic structures in the piedmont region of Pir Panjal range NW Himalaya by integrating geomorphic indicators coupled with geophysical transects (GPR). *Natural Hazards, 105*(3), 2869–2882.

Bali, B. S., Wani, A. A., Khan, R. A., & Ahmed, S. (2016). Morphotectonic analysis of the Madhumati watershed, Northeast Kashmir Valley. *Arabian Journal of Geosciences, 9*(5), 1–17.

Bali, B. S., Wani, A. A., Mir, S. A., Irfan, M., & Farooq, D. (2025b). Rupture length as a proxy for quantifying palaeo-earthquake magnitude for future seismic micro-zonation in Kashmir, NW Himalaya, India. *Natural Hazards*. https://doi.org/10.1007/s11069-025-07189-0

Bhat, G. R., Bali, B. S., Balaji, S., Iqbal, V., & Balakrishna. (2016). Earthquake triggered soft sediment deformational structures (seismites) in the Karewa formations of Kashmir Valley—An indicator for Palaeo-seismicity. *Journal of the Geological Society of India, 87*, 439–452.

Burbank, D. W., & Anderson, R. S. (2001). *Tectonic geomorphology*. Blackwell Scientific.

Cristea, A. I. (2015). Spatial analysis of channel steepness in a tectonically active region: Putna river catchment south-eastern Carpathians. *Geography Teacher, 10*, 19–27.

Dar, M.N., Bali, B.S., Mir, S.A.,Wani, A.A., 2024. Fault system dynamics of the Kashmir, NW Himalaya, India using continuous GPS observations and geomorphic evidences. *Journal of Geodynamics*, 161,102044.

DiBiase, R. A., Whipple, K. X., Heimsath, A. M., & Ouimet, W. B. (2010). Landscape form and millennial erosion rates in the sangabriel mountains. CA. *Earth and Planetary Science Letters, 289*, 134–144. https://doi.org/10.1016/J.EPSL.2009.10.036

Duvall, A., Kirby, E., & Burbank, D. (2004). Tectonic and lithologic controls on bedrock channel profiles and processes in coastal California. *Journal of Geophysical Research, 109*, F03002. https://doi.org/10.1029/2003JF000086

Flint, J. J. (1974). Stream gradient as a function of order, magnitude, and discharge. *Water Resources Research, 10*, 969–973.

Hack, J. T. (1957). Studies of longitudinal stream profile in Virginia and Maryland. *U.S. Geological Survey Professional Paper, 294-B*, 45–95.

Hack, J. T. (1973). Stream-profile analysis and stream-gradient index. *Journal of Research of the U.S. Geological Survey, 1*, 421–429.

Harel, M. A., Mudd, S. M., & Attal, M. (2016). Global analysis of the stream power law parameters based on worldwide 10Be denudation rates. *Geomorphology, 268*, 184–196. https://doi.org/10.1016/j.geomorph.2016.05.035

Homke, S., Verges, J., Emami, H., Garces, M., & Karpuz, R. (2004). Magnetostratigraphy of Miocene Pliocene Zagros foreland deposits in the front of the push-e Kush arc. Lurestan Province, Iran. *Earth and Planetary Science Letters, 225*, 397–410.

Irfan, M., Afzal, A., Bali, B. S., Topal, S., Naik, S., Ahmad, N., & Bashir, A. (2025). Exploring the Geoheritage, cultural geomorphology, and Geotourism potential of the Kashmir region, northwest Himalaya, India. *Geoheritage, 17*(2), 85.

Irfan, M., Bali, B. S., & Afzal, A. (2024a). Late quaternary tectono-geomorphological investigations of Zanskar Basin, NW Himalaya, India, using geospatial techniques. *Earth Surface Processes and Landforms, 49*(2), 684–705.

Irfan, M., Bali, B. S., & Afzal, A. (2024c). Morphological characteristics and controlling factors of the Piedmont fan Systems in the Zanskar Basin, NW Himalaya, Ladakh. *Quaternary Science Advances*. https://doi.org/10.1016/j.qsa.2024.100240

Irfan, M., Mir, S. A., Bali, B. S., Afzal, A., Bhat, N. A., Bhat, N., Bashir, A., & Nazir, M. (2024). Unveiling the geoheritage, cultural geomorphology and geotourism potential of Zanskar region, NW Himalaya, India. *Geomorphology, 462*, 109354.

Jenson, S. K., & Domingue, J. O. (1988). Extracting topographic structure from digital elevation data for geographic information system analysis. *PhotogrammEng. and Remote Sen, 54*, 1593–1600.

Kirby, E., & Whipple, K. (2001, May). Quantifying differential rock-uplift rates via stream profile analysis. *Geology*, 415–418.

Kirby, E., & Whipple, K. (2003). Distribution of active rock uplift along the eastern margin of the Tibetan plateau: Inferences from bedrock channel longitudinal profiles. *Journal of Geophysical Research, 108*(4).

Kirby, E., Whipple, K.X. 2012. Expression of active tectonics in erosional landscapes. Journal of Structural Geology. 44: 54–75 https://doi.org/10.1016/j.jsg.2012.07.009

Middlemiss, C. S. (1910). A revise of the Silurian-Trias sequence in Kashmir. *Records of the Geological Survey of India, 40*, 206–260.

Middlemiss, C. S. (1911). Sections in Pir Panjal and Sindh valley, Kashmir. *Records of the Geological Survey of India, 41*, 115–144.

Montgomery, D. R., Abbe, T. B., Buffington, J. M., Peterson, N. P., Schmidt, K. M., & Stock, J. D. (1996). Distribution of bedrock and alluvial channels in forested mountain drainage basins. *Nature, 381*, 587–589.

Pipaud, I., Loibl, D., & Lehmkuhl, F. (2015). Evaluation of TanDEM-X elevation data for geomorphological mapping and interpretation in high mountain environments — A case study from SE Tibet, China. *Geomorphology, 246*, 232–254. https://doi.org/10.1016/j.geomorph.2015.06.025

Shahzad, F., & Gloaguen, R. (2011). TecDEM: A MATLAB based toolbox for tectonic geomorphology, part 1: Drainage network preprocessing and stream profile analysis. *Computers & Geosciences, 37*, 250–260.

Shahzad, F., Mahmood, S. A., & Gloaguen, R. (2009). Drainage network and lineament analysis: An approach for Potwar plateau (northern Pakistan). *Journal of Mountain Science, 6*(1), 14–24.

Snyder, N., Whipple, K., Tucker, G., & Merritts, D. (2000). Landscape response to tectonic forcing: Digital elevation model analysis of stream profiles in the Mendocino triple junction region, northern California. *Geological Society of America Bulletin, 112*, 1250–1263.

Snyder, N. P., Whipple, K. X., Tucker, G. E., & Merritts, D. J. (2003). Channel response to tectonic forcing: Analysis of stream morphology and hydrology in the Mendocino triple junction region, northern California. *Geomorphology, 53*, 97–127.

Thakur, V. C., & Rawat, B. S. (1992). *Geologic map of Western Himalaya, 1: 1,000,000*. Wadia Institute of Himalayan Geology.

Van Laningham, S., Meigs, A., & Goldfinger, C. (2006). The effects of rock uplift androck resistance onriver morphology in a subductionzone forearc, Oregon, USA. *Earth Surface Processes and Landforms, 31*, 1257–1279.

Wani, A. A., & Bali, B. S. (2017). Geomorphic analysis of relative tectonic activity in the Sindh Basin, Jammu and Kashmir Himalaya, Northwest India. *Journal of Himalayan Geology, 38*(2), 171–183.

Wani, A. A., Bali, B. S., Bhat, G. R., & Hussain, N. (2020). Impact of tectonics on drainage network evolution of Suru basin, Kargil N/w Himalaya, Jammu and Kashmir, India. *Environmental Earth Sciences, 79*(1).

Wani, A. A., Bali, B. S., Irfan, M., & Bhat, N. A. (2025b). Quantitative assessment of relative tectonic activity and its implications for seismic Hazard analysis in the Pir Panjal and Zanskar Mountain ranges, northwest Himalaya. In S. Ghosh (Ed.), *The Himalaya Dilemma*. Springer Natural Hazards. Springer. https://doi.org/10.1007/978-3-031-95083-4_3

Wani, A. A., Bali, B. S., & Mir, S. A. (2025a). Kinematics of the Kashmir Himalaya: Inferences from geological and geodetic data. *Geodesy and Geodynamics*. https://doi.org/10.1016/j.geog.2025.02.003

Wani, A. A., Bali, B. S., & Mohammad, S. (2019). Drainage characteristics of tectonically active area: An example from Mawar Basin, Jammu and Kashmir, India. *Journal of the Geological Society of India, 93*(3), 313–320.

Wenliang, J., Zhujun, H., Jingfa, Z., & Qisong, J. (2016). Streamprofile analysis, tectonic geomorphology and neotectonic activity of the Damxung Yangbajain rift in the south Tibetan plateau. Article first published online: 17 FEB 2016. https://doi.org/10.1002/esp.3899

Whipple, K. X. (2004). Bedrock rivers and the geomorphology of active orogens. *Annual Review of Earth and Planetary Sciences, 32*, 151–185.

Whipple, K. X., & Tucker, G. E. (1999). Dynamics of the stream power river incision model: Implications for hight limits of mountain ranges, landscape response time scales, and research needs. *Journal of Geophysical Research, 104*, 17661–17674.

Whipple, K. X., & Tucker, G. E. (2002). Implications of sediment-flux-dependent river incision models for landscape evolution. *Journal of Geophysical Research: Solid Earth, 107*, ETG3–1–ETG3–20. https://doi.org/10.1029/2000JB000044

Wobus, C. W., Whipple, K. X., Kirby, E., Snyder, N. P., Johnson, J., Spyropolou, K., Crosby, B., & Sheehan, D. D. (2006). Tectonics from topography: Procedures, promise and pitfalls. In S. D. Willett, N. Hovius, M. T. Randon, & D. M. Fisher (Eds.), *Tectonics, climate and landscape evolution* (Geological Society of America Special Papers) (Vol. 398, pp. 55–74).

Landscape Evolution, Climate, Cryosphere, and Human Dimensions of the Jhelum Basin, Kashmir Himalayas

Haadiya Saleem Bhat, Shaista Mushtaq, Uzma Farooq, Saqib Iqbal Raina, Mudigandla Rajesh, and Talib Bashir Bhat

Abstract The Jhelum Basin of the Kashmir Himalayas constitutes a geomorphologically and environmentally complex mountain system where tectonics, climate, cryosphere, hydrology, and human activity intersect to shape landscape evolution and resource availability. This chapter provides a comprehensive geographical synthesis of the basin, integrating its physical setting, relief, geological framework, geomorphic development, drainage organisation, climatic regime, biophysical characteristics, cryospheric components, land surface thermal patterns, land use dynamics, and demographic trends. Spanning a wide altitudinal range from valley floors and Karewa plateaus to high-elevation glaciated terrain, the basin exhibits strong spatial heterogeneity governed by active tectonics, glacio-fluvial processes, and climatic variability. The Jhelum River and its tributaries form a structurally influenced drainage network that sustains agriculture, hydropower, ecosystems, and livelihoods across the valley. The basin's cold-temperate climate, dominated by western disturbances, regulates seasonal snow accumulation, glacier mass balance, and river discharge. Detailed assessment of glaciers and glacial lakes underscores their critical role in regional hydrology and downstream water security. Observed changes in land use/land cover and land surface temperature reveal an intensification of anthropogenic pressures,

H. S. Bhat (✉)
Indra Gandhi National Open University (IGNOU), New Delhi, India

Islamic University of Science and Technology (IUST), Awantipora, India

S. Mushtaq · U. Farooq
Department of Geography and Disaster Management, School of Earth and Environmental Sciences, University of Kashmir, Srinagar, India

S. I. Raina
Department of Geography, Jamia Millia Islamia, New Delhi, India

M. Rajesh
Division of Natural Resource Management, Faculty of Forestry, Sher-E-Kashmir University of Agricultural Sciences and Technology of Kashmir, Benhama Ganderbal J&K, Srinagar, India

T. B. Bhat
Andhra University, Visakhapatnam, India

R. Ahmed et al. (eds.), *Climate Change, Water Resources, and Natural Hazards in the Himalayas*, Advances in Geographical and Environmental Sciences,
https://doi.org/10.1007/978-981-92-0835-7_4

particularly urban expansion, deforestation, and population growth, with implications for ecosystem stability and watershed functioning. By synthesising natural and human dimensions within a unified geographical framework, this chapter highlights the Jhelum Basin as a sensitive Himalayan system where sustainable resource management and informed planning are essential for enhancing resilience under ongoing environmental and climatic change.

Keywords LULC · LST · Glaciers · Lakes · Geology · Geomorphology · Demography · Climate · Drainage · Soils · Vegetation · Jhelum basin · Kashmir Himalayas

1 Introduction

The present research work has focused on the Jhelum Basin of Kashmir Himalayas located in the North-Western Himalaya region of India. The Jhelum basin or valley of Kashmir, which is drained by the river Jhelum in length and breadth is engulfed between the youngest mountain ranges of Pir-Panjal and Zanskar that possess a dramatic landscape with snow and glaciers surging river systems (Ahmed et al., 2021b). The Jhelum River, a major tributary of the Indus River system, traverses the whole Kashmir basin and has 24 sub-watersheds. The Pir-Panjal range is drained by 10 watersheds on the river's left bank, including Ferozpur, Dodhganga, Sukhnag, Vaishav, Rambiahra, and Romshi. On the right side of the Jhelum River, 14 watersheds originate in the broader Himalayan range. These include tributaries like as Lidder, Sind, Madhumati, and Sandran, which contribute significantly to the river's flow from the larger Himalayan range (Mir et al., 2016; Mir & Gani, 2019; Mir & Jeelani, 2015). The study region has the potential to generate around 20,000 megawatts (MW) of hydropower, and numerous power plants are now operational (ul Shafiq et al., 2020). The location of the study area is depicted in Fig. 1.

2 Physical Setting

The whole Jhelum basin is located in the northern parts of India and Pakistan with a total catchment area of 33,342 km. Jhelum River is one of the largest tributaries of the Indus River system (Himayoun et al., 2020; Mahmood & Jia, 2017). The present study has been carried out on the Indian part of Jhelum River basin alternatively studied as Upper Jhelum Basin (UJB) or Jhelum basin of Kashmir Himalaya according to the scholarly literature. The physical boundary of the Jhelum River basin of Kashmir Himalayas overlaps/coincides with the Kashmir valley or Kashmir basin. The Jhelum River originates from a magnificent and perennial spring known as "Cheshma Verinag" and drains the valley of Kashmir up to 165 km (Imran & Sheikh, 2020). The river meanders towards the northwest direction from the northern slopes

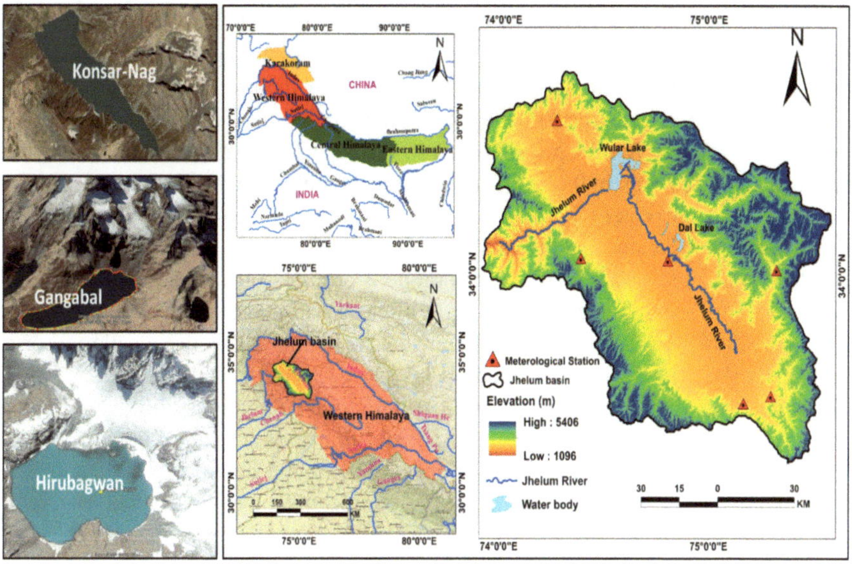

Fig. 1 Location map of the study area prepared on digital elevation model (DEM). The red color triangles in the figure show the location of the meteorological stations

of the Pir-Panjal through the valley of Kashmir to Wular Lake, which controls its flow. From the Wular lake, the Jhelum River flows towards the west direction and cuts through Pir-Panjal in a gorge at an elevation of 7000 ft. (2100 m) deep with almost vertical sides. The Kashmir Valley has an oval-shaped topology and is distinguished by a stunning environment adorned with numerous glaciers and glacial lakes spread across towering mountains. The research region is located in the Union Territory of Jammu and Kashmir, India, between the Pir Panjal Mountains to the southwest and the Greater Himalayan range to the northeast. The coordinates are 32° 20′–34° 50′ N and 73° 55′–75°35′ E (see Fig. 1). Overall, the elevation of the Jhelum basin varies from 1020 to 5200 m asl having a total catchment area of approximately 15,836 km^2 up to the Line of Control, where it enters Pakistan. The main channel of Jhelum River has a total length of 165 km (Ahmed et al., 2025; Mushtaq et al., 2024; Saleem et al., 2021, 2024a, 2024b; Wadia, 1941; Zaz et al., 2019). The location of the study area is depicted in Fig. 1.

3 Relief

Jhelum basin constitutes the important hydrological and relief features of enormous geographic significance. Based on elevation, the valley of Kashmir has been divided into four broad physiographic divisions such as mountains, foot-hills, Karewas and valley floor (Dubey et al., 2017). The region has been divided into three major physiographic zones as valley floor, Karewas and rimlands by Mohiuddin (2013) based on the geographic features, altitude, slope, climate, and pedological conditions of the region. The major elevation zones of the Jhelum basin are presented in Fig. 2. The three-dimensional 3D map Relief map of the study area generated through the SRTM-DEM in Arc-Scene is shown in the Fig. 2. The majority of the terrain/ physiographic features in the study area are the result of glacio-fluvial process. The valley of Kashmir is a composite basin with a well-developed drainage network and Jhelum serves as the master stream of the drainage. The distance between the crest of the Pir Panjal and the Great Himalayan range is around 100 km. It varies in altitude from 1549 m (Walur Lake) to 5432 m (Kolahoi Peak). The master stream Jhelum and its feeding tributaries on the left and right banks have created a valley floor or flood plain that stretches on both sides of the river, varying in width from 3.5–30 km. The surface features of the basin exhibit an inextricable juxtaposition of both deposition and erosion, having been the scene of simultaneous activity of both processes. During the Pleistocene epoch, it experienced glacial and fluvial activity cycles that corresponded to glacial and interglacial periods (Romshoo et al., 2015).

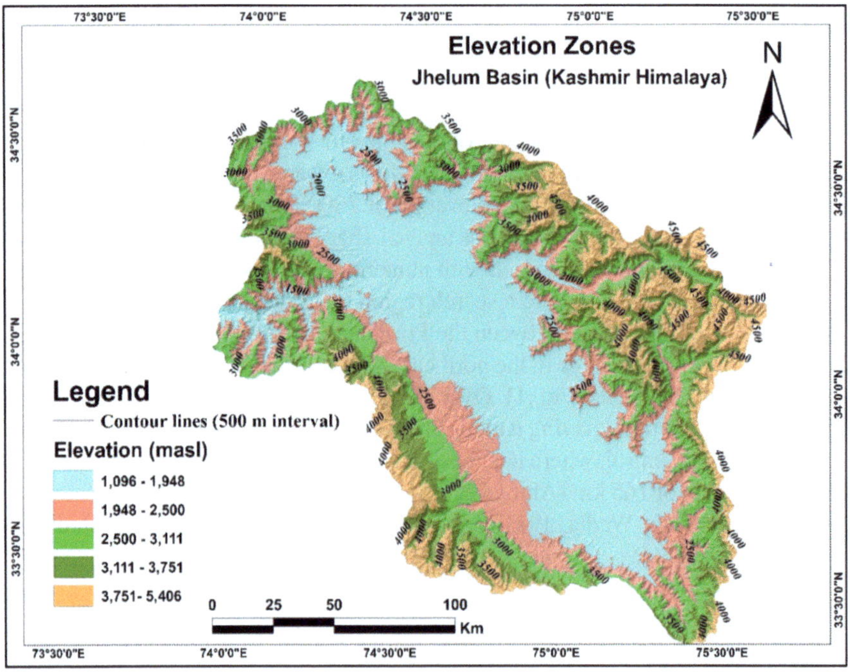

Fig. 2 Map showing elevation zones

3.1 Valley Floor

The Valley floor, like many sedimentary basins, contains an odd mix of depositional and erosional features. The bottom of the valley floor is made up of fluio-lacustrine deposits that are overlain by a repository of thick rock material from the bordering mountains. Low-lying areas that are either waterlogged or prone to frequent flooding continue to get layers of fine silt and coarse gravel. The valley floor consists of flood plains and loamy tracts that stretch from Anantnag in the south to the Baramulla in the north. The valley floor is densely populated and dominated by agricultural land with paddy and apple orchards. Most of the major cities are situated on the valley floor.

3.2 The Karewas

Karewas is the unique relief feature of the Kashmir valley locally known as "Wudar" which means elevated table land. Karewas are the lacustrine deposits that cover an area of approximately 13–26 km on the left flanks of the river Jhelum. Karewas developed during the Pleistocene Period (one million years ago), when the Kashmir Valley was completely submerged under water. When the water was drained through the Baramulla gorge at the time of Pir-Panjal rise, the deposits left over from the process are known as Karewas. The Karewas are economically and agriculturally valuable. On the Karewas, commercial and cash crops like as almonds, walnuts, apples, pears, peaches, and saffron thrive in abundance.

3.3 The Side Valleys

The valleys carved out through the tributaries of Jhelum are known as side valleys. The rate of insolation is very low and winters are harsh because of the steeper gradient. Maize is the major crop of the side valleys with paddy limited to the flat irrigated area. The region is deficient in humus content with immature soil type. Side valleys are mostly occupied by the Gujjar communities which are highly dependent on pastoral activities.

3.4 The Mountains

The valley of Kashmir is surrounded by lofty mountain ranges on the north by Greater Himalayas and on the south by Pir-Panjal mountains. The mountains of the Jhelum basin are the home of hundreds of glaciers and glacial lakes that provide continuous water supply to the downstream region. The mountains of the Jhelum basin are in the elevation zone of approximately 3400 to 5400 m asl.

4 Geology

Kashmir Valley lies between the Great Himalayan range (northeast) and the Pir-Panjal range (southwest) is classified as one of the Neogene Quaternary intermontane basins. It is located on the western side of the Himalayan orogenic system, about 100 km from the Himalaya's actively growing frontal region. Bhat (1982) proposed the first tectonic model for the formation of the Kashmir basin, which reveals that the basin is formed during the rifting process and is delimited by two large horsts and the presence of Panjal Traps in Kashmir is primarily responsible for the rifting (Shellnutt, 2018). In terms of geological importance, Kashmir valley is recognised to have complete stratigraphic record of rocks ranging in age from Archean to the Recent (Wadia, 1975; Dar & Dubey, 2015). Geologically, Jhelum basin of Kashmir Himalaya is the most important part of the Himalayan region which stretches for over 2500 km from east to west.

The oldest rocks in Kashmir valley are known as the Salkhala Series, which consist primarily of granites, gneisses, schist, carbonaceous slates, phyllites, and schist, while the most recent are known as Alluvium, which consists primarily of loam and clay in the valley floor. The study area is composed of a series of metamorphosed rocks (Salkhala's and Murree's) account for 22.53%, that are overlain by Syringhothyri's as well as Fenestella shales, which are overlain with the Triassic and Zewan Limestone (11%). Panjal Traps (24%) of volcanic (igneous) origin and Triassic limestones cover a substantial portion of the research area and are generally found in the upper mountainous parts and on the flanks of the region, whereas the lower reaches are mostly made up of Plio-Pleistocene Karewas deposits (20%) and Recent Alluvium (18%) Dar and Zeeden (2020). The major types of rock formations and lineaments are shown in Fig. 3.

In the various parts of the study region more or less complete sequence of fossiliferous Palaeozoic lithologies like Panjal volcanic series which comprises Panjal traps and agglomeratic slates, Gondwana shale, Fenstella shale, Permo-Triassic rocks, Syringothyris limestone, varved clays and conglomerate beds are found (Middlemiss, 1910; Ahmad et al., 2016) and the region is mostly covered by fluvioglacial sediments.

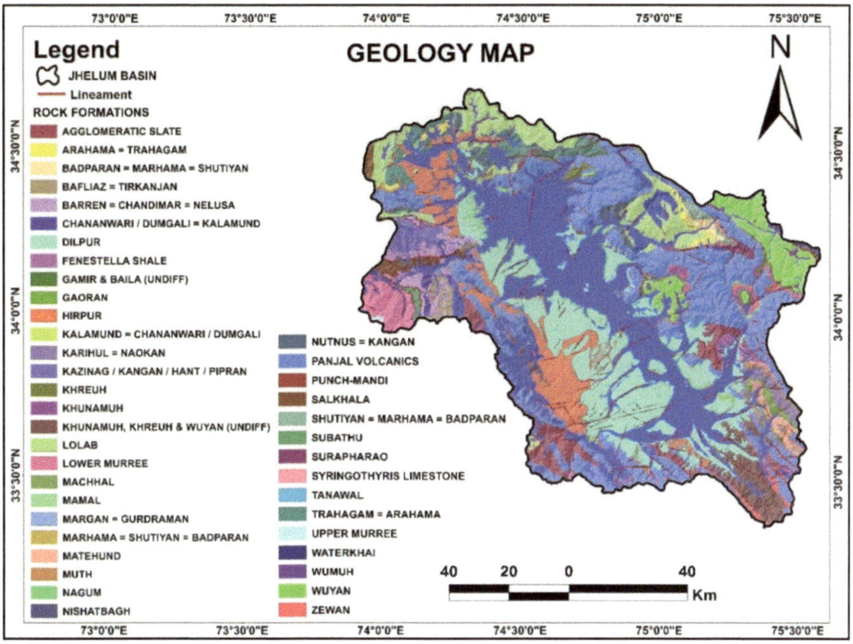

Fig. 3 Geology map of the Jhelum basin Kashmir Himalayas showing various types of rock formations

5 Geomorphic Setup of the Basin

Geomorphic detection of tectonic landforms gives an insight into regional faulting patterns and aids in determining a region's geologic history (Gürbüz & Gürer, 2008). Jhelum basin has a unique position in the Himalayas region in terms of geomorphology, as it possesses a large quantity of evidence on the evolution of its surface features. The geomorphic evolution of Kashmir valley is strikingly recent with a magnificent landscape of snow, glaciers, and a rushing river system that stretches between the Pirpanjal and the Zanskar. It is nestled between the young folded mountains and the process of uplifting is still going on in the region (Raza et al., 1978). Two tectonic episodes characterise the Orogen. The first episode pertains to the collision of India and Asia, which led in crustal shortening, thickening, and the commencement of a metamorphic effect on the sedimentary series of Indian continental margins. The second episode resulted into the extensional tectonic structure related to the exhumation of high-grade metamorphic rock of High Himalayan Crystalline which led to compression of beds due to ductile normal shearing, doming and normal faulting. The geomorphology of the Jhelum basin is shown in Fig. 4. The Kashmir basin has been tilted northeast due to tectonic activity, resulting in an asymmetrical position on the valley floor, closer to the Great Himalayan flank than the Pir Panjal flank

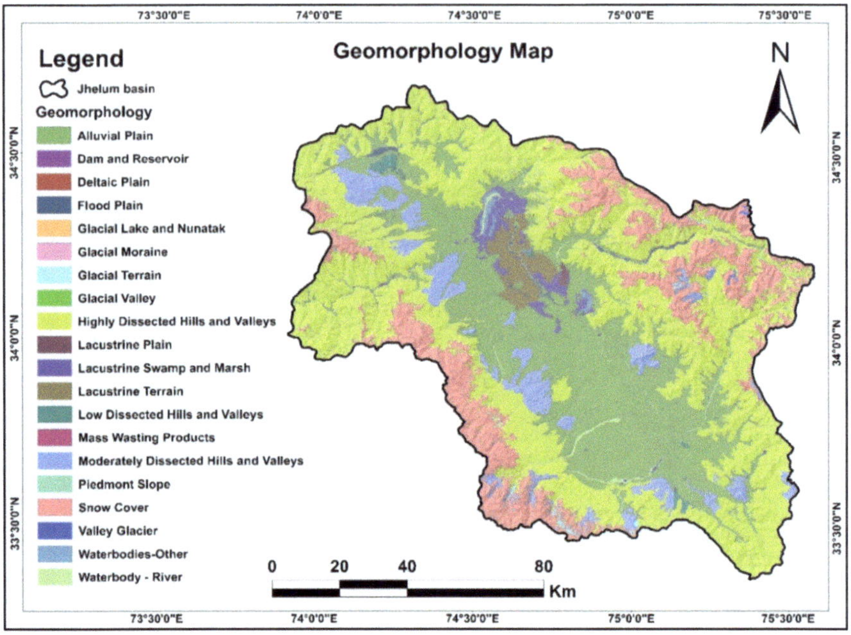

Fig. 4 Geomorphology of the study area

(Romshoo et al., 2021). The variation in the altitude of the Jhelum basin, which varies between 1079–5320 masl implies a wide range of its geomorphic properties.

6 Drainage

The Kashmir basin is flanked by two linear mountain ranges, the Pir Panjal and the Great Himalaya, with the Jhelum River providing an efficient drainage system. The surrounding mountain ranges are snow-covered nearly all year, providing fertile ground for the growth of several streams that have established themselves as separate entities inside the valley. The drainage system of the Kashmir valley is made up of tributary drainage basins and the river Jhelum. Various drainage patterns can be seen in Jhelum basin including dendritic, radial, trellis, rectangular, parallel and barbed patterns. In areas with homogeneous lithology, low-order streams frequently intersect with higher-order tributaries at angles less than 90°, resulting in a dendritic drainage pattern. This pattern is predominantly determined by lithological properties, with negligible structural influence (Yousuf et al., 2021). Antecedent and superimposed drainage patterns are common in places with less resistant rocks, as well as flat, low-lying horizontal igneous, sedimentary, and metamorphic formations across the basin. The Jhelum Basin is geologically young and has tremendous tectonic activity as a result of the Himalayas' continuous ascent. Past rapid uplift episodes have caused changes in the drainage pattern (Basavaiah et al., 2010), and contemporary tectonic processes are still shaping and defining the pattern, morphology, and structure of the drainage system (Yousuf et al., 2021).

The Jhelum River is the only major river that drains the Kashmir basin. It runs through the centre of the valley, fed on both sides by various tributaries, and drains the basin through a short gorge near Baramulla. Jhelum basin is comprised of 24 watersheds drained by 24 left and right bank tributaries. Bringi, Aripat, Kuthar, Lider, Dachigam, Erin, Aripal, Sind, and Madhumati feed the river on the north, whereas Veshaw, Sandran, Romushi, Rambiara, Shaliganga, Doodhganga, Sukhnag, Ningal, Ferozpor, Garzan, and Vijay-Dhakil feed it on the south (Fig. 5).

From the east, Pohru and Mawar tributaries join it. Among these streams, Lidder and Sind are in the sixth order, whereas Erin, Madhumati, Aripal, Dachigam, Bringi, Aripat, Sandran and Kuthar, are examples of smaller rivers with fifth-order streams. The majority of these rivers have substantially narrower river courses with steep basins and gorges. The drainage network of the Jhelum basin is depicted in Fig. 5 with Jhelum River shown in light blue colour. Jhelum river originates from a spring locally known as "Cheshma Verinag" after that it joins with the tributary stream of large size (sixth order) near the Khanbal area of Anantnag and becomes a sizable river. It flows from several districts and gets fed by various catchments before draining into the Wular lake. From the Wular lake the river Jhelum takes the southwesterly direction of Baramulla and makes its way through a narrow gorge across the mountains.

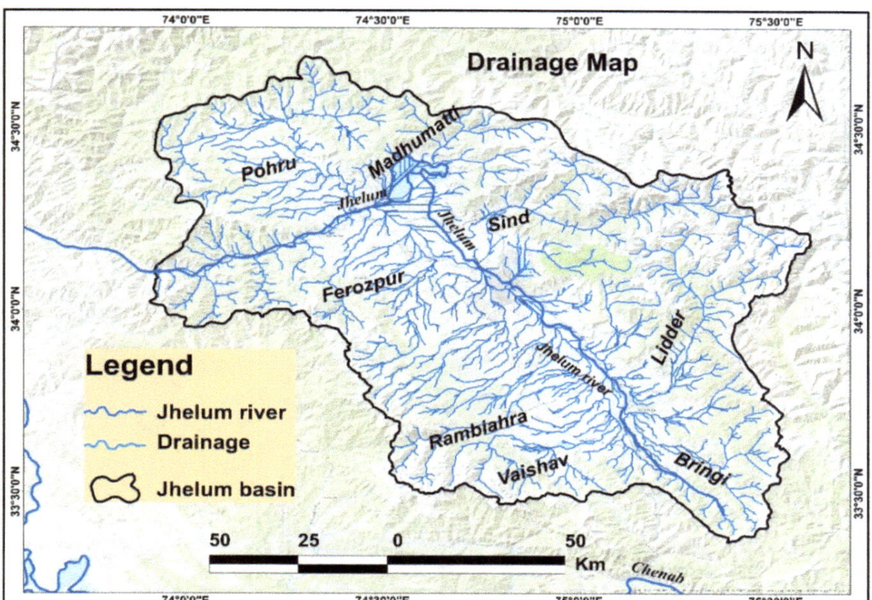

Fig. 5 Drainage map of the study area

7 Climate Setting

The study region is characterized by a cold temperate climatic regime, with pronounced seasonal contrasts. The Kashmir Valley typically experiences cold and moist winters marked by moderate to heavy snowfall, while summers remain comparatively mild. A significant proportion of the annual precipitation occurs during winter, primarily associated with Western Disturbances that transport moisture from the Mediterranean region. Precipitation distribution shows strong spatial variability across the valley. For example, Srinagar records an average annual rainfall of approximately 650 mm, whereas higher-altitude locations such as Pahalgam and Sonamarg receive more than 1500 mm annually (Fig. 6). Mean seasonal temperatures further reflect this climatic pattern, with winter averages around 2.5 °C and summer averages near 19.8 °C (Dar et al., 2014). Glaciers in Jammu and Kashmir, including those in Ladakh, are primarily nourished by snowfall associated with Western Disturbances, although limited monsoonal input occurs during July and August. The Jhelum River and its tributaries maintain perennial flow, supported by sustained meltwater contributions from high-altitude snowfields and glaciers. Peak discharge generally occurs in May and June, coinciding with intensified snow and ice melt during the early summer season. The downstream topographic gradient further enhances the river's suitability for hydropower development at multiple sites.

The climate of the Jhelum basin is sub-Mediterranean, with four distinct seasons depending on temperature and precipitation: winter (December–February), spring (March–May), summer (June–August), and autumn (September–November) (Bagnolus & Meher-Homji, 1959). However, there are micro-level variations in the basin's overall meteorological and climatic characteristics. The surrounding mountains, with their snow-capped peaks, have a significant impact on local weather

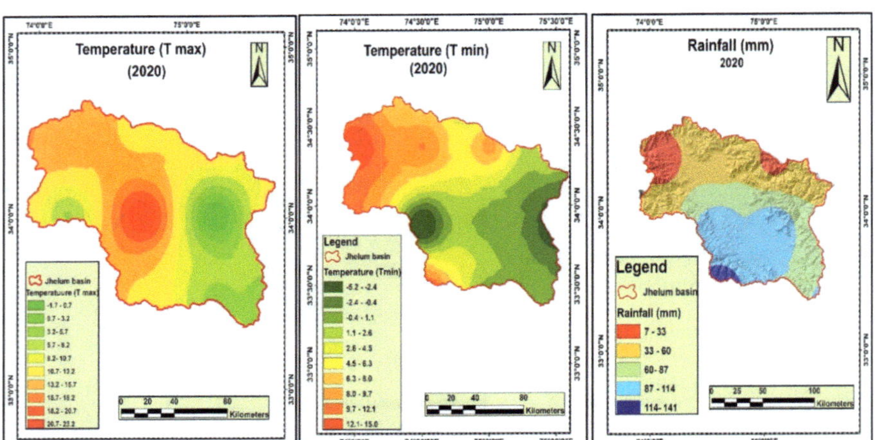

Fig. 6 Showing temperature (Tmax, Tmin, and Tmean) and precipitation of the Kashmir valley for the year 2020

patterns. They shield the Valley from both the bitter cold of the north and the searing heat of the south, contributing greatly to the Valley's well-known moderate climate. The weather and climate of the Jhelum basin is governed by several factors such as relief, monsoon winds, altitude and forests. A brief detail of the four seasons of Kashmir valley is given below:

1. **Winter season**: The winter season lasts from Mid-November to Mid-March with mean maximum (Tmax) and mean minimum (Tmax) temperatures of about 14 °C and 1.5 °C which decrease to −1.40 °C (mean minimum) to 8 °C (mean maximum). The winters in the region are mostly dominated by winds called western disturbances. The coldest month of the year is January locally known as a period of *Chillai-Kalan*. During the winter season relative humidity remains around 90% with low sunshine. Most of the water bodies including Dal and Wular Lake freezes during these months. The region receives precipitation mostly in the form of snow in the winter season with maximum snow in January.
2. **Spring season**: The weather begins to improve as March progresses, and the temperatures begin to rise slowly. In March, the snow begins to melt and the day temperature varies between 10 and 16 °C. Temperatures, on the other hand, vary greatly from one location to another and from year to year. Sunnier days and cooler nights characterise the weather throughout this season. With the continuous rise in temperature throughout this season, lush green grass emerges on the surface and leaves grow on the dormant vegetation.
3. **Summer season**: This season starts in the mid-May and ends in mid-September. The study region experiences hot and humid temperatures during the summer season, which extends from June to August. The temperature rises during the day, reaching 37 °C at times. The month of July is the hottest of the year, with high temperatures and humidity. Summer monsoon season spans from June to September across India's subcontinent, while the Valley of Kashmir receives significantly less rainfall during this time. The hot weather in June and July helps paddy and vegetable crops to sprout and mature quickly. The ripening of Cherry, Peaches, Pears, and early kinds of apples is also aided by these conditions.
4. **Autumn season**: The autumn season in the valley of Kashmir lasts from Mid-September to Mid-November, is marked by calm and undisturbed weather, clear skies, most sunshine, a wide diurnal temperature range minimal rain or snow as well as the most enjoyable weather. The average maximum and minimum temperatures in September are recorded around 25 °C and 11 °C, respectively. The temperature, on the other hand, drops sharply when clouds appear. In October, the temperature difference between day and night is extremely noticeable.

8 Vegetation

The vegetation of the Jhelum basin is determined by various factors such as climate, soils and altitude of place. Mostly deciduous type of vegetation is found in the valley of Kashmir. Chinar, deodar, poplar, pine, fir, mulberry, kail, partal, walnut and various other types of fruit-bearing trees grow throughout the valley. The vegetation of the study area has been divided into three major types found in different elevation zones: 1) temperate forests (1600–2700 masl) 2) Kashmir subalpine forests (2700–3500 masl) and 3) Kashmir alpine forests (above 3500). The important plant species of the temperate forests are silver fir, spruce, blue pine, yew and little deodar with various shrubs and bushes, while silver fir dominates in the lower sections, and birch dominates at higher elevations in Kashmir subalpine forests. Kashmir Alpine forests are generally found above the tree line. Alpine scrubs with stunted birches, junipers, and certain willow species make up the vegetation. Furthermore, vast areas of grasslands locally known as "Margs and Dhoks" are found in this region.

9 Soils

The unconsolidated mineral or organic material on the earths near surface that acts as a natural medium for land plant growth is known as soil. The Kashmir valley, is located in a temperate zone and has an alluvial soil that is rich in nitrogen and organic matter. The presence of high nitrogen content in the alluvial soils of the valley is due to plant residue, agricultural stubble, natural flora, and animal excretion. Slit (Gurti), loam (Bahil), sandy (Sekil), vegetable soil (Surzamin), karewa soil (Wuder), peaty soil (Nambal) Lemb, Floating Garden soils and Karewa soils are some of the major soil types found in the valley. The soils of Kashmir were studied by Raza et al. (1978) using the following general physiographic divisions as: (a) The valley basin and the side valleys of the Jhelum River up to an altitude of 1850 m (b) The highlands, found mainly within the elevation zones of 1850 and 3350 m and (c) The Karewa uplands. The soil map reclassified by Dar et al. identifies six major soil types within the basin. These include impermeable coarse loam exposed on rock outcrops (10.23%), clayey skeletal soils (2.66%), moderately well-drained sandy loam (1.4%), moderately well-drained calcareous fine silt (10.2%), well-drained fine clay (4.29%), and fine loam, which accounts for the majority of the basin (71.16%).

10 Glaciers

Glaciers in the Jhelum basin of Kashmir Himalaya serve as a natural reserve of water for the downstream region and they have extremely important socioeconomic and scientific value. Jhelum basin is home to hundreds of glaciers and glacial lakes as reported by Romshoo et al. (2020), Ahmed et al. (2021a, 2021b, 2022a, 2022b), Ahmad et al. and forms a significant part of the hydrological scenario of the Upper Indus Basin. The glaciers in the Jhelum basin provide meltwater supply to the tributaries feeding the Jhelum River and support agriculture, tourism, hydropower generation, and other domestic water uses (Dar et al., 2014). The glaciers in the study area are predominately fed by the westerlies. Romshoo et al. (2020) have reported 147 glaciers in the Jhelum basin with sizes ranging between 0.3 to 13.72 km^2. Whereas, Ahmad et al. have identified 307 glaciers in the Jhelum basin having a size greater than 0.01 km^2. Kolahoi is the largest glacier in the study region with an area of 10.55 km^2 located in the Lidder catchment. The Sind, Lidder, and Veshav sub-basins of the Jhelum are the most glacier-rich and wide-ranging in terms of both number and area covered, with the most glaciers and accounting for 86 percent of the total glacier surface.

Out of the 24 sub-catchments of Jhelum basin glaciers are found in only four catchments namely Lidder, Sind, Vishav, and Doodhganga (Romshoo et al., 2020). All glaciers in the Jhelum basin are located between 3500 and 5000 m above sea level, because of local changes in topography and climatic conditions. The glaciers in the study area have been mapped by various national and international organisations for example, Global Land Ice Measurements from Space (GLIMS) Randolph Glacier Inventory V6 have reported 206 glacier polygons in the Jhelum basin. ICIMOD and GSI have reported 388 and 133 glacier polygons in the region (Fig. 7).

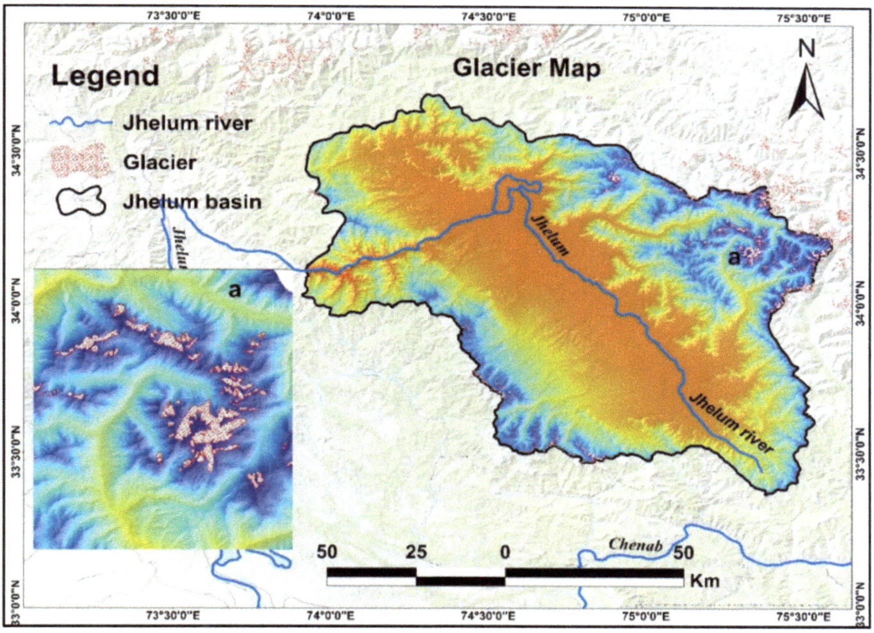

Fig. 7 Glacier inventory of Jhelum basin

11 Glacial Lakes

We performed a detailed examination of glacial lakes in the Jhelum Basin using high-resolution satellite data from 2018 and 2019. A total of 393 glacial lakes with a size equal to or higher than 1000 m^2 were discovered, covering an area of 21.55 km^2. Among these lakes, 102 were classified as proglacial lakes, covering approximately 9.12 km^2, and 278 as disconnected glacial lakes, covering 12.28 km^2 (Fig. 8). Furthermore, 13 supraglacial lakes were discovered, accounting for 0.15 km^2 of the total lake area. Our data found that disconnected glacial lakes made up the largest fraction in terms of both number (70.73%) and total area (56.98%), followed by proglacial lakes (number: 25.92%, area: 43.32%). Supraglacial lakes, while small in quantity and extent, are predicted to become more significant in the future as climatic warming continues (Ahmed et al., 2026a, 2026b).

Further investigation revealed 50 glacial lakes larger than 0.1 km^2, with only Gangabal and Konsar Nar above 1 km^2. The Sind and Vaishav watersheds of the Jhelum Basin contain the larger lakes Gangabal (1.65 km^2) and Konsar Nar (1.37 km^2). The distribution and properties of glacial lakes have far-reaching consequences for both local and downstream regions. Proglacial lakes, which are directly or

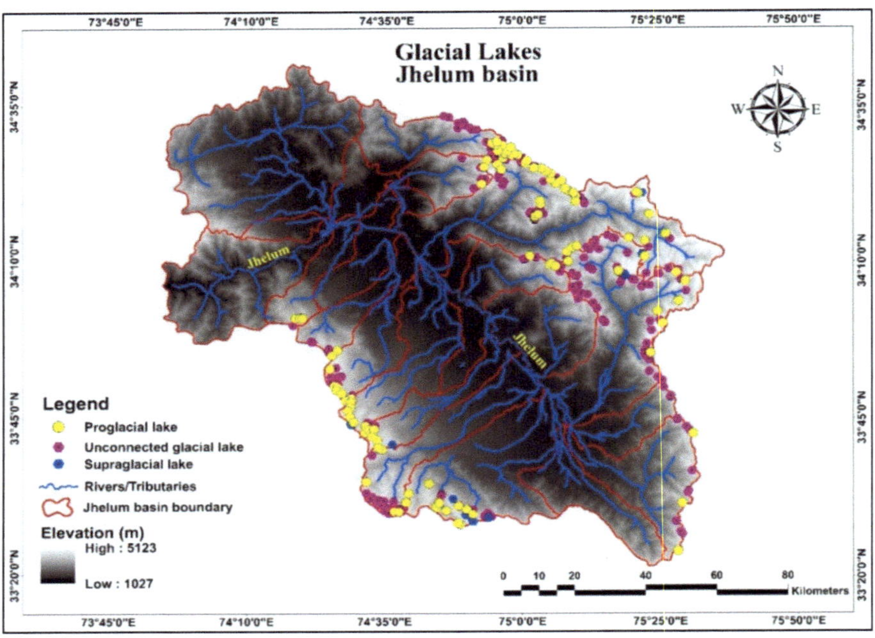

Fig. 8 Glacial lake distribution in the Jhelum basin

indirectly connected to glaciers, are more vulnerable to glacier retreat and accompanying hazards such as GLOFs. Their presence and behaviours pose dangers to downstream communities and infrastructure, demanding risk assessment and management solutions.

12 LST Distribution

Previous research has shown that plant cover has a considerable influence on the distribution of land surface temperature (LST) (Song et al., 2018; Weng et al., 2004). Studies have consistently found a negative correlation between the Normalised Difference Vegetation Index (NDVI) and LST geographical fluctuation. Vegetated regions absorb heat and lower surface radiant temperatures through transpiration, resulting in less radiation. Changes in land-use characteristics between 2000 and 2020 resulted in a considerable increase in the research area's mean surface temperature, which rose from 13.49 to 15.40 °C. This movement is paralleled in NDVI values, with the greatest NDVI value falling from 0.77 in 2000 to 0.52 by 2020. Conversely, the lowest NDVI values have risen from − 0.28 in 2000 to − 0.48 in 2020. These patterns indicate that urbanisation and the extension of barren lands have contributed to rising surface temperatures by replacing natural vegetation with non-evaporative surfaces such as metal and concrete. Thus, there is a negative correlation between vegetation (NDVI) and land surface temperature. The distribution of LST in the Kashmir valley changed significantly between 2000 and 2020. The LST maps were divided into five temperature ranges: very low (< 5 °C), low (5–10 °C), medium (10–15 °C), high (15–22 °C), and very high (> 22 °C). High surface temperatures (over 22 °C) covered 116.60 km^2 in 2000, with only 0.85% falling into the very high category. The majority of the region was classified as high-temperature zones (Fig. 9).

13 Demography and Landuse

According to the 2011 Census of India, the overall population of Kashmir Valley is 69 lakhs. The population of the valley has shown a substantial increase from 24 lakhs in 1971 to 69 lakhs in 2011. Thus, showing an increase of 183% at the rate of 4.5% per year over 40 years resulting in a higher population density. The landuse and landcover status of the study area was also analysed for the year 2020. Land use refers to human activities such as agriculture, settlement, and industry, whereas land cover refers to features on the surface of the earth such as lakes, mountains, vegetation, and rocks. The landscape in the Jhelum basin has changed dramatically in recent decades as land has been converted to new uses without regard for its inherent land use suitability. The major LULC classes in the study area by percentage are Forest area (29.99%), water aqua vegetation (18.19%), Agriculture/horticulture (14.67%),

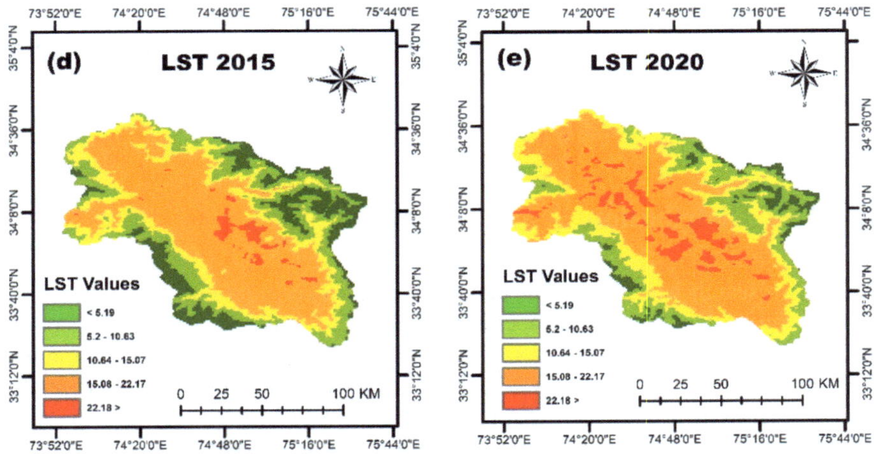

Fig. 9 LST distribution in the Jhelum basin

Pasture/Plantation (14.65%), Barren land (7.67%) and Built-up (1.49%). The LULC map of the Jhelum basin is depicted in Fig. 10.

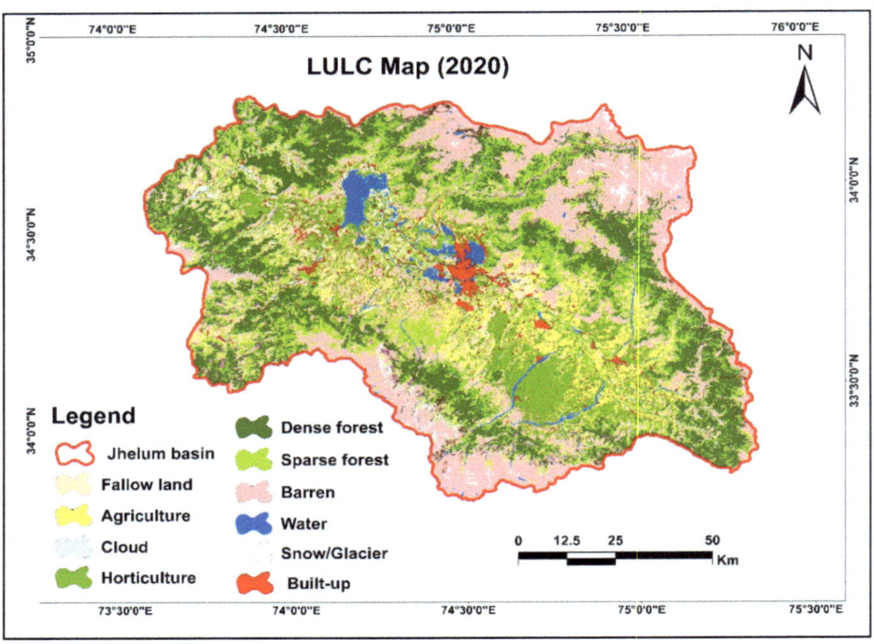

Fig. 10 Landuse and landcover status of Jhelum basin in 2020

Changes in the land system within the Jhelum basin have far-reaching consequences for watershed-scale hydrological processes, affecting not just the basin but also downstream regions of the Indus (Rather et al., 2016). Deforestation of natural forests in the Jhelum's highland catchment area, largely for timber and fuel wood, has reduced forest ecosystems' water-retention capacity.

14 Discussion

The extensive geographical overview offered in this study provides light on the complex dynamics of the Jhelum Basin in the Kashmir Himalayas. This discussion explores into the study's primary findings and implications, emphasising the need of understanding the region's geographic peculiarities. The physical setting of the Jhelum Basin is one of the study's main focus points. The valley, located between the Pir-Panjal and Zanskar mountain ranges, features a stunning scenery characterised by snow-capped peaks, glaciers, and a network of rivers and streams. This physical environment determines not only the region's hydrology, but also its weather patterns and flora distribution. The geomorphology of the Jhelum Basin is heavily influenced by its relief characteristics. The region's topography is diverse, with mountains, foothills, Karewas, and valley floors. The existence of Karewas, lacustrine deposits developed during the Pleistocene Period, contributes to the basin's distinct geomorphic characteristics. Furthermore, the juxtaposition of depositional and erosional structures on the valley floor emphasises the dynamic nature of the fluvial-influenced environment.

Geologically, the Jhelum Basin provides vital information about the tectonic history and rock formations of Kashmir. The existence of Neogene Quaternary intermontane basins and a variety of lithologies ranging from granites to alluvium highlights the area's geologic complexity. Understanding the geological makeup is critical for determining geological dangers and resource potential in the region. The discussion also emphasises the significance of drainage patterns in the Jhelum Basin. The Jhelum River, fed by multiple tributaries and glaciers, is a vital resource for the region, providing water for agriculture, hydropower generation, and ecological sustenance. The study illustrates the basin's various drainage patterns, which are impacted by lithology, tectonic activity, and glaciation.

Climate appears as another important factor influencing the spatial dynamics of the Jhelum Basin. The region's agricultural techniques, water supply, and tourism are all influenced by the cold temperate environment, which has distinct seasons. The study emphasises the influence of Western disturbances and monsoon patterns on precipitation and temperature variability in the Basin. Vegetation distribution and soil characteristics are also explored, with a focus on how they interact with climate, terrain, and land use practices. The existence of deciduous forests, alpine plants, and varied soil types demonstrates the region's biological richness. Unsustainable land use practices, such as deforestation and shifting agriculture patterns, jeopardise ecosystem integrity and water resource management.

Furthermore, the study emphasises glaciers' importance as natural storage of water, which is critical for downstream water supply and other socioeconomic activity. The inventory of glaciers and their contribution to the hydrological regime highlight their relevance in terms of climate change and water resource management. Finally, the topic delves into the population patterns and land use changes found in the Jhelum basin. Rapid population increase and land-use changes, particularly from agriculture to horticulture, have implications for water availability, soil conservation, and ecosystem services. Understanding demographic and land use patterns is critical for regional sustainable development planning.

15 Conclusion

The geographical dynamics of the Jhelum basin in the Kashmir Himalayas have important policy implications and management issues for sustainable development. With its diversified geography and critical water resources, effective regulations are required to balance conservation efforts and socioeconomic growth. Management plans must consider the effects of urbanisation, deforestation, and population growth on land use and water supply. Furthermore, proactive steps are required to mitigate the risks connected with glacial lakes and changing climate patterns. Collaboration among government agencies, local communities, and environmental organisations is critical to safeguarding the region's long-term resilience and prosperity. Policymakers in the Jhelum basin can promote sustainable development practices that benefit both current and future generations by focusing on environmental conservation, equitable resource distribution, and community participation.

Acknowledgements **Conflict of Interest Disclosure** The authors declare no conflict of interest.

Ethics Approval This study was conducted in accordance with relevant ethical guidelines and regulations. Ethical approval was not required as this research did not involve human or animal subjects.

Permission to Reproduce Material from Other Sources Permission to reproduce material from other sources was not required for this study.

Funding No funding was received to carry out this study.

Data Availability NA

References

Ahmad, S., Alam, A., & Ahmad, B. (2016). Comment on: "Earthquake geology of Kashmir Basin and its implications for future large earthquakes" by Shah (2013); "Kashmir Basin fault and its tectonic significance in NW Himalaya, Jammu and Kashmir, India" by Shah (2015). *International Journal of Earth Sciences, 105*, 681–687.

Ahmed, R., Ahmad, S. T., Wani, G. F., Mir, R. A., Ahmed, P., & Jain, S. K. (2022a). High resolution inventory and hazard assessment of potentially dangerous glacial lakes in upper Jhelum basin, Kashmir Himalaya, India. *Geocarto International, 37*(25), 10681–10712.

Ahmed, R., Rather, A., Banerjee, P., Wani, G., & Saleem, S. (2025). Impact of climate variability on the feeding glaciers of potentially dangerous glacial lakes in the Jhelum basin of Kashmir Himalaya, India. *Natural Hazards Review.* https://doi.org/10.1061/NHREFO.NHENG-2140

Ahmed, R., Rawat, M., Wani, G. F., Ahmad, S. T., Ahmed, P., Jain, S. K., Farooq, M., et al. (2022b). Glacial Lake outburst flood hazard and risk assessment of Gangabal lake in the upper Jhelum Basin of Kashmir Himalaya using geospatial technology and hydrodynamic modeling. *Remote Sensing, 14*(23), 5957.

Ahmed, R., Wani, G. F., Ahmad, S. T., Mir, R. A., Almazroui, M., Jain, S. K., & Ahmed, P. (2021a). Spatiotemporal dynamics of glacial lakes (1990–2018) in the Kashmir Himalayas, India using remote sensing and GIS. *Discover Water, 1*, 1–17.

Ahmed, R., Wani, G. F., Ahmad, S. T., Sahana, M., Singh, H., & Ahmed, P. (2021b). A review of glacial lake expansion and associated glacial lake outburst floods in the Himalayan region. *Earth Systems and Environment, 5*(3), 695–708.

Ahmed, R., Bhardwaj, A., Sam, L., & Van Tricht, L. (2026a). Ice-rock avalanches in a warming Himalaya indicate pathways toward effective preparedness. *Communications Earth & Environment, 7*(1), 217.

Ahmed, R., Sam, L., Bhardwaj, A., Kulkarni, A. V., & Yamasaki, S. (2026b). Increasing global threat of outburst floods from overlooked small alpine lakes. *Nature Sustainability*, 1–10.

Bagnolus, F., & Meher-Homji, V.M. (1959). *Bioclimatic types of South East Asia* (Vol. 277). Travaux de la Section Scientific at Technique Institute Franscis de Pondicherry.

Basavaiah, N., Appel, E., Lakshmi, B. V., Deenadayalan, K., Satyanarayana, K. V. V., Misra, S., Malik, M. A., et al. (2010). Revised magnetostratigraphy and characteristics of the fluviolacustrine sedimentation of the Kashmir basin, India, during Pliocene-Pleistocene. *Journal of Geophysical Research: Solid Earth, 115*(B8).

Bhat, M. I. (1982). Thermal and tectonic evolution of the Kashmir basin Vis-a-Vis petroleum prospects. *Tectonophysics, 88*(1–2), 117–132.

Dar, J. A., & Dubey, R. K. (2015). Probabilistic seismic hazard analyses (PSHA) and liquefaction susceptibility evaluation of Kashmir Valley, India. *Proceedings of the National Academy of Sciences, India Section A: Physical Sciences, 85*, 177–186.

Dar, R. A., & Zeeden, C. (2020). Loess-palaeosol sequences in the Kashmir Valley, NW Himalayas: A review. *Frontiers in Earth Science, 8*, 113.

Dar, R. A., Rashid, I., Romshoo, S. A., & Marazi, A. (2014). Sustainability of winter tourism in a changing climate over Kashmir Himalaya. *Environmental Monitoring and Assessment, 186*, 2549–2562.

Dubey, R. K., Dar, J. A., & Kothyari, G. C. (2017). Evaluation of relative tectonic perturbations of the Kashmir Basin, northwest Himalaya, India: An integrated morphological approach. *Journal of Asian Earth Sciences, 148*, 153–172.

Gürbüz, A., & Gürer, Ö. F. (2008). Anthropogenic affects on Lake sedimentation process: A case study from Lake Sapanca, NW Turkey. *Environmental Geology, 56*(2), 299–307.

Himayoun, D., Mohsin, F., & Roshni, T. (2020). Efficacy in simulating the peak discharge response using soft computing techniques in the Jhelum River basin, India. *International Journal of River Basin Management, 18*(1), 81–93.

Imran, M., & Sheikh, A. K. P. (2020). Forecasting water level of Jhelum River of Kashmir Valley India, using prediction and earlywarning system. *Geography, Environment, Sustainability, 13*(2), 35–42.

Mahmood, R., & Jia, S. (2017). Spatial and temporal hydro-climatic trends in the transboundary Jhelum River basin. *Journal of Water and Climate Change, 8*(3), 423–440.

Middlemiss, C. S. (1910). Revision of Silurian-Trias sequence in Kashmir. *Records of the Geological Survey of India, 40*(3), 206–260.

Mir, R. A., & Gani, K. M. (2019). Water quality evaluation of the upper stretch of the river Jhelum using multivariate statistical techniques. *Arabian Journal of Geosciences, 12*, 1–19.
Mir, R. A., & Jeelani, G. H. (2015). Textural characteristics of sediments and weathering in the Jhelum River basin located in Kashmir Valley, western Himalaya. *Journal of the Geological Society of India, 86*, 445–458.
Mir, R. A., Jeelani, G., & Dar, F. A. (2016). Spatio-temporal patterns and factors controlling the hydrogeochemistry of the river Jhelum basin, Kashmir Himalaya. *Environmental Monitoring and Assessment, 188*, 1–24.
Mohiuddin, S. (2013). Identification of backward regions of Kashmir Valley (India)—A spatio-temporal approach. *Journal of Geography and Regional Planning, 6*(8), 294–301.
Mushtaq, S., Saleem, S., Ahmed, R., Tass, M. S., Rather, J. A., & Rather, G. M. (2024). Spatio-temporal analysis of land use land cover changes in South Kashmir region of North-Western Himalayas using Landsat data. *Discover Geoscience, 2*(1), 37.
Rather, M. I., Rashid, I., Shahi, N., Murtaza, K. O., Hassan, K., Yousuf, A. R., & Shah, I. Y. (2016). Massive land system changes impact water quality of the Jhelum River in Kashmir Himalaya. *Environmental Monitoring and Assessment, 188*(3), 185.
Raza, M. A., Ahmad, A. A., & Mohammad, A. (1978). The Land. In *The valley of Kashmir—A geographical interpretation* (Vol. 1).
Romshoo, S. A., Bhat, M. A., & Beig, G. (2021). Particulate pollution over an urban Himalayan site: Temporal variability, impact of meteorology and potential source regions. *Science of the Total Environment, 799*, 149364.
Romshoo, S. A., Dar, R. A., Rashid, I., Marazi, A., Ali, N., & Zaz, S. N. (2015). Implications of shrinking cryosphere under changing climate on the streamflows in the Lidder catchment in the upper Indus Basin, India. *Arctic, Antarctic, and Alpine Research, 47*(4), 627–644.
Romshoo, S. A., Fayaz, M., Meraj, G., & Bahuguna, I. M. (2020). Satellite-observed glacier recession in the Kashmir Himalaya, India, from 1980 to 2018. *Environmental Monitoring and Assessment, 192*, 1–17.
Saleem, S., Ahmed Hajam, F., & Rather, J. A. (2021). Spatio-temporal analysis of land use land cover changes in Sind catchment of the Kashmir Valley, India. *Geo-Eye, 10*(2), 35–42.
Saleem, S., Rather, J. A., Ahmad, S., Ahmed, R., & Hajam, L. A. (2024a). Assessment of livestock carrying capacity in the alpine grasslands of the Kashmir Himalayas. *Pastoralism: Research, Policy and Practice, 14*, 13541.
Saleem, S., Rather, J. A., Ahmed, S., Mushtaq, S., Ahmed, R., & Malik, I. H. (2024b). Above ground biomass estimation for alpine grasslands of Kashmir Himalayas using remote sensing and field-data. *Rangeland Ecology and Management, 96*, 117–127.
Shellnutt, J. G. (2018). The panjal traps. *Geological Society, London, Special Publications, 463*(1), 59–86.
Song, Z., Li, R., Qiu, R., Liu, S., Tan, C., Li, Q., Ma, M., et al. (2018). Global land surface temperature influenced by vegetation cover and PM2. 5 from 2001 to 2016. *Remote Sensing, 10*(12), 2034.
ul Shafiq, M., Ashraf, I., ul Islam, Z., Ahmed, P., & Dimri, A. P. (2020). Response of streamflow to climate variability in the source region of Jhelum River basin in Kashmir valley, India. *Natural Hazards, 104*, 611–637.
Wadia, D. N. (1941). Pleistocene ice age deposits of Kashmir. *Proceedings of the National Institute of Sciences of India, 7*, 49–59.
Wadia, D.N., (1975). *Geology of India.* Tata McGraw Hill Co. 344pp.
Weng, Q., Lu, D., & Schubring, J. (2004). Estimation of land surface temperature–vegetation abundance relationship for urban heat Island studies. *Remote Sensing of Environment, 89*(4), 467–483.
Yousuf, M., Bukhari, S. K., & Bhat, G. R. (2021). Using paleo-liquefaction features to determine the likely source, magnitude and ground accelerations of pre-historic earthquakes in the Kashmir Basin (northwestern Himalaya), India. *Engineering Geology, 293*, 106302.
Zaz, S. N., Romshoo, S. A., Krishnamoorthy, R. T., & Viswanadhapalli, Y. (2019). Analyses of temperature and precipitation in the Indian Jammu and Kashmir region for the 1980–2016 period: Implications for remote influence and extreme events. *Atmospheric Chemistry and Physics, 19*(1), 15–37.

Water Resources, Wetlands, and Hydroclimatic Variability

Quantification of Long-Term Groundwater Storage Change in Transboundary Aquifers of Indus River Basin in India Using Remotely Sensed Data

Ankita Sharma, Chandan K. Singh, and Mukesh K. Verma

Abstract Groundwater overexploitation in the Indus Basin transboundary aquifers has a significant implication for water security in South Asia. To measure changes in groundwater storage in these aquifers over a given time frame, Gravity Recovery and Climate Experiment (GRACE) satellite data has been utilized to provide estimates of Terrestrial Water Storage (TWS) anomalies and its Spatial-temporal variation. GRACE-derived data of TWS for the Indus River Basin area from 200 to 2024 was used for the present investigation. Rescaling method is used to resample the GRACE data, and uncertainty is determined to minimize error and increase accuracy. The monthly long-term variation in TWS has been observed for the entire Indus River basin and compared with field ground water level data. Soil moisture data, Global Land Data Assimilation System (GLDAS), has also been utilized to compute groundwater storage time series data and their uncertainty. The result shows that most of the aquifers in the Indus River Basin have been over exploited due to excessive irrigation from ground water. The study concludes that if the rate of extraction of ground water remains constant are increases, then aquifer would be vulnerable considering the changing climate pattern and increasing demand for irrigation from ground water resources.

Keywords Terrestrial water storage (TWS) · Groundwater storage (GWS) · Soil moisture (SM) · Specific yield (Sy) · GRACE · GLDAS · LAND · TELLUS · Gravity · CSR

A. Sharma (✉) · C. K. Singh · M. K. Verma
Department of Civil Engineering, National Institute of Technology Raipur, Raipur, India
e-mail: asharma001.mtech2024@civil.nitrr.ac.in

C. K. Singh
e-mail: cksingh.ce@nitrr.ac.in

R. Ahmed et al. (eds.), *Climate Change, Water Resources, and Natural Hazards in the Himalayas*, Advances in Geographical and Environmental Sciences,
https://doi.org/10.1007/978-981-92-0835-7_5

1 Introduction

The Indus River Basin is one of the most important transboundary water systems in South Asia, covering large portions of India, Pakistan, Afghanistan, and China. In India, the basin spans across several northern and northwestern states including Jammu and Kashmir, Ladakh, Himachal Pradesh, Punjab, Haryana, and parts of Rajasthan. These areas are densely cultivated and highly dependent on groundwater for irrigation, primarily due to unreliable canal water deliveries, erratic monsoon rainfall, and increasing agricultural water demands. In states such as Punjab and Haryana, large-scale groundwater extraction often incentivized by subsidized electricity and minimum support price policies has led to significant and sustained declines in water tables (Basharat, 2020; Fallatah et al., 2017; Pham-Duc et al., 2020).

This rising dependence on groundwater, alongside population growth, urban expansion, and changing climate patterns, has placed extreme pressure on aquifers throughout the Indus River Basin (Frappart, 2020; Gonçalvès et al., 2020; Mehmood et al., 2022). However, the management of this groundwater crisis is hindered by major data limitations. Conventional monitoring through observation wells is often sparse, unevenly distributed, and resource-intensive, making it difficult for policymakers and regulatory agencies to track trends and implement timely, location-specific interventions (Almulla et al., 2020; Fallatah et al., 2017).

To overcome these data constraints, satellite-based observations have become valuable tools in groundwater research and planning. The Gravity Recovery and Climate Experiment (GRACE) mission, in particular, has provided large-scale estimates of terrestrial water storage by measuring temporal changes in the Earth's gravity field (Mehmood et al., 2022). Used GRACE data, in combination with hydrological models such as the Global Land Data Assimilation System (GLDAS) and the WaterGAP Hydrological Model (WGHM), to assess changes in groundwater storage across the entire Indus River Basin between 2002 and 2025. Their study introduced a zonation framework dividing the basin into upper, middle, and lower segments to account for the varying topographic and climatic conditions. This tiered approach enabled more targeted insights that are useful for policy formulation and regional water management.

However, the study by (Mehmood et al., 2022) does not address data uncertainty or validate the satellite-derived estimates against field-based observations, two components that are essential for building policy trust in remotely sensed datasets. By Ali et al. (2022) Addressed this shortcoming by applying the Generalized Three-Cornered Hat method and Empirical Orthogonal Function analysis to evaluate uncertainty and spatial variability in GRACE-derived groundwater storage estimates within the lower portion of the Indus River Basin. Their methodology also included cross-validation with other global hydrological models and in situ datasets, enhancing the credibility of the satellite-based outputs.

Building upon these previous works, this study focuses on groundwater trends in the Indian section of the Indus River Basin using a zonal framework that separates the region into upper, middle, and lower zones. This division supports the generation of location-specific insights and accounts for the hydrological and agro-ecological variability across the basin. Furthermore, this study incorporates uncertainty quantification and correlation analysis techniques to evaluate the robustness of GRACE-based groundwater estimates. The resulting insights are intended to guide evidence-based water governance in India, including the design of regional abstraction controls, improvements in irrigation efficiency, and targeted groundwater recharge initiatives across vulnerable areas of the basin.

The Indus River Basin spans multiple geopolitical boundaries and serves as a lifeline for millions across South Asia. In the Indian context, it extends over critical agrarian regions such as Jammu and Kashmir, Ladakh, Himachal Pradesh, Punjab, Haryana, and parts of Rajasthan. Agriculture in these areas is predominantly irrigated, placing immense pressure on both surface water and groundwater resources. With increasing canal water unreliability, declining rainfall consistency, and expanding cultivation of high water-demand crops, farmers have turned extensively to groundwater, often supported by state subsidies and inadequate abstraction regulations. This widespread extraction has led to unsustainable groundwater declines, particularly in states like Punjab and Haryana (Basharat, 2020; Fallatah et al., 2017; Frappart, 2020).

2 Study Area

The Indus River Basin occupies a vast region of South Asia, extending across China, India, Pakistan, and Afghanistan, with a total drainage area ranging from approximately 1.12 to 1.17 million km^2. The basin extends between approximately 24°N–37°N latitudes and 67°E–82°E longitudes and encompasses diverse physiographic settings, including high mountain ranges, plateaus, foothills, and extensive alluvial plains. The Indus River originates from the Bokhar Chu Glacier in the Kailash Range near Lake Mansarovar on the Tibetan Plateau at an elevation of about 5,182 m above mean sea level. From its source, the river flows through the Himalayan and Karakoram mountain systems before traversing the plains of Pakistan and ultimately discharging into the Arabian Sea through the Indus Delta southeast of Karachi. The basin is bounded by the Tibetan Plateau and Kailash Range in the east and by the Sulaiman and Kirthar mountain ranges in the west. The Indus River system comprises several major tributaries, including the Jhelum, Chenab, Ravi, Beas, and Sutlej rivers, which collectively contribute to the hydrological regime of the basin. Water availability within the basin is primarily governed by glacier melt, seasonal snowmelt, and monsoonal precipitation, resulting in significant spatial and temporal variability in water resources. Due to its transboundary nature, large population dependence, extensive agricultural activities, and increasing pressure on groundwater resources, the Indus River Basin represents a critically important region for investigating long-term terrestrial water storage and groundwater storage dynamics. Understanding changes in groundwater storage within the basin is essential for sustainable water-resource management and for assessing the impacts of climatic variability and anthropogenic water extraction on regional water security (CWC, 2026).

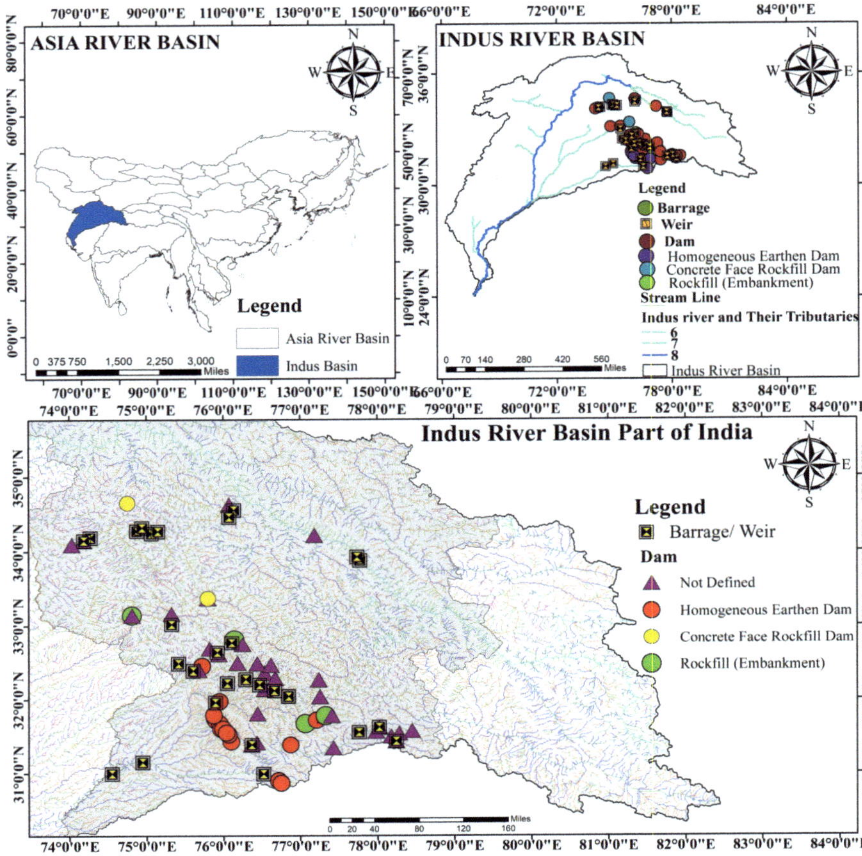

Fig. 1 Transboundary Indus River basin with its tributaries

3 Data Source

The monthly terrestrial water storage (TWS) data from the Gravity Recovery and Climate Experiment (GRACE) Release-06 (RL06) were extracted for the period spanning 2002 to 2024. The data, provided by the National Aeronautics and Space Administration (NASA), have a global coverage with a 1° × 1° spatial resolution. Specifically, the GRACE land mass datasets from the Centre for Space Research (CSR) were sourced from the NASA website (https://www.jpl.nasa.gov/), which is supported by the NASA Making Earth System Data Records for Use in Research Environments (MEASUREs) Program. This access was granted on December 17, 2024 (Landerer, 2024; Landerer & Swenson, 2012; Mehmood et al., 2022). The total land water storage anomalies are derived from the NOAH model within the Global Land Data Assimilation System (GLDAS). GLDAS estimates land water content by integrating various land surface models with data assimilation techniques. The combined land water anomalies—calculated as the sum of soil moisture, snow water, and canopy water—are suitable for comparing and evaluating satellite observations from the GRACE and GRACE-FO missions over terrestrial regions. These anomalies are calculated on a monthly basis using the same days covered by each GRACE and GRACE-FO monthly measurement. The results are available in NetCDF format on a global $1^\circ \times 1^\circ$ latitude/longitude grid. At present, the time span used to generate these monthly anomalies aligns with that of the GRACE-FO Level-2 RL06.3 JPL solutions (Rodell, 2020).

4 Methodology

To Estimate long-term groundwater storage changes the GRACE data, which measures temporal changes in the Earth's gravity field to provide estimates of Terrestrial Water Storage (TWS), was used for the period of 2003 to 2024. Monthly averages of all datasets were computed to produce a month-wise mean TWS distribution using Python. In addition, total land water storage anomalies were downloaded from the Global Land Data Assimilation System (GLDAS) NOAH model for the period 2003–2017, and corresponding month-wise averages were calculated and visualized using Python. Groundwater storage anomaly (GWC) was then derived using the equation:

$$\Delta GWC = TWS - \Delta SM - \Delta SWE - \Delta CWS \quad (1)$$

where ΔSM, ΔSWE, and ΔCWS are the anomalies for soil moisture, snow water equivalent, and canopy water storage, respectively, and ΔGWC represents the groundwater storage change (groundwater anomaly), while TWS denotes the total water storage(Frappart & Ramillien, 2018; Mehmood et al., 2022; Voss et al., 2013).

5 Result and Discussion

See Fig. 2.

6 Observed Trends and Interpretation

The y-axis represents the WET in centimeters (cm), while the x-axis denotes time. The blue bars represent monthly WET values, oscillating around a general downward trend. The red curve is a **polynomial regression line**, specifically a quadratic equation, that models the long-term trend. The equation for this curve, $y(t) = -0.000343\,t - 0.0135\,t + 10.7097$, where t is time in months, indicates a **non-linear,**

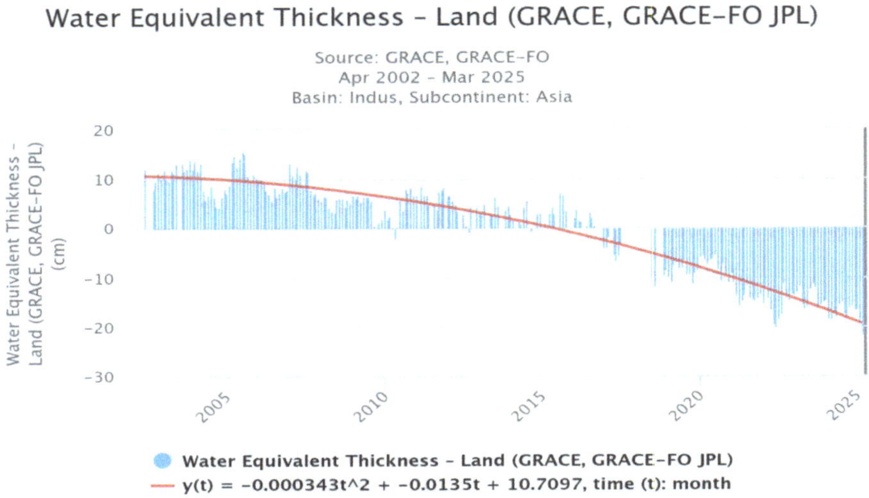

Fig. 2 Water equivalent thickness in the Indus Basin (2002–2025)

accelerating decline in WET over the period (Fig. 2). The negative coefficient of the t2 term signifies that the rate of water loss is increasing over time. This suggests a **significant and worsening depletion of terrestrial water storage** in the Indus basin. The initial positive WET values (above 0 cm) transition to a consistent negative trend, with values dropping to approximately − 20 cm by 2025 (Fig. 2). This persistent and accelerating decline is indicative of **unsustainable water management practices**, such as excessive groundwater extraction for agriculture, and potential impacts from **climate change**, leading to reduced glacial meltwater inflow and altered precipitation patterns. The significant water loss poses a serious threat to regional water security and ecological stability.

Northern and Eastern Regions: The northern and eastern parts of the basin generally exhibit higher soil moisture, especially during the months of March, April, May, and June (Fig. 3). This is indicated by the green and light-yellow colors in these areas. This period of increased moisture is likely a result of winter snowmelt and the onset of pre-monsoon rainfall.

Southern and Western Regions: These areas, in contrast, consistently show lower soil moisture throughout the year, as seen by the prevalence of red and orange hues. The soil moisture in these regions is particularly low from July to October, which corresponds to the peak of the agricultural season and high water demand for irrigation.

Overall Seasonal Pattern: The basin as a whole experience a cyclical change in soil moisture. Moisture levels tend to increase from March to June, peaking in the northern and eastern parts. After June, there is a general decrease in moisture across the basin, with the southern and western areas showing the most significant decline.

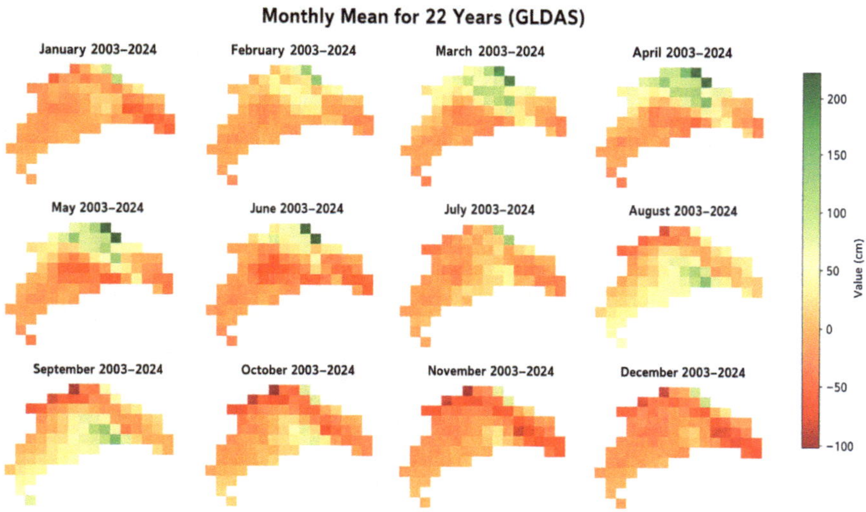

Fig. 3 Mean monthly land water anomalies 2003–2024

This trend aligns with the high irrigation demands during the agricultural growing season, which deplete the soil moisture faster than it can be replenished by rainfall.

The maps show a clear seasonal pattern in water storage. During the colder months, from October to March, the northern and western parts of the basin exhibit a negative TWS anomaly, indicated by red and orange hues (Fig. 4). This is likely due to a combination of factors, including reduced precipitation and the accumulation of snow in the mountainous regions, which is not yet melted and contributing to the liquid water storage. Conversely, the southern and eastern parts of the basin show a slight positive anomaly during this period.

From April to September, the pattern shifts significantly. The northern parts of the basin, particularly in the Himalayas, show a strong positive TWS anomaly, represented by the deep blue colors. This gain is primarily attributed to snowmelt and increased monsoon rainfall during the summer months. The southern and eastern regions, which are densely cultivated, show a negative TWS anomaly during this period. This is likely due to the high demand for groundwater for irrigation during the agricultural growing season, which outpaces natural recharge from rainfall and snowmelt.

This Fig. 5 presents the spatial differences between GRACE-derived groundwater storage anomalies and GLDAS-simulated values, averaged month-wise for the period under study. The basin-wide patterns reveal strong geographical contrasts, reflecting both climatic and anthropogenic influences.

The groundwater storage anomaly derived from the residual difference between GRACE-derived Terrestrial Water Storage (TWS) and GLDAS land-water storage components provides insight into the spatial distribution of groundwater variability

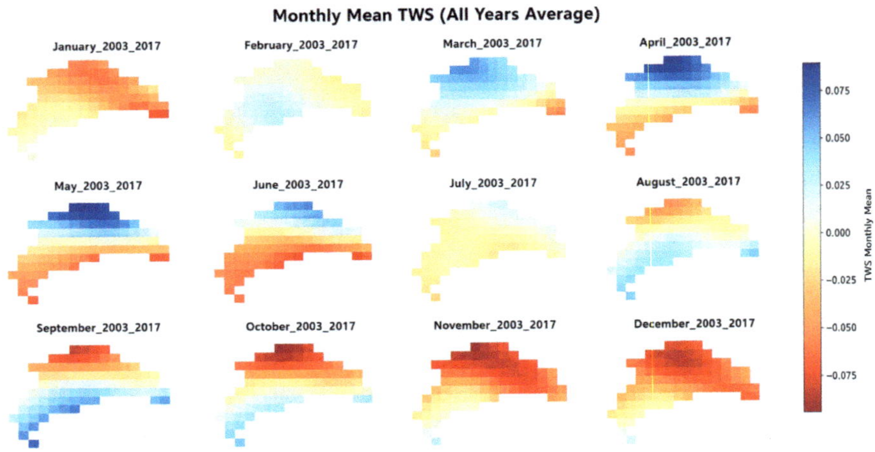

Fig. 4 Monthly mean terrestrial water storage (TWS) for the period 2003–2017

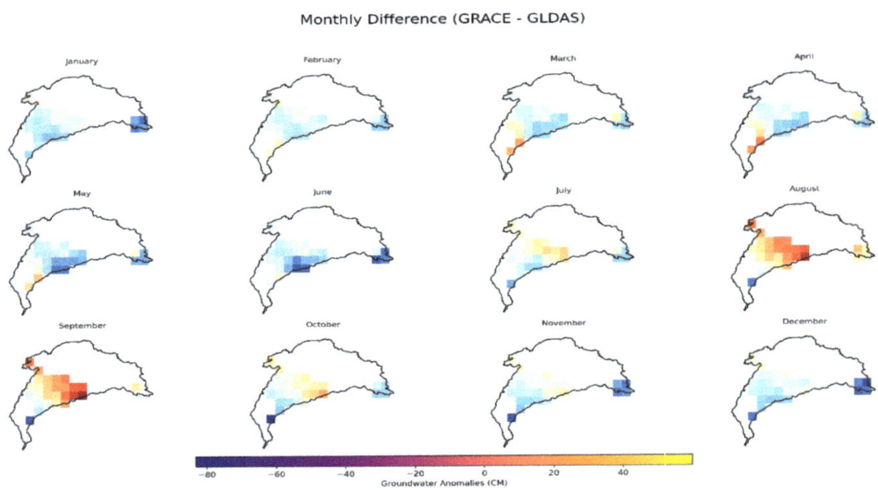

Fig. 5 Spatial distribution of monthly groundwater anomaly differences (GRACE—GLDAS) over the Indus basin

across the Indus River Basin. The monthly anomaly maps reveal considerable regional differences in groundwater storage behaviour, indicating that groundwater dynamics vary substantially across different physiographic settings of the basin. The GRACE–GLDAS residual highlights localized groundwater responses that are not readily apparent from total water storage alone and provides a clearer representation of regional groundwater storage conditions.

Northern Region (Upper Indus – Himalayan Foothills and Upstream Catchments): The northern region exhibits pronounced fluctuations in groundwater storage anomalies throughout the year. Negative anomalies become increasingly widespread during March, April, and May, with several areas displaying strong groundwater deficits during the late pre-monsoon period. During the monsoon and post-monsoon months, positive anomalies expand across large parts of the region, indicating a marked shift in groundwater storage conditions. The observed transition from predominantly negative to positive anomalies suggests a strong response of groundwater storage to seasonal hydrological inputs. The magnitude of anomaly variations observed in this region highlights the influence of climatic and topographic controls on groundwater storage dynamics within the upper basin.

North-western Plains (Punjab and Haryana Alluvial Plains): The north-western plains display a complex groundwater anomaly pattern characterized by the coexistence of both positive and negative anomalies throughout the year. Negative anomalies are more apparent during the dry pre-monsoon months, whereas localized positive anomalies become increasingly evident during the monsoon and post-monsoon period. The heterogeneous anomaly distribution indicates substantial spatial variability in groundwater storage conditions across the region. The persistence of contrasting anomaly patterns suggests that groundwater responses vary considerably within the alluvial plains and may reflect differences in recharge availability, groundwater utilization, and local hydrogeological conditions.

Southern and South-western Region: The southern and south-western sectors exhibit comparatively moderate groundwater anomaly fluctuations relative to the northern region. Positive anomalies occur across extensive portions of the region during several months of the year, particularly during the monsoon and post-monsoon period. However, localized negative anomalies are also observed intermittently, indicating that groundwater storage conditions are spatially variable rather than uniformly positive. The persistence of positive anomalies across large areas suggests relatively favourable groundwater storage conditions in certain sectors, whereas localized deficits indicate the continued influence of climatic and hydrogeological variability.

Overall, the GRACE–GLDAS residual analysis reveals marked spatial heterogeneity in groundwater storage behaviour across the Indus River Basin. The northern region is characterized by substantial seasonal shifts in groundwater anomalies, the north-western plains exhibit mixed groundwater responses, and the southern and south-western region generally displays comparatively stable groundwater conditions with localized variability. A notable outcome of the analysis is the uneven spatial distribution of groundwater storage anomalies across the basin, indicating that groundwater recharge and depletion processes operate differently among regions. These findings demonstrate that groundwater storage behaviour within the Indus River Basin is strongly influenced by regional hydro-climatic and physiographic controls rather than a uniform basin-wide response.

7 Conclusion

Groundwater Storage Anomaly: GRACE–GLDAS Residual Analysis The groundwater storage anomaly derived from the residual difference between GRACE-derived Terrestrial Water Storage (TWS) and GLDAS land-water storage components provides insight into the spatial distribution of groundwater variability across the Indus River Basin. The monthly anomaly maps reveal substantial regional differences in groundwater storage behaviour, indicating that groundwater dynamics vary considerably across different physiographic regions of the basin. Unlike total terrestrial water storage, the GRACE–GLDAS residual highlights localized groundwater responses and emphasizes the heterogeneous nature of groundwater storage conditions throughout the basin. Northern Region (Upper Indus – Himalayan Foothills and Upstream Catchments): The northern region is characterized by pronounced fluctuations in groundwater storage anomalies throughout the year. Negative anomalies become increasingly widespread during March, April, and May, with several parts of the region exhibiting strong groundwater deficits during the late pre-monsoon period. During the monsoon and post-monsoon months, positive anomalies expand across substantial portions of the region, indicating a noticeable shift in groundwater storage conditions. The observed transition from negative to positive anomalies suggests a strong response of groundwater storage to seasonal hydrological inputs. The magnitude and spatial extent of anomaly variations in this region highlight the importance of climatic and topographic controls on groundwater dynamics within the upper basin. North-western Plains (Punjab and Sindh Headwaters): The north-western plains exhibit a mixed groundwater anomaly pattern throughout the year. Both positive and negative anomalies are observed across different parts of the region, reflecting considerable spatial variability in groundwater storage conditions. Negative anomalies are more evident during the dry season, whereas localized positive anomalies become increasingly apparent during the monsoon and post-monsoon months. The coexistence of contrasting anomaly patterns within the same region suggests that groundwater storage responses are not spatially uniform and may reflect differences in recharge availability, groundwater utilisation, and local hydrogeological characteristics. Compared with the northern region, groundwater recovery appears more heterogeneous and spatially fragmented. Southern and South-western Region: The southern and south-western sectors display relatively moderate groundwater anomaly fluctuations compared with the northern region. Positive anomalies are observed across large parts of the region during several months of the year, particularly during the monsoon and post-monsoon period. However, localized negative anomalies also occur intermittently, indicating that groundwater storage conditions are not uniformly positive throughout the region. The persistence of positive anomalies over extensive areas suggests comparatively favourable groundwater storage conditions in certain sectors, while the occurrence of localized deficits highlights the influence of regional climatic and hydrogeological variability. Overall, the GRACE–GLDAS residual analysis reveals marked spatial heterogeneity in groundwater storage behaviour across the Indus River Basin. The northern region exhibits substantial seasonal shifts in groundwater anomalies, the north-western plains demonstrate a complex mixture of positive and negative storage responses, and the southern and south-western region generally displays more stable groundwater conditions with localized variability. A key finding of this analysis is that groundwater storage responses differ considerably between regions, indicating that groundwater recharge and depletion processes are spatially uneven across the basin. These results demonstrate that basin-wide groundwater assessments may mask important regional differences in groundwater storage dynamics and emphasize the need for region-specific groundwater management approaches.

Acknowledgements **Conflict of Interest Disclosure** The authors declare no conflict of interest.

Ethics Approval This study was conducted in accordance with relevant ethical guidelines and regulations. Ethical approval was not required as this research did not involve human or animal subjects.

Permission to Reproduce Material from Other Sources Permission to reproduce material from other sources was not required for this study.

Funding No funding was received to carry out this study.

Data Availability NA

References

Ali, S., Wang, Q., Liu, D., et al. (2022). Estimation of spatio-temporal groundwater storage variations in the lower transboundary Indus Basin using GRACE satellite. *Journal of Hydrology, 605*, 127315. https://doi.org/10.1016/j.jhydrol.2021.127315

Almulla, Y., Ramirez, C., Pegios, K., et al. (2020). A GIS-based approach to inform agriculture-water-energy nexus planning in the North Western Sahara aquifer system (NWSAS). *Sustainability, 12*(17), 17. https://doi.org/10.3390/su12177043

Basharat, M. (2020). Study quantum of transboundary groundwater drainage from India to Pakistan. *Global Journal of Engineering Sciences, 5*(5). https://doi.org/10.33552/gjes.2020.05.000624

Central Water Commission Board, Indus Basin Organization. (2026). *CWC*, https://doi.org/https://www.cwc.gov.in/ibo/home.

Fallatah, O. A., Ahmed, M., Save, H., & Akanda, A. S. (2017). Quantifying temporal variations in water resources of a vulnerable middle eastern transboundary aquifer system. *Hydrological Processes, 31*(23), 4081–4091. https://doi.org/10.1002/hyp.11285

Frappart, F. (2020). Groundwater storage changes in the major north African transboundary aquifer systems during the GRACE era (2003–2016). *Water, 12*(10), 10. https://doi.org/10.3390/w12102669

Frappart, F., & Ramillien, G. (2018). Monitoring groundwater storage changes using the gravity recovery and climate experiment (GRACE) satellite Mission: A review. *Remote Sensing, 10*(6), 829. https://doi.org/10.3390/rs10060829

Gonçalvès, J., Deschamps, P., Hamelin, B., Vallet-Coulomb, C., Petersen, J., & Chekireb, A. (2020). Revisiting recharge and sustainability of the North-Western Sahara aquifers. *Regional Environmental Change, 20*(2), 47. https://doi.org/10.1007/s10113-020-01627-4

Landerer, F. (2024). CSR TELLUS GRACE-FO Level-3 monthly land water-equivalent-thickness surface mass anomaly release 6.3 version 04. Ver. RL06.3v04. PO.DAAC, CA, USA. Dataset Accessed [YYYY-MM-DD] at: https://doi.org/10.5067/GFLND-3C634.

Landerer, F. W., & Swenson, S. C. (2012). Accuracy of scaled GRACE terrestrial water storage estimates. *Water Resources Research, 48*(4). https://doi.org/10.1029/2011WR011453

Mehmood, K., Tischbein, B., Flörke, M., & Usman, M. (2022). Spatiotemporal analysis of groundwater storage changes, controlling factors, and management options over the transboundary Indus Basin. *Water, 14*(20), 20. https://doi.org/10.3390/w14203254

Pham-Duc, B., Sylvestre, F., Papa, F., Frappart, F., Bouchez, C., & Crétaux, J.-F. (2020). The Lake Chad hydrology under current climate change. *Scientific Reports, 10*(1), 5498. https://doi.org/10.1038/s41598-020-62417-w

Rodell, M. (2020). TELLUS_GLDAS-NOAH-3.3_TWS-ANOMALY_MONTHLY. Ver. 3.3. PO.DAAC, CA, USA. Dataset accessed [YYYY-MM-DD] at: https://doi.org/10.5067/GGDAS-3NH33.

Voss, K. A., Famiglietti, J. S., Lo, M. H., de Linage, C., Rodell, M., & Swenson, S. C. (2013). Groundwater depletion in the Middle East from GRACE with implications for transboundary water management in the Tigris-Euphrates-Western Iran region. *Water Resources Research, 49*(2), 904–914. https://doi.org/10.1002/wrcr.20078

Spatio-Temporal Variability and Trends of Rainfall in Bihar, India: Implications for Climate Extremes and Water Resources

Ravi Ranjan Kumar, Shahid Saleem, Javeed A. Rather, Talib Bashir Bhat, and Atiqulla Maalik

Abstract The trend analysis and variability of the annual and seasonal rainfall at the spatial and temporal scales were carried out for fifteen places in Bihar, India. Nonparametric tests such as the Mann-Kendall (MK) test and Sen's slope estimator were employed to determine the magnitude of the trends, while the Kriging interpolation technique mapped spatial variations. Although rainfall trends varied across sites, a significant majority exhibited increasing trends across annual, pre-monsoon, monsoon, and post-monsoon Seasons. The magnitude of the significant increases was estimated at 8.38 to 24.82 mm yr^{-1} for annual rainfall, 1.22 to 5.27 mm yr^{-1} for pre-monsoon rainfall, and 6.82 to 16.73 mm yr^{-1} for monsoon rainfall. The most likely years of change in observed annual and seasonal rainfall were shifted from 1988 to 2019 and 1997 to 2005, respectively. At spatial scales, the northeastern region of the state observed higher variations and high rainfall in the annual and monsoon seasons. During pre-monsoon and post-monsoon seasons, spatial variability was observed to be higher in the western and northern regions of Bihar, respectively. In contrast, high rainfall was observed in the eastern region of Bihar. Overall, this study shows that wide variation was noticed at different spatial scales in the case of annual and seasonal rainfall.

Keywords Mann-Kendall · Sen's slope · Trend analysis · Interpolation · Rainfall

R. R. Kumar
Horticulture College, Chaibasa, India

S. Saleem (✉) · J. A. Rather · A. Maalik
Department of Geography and Disaster Management, School of Earth and Environmental Sciences, University of Kashmir, Srinagar, India

J. A. Rather
e-mail: drjarather@uok.edu.in

T. B. Bhat
Division of Natural Resource Management, Faculty of Forestry, Sher-E-Kashmir University of Agricultural Sciences and Technology of Kashmir, Srinagar, India

R. Ahmed et al. (eds.), *Climate Change, Water Resources, and Natural Hazards in the Himalayas*, Advances in Geographical and Environmental Sciences,
https://doi.org/10.1007/978-981-92-0835-7_6

1 Introduction

Climate change represents one of the most critical environmental challenges of our time, evidenced by rising temperatures Bellard et al., 2012; Fischer & Knutti, 2013; Lindsey & Dahlman, 2020; Morak et al., 2013 and an increase in extreme events such as cyclones and floods (Holland & Bruyère, 2014; Kundzewicz et al., 2014; Moon et al., 2019; Walsh et al., 2016). reductions in snow cover (Mir et al., 2017; Wang et al., 2014), and escalating wildfires and desertification (Huang et al., 2020), with vulnerable regions being particularly impacted (IPCC 6th Assessment Report, 2022). In India, while temperatures have shown a consistent upward trend (Dash et al., 2007; Kumar et al., 1994; Srivastava, 2001), rainfall patterns display marked spatial variability, with some areas experiencing increases and others declines (Guhathakurta et al., 2011; Kripalani et al., 2003; Misra et al., 2018; Singh & Sontakke, 2002). Focusing on Bihar, this study examines the trends and variability of rainfall, aiming to better understand regional climatic impacts. Similar patterns of climatic variability and environmental change have also been observed in Himalayan regions, where shifts in land use, vegetation, and ecological conditions reflect broader climate-driven transformations (Mushtaq et al., 2024; Saleem et al., 2021). Additionally, changing climatic conditions have implications for ecosystem productivity and resource availability, as evidenced in studies on biomass dynamics and carrying capacity in mountain environments (Saleem, Rather, Ahmad, Ahmed, & Hajam, 2024; Saleem, Rather, Ahmed, Mushtaq, Ahmed, & Malik, 2024). The study of rainfall variability has become increasingly important in recent years due to the changing climate and its potential impact on global water resources (Dhakal et al., 2015; Maussion et al., 2013; Pascale et al., 2015). Understanding the spatial and temporal patterns of rainfall variability is crucial for developing effective strategies for water management and adaptation to climate change. The effects of climate change and its influence on the pattern of rainfall, temperature, humidity, wind speed, and direction are becoming increasingly obvious Pascale et al. (2015), and they pose a significant risk to the viability and consistency of all ecosystems (Wang et al., 2016). The consequence of climate change is already being felt in every region of the world. However, nations like India, are more susceptible than others because of their large populations, excessive demands placed on natural resources, and relatively ineffective coping mechanisms.

Much attention has been given to understanding past and recent climate change through the improvement and expansion of many datasets and the use of more sophisticated data analyses around the world (Ahmed et al., 2026). Pal and Mishra (2017) state that changes in the global climate could affect long-term patterns of rainfall, which could affect the amount of water available and make it more likely that droughts and floods will happen more often. Rainfall is the most critical essential physical parameter of climate because it affects a region's environment, Talaee et al. which in turn affects the amount of food that can be grown there. The distribution of rainfall (spatial and temporal) is essential for hydrology, water resource management, agriculture, and climate change policy and planning (Hammouri et al., 2015). The uneven distribution of rainfall results in a mismatch between water availability and demand;

irrigation structures are required to redistribute water according to the needs of a particular region. The frequency of extreme rainfall events is also a crucial design parameter for dams, spillways, flood protection structures, and drainage networks. Therefore, it is essential to learn about the spatial and temporal patterns of rainfall and their variability to understand the balance of water dynamics for water resources management and to plan strategies for solving many problems, such as predicting those caused by heavy rainfall. Studying the spatial pattern and variability of precipitation can help people plan for future water resources and contribute to efforts to keep an ecological system stable.

Bihar, where agriculture is the mainstay for millions, is highly sensitive to variations in rainfall. With nearly two-thirds of India's crop production depending on rainfall, particularly from the southwest monsoon that provides about 80% of the total rainfall (Rajendran et al., 2022), any deviation in rainfall patterns can significantly impact the region's agricultural output. Although national studies have reported no clear trend in mean annual rainfall (Das & Bhattacharya, 2018; Jain et al., 2012; Pingale et al., 2014), the eastern region, including Bihar, has received less attention. This study examines rainfall trends and variability across different parts of Bihar using nonparametric methods the Mann-Kendall test, Sen's slope estimator, and the MWP test which are advantageous for handling missing data, requiring fewer assumptions, and being independent of data distribution (Ali et al., 2019; Sarkar et al., 2022).

2 Methods and Materials

2.1 Study Area

The study area encompasses 15 locations within the state of Bihar, situated between latitudes 24°–27° N and longitudes 82°–88° E (Fig. 1). These 15 locations were selected to represent the diverse climatic and geographical conditions across Bihar, ensuring a comprehensive analysis of rainfall variability in the region. Bihar, located in the Indo-Gangetic plains, covers a total geographical area of 94,163 km^2. The climate of the region is characterized as subtropical to sub-humid, and the state is divided into three distinct agro-climatic zones: Zone-I (North-West Alluvial Plains), Zone-II (North-East Alluvial Plains), and Zone-III (South Alluvial Plains). The state receives an average annual rainfall of approximately 1235 mm, with the southwest monsoon season (June to September) contributing 75% of the total precipitation. Due to the considerable spatio-temporal variability of the Indian Summer Monsoon Rainfall (ISMR), Bihar frequently experiences both floods and droughts.

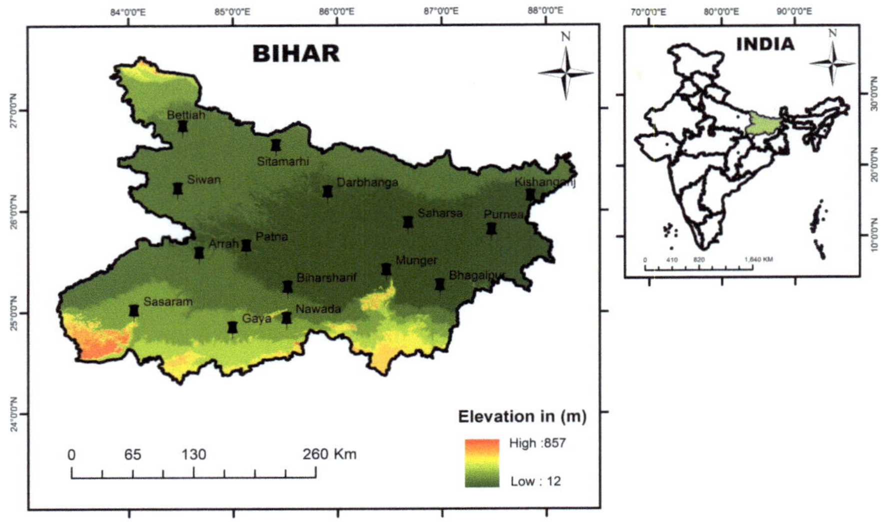

Fig. 1 Geographical location map of the study area

2.2 *Data and Preprocessing*

The data used in the present study consists of the gridded data of monthly rainfall (mm) for 15 locations selected to represent the diverse climatic and geographical conditions in Bihar during the period from January 1981 to December 2021. Further, the time series data sets were segregated into five data sets, viz. annual and seasonal, as characterized by the India Meteorological Department (IMD) guidelines: winter (Dec–Feb), pre-monsoon (Mar–May), monsoon (Jun–Sept), and post-monsoon (Oct–Nov). The data collected from NASA's prediction of worldwide energy resources (https://power.larc.nasa.gov/data-access-viewer/) provides gridded data at a moderate spatial resolution. Although this dataset offers extensive coverage, potential limitations include the spatial resolution that may not capture very localized rainfall variations. ArcGIS 10.8 and R.4.1.2 software have been used for mapping and data analysis purposes, whereas the Kriging technique was used for interpolating point values (Kumar et al., 2021). Kriging results help determine how average rainfall changes over time and space in study areas where data is unavailable or is not of good quality. To assess rainfall trends, nonparametric methods—namely, the Mann-Kendall test, Sen's slope estimator, and the Pettitt test—were utilized. These methods were chosen over parametric alternatives because they handle missing data more effectively, require fewer assumptions about data distribution, In particular, the Pettitt test is relevant for detecting abrupt change points in the time series, which is crucial for understanding shifts in rainfall patterns.

2.3 Trend Analysis

The statistical definition of a trend is a significant ± change in a random variable over time, which may be assessed using parametric and nonparametric tests. In this research, rainfall variables were evaluated using a nonparametric test. The MK test, Sen's Slope estimator, and MWP test were used to find trends in annual and seasonal rainfall time series data. Standard mathematical processes were utilized in the computing of descriptive statistics for the time series data. These statistics included the mean, standard deviation, coefficient of variation (CV), and skewness.

2.4 Mann–Kendall Test

The MK test (Kendall, 1975; Mann, 1945) has been used for identifying trends in rainfall data. The MK test statistics (S) can be computed as follows.

Null hypothesis (H_0): There is the absence of a trend.
Alternative hypothesis (H_1): There is the presence of a trend.

The MK test statistic (S):

$$S = \sum_{i=1}^{n-1} \sum_{j=i+1}^{n} \operatorname{sgn}(X_j - X_i), \tag{1}$$

where n is the number of the data points, X_i is the observed value at time i $(i = 1,2,..,n)$ and X_j is the observed value at time j $(j = i + 1, i + 2, \ldots, n)$. The value of $sgn(X_j - X_i)$ is computed according to the $sgn(X_j - X_i)$ as follows:

$$\operatorname{sgn}(X_j - X_i) = \begin{cases} +1, & (X_j - X_i) > 0 \\ 0, & (X_j - X_i) = 0 \\ -1, & (X_j - X_i) < 0 \end{cases}. \tag{2}$$

For samples with a size of 10 or more, the test is done using a normal approximation (Z statistics) with the following mean and variance:

$$E(S) = 0, \tag{3}$$

$$Var(S) = \frac{1}{18}\left[n(n-1)(2n+5) - \sum_{p=1}^{q} t_p(t_p - 1)(2t_p + 5) \right], \tag{4}$$

where q is the number of groups with identical values and t_p is the number of data values in the pth group. Test Z according to the following:

$$Z = \begin{cases} \frac{S-1}{\sqrt{Var(S)}} & if\ S > 0 \\ 0 & if\ S = 0 \\ \frac{S+1}{\sqrt{Var(S)}} & f\ S < 0 \end{cases}. \quad (5)$$

A statistically significant trend is assessed using the Z value, with a positive value indicating an increasing trend and a negative value indicating a decreasing trend.

2.5 Sen's Slope Estimator

Sen's nonparametric method has been used to estimate the true slope of an existing trend (as a change per year). Sen's nonparametric method has been used. Sen's method (Sen, 1968) can be used in cases where the trend can be assumed to be linear.

$$Q = Median\left(\frac{X_j - X_i}{j - i}\right), \quad (6)$$

Where X_j is the observed value at time j and X_i is the observed value at the time i. Q is the magnitude of the trend estimate. Sen's slope shows that positive values in the time series tend to go up, while negative values tend to go down.

2.6 Mann-Whitney-Pettitt (MWP) Test

The MWP test (Pettitt, 1979) distinguishes the most likely change year in the yearly time series sequence. The MWP test statistics can be computed as follows;

Assume a time series $\{X_1, X_2,\ldots, X_n\}$ with a length n. Two samples $\{X_1, X_2,\ldots, X_t\}$ and $\{X_{t+1}, X_{t+2},\ldots, X_n\}$, can then be derived by dividing the time series at time t. Where t is the time of the most likely change point. An index, U_t, is derived as:

$$U_t = \sum_{i=1}^{t} \sum_{j=1}^{n} \text{sgn}(X_i - X_j), \quad (7)$$

$$\text{sgn}(X_i - X_j) = \begin{cases} +1 & if\ (X_i - X_j) > 0 \\ 0 & if\ (X_i - X_j) = 0 \\ -1 & if\ (X_i - X_j) < 0 \end{cases}. \quad (8)$$

A plot of U_t against t for a time series with no change point would result in a continually increasing value of U_t. However, if there is a change point (even a local change point), then U_t would increase up to the change point and then begin to decrease. In a time series, this increase, followed by a decrease, may occur several times, indicating several local change points. So there is still the question of determining the most significant change point. We can identify the most significant change point K_t where the value of $|U_t|$ is maximum.

$$K_t = {}_{1\le t\le T}{}^{\max}|U_t|. \tag{9}$$

The approximated significance probability *P(t)* for a change point is given as:

$$P = 1 - \exp\left[\frac{-6K_t^2}{n^3 + n^2}\right]. \tag{10}$$

The change point is statistically significant at the time t with a significance level of α when probability *P(t)* exceeds 1—α.

3 Results and Discussions

The analysis of annual and seasonal rainfall patterns over 41 years highlights significant spatial and seasonal variability such as mean (mm), CV (%), and skewness, are presented in (Tables 1 and 2), across the regions. The average annual (1396.95 mm), pre-monsoon (200.07 mm), monsoon (1101.16 mm), and post-monsoon (80.35 mm) rainfall over 41 years was recorded maximum in Kishanganj, while the average minimum rainfall was recorded in Sasaram region for annual (894.78), pre-monsoon (38.76 mm), and post-monsoon (41.95 mm) rainfall. During the winter season, the average maximum rainfall was recorded in Sasaram (29.03 mm), and the minimum rainfall was recorded in Purnea (15.07 mm). The probability distribution analysis of the time series data reveals slight positive skewness for winter, monsoon, and annual rainfall, suggesting moderate variability in these periods. In contrast, the pre-monsoon and post-monsoon seasons exhibit pronounced positive skewness, indicating a higher frequency of extreme rainfall events or greater variability in these periods. This implies that, although annual rainfall trends may appear relatively stable, seasonal rainfall particularly during pre- and post-monsoon periods shows more dramatic fluctuations. Tables 1 and 2 provide detailed statistical parameters that support these findings, highlighting the seasonal dependency of rainfall variability and its potential implications for agricultural planning and water resource management in the region.

3.1 Variability and Trend of Annual Rainfall

Spatio-temporal variability of annual rainfall in Bihar as illustrated Fig. 2a and b, which ranges from 890 mm to 1400 mm with a moderate coefficient of variation (CV) between 18.69 and 35.00%. Annual rainfall is higher in the northeastern region and lower in the south-western region. Moreover, the CV analysis shows greater

Table 1 Geographical location with corresponding rainfall statistics

	Coordinates		Annual rainfall			Winter rainfall		
Locations	Latitude (N)	Longitude (E)	Mean (mm)	CV (%)	Skew	Mean (mm)	CV (%)	Skew
Arrah	25.555	84.675	945.325	22.511	0.267	26.537	81.518	0.436
Bhagalpur	25.238	86.982	964.811	24.051	0.432	18.036	78.023	0.715
Biharsharif	25.218	85.521	968.842	20.270	−0.022	26.794	84.129	0.598
Bettiah	26.807	84.521	1037.209	34.999	0.802	27.223	87.889	0.997
Darbhanga	26.158	85.905	994.047	26.620	0.732	22.582	82.775	0.354
Gaya	24.820	84.993	970.315	20.375	−0.001	28.015	85.217	0.706
Kishanganj	26.119	87.851	1396.950	32.712	0.531	15.347	85.984	1.654
Munger	25.387	86.465	947.633	22.029	0.435	20.616	82.739	0.665
Nawada	24.912	85.510	968.842	20.270	−0.022	26.794	84.129	0.598
Patna	25.625	85.125	968.842	23.598	−0.022	26.794	81.897	0.598
Purnea	25.783	87.476	1135.898	29.708	0.403	15.070	79.123	0.897
Sasaram	24.989	84.048	894.788	18.696	−0.136	29.034	87.554	1.009
Saharsa	25.852	86.673	972.591	34.552	1.280	15.846	81.378	0.578
Sitamarhi	26.616	85.410	1009.680	29.546	0.799	22.583	77.424	0.495
Siwan	26.191	84.470	962.670	25.002	0.637	24.394	80.067	0.261

CV = Coefficient of variation, Skew = Skewness, mm = millimetre

variability in the northeastern and north-western regions, whereas the central and south-western areas exhibit more stable rainfall patterns. Consequently, the north-eastern region receives considerably more rainfall than the central region, possibly due to rapid urbanization and increased greenhouse gas emissions in the central area that alter local climatic conditions. The trend analysis reveals a mixture of increasing and decreasing trends across Bihar. Notably, several locations—namely Bhagalpur, Bettiah, Kishanganj, Purnea, Saharsa, and Sitamarhi—exhibit highly significant increasing trends, as clearly visualized in Fig. 3. The magnitude of these increases varies widely: Bettiah shows an increase of 17.93 mm year^{-1}, Sitamarhi 8.38 mm year^{-1}, Kishanganj 24.82 mm year^{-1}, Bhagalpur 10.97 mm year^{-1}, Saharsa 13.33 mm year^{-1}, and Purnea 15.10 mm year^{-1}. Furthermore, the analysis (see Table 3) suggests that the most probable change point in the observed rainfall occurred between 1988 and 2019.

3.2 *Variability and Trend of Winter Rainfall*

The seasonal spatial and temporal variability of winter rainfall is depicted in Fig. 4a and b. The region's winter rainfall ranged from 15.070 mm in Purnea to 29.034 mm in Sasaram, with a very high CV (> 70%). The amount of winter rainfall was higher

Table 2 Geographical location with corresponding rainfall statistics

Locations	Pre-monsoon rainfall			Monsoon rainfall			Post-monsoon rainfall		
	Mean (mm)	CV (%)	Skew	Mean (mm)	CV (%)	Skew	Mean (mm)	CV (%)	Skew
Arrah	50.921	79.576	1.393	823.504	24.745	0.467	44.361	93.082	1.332
Bhagalpur	101.828	70.777	1.065	781.132	22.370	0.097	63.813	83.670	1.109
Biharsharif	52.212	77.055	1.679	838.285	22.144	0.058	51.550	87.398	1.235
Bettiah	87.496	82.064	1.277	877.750	33.204	0.506	44.738	101.431	1.599
Darbhanga	64.510	85.664	1.634	862.687	27.761	0.721	44.263	97.588	1.439
Gaya	50.902	78.456	1.821	842.431	22.420	0.060	48.965	90.383	1.208
Kishanganj	200.078	51.512	0.781	1101.166	31.066	0.676	80.357	98.341	1.047
Munger	64.462	82.506	1.286	812.997	22.816	0.173	49.552	90.222	1.342
Nawada	52.212	77.055	1.679	838.285	22.144	0.058	51.550	87.398	1.235
Patna	52.212	81.344	1.679	838.285	25.750	0.058	51.550	93.534	1.235
Purnea	149.808	52.967	0.715	911.048	28.309	0.443	59.973	99.003	1.109
Sasaram	38.761	76.528	1.542	785.038	20.257	0.131	41.952	88.431	1.415
Saharsa	108.894	71.466	1.146	797.838	32.809	1.159	50.010	99.935	1.543
Sitamarhi	79.662	82.849	1.408	863.728	29.133	0.697	43.703	98.182	1.813
Siwan	56.779	80.428	1.546	839.132	26.755	0.785	42.360	95.705	1.320

CV = Coefficient of Variation, Skew = Skewness, mm = millimetre

in the north-western to south-western regions of Bihar. In addition, the CV of winter rainfall showed that the variability was high in the north-western region, the south-western region, and some parts of the northeastern region. The results of the MK test with Sen's slope estimator at the spatial distribution of temporal trends showed statistically insignificant decreasing trends during the study period (Table 3 and Fig. 8d). The magnitude of the decreasing trends was quite lower in all the study areas and the most likely year of change in the observed rainfall was shifted from 1997 to 1998 (Figs. 5, 6 and 7).

3.3 *Variability and Trend of Pre-monsoon Rainfall*

Spatio-temporal variability of pre-monsoon rainfall has represented in Fig. 5a and b. The pre-monsoon rainfall of the region varied from 38.761 mm in Sasaram to 200.078 mm in Kishanganj, with a high CV ranging from 51.512 to 85.664%. The amount of rainfall was observed to be higher in the northeastern region, while a lower amount of rainfall was observed in the regions of the central and western. The variability of rainfall was higher in the central and north-western regions, whereas lower variability was observed in the northeastern region. After the trend analysis, the test results showed statistically highly significant increasing trends except for

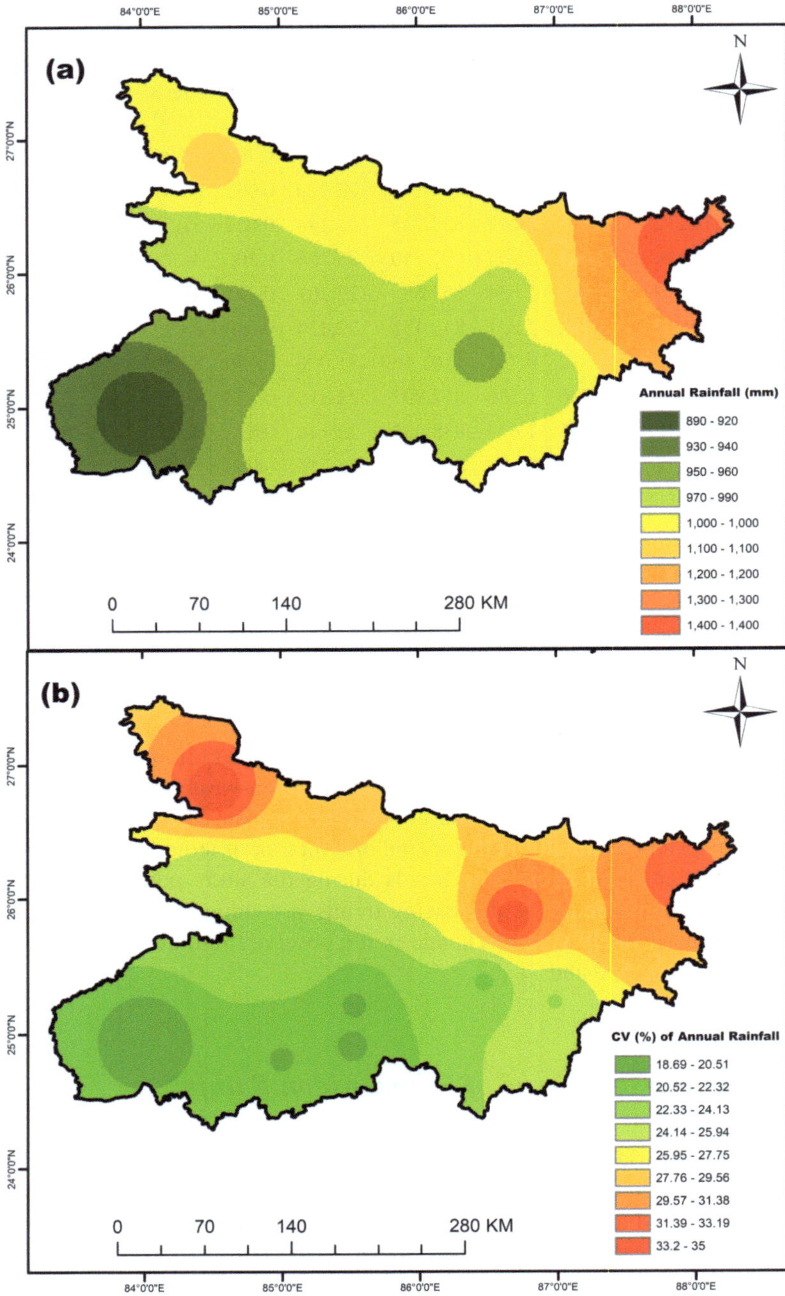

Fig. 2 **a** and **b** Distribution of annual rainfall in Bihar **a** spatial variability and **b** co-efficient of variation

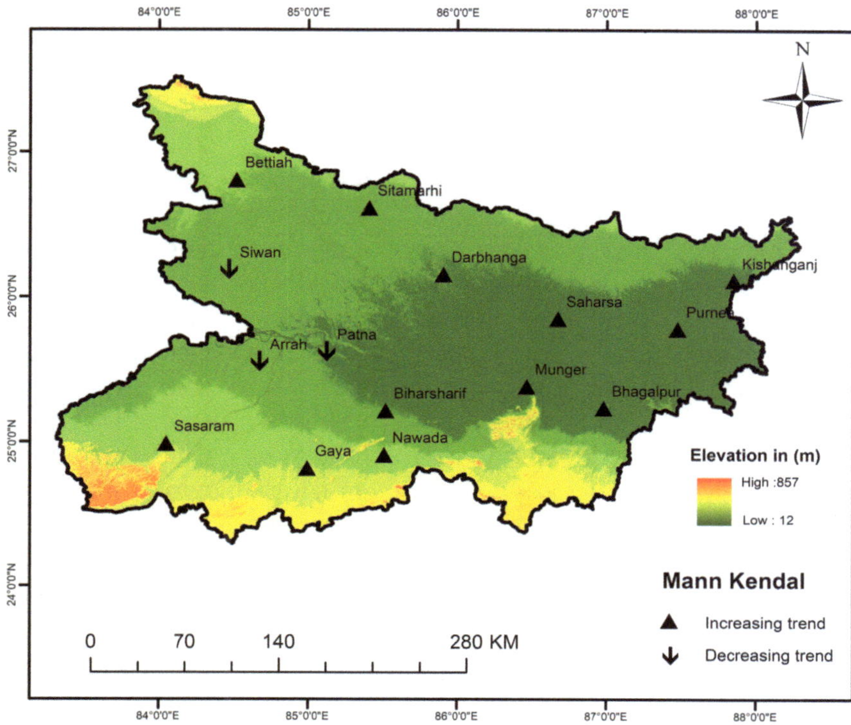

Fig. 3 Trends in annual rainfall in Bihar

Sasaram, which can be easily visualized in Fig. 8a. The significant increase in pre-monsoon rainfall trends may aid in early Kharif crop cultivation and reduce the need for irrigation. The magnitude of the significant increasing trend varied from 1.217 mm year^{-1} in Arrah to 5.274 mm year^{-1} in Kishanganj from 1981 to 2021, and the most probable year of change in the observed pre-monsoon rainfall was shifted to 1998 (Table 3).

3.4 Variability and Trend of Monsoon Rainfall

The spatial and temporal variation in monsoon rainfall has represented in Fig. 6a and b. The amount of monsoon rainfall in the Bihar region varied from 781.132 mm in Bhagalpur to 1101.166 mm in Kishanganj, with a medium CV ranging from 22.144% to 33.204%. The highest rainfall was observed in the region of north-eastern Bihar, while a lower amount of monsoon rainfall was observed in the central and south-western regions. The variability of monsoon rainfall was higher in the

Table 3 The trend results of annual and seasonal rainfall at different locations in Bihar

Locations	Annual rainfall			Winter rainfall			Pre-monsoon rainfall		
	MK test (Z)	Sen's slope	MWP test	MK test (Z)	Sen's slope	MWP test	MK test (Z)	Sen's slope	MWP test
Arrah	−0.382	−1.718	1988	−1.332	−0.310	1997	2.453	1.217	1998
Bhagalpur	3.607	10.970	2003	−1.168	−0.207	1998	3.754	3.516	1998
Biharsharif	0.236	0.753	2019	−1.219	−0.377	1998	2.819	1.318	1998
Bettiah	3.719	17.930	2003	0.045	0.000	2018	3.475	3.120	2003
Darbhanga	0.348	1.479	1988	−0.852	0.000	2008	3.160	2.109	1998
Gaya	0.360	0.962	2005	−1.014	−0.277	1997	2.430	1.054	1998
Kishanganj	4.224	24.815	1997	−0.330	0.000	1982	4.022	5.274	1997
Munger	1.224	3.920	2005	−1.074	−0.235	1997	3.286	2.176	1998
Nawada	0.236	0.753	2019	−1.219	−0.377	1998	2.819	1.318	1998
Patna	−0.101	−0.550	1988	−1.287	−0.310	1997	2.693	1.388	1998
Purnea	3.191	15.100	2003	−1.658	−0.229	1997	3.486	3.514	2006
Sasaram	0.438	0.950	2010	−1.230	−0.346	1997	1.926	0.649	1996
Saharsa	3.303	13.329	2003	−1.552	−0.235	1997	3.225	3.369	2001
Sitamarhi	2.067	8.377	2003	−1.298	−0.285	2008	3.485	2.784	2000
Siwan	−0.180	−0.682	1988	−0.782	0.000	1997	3.038	1.572	1998

northeast and northwest regions, while lower variability was observed in the southwestern region. After the trend analysis of monsoon rainfall, the results revealed that a mixture of increasing and decreasing trends was detected in different regions (Fig. 8b). The significant increasing trend was found in some places like Bhagalpur (6.818 mm year^{-1}), Bettiha (14.580 mm year^{-1}), Kishanganj (16.725 mm year^{-1}), Purnea (9.765 mm year^{-1}), and Saharsa (9.708 mm year^{-1}) from 1981 to 2021. The most likely years of change in observed monsoon rainfall were 2003 and 2005 (Table 4). These findings suggest that the monsoon rainfall in the Bihar region is highly variable, and the trend analysis results show that the rainfall patterns have been changing over the years. The observed increasing trend in monsoon rainfall in some regions of Bihar is consistent with previous studies that have reported increasing trends in monsoon rainfall in different parts of India. These findings can be attributed to several factors, including climate change, land use change, and urbanization. However, the decreasing trend observed in other regions could be due to natural variability or other local factors.

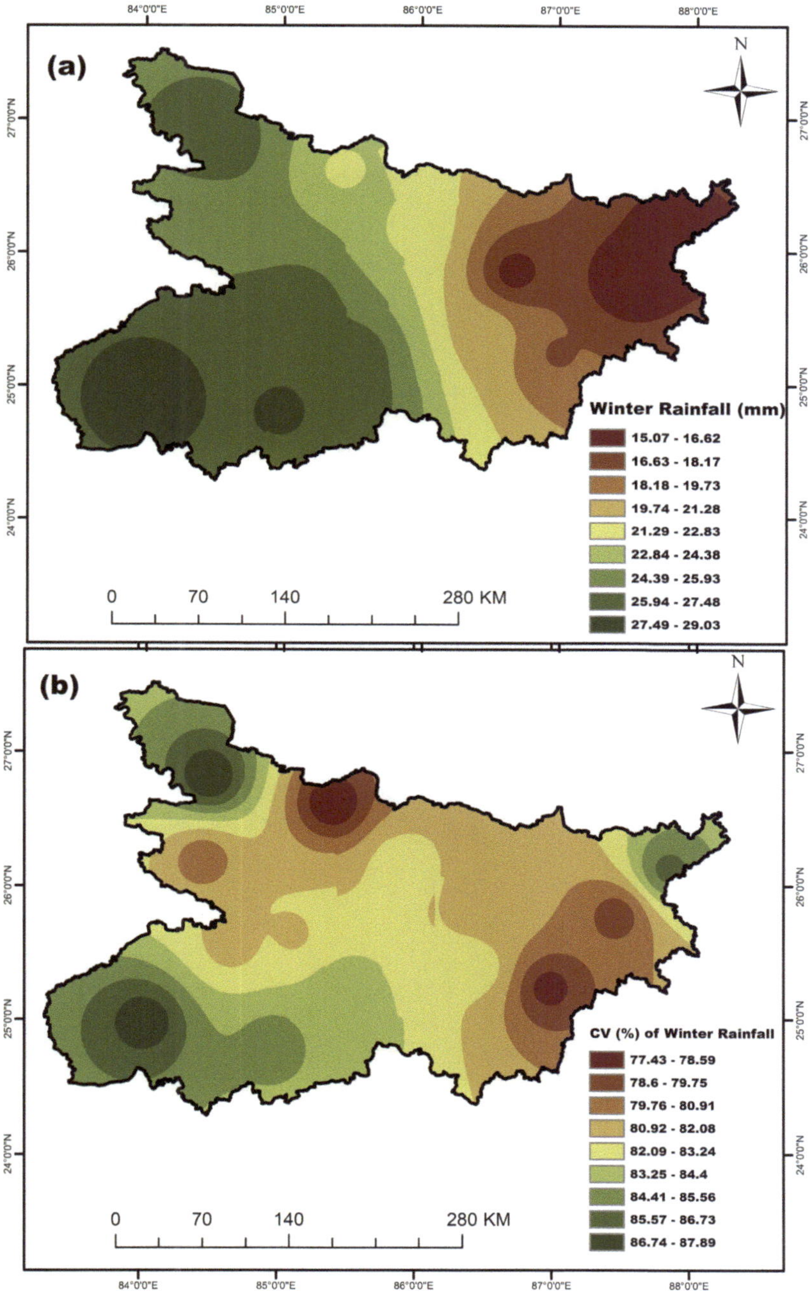

Fig. 4 **a** and **b** Distribution of winter rainfall in Bihar **a** spatial variability and **b** co-efficient of variation

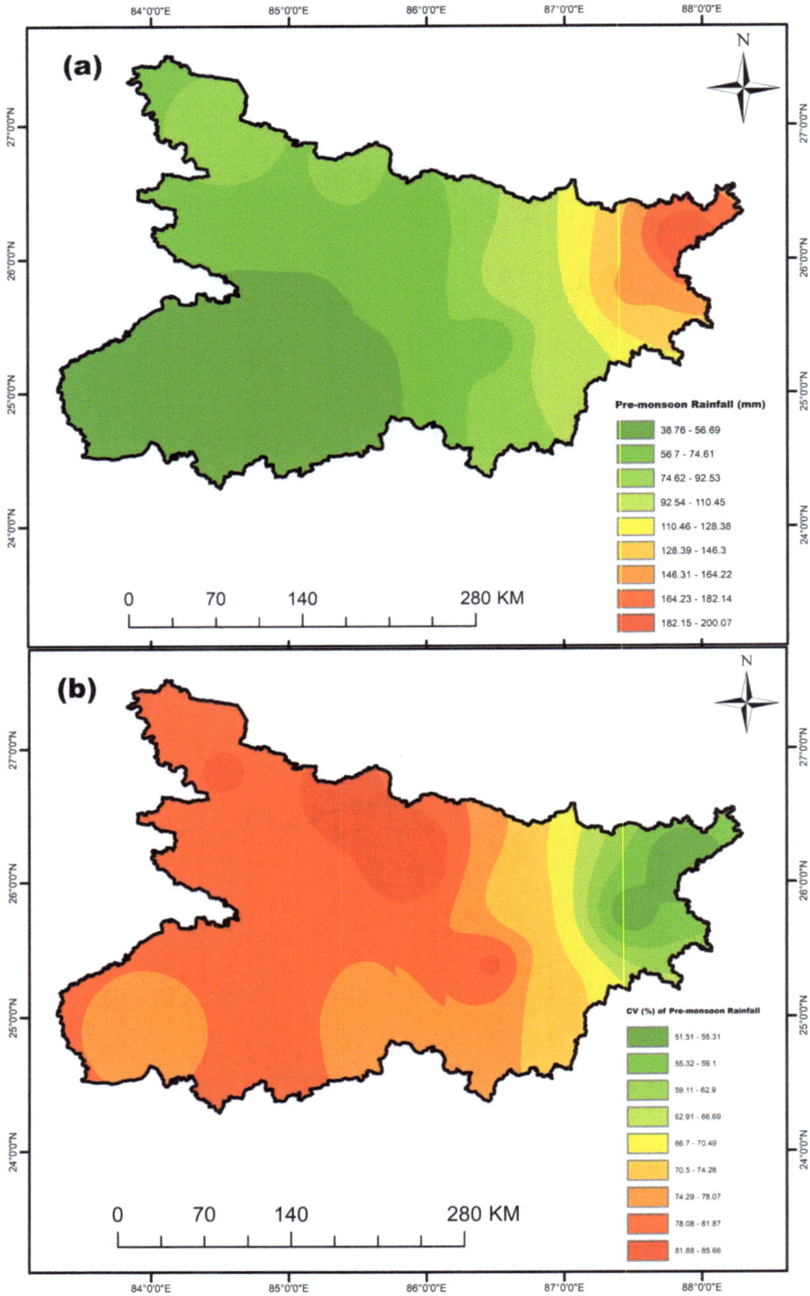

Fig. 5 **a** and **b** Distribution of pre-monsoon rainfall in Bihar **a** *spatial variability and* **b** *co-efficient of variation*

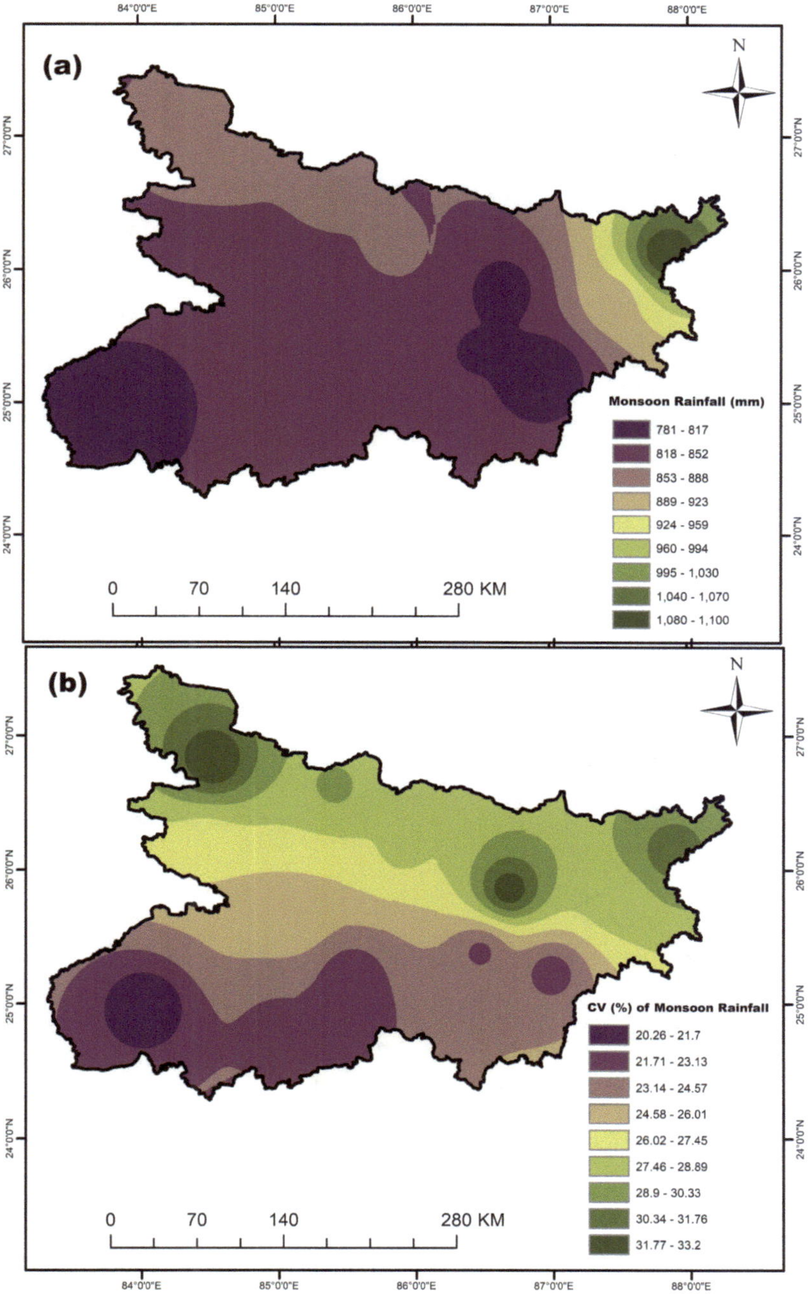

Fig. 6 **a** and **b** Distribution of monsoon rainfall in Bihar **a** *spatial variability and* **b** *co-efficient of variation*

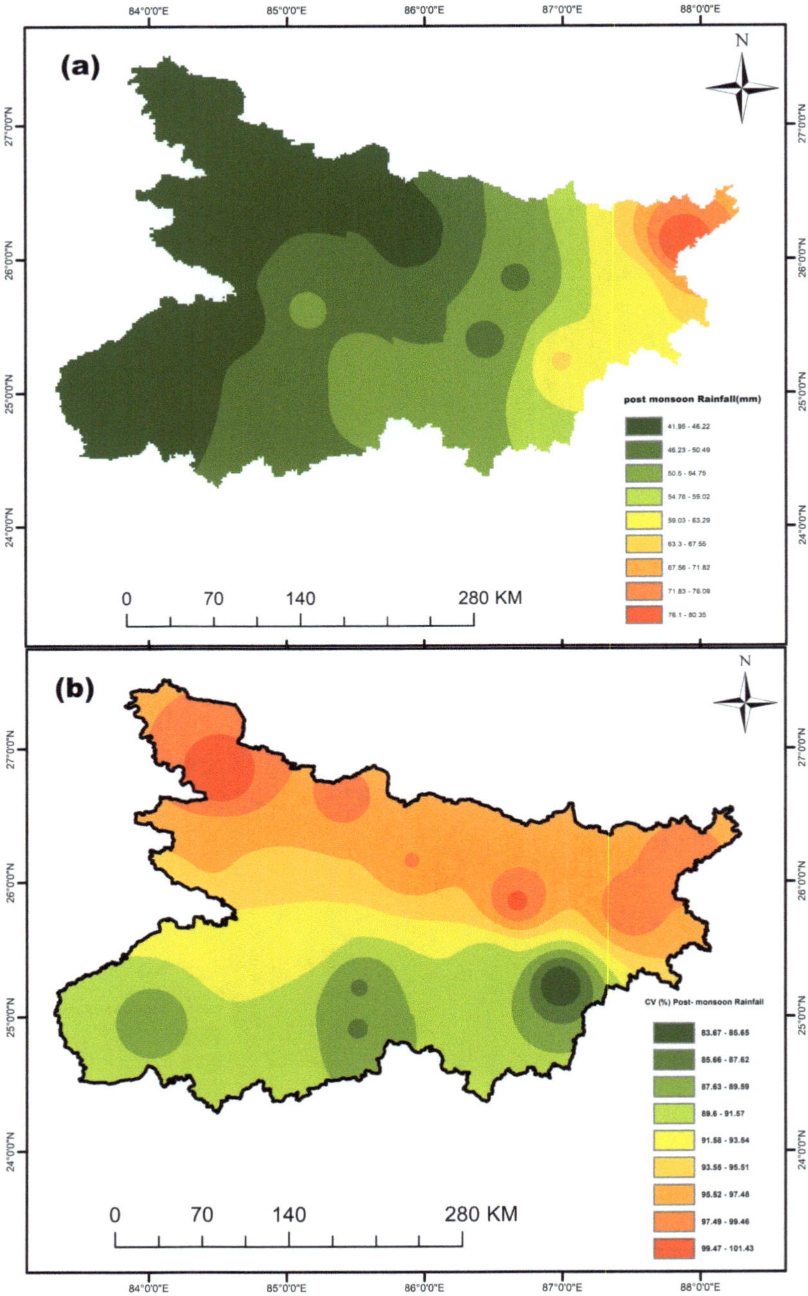

Fig. 7 **a** and **b** Distribution of post-monsoon rainfall in Bihar **a** *spatial variability and* **b** *co-efficient of variation*

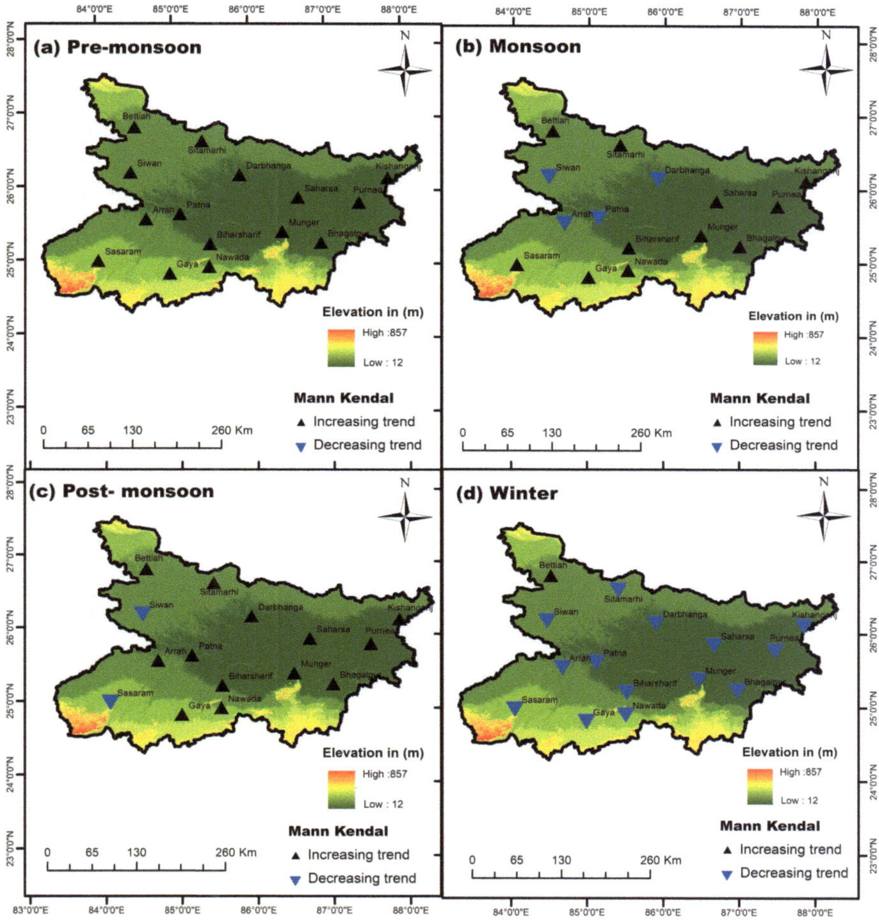

Fig. 8 **a–d** The trend of seasonal rainfall in *Bihar* **a** *pre-monsoon,* **b** *monsoon,* **c** *post-monsoon and* **d** *winter*

3.5 *Variability and Trend of Post-monsoon Rainfall*

The amount of post-monsoon rainfall in the Bihar region varied from 41.951 mm in Sasaram to 80.357 mm in Kishanganj, with a very high CV ranging from 83.67% to 101.43%. The highest post-monsoon rainfall was observed in the eastern region of Bihar, while the lowest rainfall was observed in the western area (Fig. 7a). The variability of post-monsoon rainfall was higher in the northern region, while lower variability was observed in the southern region (Fig. 7b). The trend analysis of post-monsoon rainfall found the majority of the regions showed non-significant increasing trends. However, Kishanganj showed a significant increasing trend at a rate of 1.545 mm year^{-1} during the study period (Fig. 8c). The most probable

Table 4 The trend of seasonal rainfall at different locations in Bihar

Locations	Monsoon rainfall			Post-monsoon rainfall		
	MK test (*Z*)	Sen's slope	MWP test	MK test (*Z*)	Sen's slope	MWP test
Arrah	−0.685	−2.466	2008	0.101	–0.753	2014
Bhagalpur	2.889	6.818	2003	1.868	0.974	1997
Biharsharif	0.258	0.880	2005	0.630	0.341	1992
Bettiah	3.640	14.580	2003	0.653	0.199	1994
Darbhanga	−0.056	−0.006	1988	0.349	0.000	1994
Gaya	0.337	0.905	2005	0.258	0.000	1992
Kishanganj	3.966	16.725	2003	2.139	1.545	1997
Munger	0.932	2.501	2005	1.114	0.527	1994
Nawada	0.258	0.880	2005	0.630	0.341	1992
Patna	−0.326	−1.485	2008	0.236	0.000	1993
Purnea	2.965	9.765	2003	1.812	0.983	1997
Sasaram	0.348	0.822	2010	−0.078	0.000	2003
Saharsa	2.786	9.708	2003	1.655	0.674	1997
Sitamarhi	1.427	5.273	2003	0.157	0.000	1994
Siwan	−0.663	−1.960	1988	−0.248	0.000	2014

year of change in the observed post-monsoon rainfall was shifted from 1992 to 1997 (Table 4).

4 Conclusions

The study provides valuable insights into the spatio-temporal trends and variability of rainfall in Bihar from 1981 to 2021. The finding shows a mixture of increasing and decreasing rainfall trends at different places. In the case of pre-monsoon rainfall, all the study areas showed statistically significant increasing trends except for Sasaram, and the magnitude of the significant increasing trend varied from 1.217 mm year^{-1} to 5.274 mm year^{-1}. During annual, monsoon, and post-monsoon rainfall, the majority of the places also showed increasing trends. However, during winter rainfall, a decreasing trend was detected during the study period. The magnitude of significant increasing trends in annual and monsoon rainfall varied from 8.377 mm year^{-1} to 24.815 mm year^{-1} and 6.818 mm year^{-1} to 16.725 mm year^{-1}, respectively. The most likely years for annual and seasonal rainfall changes were shifted between 1988 to 2019 and 1997 to 2005, respectively. In the case of spatial variability, the northeastern region showed high variation and high rainfall during the annual and monsoon seasons, whereas it was quite low in the south-western region of Bihar. The amount of rainfall was shown to be higher in the eastern region during pre-monsoon and pot-monsoon, whereas the variability of rainfall was shown to be higher in the western region for pre-monsoon and the northern region for post-monsoon. Overall, most of the place's increasing trend in rainfall magnitude was found in Bihar from 1981–2021, except for winter rainfall. However, the spatial variability was quite high, and it has significant implications for water resource management and agricultural planning, underscoring the need for region-specific adaptation strategies. Future research should address these limitations by incorporating finer-scale data, exploring the underlying drivers of rainfall variability in greater depth, and assessing its impacts across different sectors such as agriculture, water resource management, and urban development to better inform effective adaptation strategies.

Acknowledgements We are grateful to the USGS, Open-source Data used in this study.

Funding The authors declare that no funds, grant or other support were received during the preparation of this manuscript.

Author Contributions All authors contributed to the study conception and design. Material preparation, data collection and analysis were performed by Ravi Ranjan Kumar, and Shahid Saleem. The first draft of the manuscript was written by Ravi Ranjan Kumar, Shahid Saleem and Javeed A Rather all authors commented on previous versions of the manuscript. All authors read and approved the final manuscript.

Data Availability All data are available online as provided in references.

Code Availability The codes written for the current study are available on reasonable request from the corresponding author.

Consent to Participate Not applicable.

Consent for Publication Not applicable.

Conflict Interest The authors declare that they have no conflict of interest.

Ethical Approval The authors confirm that this work is original and has not been published elsewhere nor is it currently under consideration for publication elsewhere.

References

Ahmed, R., Sam, L., Bhardwaj, A., Kulkarni, A. V., & Yamasaki, S. (2026). Increasing global threat of outburst floods from overlooked small alpine lakes. *Nature Sustainability*, 1–10.

Ali, R., Kuriqi, A., Abubaker, S., & Kisi, S. (2019). Long-term trends and seasonality detection of the observed flow in Yangtze River using Mann-Kendall and Sen's innovative trend method. *Water, 11*, 185. https://doi.org/10.3390/w11091855

Bellard, C., Bertelsmeier, C., Leadley, P., Thuiller, W., & Courchamp, F. (2012). Impacts of climate change on the future of biodiversity. *Ecology Letters, 15*(4), 365–377.

Das, J., & Bhattacharya, S. K. (2018). Trend analysis of long-term climatic parameters in Dinhata of Koch Bihar District, West Bengal. *Spatial Information Research, 3*, 1–10. https://doi.org/10.1007/s41324-018-0173-3

Dash, S. K., Jenamani, R. K., Kalsi, S. R., & Panda, S. K. (2007). Some evidence of climate change in twentieth-century India. *Climatic Change, 85*(3–4), 299–321.

Dhakal, N., Jain, S., Gray, A., Dandy, M., & Stancioff, E. (2015). Nonstationarity in seasonality of extreme precipitation: A nonparametric circular statistical approach and its application. *Water Resources Research, 51*(6), 4499–4515. https://doi.org/10.1002/2014WR016292

Fischer, E. M., & Knutti, R. (2013). Robust projections of combined humidity and temperature extremes. *Nature Climate Change, 3*(2), 126–130.

Guhathakurta, P., Sreejith, O. P., & Menon, P. A. (2011). Impact of climate change on extreme rainfall events and flood risk in India. *Journal of Earth System Science, 120*(3), 359–373.

Hammouri, N., Al-Qinna, M., Salahat, M., Adamowski, J., & Prasher, S. O. (2015). Community based adaptation options for climate change impacts on water resources: The case of Jordan. *Journal of Water and Land Development, 26*, 3–17.

Holland, G., & Bruyère, C. L. (2014). Recent intense hurricane response to global climate change. *Climate Dynamics, 42*, 617–627.

Huang, J., Yu, H., Dai, A., Wei, Y., & Kang, L. (2020). Drylands face potential threat under 2°C global warming target. *Nature Climate Change, 7*(2), 123–129.

IPCC. (2022). *Sixth assessment report of the intergovernmental panel on climate change.*

Jain, S. K., Kumar, V., & Saharia, M. (2012). Analysis of rainfall and temperature trends in Northeast India. *International Journal of Climatology, 33*(4), 968–978. https://doi.org/10.1002/joc.3483

Kendall, M. G. (1975). *Rank correlation methods.* Griffin.

Kripalani, R. H., Kulkarni, A., Sabade, S. S., & Khandekar, M. L. (2003). Indian monsoon variability in a global warming scenario. *Natural Hazards, 29*(2), 189–206.

Kumar, A., Giri, R. K., Taloor, A. K., & Singh, A. K. (2021). Rainfall trend, variability and changes over the state of Punjab, India 1981–2020: A geospatial approach. *Remote Sensing Applications: Society and Environment, 23*, 100595.

Kumar, K. R., Kumar, K. K., & Pant, G. B. (1994). Diurnal asymmetry of surface temperature trends over India. *Geophysical Research Letters, 21*(8), 677–680.

Kundzewicz, Z. W., Kanae, S., Seneviratne, S. I., Handmer, J., Nicholls, N., Peduzzi, P., & Sherstyukov, B. (2014). Flood risk and climate change: Global and regional perspectives. *Hydrological Sciences Journal, 59*(1), 1–28.

Lindsey, R., & Dahlman, L. (2020). Climate change: Global temperature. *Climate.gov, 16*, 1–5.

Mann, H. B. (1945). Nonparametric tests against trend. *Econometrica, 13*, 245–259.

Maussion, F., Scherer, D., Mölg, T., Collier, E., Curio, J., & Finkelnburg, R. (2013). Precipitation seasonality and variability over the Tibetan plateau as resolved by the high Asia reanalysis. *Journal of Climate, 27*(6), 1910–1927. https://doi.org/10.1175/JCLI-D-13-00282.1

Mir, R. A., Jain, S. K., Thayyen, R. J., & Saraf, A. K. (2017). Assessment of recent glacier changes and its controlling factors from 1976 to 2011 in Baspa basin, western Himalaya. *Arctic, Antarctic, and Alpine Research, 49*(4), 621–647.

Misra, A. K., Goswami, P., & Prajapat, K. (2018). Impact of climate change on rainfall pattern and water resources in India. *Journal of Earth System Science, 127*(3), 1–10.

Moon, I. J., Kim, S. H., & Chan, J. C. (2019). Climate change and tropical cyclone trend. *Nature, 570*(7759), E3–E5.

Morak, S., Hegerl, G. C., & Christidis, N. (2013). Detectable changes in the frequency of temperature extremes. *Journal of Climate, 26*(5), 1561–1574.

Mushtaq, S., Saleem, S., Ahmed, R., Tass, M. S., Rather, J. A., & Rather, G. M. (2024). Spatio-temporal analysis of land use land cover changes in South Kashmir region of North-Western Himalayas using landsat data. *Discover Geoscience, 2*(1), 37.

Pal, A. B., & Mishra, P. K. (2017). Trend analysis of rainfall, temperature and runoff data: A case study of Rangoon watershed in Nepal. *International Journal of Students' Research in Technology & Management, 5*(3), 21–38.

Pascale, S., Lucarini, V., Feng, X., Porporato, A., & Hasson, S. (2015). Analysis of rainfall seasonality from observations and climate models. *Climate Dynamics, 44*(11–12), 3281–3301. https://doi.org/10.1007/s00382-014-2256-6

Pettitt, A. N. (1979). A non-parametric approach to the change point problem. *Journal of the Royal Statistical Society. Series C, Applied Statistics, 28*, 126–135. https://doi.org/10.2307/2346729

Pingale, S. M., Khare, D., Jat, M. K., & Adamowski, J. (2014). Spatial and temporal trends of mean and extreme rainfall and temperature for the Rajasthan, India. *Atmospheric Research, 138*, 73–90.

Rajendran, K., Surendran, S., Varghese, S. J., & Sathyanath, A. (2022). Simulation of Indian summer monsoon rainfall, interannual variability and teleconnections: Evaluation of CMIP6 models. *Climate Dynamics, 58*(9–10), 2693–2723.

Saleem, S., Ahmed Hajam, F., & Rather, J. A. (2021). Spatio-temporal analysis of land use land cover changes in Sind catchment of the Kashmir Valley, India. *Geo-Eye, 10*(2), 35–42.

Saleem, S., Rather, J. A., Ahmad, S., Ahmed, R., & Hajam, L. A. (2024). Assessment of livestock carrying capacity in the alpine grasslands of the Kashmir Himalayas. *Pastoralism: Research, Policy and Practice, 14*, 13541.

Saleem, S., Rather, J. A., Ahmed, S., Mushtaq, S., Ahmed, R., & Malik, I. H. (2024). Above ground biomass estimation for alpine grasslands of Kashmir Himalayas using remote sensing and field-data. *Rangeland Ecology and Management, 96*, 117–127.

Sarkar, K. A., Kumar, R. R., Dhakre, D. S., & Bhattacharya, D. (2022). Analysis of spatio-temporal trend of solar radiation in the agro climatic zone-I of Bihar. *The Pharma Innovation Journal, SP-11*(9), 431–438.

Sen, P. K. (1968). Estimates of the regression coefficient based on Kendall's tau. *Journal of the American Statistical Association, 63*, 1379–1389.

Singh, N., & Sontakke, N. A. (2002). On climatic fluctuations and environmental changes of the Indo-Gangetic plains, India. *Climatic Change, 52*(3), 287–313.

Srivastava, P. (2001). Paleoclimatic implications of pedogenic carbonates in Holocene soils of the Gangetic Plains, India. *Palaeogeography, Palaeoclimatology, Palaeoecology, 172*(3–4), 207–222.

Walsh, K. J., McBride, J. L., Klotzbach, P. J., Balachandran, S., Camargo, S. J., Holland, G., & Sugi, M. (2016). Tropical cyclones and climate change. *WIREs Climate Change, 7*(1), 65–89.

Wang, H., Zhou, S., Li, X., Liu, H., Chi, D., & Xu, K. (2016). The influence of climate change and human activities on ecosystem service value. *Ecological Engineering, 87*, 224–239. https://doi.org/10.1016/j.ecoleng.2015.11.027

Wang, W., Huang, X., Deng, J., Xie, H., & Liang, T. (2014). Spatio-temporal change of snow cover and its response to climate over the Tibetan plateau based on an improved daily cloud-free snow cover product. *Remote Sensing, 7*(1), 169–194.

A Perspective on Himalayan Wetland as Asset and Threat to Community

Lakpa Tamang, Rajesh Joshi, and Prakash Chhetri

Abstract Himalayan wetland support souls of million populations from glaciated valleys to river mouths. The wetland acts as the mediator between the resource and beneficiaries. The life sustaining resource water, begin to descends from these wetlands in the forms of channel and feed million of population in a process of one complete hydrological cycle. Further, wetland functions as carbon storage unit, water filtration, supports diverse biodiversity, livelihood etc. The central discussion of this paper circles around the wetland resources of Sikkim Himalaya and the related hazards over the decades. From 1968 to 2023, 64 lives have been taken by such hazards which are likely to intensify in coming years with changing climatic patterns over Himalayan belt. Wetland resource map, stream order map and NDSI map of Sikkim has been prepared to understand its importance and urging different stakeholders for the urgent monitoring and preservation need of mountain wetlands.

Keywords Wetland · Hazards · Teesta basin

1 Introduction

Wetlands are specialized ecosystems where soil is covered or saturated by the water and supports a unique vegetational growth (Moor et al., 2017). Wetlands are "Areas of marsh, fen, peatland or water, whether natural or artificial, permanent or temporary, with water that is static or flowing, fresh, brackish or salt, including areas of marine water the depth of which at low tide does not exceed six meters" as defined by Ramsar convention (Matthews, 1993; Chakraborty et al., 2023). Wetlands can be formed in diverse environmental condition from tropics to arctic irrespective of altitudes and latitudes. Globally wetland covers 6% of total land surface area (Maltby & Barker, 2009). India's 4.5% of geographical area is covered by more than 7 lakhs of wetlands. Wetland offer range of ecosystem services from breeding grounds to many animal

L. Tamang (✉) · R. Joshi · P. Chhetri
Sikkim Regional Centre, G.B Pant National Institute of Himalayan Environment, Gangtok, India
e-mail: dorjeelakpatamang@gmail.com

R. Ahmed et al. (eds.), *Climate Change, Water Resources, and Natural Hazards in the Himalayas*, Advances in Geographical and Environmental Sciences,
https://doi.org/10.1007/978-981-92-0835-7_7

species, groundwater aquifer recharge to aesthetic values to the landscape for tourist attraction (Sharma & Naik, 2024). As per the available inventories, only 0.8% of the country's total wetlands (lakes) is located in high-altitude (above 3000 m) areas of the Indian Himalayan region (Sharma et al., 2020).

In Sikkim a total of 526 lacustrine wetlands have been geo-tagged by state government. Some of these wetlands cover 0.08 ha area while some expands over 132.38 ha of area (O'Neill, 2019). It has been observed that over the decade's glacier in Sikkim Himalay is receding at much faster rate and leading to formation and concentration of many vulnerable lakes in high altitude areas (Shukla et al., 2018). During 2003, 266 lakes having an area more than 0.1 km^2 were identified and designated 14 lakes as potential threat with 3 lakes having past GLOF instances in Sikkim (Shukla et al., 2018). The recent 2023 Sikkim GLOF is the wakening call for the community about the losing water reserves from the mountain top and proper developmental activity needs to be carried out along river channel after area vulnerable assessment to cope with yearly unfolding flood hazards (Ahmed et al., 2025). With glacier retreat number of new lakes have emerged in high elevated areas having potential to recharge ground water table, to support stream flow and to swipe the valley bottom if the pressure is release at once due to sudden cloud burst, moraine dam failure etc. (Gurung et al., 2017). Therefore, these are emerging challenges and need to address in our developmental activities to minimize the impact in this eco-sensitive zone.

2 Study Area

This study is conducted in the eastern Himalayan state of Sikkim, located in north-eastern India between 27.08°–28.13° N latitude and 88.08°–88.92° E longitude (Fig. 1). The state occupies an area of roughly 7096 km^2 and extends approximately 114 km along a north–south axis and about 64 km east–west. Elevation varies dramatically across this relatively small spatial extent, rising from nearly 250 m in the southern foothills to 8598 m in the northern high mountain region (Subba, 2008).

A pronounced altitudinal gradient defines the regional landscape, with nearly half of the territory characterized by steep slopes and dissected, rugged terrain. This sharp elevational transition exerts strong control over climatic conditions. Lower elevations in the southern part of the state experience subtropical conditions, whereas the northern high-altitude areas are dominated by alpine environments and tundra-type vegetation (Aggarwal et al., 2017). From a physiographic perspective, northern Sikkim forms part of the Greater Himalaya, while the southern sector is associated with the Lesser Himalayan domain.

The state experiences four distinct seasons—spring, summer, autumn, and winter—with the southwest monsoon delivering the majority of annual precipitation between June and September. Although the terrain is mountainous and topographically complex, agriculture remains the backbone of the state's economy, supporting

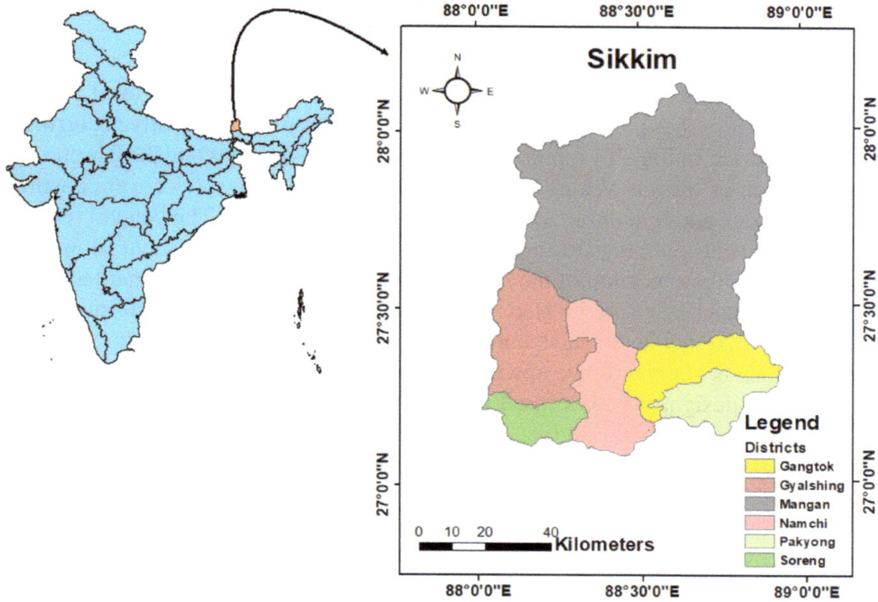

Fig. 1 Location map of the study area

over 60% of the population (Mishra et al., 2021). Recent modifications in cropping practices, however, have encouraged a gradual shift toward alternative livelihood options, some of which raise sustainability concerns (Tamang & Joshi, 2025). Topographically, the northern region is dominated by high-elevation glaciated peaks and snow-covered mountain systems, while comparatively lower valleys and floodplains occur toward the southern margin, reflecting the overall north–south elevation gradient.

3 Methodology

The present study is entirely based on analysis of remotely sensed secondary data sources. Various secondary sources platform has been accessed like Google scholar, Sodhganga, government published reports etc. for this study. Various state, central government and international organization portal has also been accessed like ENVIS, USGS, Esri LULC etc. To map wetlands and snow cover areas of Sikkim Landsat 8 data has been used. To prepare NDSI, drainage map of study area Arc GIS 10.2 software has been used. This study is totally based on explanatory methodology to analyze the findings and to draw meaningful conclusion.

4 Results

The above NDSI of Sikkim has been calculated using USGS Landsat 8 level 2 image. The above figure shows the snow cover areas of Sikkim which is mostly occurring in Mangan district. In Fig. 2, 1 represents the non-snow cover areas and the 2 represents the extension of snow cover area. Out of total area of Sikkim snow cover constitute 14.64% of area and remaining 85.36% is non-snow cover area. The general regional elevation is from south towards north which is why the major permanent snow cover areas lie in northern portion of the state. The elevation map indicates that larger and maximum glacier lakes are located at 4600 to 6000 m. The sensitive elevation zone of 3100 to 4500 m where sparse human settlements are also located with retreating glacier which is likely to move upward and there is chances of Lake Eutrophication. Some tourist designated lakes like Tsomgo, Aritar Lake already shows sign of high nutrient influx from various human induced sources (Fig. 3).

The wetlands are unevenly distributed across Sikkim the major ones are located in the north-east and north-west portion of the state which are glacier feed and only handful of wetlands located in lower ridges are spring-stream feed. The total wetlands cover only 0.47% of the state area which is the direct supplier of fresh water and provide various ecosystem services in the region. Among the Himalayan states Sikkim is dotted with third highest lakes with 2.63% of area cover (Panigrahy et al., 2010).

The drainage map of Sikkim has been prepared using Strahler stream order in Arc GIS software. Figure 4, captures the close association of wetlands and drainage network. Stream order length wise stream 1 covers 50.98% of length, stream 2 covers 25.05%, stream 3 covers 11.43%, stream 4 covers 6.88%, stream 5 covers 3.66% and stream 6 covers only 1.99% of stream length with carrying highest water volume cumulated from all other lower stream order. If observe closely then majority of lower order streams are originating from glacier feed wetlands only. Few lower order streams 1–3 orders are mainly originating from spring feed in the mid-southern portion of state which ultimately joins with higher stream order coming from glaciers. The higher order stream flowing through the heart of the state is Teesta River which forms the major drainage system of the state (Fig. 5).

Teesta River forms the major regional drainage basin of the western portion of Eastern Himalaya. It originates from Teesta Khangse Glacier in Mangan District and joined by several tributaries originating from existing wetlands as we have seen in Fig. 4. The number of lower stream order forming many small tributaries coming together on-stream order 5 and joined to form River Teesta which is in 6 order of hierarchy. The base level flow of Teesta is maintained by the continuous water release from wetlands present in the upward basin area. The presence of wetland as a storage unit in upper reaches is very significant to contribute river base level flow especially during winter. Teesta River is 414 km long with total catchment area of 12,444 km^2 in which 6930 km^2 of catchment area fall within Sikkim (Syed et al., 2017). In general, wetland act as the moderator in between receiving melt water from glaciers and releasing exceeding threshold water as outflow or channel headwater.

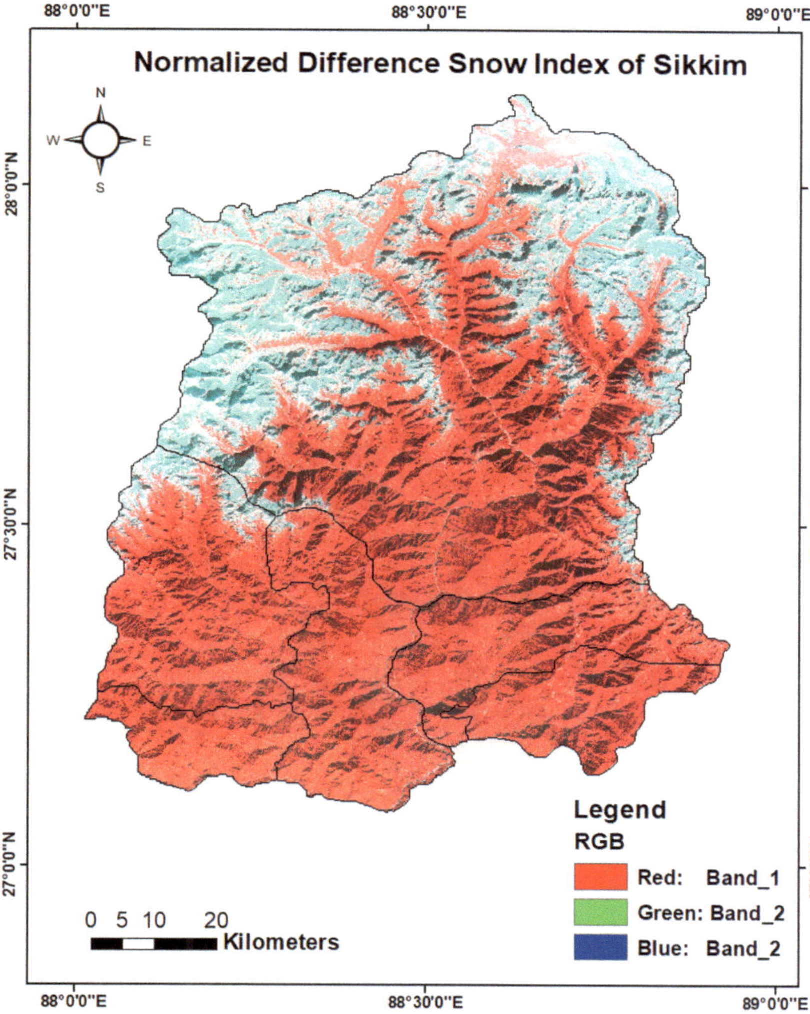

Fig. 2 NDSI map

5 Discussions

During 2017 there were 466 High altitude glacial lakes in Sikkim and over the period from 1975 to 2017 lake has increased by 9% with an area expansion by 24% as stated by Shukla et al. (2018). NDSI calculation performed in Arc Gis software of Landsat 8 data of February 2024 gives the value of snow cover area as 14.64%. This much of snow or glacier cover is sufficient to sustain 544 mapped wetlands with 3311 hectares of area expansion using Esri LULC map and landsat 8 images of 2024.

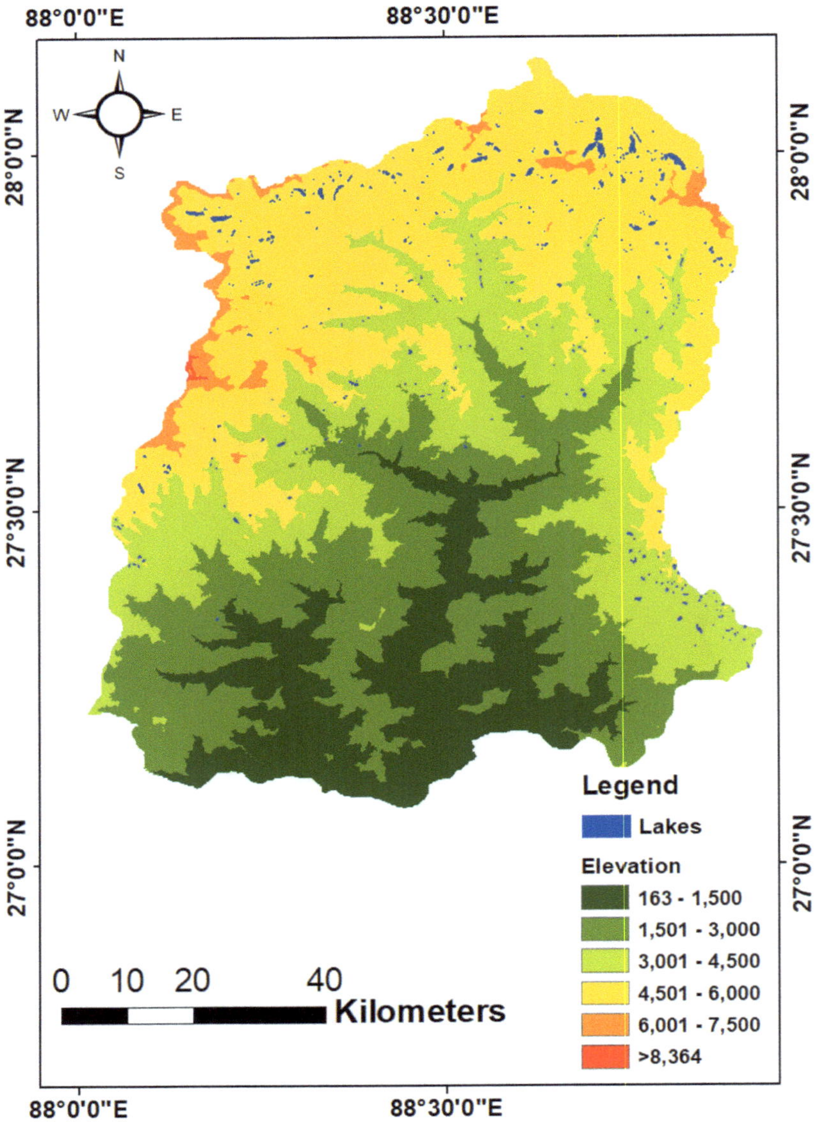

Fig. 3 Elevation map

The total state wetland cover is only 0.47% of area but enough to sustain the major Teesta River system and support the population residing within Teesta River basin. The total catchment area of Teesta River is 12,159 km^2 in which greater portion falls in Sikkim with 56.99% of area, 26.53% lies in West Bengal and remaining 16.48% of catchment area lies in Bangladesh. The entire Teesta basin support 30 million

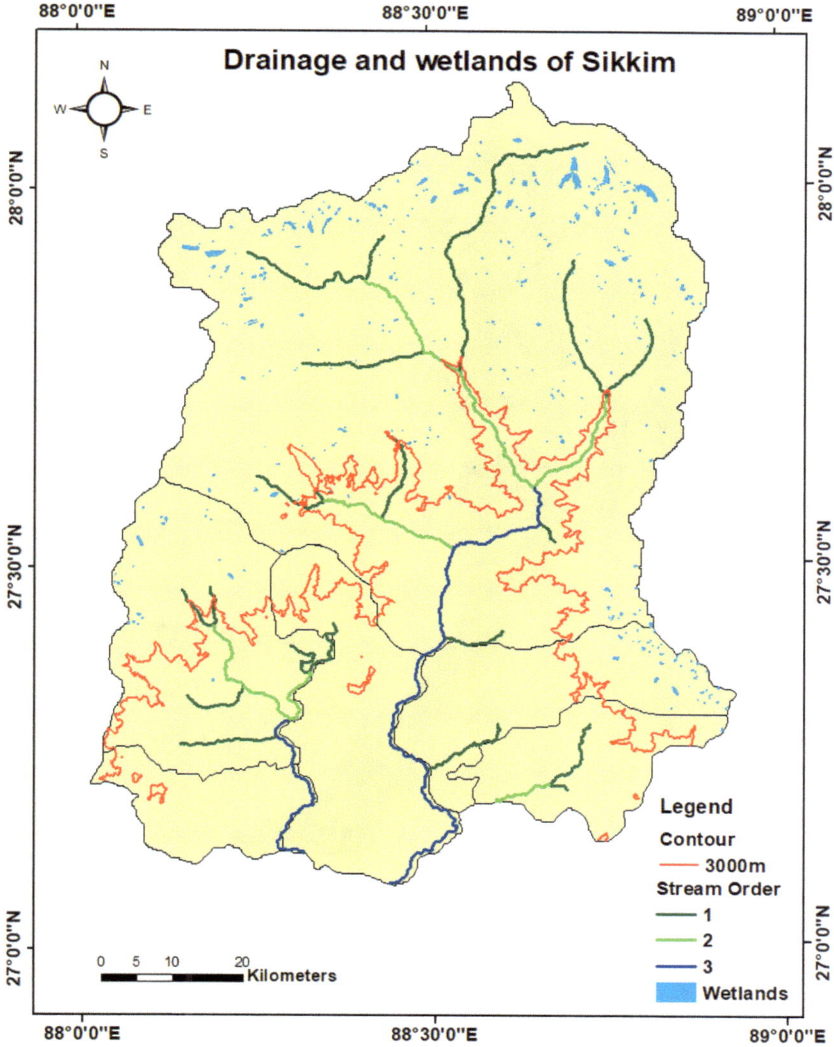

Fig. 4 Drainage and wetland distribution

population out of which 2% of inhabitants are in Sikkim, 27% of inhabitants are in West Bengal and 71% of inhabitants are in Bangladesh (Syed et al., 2017). More than half of the basin population are directly dependent on Teesta water to carry their agricultural activities. For the Hydropower generation along the Teesta River basin 5353 MW plant is installed on Sikkim side and 2841 MW plant is installed on West Bengal side as of 2019 (Rahaman, 2020). There is many more ecosystem services as provided by wetlands which still needs to be assess and documented.

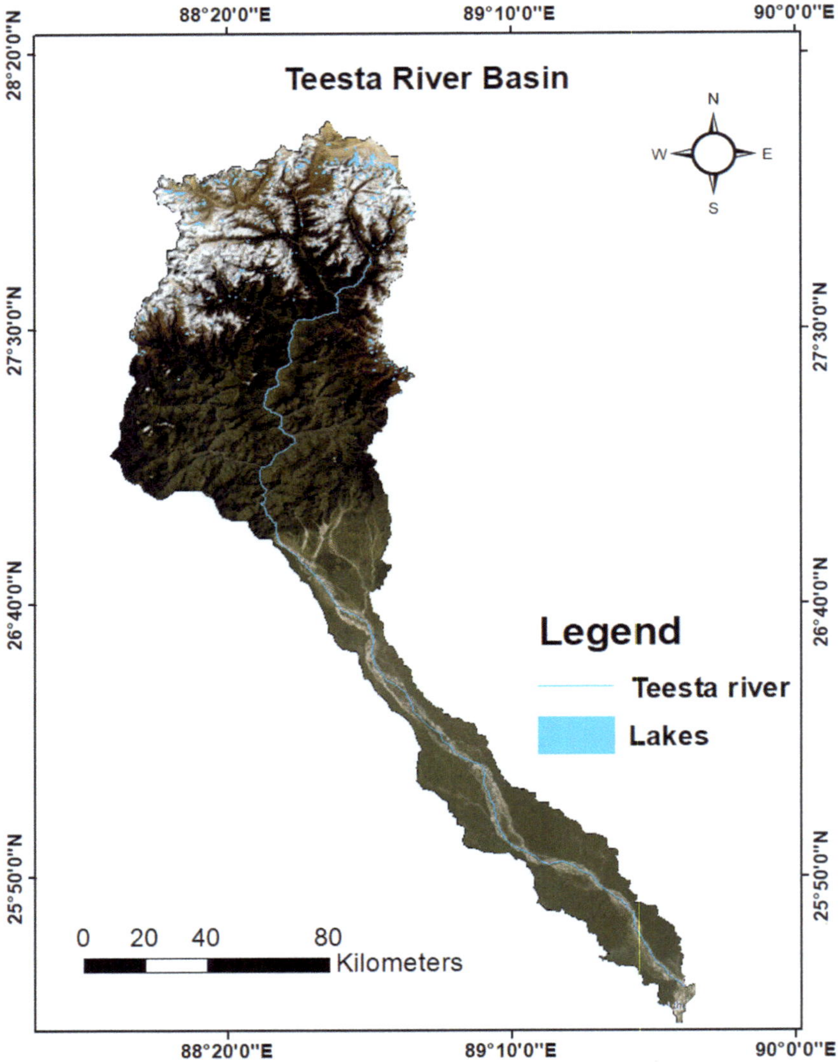

Fig. 5 Teesta river basin

Despite a number of direct and unexplored advantages of wetland it also poses as a serious threat to the community. There are number of growing instances of Glacial Lake outburst as reported from Hindu Kush Himalayan region mainly due to glacier retreat. Furthermore, there are number of Flash floods and GLOF has also been reported in head ward region of Teesta basin and cause havoc among Teesta basin residents (Table 1).

Table 1 Hazards related to wetland reported in Sikkim

Events	Year	Place	Casualties
Flash floods	1968	Teesta basin	–
	2001	Tir Khola	–
	2003, 2010	Places not specified	1 death
	2012	Chungthang	22 deaths
	2013	Rolep	–
GLOF	2023	South Lhonak lake	41 deaths

The above figure indicates the number of wetland related hazards occurred in Sikkim over past 55 years and are more likely to occur in near future (Prakash, 2011). In Sikkim Himalaya the major casualties were claimed by recent 2023 GLOF so far. Till now 91 deaths have been reported out of which 41deaths from Sikkim, 46 deaths from West Bengal and 4 deaths from Bangladesh (Dahal et al., 2024). These hazards occur due to sudden release of huge volume of water from the wetlands. It occurs when the moraine dam wetlands collapsed, when glacier breaks off and fall into lake or due to cloud burst over these wetlands. Lakes in Sikkim has increased by 9% within four decades and 14 lakes have found to be potential threat in near future (Shukla et al., 2018). When hazards like flash flood and GLOF origin at the source of river channel it's hard to estimate impact because it not only impacts the flooded area but the functioning of whole basin area. These hazards cannot be averted but can be minimized with proper systematic monitoring of high altitude wetland area with proper developmental plan along river channel. In every Monsoon season there is a sliding of NH-10 road which runs parallel to river Teesta has great influence on State economy.

Few of the wetlands are slowly deteriorated due to increased tourism activity, tourist influx and improper extraction and utilization of wetland resources. It is really necessary to sensitize local community about the importance of wetlands as the landscape unit contributing in the total regional functioning of ecosystem. GLOF and Flash flood are the ill effects of our activities on a global scale but the potential threat we are creating by polluting lakes, disturbing wetland ecosystem functioning at the source can be controlled. We need decentralized power thread from locals to international bodies for its conservation and effective management. By monitoring human activities around wetlands in high elevated areas can prevent emerging problem of lake eutrophication, associated health hazards and stand to sustainable water supply. Therefore, it is prime time to do hydrological assessment of high altitude lakes to understand its changing nature in respect to threat and freshwater depletion from mountains.

6 Conclusion

Wetlands are the most productive ecosystem in the world. It offers various ecosystem services like ground water recharge, carbon sequestration, sustain base flow of river channels, supports biodiversity, bio geo-chemical cycle and hydrological cycle. Sikkim covers 2.63% of wetland area among the Himalayan states. The concentration of high-altitude lakes is likely to increase due to rapid glacier retreat by about 9% in Sikkim Himalaya as per earlier findings. The wetlands are the source of resource regeneration and also hold potential threat if not consider during developmental plans along downstream river channels. Similarly, a number of wetland origin hazards in Sikkim have already claimed many lives. Therefore, it is really necessary to monitor these resources in order to minimize the damage and to sensitize the local people about its importance for the sustainable use.

Acknowledgments I would like to thank Director of G.B Pant National Institute of Himalayan Environment, for the opportunity and the facilities to carry my research. I like to extend my gratitude to funding agency NMHS under MOEF&CC, Government of India. Lastly, my sincere thanks to PI of the project and director Sikkim Regional Centre for the support and guidance.

References

Aggarwal, S., Rai, S. C., Thakur, P. K., & Emmer, A. (2017). Inventory and recently increasing GLOF susceptibility of glacial lakes in Sikkim, eastern Himalaya. *Geomorphology, 295*, 39–54.

Ahmed, R., Sam, L., & Bhardwaj, A. (2025). It is time to build GLOF-resilient communities globally before another GLOF disaster strikes. *NPJ Natural Hazards, 2*(1), 41.

Chakraborty, S. K., Sanyal, P., & Ray, R. (2023). Diversity and classification of wetlands in international and national perspectives. In *Wetlands ecology: Eco-biological uniqueness of a Ramsar site (East Kolkata wetlands, India)* (pp. 167–226). Springer International Publishing.

Dahal, D. R., Mehta, M., Bhatt, C. M., & Luirei, K. (2024). Flash floods and their cascading tumults: An example from Teesta River valley, Eastern Himalaya, Sikkim, India. *Current Science, 126*(10), 1280.

Gurung, D. R., Khanal, N. R., Bajracharya, S. R., Tsering, K., Joshi, S., Tshering, P., Chhetri, L. K., Lotay, Y., & Penjor, T. (2017). Lemthang Tsho glacial Lake outburst flood (GLOF) in Bhutan: Cause and impact. *Geoenvironmental Disasters, 4*(1), 17.

Maltby, E., & Barker, T. (2009). 43 conclusions: Wetlands for the future. In *The Wetlands handbook* (p. 983). Wiley.

Matthews, G. V. T. (1993, Mar). *The Ramsar convention on wetlands: Its history and development*. Ramsar Convention Bureau.

Mishra, P. K., Rai, A., Abdelrahman, K., Rai, S. C., & Tiwari, A. (2021). Analysing challenges and strategies in land productivity in Sikkim Himalaya, India. *Sustainability, 13*(19), 11112.

Moor, H., Rydin, H., Hylander, K., Nilsson, M. B., Lindborg, R., & Norberg, J. (2017). Towards a trait-based ecology of wetland vegetation. *Journal of Ecology, 105*(6), 1623–1635.

O'Neill, A. R. (2019). Evaluating high-altitude Ramsar wetlands in the eastern Himalayas. *Global Ecology and Conservation, 20*, e00715.

Panigrahy, S., Patel, J. G., Singh, T. S., & Murthy, T. V. R. (2010). National Wetland inventory and assessment. In *High altitude Himalayan lakes* (pp. 1–20). SAC/ESPA/NWIA/IN.

Prakash, S. (2011). Historical records of socio-economically significant landslides in India. *Journal of South Asia Disaster Studies, 4*(2), 177–204.

Rahaman, M. M. (2020). Hydropower development along Teesta River basin: Opportunities for cooperation. *Water Policy, 22*(4), 641–657.

Sharma, L. K., & Naik, R. (2024). Wetland ecosystems. In *Conservation of saline wetland ecosystems: An initiative towards UN decade of ecological restoration* (pp. 3–32). Springer Nature Singapore.

Sharma, S., Gosavi, V. E., Kanwal, K. S., Agnihotri, V., Chandra Sekar, K., Arya, S. C., & Rawal, R. S. (2020). *High altitude wetlands in the Indian Himalaya: Conservation and management.* G. B. Pant National Institute of Himalayan Environment.

Shukla, A., Garg, P. K., & Srivastava, S. (2018). Evolution of glacial and high-altitude lakes in the Sikkim, eastern Himalaya over the past four decades (1975–2017). *Frontiers in Environmental Science, 6*, 81.

Subba, J. R. (2008). *History, culture and customs of Sikkim.* Gyan Publishing House.

Syed, A., Haq, A., Uzzaman, A., Goodrich, C.G., Mallick, D., Mini, G., Sharma, G., Nyima, K., Mamnun, N., Varma, N., & Singh, P. (2017). *The Teesta Basin: Enough water for power and agriculture for all* (HI-AWARE Working Paper, 12).

Tamang, L., & Joshi, M. (2025). Transition in traditional cropping pattern led to engagement in alternative economic activity. In *Climate change impact on Himalayan biodiversity* (pp. 295–304). Springer Nature Switzerland.

Natural Hazards, Terrain Instability, and Risk Reduction

Understanding Himalayan Glacial Lake Outburst Floods through Ecological, Geomorphological, and Socio-Economic Lenses

Rayees Ahmed, Shahid Saleem, Abid Farooq Rather, Syed Hameem, Abrar Farooq, Gowhar Farooq Wani, Saqib Iqbal Raina, and Haadiya Saleem Bhat

Abstract Glacial lake outburst floods (GLOFs) are emerging as one of the most consequential climate-driven hazards in high-mountain regions, with the Himalayas at the global epicentre of risk. Accelerated glacier retreat, rapid expansion of moraine- and ice-dammed lakes, and increasing slope instabilities are fundamentally reshaping hazard regimes, exposing millions of people, critical infrastructure, and fragile ecosystems to low-frequency but high-impact disasters. Here, we synthesise current scientific evidence on Himalayan GLOFs through an integrated ecological, geomorphological, and socio-economic lens, linking physical triggering mechanisms to downstream impacts and risk governance challenges. We show that GLOFs not only drive profound geomorphic reorganisation of mountain valleys and long-term ecosystem degradation, but also generate cascading socio-economic consequences that undermine livelihoods, development trajectories, and regional stability. Despite major advances in remote sensing, hydrodynamic modelling, and hazard mapping, a persistent gap remains between scientific knowledge and actionable risk reduction. We identify community-centred, non-structural interventions—early warning systems, land-use planning, local capacity building, and knowledge co-production—as critical yet under-implemented components of effective GLOF risk management. Crucially, we argue that integrating disaster risk reduction into development planning and strengthening transboundary cooperation are essential to address the cross-border nature of Himalayan cryospheric hazards. By bridging physical science, ecology, and social vulnerability, this synthesis provides a holistic framework for managing GLOF

R. Ahmed (✉) · S. Saleem · A. F. Rather · S. Hameem · A. Farooq · G. F. Wani
Department of Geography and Disaster Management, School of Earth and Environmental Sciences, University of Kashmir, Srinagar, India

S. I. Raina
Department of Geography, Jamia Millia Islamia, New Delhi, India

H. S. Bhat
Indra Gandhi National Open University (IGNOU), New Delhi, India

Islamic University of Science and Technology (IUST), Awantipora, India

R. Ahmed et al. (eds.), *Climate Change, Water Resources, and Natural Hazards in the Himalayas*, Advances in Geographical and Environmental Sciences,
https://doi.org/10.1007/978-981-92-0835-7_8

risk in a warming world and offers policy-relevant insights for enhancing resilience across high-mountain socio-ecological systems.

Keywords GLOFs · Climate change · Glacier melting · Risk management · Disaster risk reduction

1 Introduction

High mountain regions operate as highly dynamic systems controlled by the interaction of multiple geomorphic, climatic, and cryospheric processes. Due to this complexity, these environments are particularly sensitive to external disturbances. Under present conditions, climate change and climate variability have emerged as major concerns for the scientific community (Gadek, 2014; Ahmed, 2024); Shamim et al., 2025). The increase in global temperature has clearly affected major Earth systems, including the oceans and the cryosphere, and observations confirm that this warming trend is ongoing and expected to continue in the future (IPCC, 2013). Since the beginning of the industrial era, the global mean temperature has risen by approximately 1 °C (Ahsan et al., 2022; Bajracharya et al., 2008; IPCC 1990; Mool & Shrestha, 2008; Ahmed et al., 2022a, b, c; Mir et al., 2025). In India, the mean annual temperature increased by about 0.56 °C between 1901 and 2009 (Attri & Tyagi, 2010), and projections indicate a further rise of 3.5 to 5.5 °C by the year 2100 (Lal, 2002). One of the most visible indicators of global warming is the widespread decline in snow cover and the retreat of glaciers. During the past century, glacier thinning and recession have promoted the formation and expansion of glacial lakes along glacier margins and moraine systems across high mountain regions worldwide, thereby increasing the risk of glacier-related hazards. Among these hazards, Glacial Lake Outburst Floods represent one of the most significant threats. As glaciers retreat, meltwater accumulates behind natural dams composed of moraines, ice, landslide debris, or bedrock. Lakes impounded by unconsolidated moraines and ice are considered particularly dangerous because these materials are comparatively weak and unstable. Failure of such dams can result in the sudden release of large volumes of water and debris. These outburst events can discharge millions of cubic meters of water within a short time, producing destructive floods that endanger downstream populations and infrastructure (Allen et al., 2016; Ahmed et al., 2022a; Taylor et al., 2023). Many newly developing glacial lakes are located beneath hanging glaciers and seracs or below steep and potentially unstable rock slopes (Frey et al., 2010). This setting increases their susceptibility to external triggers, especially since climate change is also expected to reduce slope stability. Under warming conditions in high mountain environments, the likelihood of rockslides and avalanches is expected to increase (Khan et al., 2025; Stoffel & Huggel, 2012). Glacial lakes classified as high hazard, particularly those that are newly formed or rapidly expanding, are especially vulnerable to such mass movement processes. Among the various triggering factors of GLOFs, including landslides, rock avalanches, snow avalanches, ice calving, slope instability, and cloudbursts, ice avalanches and calving events are becoming increasingly frequent due to continued glacier melt and retreat.

GLOF is generally defined as a type of flood that occurs when water that has been trapped by a glacier or a moraine is suddenly released. As the glaciers retreat, the meltwater accumulates behind the glacier ice or loosely dammed moraines made up of sand, ice, stone, and ice residues, etc. Over the past few centuries, human activities and settlements have gradually advanced towards the ecologically fragile regions especially towards the river headwaters, accepting the fact that the environment hosts sources of several natural hazards (Veh et al., 2019). One commonly held view is that continuous global atmospheric warming will manifest itself most dramatically in high mountains. Snow and ice-related disasters may thus become more common, with potentially greater magnitudes. The Himalayan region is highly dependent on meltwater supplies, yet it is also frequently vulnerable to various natural hazards. The Himalayan region, encompassing vast areas of central Asia, is host to thousands of glaciers and glacial lakes, constituting the third-largest snow and ice deposits globally. With approximately 15,000 glaciers and 9000 glacial lakes, the Himalayan region is a significant contributor to the world's cryosphere (Mool et al., 2005) This area experiences persistent snowfields, and during the winter season, substantial snowfall blankets most high-elevation regions (Kumar et al., 2005). Climate change is exerting a profound impact on the Himalayas, a geologically young and fragile mountain range highly sensitive to even minor climate variations. Observations indicate regional warming in the Himalayas ranging from 0.15 to 0.60 °C per decade, surpassing the mean global warming rate of 0.74 °C per hundred years. This warming trend in the Himalayas has led to noticeable glaciological changes, manifesting in the formation, expansion, and disappearance of various glacial lakes. The transformation of the Himalayan landscape due to climate change has given rise to concerns, particularly regarding the dangerous glacial lakes formed or expanded in the process. These lakes being a significant threat to the downstream communities and infrastructure, emphasize the need for proactive measures to address the potential hazards associated with them. GLOFs are a type of hazard that has sparked a lot of scientific interest in the last two decades. They happen rarely, but they usually have devastating and disastrous repercussions for downstream people and infrastructure. In this study we aim to systematically summarise the scholarly literature to provide the hazard and risk management strategies in the Himalayan region. Accordingly, this paper provides a comprehensive synthesis of Glacial Lake Outburst Flood (GLOF) hazards in the Himalayan region by integrating physical, ecological, geomorphological, and socio-economic perspectives. Specifically, the study aims to: (i) examine the dominant climatic and geomorphic drivers of glacier retreat and glacial lake evolution under a warming climate; (ii) synthesise current understanding of GLOF triggering mechanisms and their cascading downstream impacts on landscapes, ecosystems, and human systems; (iii) evaluate advances in hazard mapping, modelling approaches, and risk assessment frameworks for identifying vulnerable glacial lakes and downstream exposure; (iv) assess the ecological, geomorphological, and long-term socio-economic consequences of GLOF events; (v) highlight the role of community-centred, non-structural risk reduction strategies, including early warning systems, contingency planning, capacity building, and land-use planning; (vi) examine the integration of disaster risk reduction into development planning and natural resource management; and (vii) emphasise the need for regional and transboundary coordination to manage the growing GLOF risk in the Himalayas. By adopting a holistic and interdisciplinary framework, this study seeks to bridge the gap between scientific understanding and actionable risk management, offering policy-relevant insights to enhance resilience in high-mountain environments.

1.1 Climate Change-Induced Glacier Hazards in the Himalayas

The Himalayas, often referred to as the "Third Pole," are home to some of the world's most significant ice reserves outside the polar regions (Ahmed et al., 2021a; Allen et al., 2022; Mir et al., 2022; Rawat et al., 2023; Gupta et al., 2022). These glaciers play a crucial role in feeding major river systems across South and East Asia, making them vital for the freshwater supply of nearly 10% of the global population (Barua et al., 2022). However, the impacts of climate change have increasingly threatened these glaciers, leading to severe hazards such as GLOFs, which pose a significant risk to downstream communities and infrastructure (Mohanty & Maiti, 2021). The rapid warming of the Earth's climate has accelerated the melting of Himalayan glaciers, resulting in the formation and expansion of glacial lakes, particularly in the Indian Himalayan Region (Nie et al., 2018; Mir et al., 2024). The melting ice increases the volume of these lakes, many of which are dammed by unstable moraines that can fail catastrophically, leading to GLOFs (Ahmed et al., 2022a, b; Sharma et al., 2024). These floods are capable of causing widespread devastation, as evidenced by discharge rates that can exceed 30,000 m^3/s and run-out distances over 200 km (Richardson & Rennolds, 2000). In addition to GLOFs, the region is also susceptible to other glacier-related hazards, including ice avalanches, debris flows, and landslides. The variability and unpredictability of these events, exacerbated by climate change, make effective risk management a complex challenge.

2 GLOF Triggering Mechanism and Downstream Impacts

GLOFs result from a combination of natural processes and human influences. A key natural factor is the ongoing rise in temperature, which accelerates glacier melt and promotes the formation and enlargement of glacial lakes. As glaciers thin and retreat, the steepening of their termini can destabilize moraine dams that retain accumulated meltwater. In addition, external triggers such as earthquakes or volcanic eruptions may suddenly introduce large masses of ice or rock into these lakes, generating displacement waves capable of overtopping or breaching the moraine barriers.

Human interventions—including the construction of roads and hydropower projects in glaciated regions—can further intensify these risks by disturbing natural drainage systems and increasing stress on moraine dams (Ahmed et al., 2023; Dasgupta, 2021; Trivedi et al., 2021). Once initiated, a GLOF can have severe downstream consequences. Floodwaters may rapidly inundate valleys, destroy infrastructure, cause loss of life, and induce long-lasting geomorphic changes. The sudden release of large volumes of water combined with entrained debris often results in significant sediment deposition, riverbed aggradation, and ecological alterations downstream. These impacts can degrade water quality, disrupt biodiversity, and affect socio-economic activities such as agriculture and hydropower production (Ahmed et al., 2022a, b; Emmer et al., 2020; Richardson & Reynolds, 2000).

Therefore, communities located downstream of glacial lakes must implement comprehensive and effective strategies to reduce potential losses. Addressing GLOF hazards requires an integrated approach incorporating multiple components. These include systematic hazard and risk assessment to delineate vulnerable areas, development of dependable early warning systems to enable timely evacuation, and the adoption of both structural and non-structural measures.

Structural measures may consist of engineering interventions such as controlled lake drainage, strengthening of moraine dams, and construction of protective infrastructure. Non-structural approaches involve land-use regulation, community awareness initiatives, capacity development, and preparation of emergency response plans. A schematic illustration outlining the processes responsible for GLOF initiation and their downstream impacts is provided in Fig. 1.

2.1 Ecological Impacts of Glacial Lake Outburst Floods

GLOFs exert profound impacts on fragile high-mountain ecosystems, often causing irreversible ecological disturbances. The sudden release of large volumes of water and debris can drastically alter river morphology, erode mountain slopes, and destroy vegetation and soil cover along flood paths. These events disrupt aquatic habitats by increasing sediment loads, reducing water quality, and altering flow regimes, which threaten endemic fish and macroinvertebrate species. The deposition of silt and debris downstream can smother riparian zones and agricultural lands, reducing soil fertility and hindering natural regeneration. Furthermore, GLOFs may trigger cascading environmental effects such as landslides, channel avulsions, and habitat fragmentation, thereby weakening the ecological balance of alpine and sub-alpine ecosystems (Fig. 2). Over the long term, these disturbances diminish ecosystem resilience, reduce biodiversity, and compromise the ecological services that mountain communities depend on for their livelihoods and sustainability.

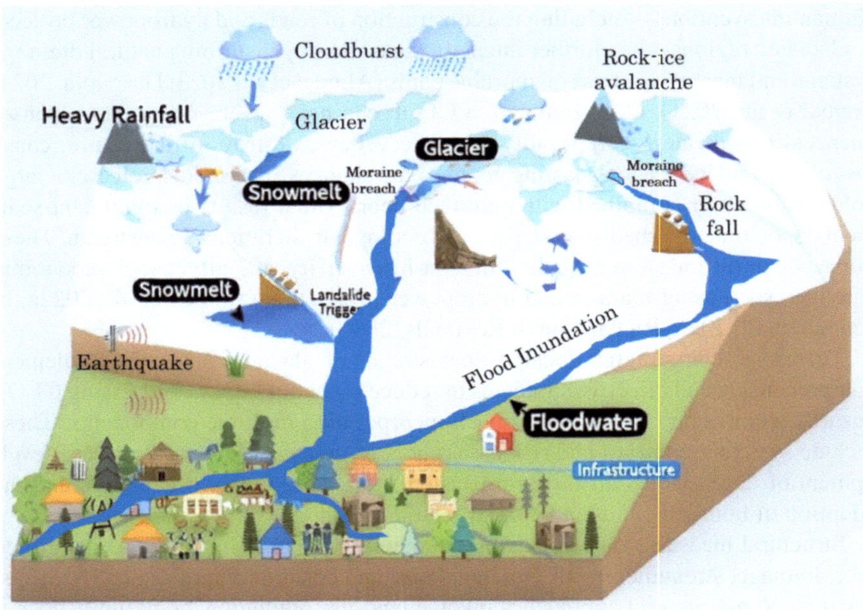

Fig. 1 Illustrates the key processes leading to GLOFs in mountainous regions. It shows how factors like ice avalanches, landslides, and earthquakes can destabilize a glacial lake, triggering a flood that causes downstream damage through erosion and debris deposition. Extreme weather conditions, such as heavy snowfall and rainfall, further exacerbate the risks

2.2 Geomorphological Impacts of Glacial Lake Outburst Floods

GLOFs bring about significant geomorphological transformations in high-mountain environments. The immense force of floodwaters and the accompanying debris flow reshape entire landscapes by eroding valley floors, scouring slopes, and modifying river channels. These events often lead to the widening and deepening of riverbeds, removal of vegetation cover, and exposure of unstable sediments, making surrounding terrain more prone to future erosion and landslides. The sudden transport and deposition of large volumes of sediment downstream alter floodplain morphology and can lead to the formation of new alluvial fans, terraces, or even secondary lakes. In some cases, GLOFs trigger cascading geomorphic hazards such as landslides, debris avalanches, or dam breaches, further amplifying landscape instability (Fig. 2). Over time, these processes not only reshape the topography of glaciated valleys but also disrupt the natural geomorphic equilibrium, leaving long-term imprints on the mountain environment that can influence hydrology, sediment transport, and ecosystem dynamics for decades.

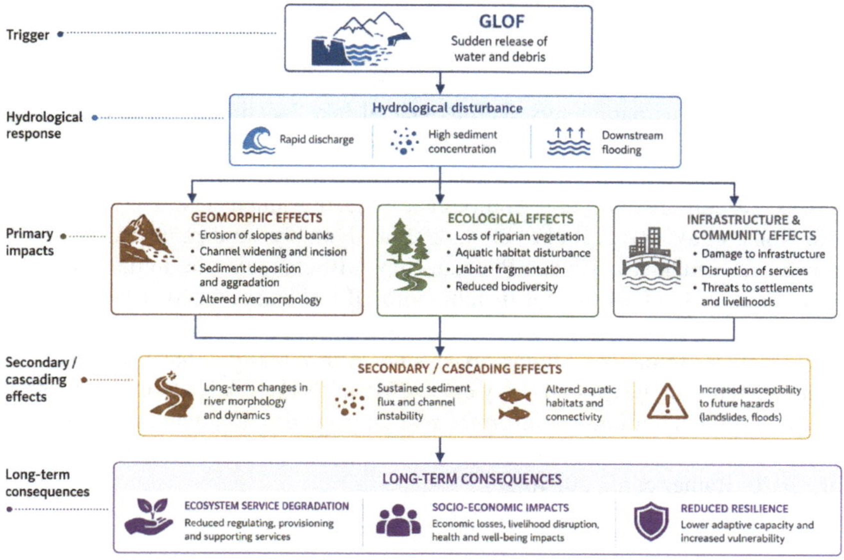

Fig. 2 Impacts of Glacial Lake Outburst Floods (GLOFs) on high-mountain environments, illustrating the cascade from sudden release of water and debris through immediate ecological and geomorphological effects to long-term consequences

2.3 Long-Term Impacts of Glacial Lake Outburst Floods

The long-term impacts of GLOFs extend far beyond the immediate destruction, leaving lasting imprints on mountain landscapes, ecosystems, and communities. Over time, GLOF-induced geomorphic changes—such as altered river courses, unstable slopes, and sediment accumulation—can permanently transform catchment hydrology and increase susceptibility to future hazards like floods and landslides. The loss of vegetation and soil fertility slows ecological recovery, leading to reduced biodiversity and weakened ecosystem resilience. For local communities, the economic repercussions are enduring, as agricultural productivity declines, infrastructure rebuilding is slow, and traditional livelihoods are disrupted. Socially, recurring GLOF threats can lead to long-term displacement, changes in settlement patterns, and psychological stress among affected populations. At a broader scale, repeated GLOF occurrences undermine regional development goals and increase adaptation costs, posing significant challenges for sustainable mountain development in a warming climate.

3 Hazard Mapping

Hazard mapping is a crucial tool for mitigating GLOF risks, involving the identification of various parameters such as glacial lake locations, moraine dams, and downstream vulnerabilities (Frey et al., 2018). Remote sensing techniques and hydrodynamic models aid in this process, integrating physical and socio-economic factors to assess risk comprehensively (Ahmed et al., 2021b; Ahmed et al., 2025a; Ives et al., 2010; Koike et al., 2012; Rather et al., 2024a). Notably, studies in regions like the Hindu Kush Himalayas, Tibetan Plateau, Nepal, Bhutan, and the Indian Himalayan Region have yielded significant insights into GLOF hazards and risk assessments (Ahmed, 2025b; Bajracharya et al., 2008; Sattar et al., 2021; Wang et al., 2022). However, while numerous studies exist, only a few findings have translated into actionable plans on the ground (Frey et al., 2016). It's imperative to utilize these risk maps and research findings for informed planning and development activities, particularly in ecologically fragile areas, to minimize potential GLOF impacts (Emmer et al., 2020; Rather et al., 2024b).

4 Developing a GLOF Hazard and Risk Management Framework

The creation of a thorough analytical framework is critical for assessing and reducing the risks associated with glacial lake outburst floods. This framework is built on a thorough review of the literature, which incorporates current knowledge and methods. The first part entails locating and inventorying glacial lakes with field surveys and remote sensing data to create a comprehensive database (Fig. 3).

The dynamics of these lakes, including changes in volume, area, and potential triggers for outburst occurrences, are then assessed using modelling techniques and remote sensing technology. Hazard assessment follows, which includes identifying vulnerable downstream places, calculating flood volumes, and analysing potential flood courses using hydrological and hydraulic modelling approaches in conjunction with Geographic Information Systems (GIS). Risk mapping integrates vulnerability and exposure evaluations with hazard data, taking into account infrastructure, population density, and land use in downstream areas, to prioritise risk reduction initiatives and spatially portray probable effect zones. Stakeholder involvement is critical throughout the process for improving data quality, evaluating models, and incorporating local knowledge. The developed methodological framework is an invaluable resource for planners, policymakers, and disaster management authorities, allowing for targeted policy interventions such as land-use planning, early warning systems, and infrastructure development in high-risk zones based on spatial visualisation of potential risk areas. Regular updates based on ongoing observation and study enhance adaptive management tactics.

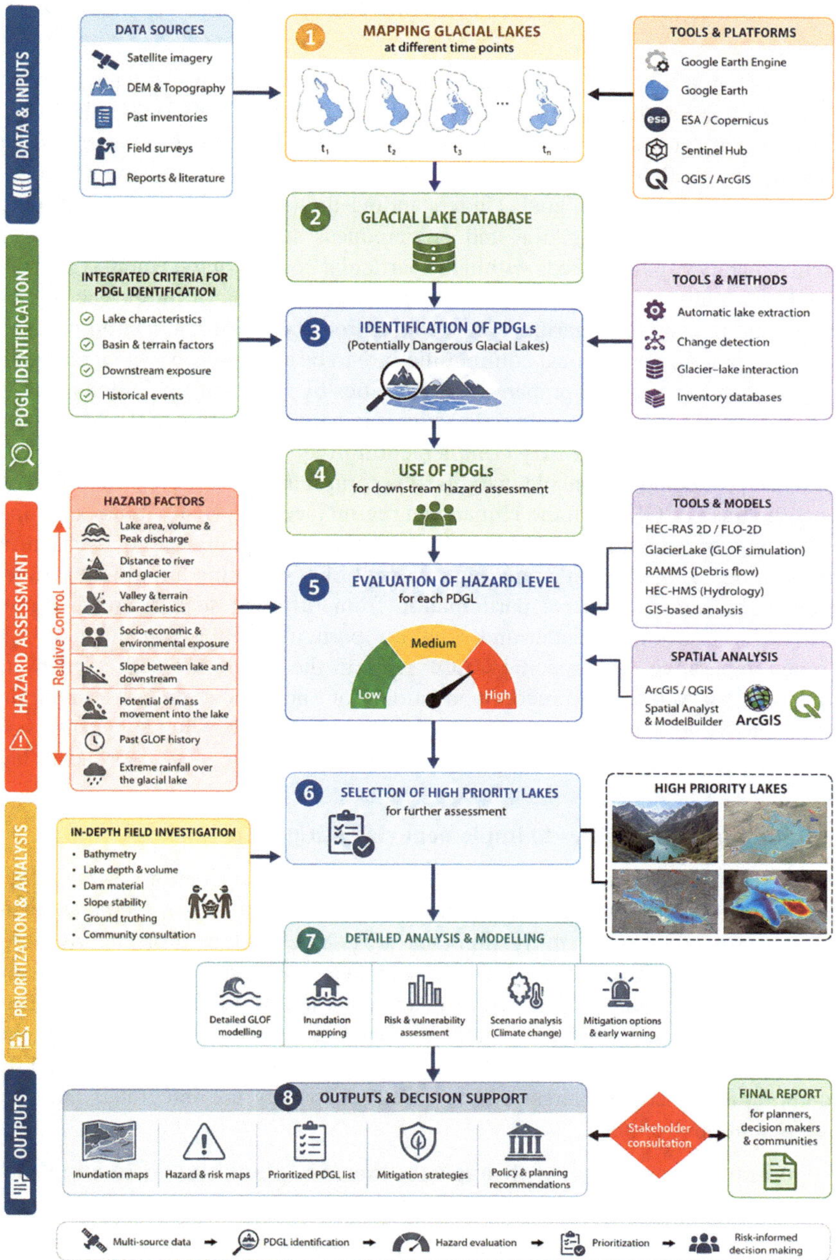

Fig. 3 Framework for the hazard and risk mapping of glacial lakes

5 Reducing GLOF Risks: A Community-Centric Approach

The primary goal of the GLOF Risk Reduction Initiative in the Himalayan region should be on strengthening the current efforts to reduce the region's vulnerability to GLOFs, with an emphasis on community-based and non-structural interventions. Adopting a methodology that entails the careful preparation of assessment reports is essential for achieving this goal. Understanding the socioeconomic impact, determining practical risk mitigation and preparedness strategies, and evaluating the current capacity gaps and needs within the particular context of the Himalayan region will all be greatly aided by these reports. Important aspects of the project include encouraging information sharing, facilitating online discussions, and building a local network of practitioners. Local communities are to be empowered to actively participate in risk mitigation and preparedness activities by means of the initiative, which will assist project countries in obtaining a thorough understanding of the hazards and promote practical measures. By complementing the existing structural mitigation efforts and producing actionable insights, this comprehensive approach significantly lowers the risks of GLOF in the Himalayan region (see Fig. 4). To create a comprehensive plan for reducing GLOF risk, sociological or community-based interventions must be added to the region's ongoing interventions. Adopting a multi-stakeholder approach that includes active participation from different sectors and integrates cross-cutting issues like gender and disability particular to the Himalayan context becomes imperative. The specific GLOF risks in the region are addressed by the actionable observations and recommendations for "non-structural" or community-centric approaches. In order to address GLOF risks, the following suggestions and practical observations about "non-structural" or community-centric approaches have been developed.

(a) **Enabling communities to implement risk mitigation**

In order to prevent adverse negative effects, a thorough evaluation of the valley's topography, community settlements, and river characteristics is imperative for the effective assessment of risk mitigation measures. Assets can be protected by simple mitigation techniques like makeshift embankments and natural barriers like plantations and boulders, but caution must be taken to avoid obstructing water flow during GLOF or flash flood events. Community ownership is ensured by connecting plantation initiatives to socioeconomic needs, and formal engagement and acceptability of community-based measures are increased when they are integrated into local development programmes.

(b) **Community preparation through contingency planning: a stitch in time**

Community preparedness necessitates contingency planning, which includes cooperative disaster management plans and public awareness campaigns. Because of migratory patterns, it must take into account the special needs of women, children, and the elderly in mountain communities. Because it is difficult to mobilise outside assistance in mountainous areas, extra care is required. For sustainability, planning should be anchored in local administrative units and should involve task forces, capacity building, and frequent drills. Including social, cultural, and religious institutions and incorporating traditional knowledge improves preparedness at the community level.

(c) **Advocacy and awareness for a vigilant and secure community**

Public awareness and advocacy are crucial for community security. Communities need to be aware of the risks and prepared to mitigate them. Awareness is improved by having experience with GLOFs or flash floods. A culture of risk reduction is aided by innovative projects and consistent awareness campaigns. At all levels, persistent policy advocacy promotes risk-reduction strategies and wins support from stakeholders. A large stakeholder network encourages cooperative action and resource sharing.

(d) **Making early warnings "very early"**

A quick Early Warning System (EWS) is necessary to guarantee timely warnings for hydro-meteorological hazards, especially in hilly areas with sudden events like GLOFs and flash floods. While remote sensing and other tools are used for monitoring, there is currently no real-time, 24*7 EWS deployment. Current EWSs mostly run autonomously, without administration integration or community focus, which causes information to be disseminated slowly. A community-based, easily maintainable EWS that prioritises community-based integration into administrative systems should be put in place to address these issues. Last-mile connectivity and guidelines for improved community understanding and response are essential. The difficulties presented by dispersed mountain communities can be addressed through coordination procedures between technical agencies and civil administration. It's imperative to have a comprehensive EWS that takes local customs, stakeholder roles, user data, and regulatory frameworks into account. Creating an EWS that is tailored to the vulnerabilities, topography, and socioeconomic circumstances of mountain communities guarantees prompt and effective alerts, promoting increased community acceptance and responsibility.

(e) **Developing capacities to improve preparedness**

Enhancing preparedness through capacity building is critical, following the military adage that "the more you sweat in peacetime, the less you bleed in war." Proactive training and capacity-building activities are critical for empowering communities, local governments, and key stakeholders to mitigate, prepare for, and respond to natural disasters. Regular programmes focusing on contingency planning, disaster preparedness, and first responder training, particularly search and rescue and first aid, are essential. Given the difficult mountain terrain and limited access routes, increasing local capacity for immediate response is critical to maximising external assistance.

(f) **Land use planning—making risks 'visible'**

Critical infrastructure and community assets in mountain valleys, where flat ground is uncommon, are frequently placed near rivers, leaving them subject to hydrometeorological hazards. To overcome this issue, community-level land use planning is required. This method aids in identifying hazard-prone areas and guides the placement of low-value assets in risk zones to reduce losses. While limited land may compel construction in hazard-prone places, community awareness and visible activities can assist people realise the need of smart asset placement. This approach is necessary for protecting socioeconomic assets such as critical development projects that are important to the economies of mountain countries like Nepal and Bhutan. In Bhutan, for example, preserving hydropower projects, which contribute considerably to national earnings, becomes critical to enable a faster recovery from a catastrophic event.

(g) **Making knowledge 'actionable' through networking**

Addressing the effects of climate change in sensitive mountain ecosystems requires effective knowledge networking. It is a two-way process: disseminating research findings for practical application and incorporating experiential insights into research outcomes. Despite countless studies on glacial lakes and GLOF risk, there is still a need to translate this information into practical measures. Dispersed information must be collected and synthesised for accessibility and application in disaster risk management. The effectiveness of risk-reduction strategies is improved by demystifying technical components and implementing new knowledge-sharing platforms. Collaboration at the regional level is critical for exchanging knowledge and managing emerging risks such as GLOFs. This networking technique enables targeted research, which contributes to community resilience and effective risk reduction.

6 Integrating Disaster Risk Reduction (DRR) into Development Planning

Integrating DRR into development planning is crucial for building resilient and sustainable communities, particularly in sensitive areas. Development plans, whether at the national or municipal level, should not be stand-alone activities but must incorporate DRR components. The enormous expenditures in socioeconomic and infrastructure assets necessitate a proactive approach to addressing hydrometeorological threats caused by climate change. Integrating risk reduction measures improves the safety and lifespan of projects like dams, hydroelectric projects, and bridges. Land use planning principles should guide the strategic location of vital infrastructure, taking into account GLOF shadow areas and applying risk-reduction measures to improve hazard resistance. Communicating and implementing DRR at the community level is also critical for promoting sustainable development.

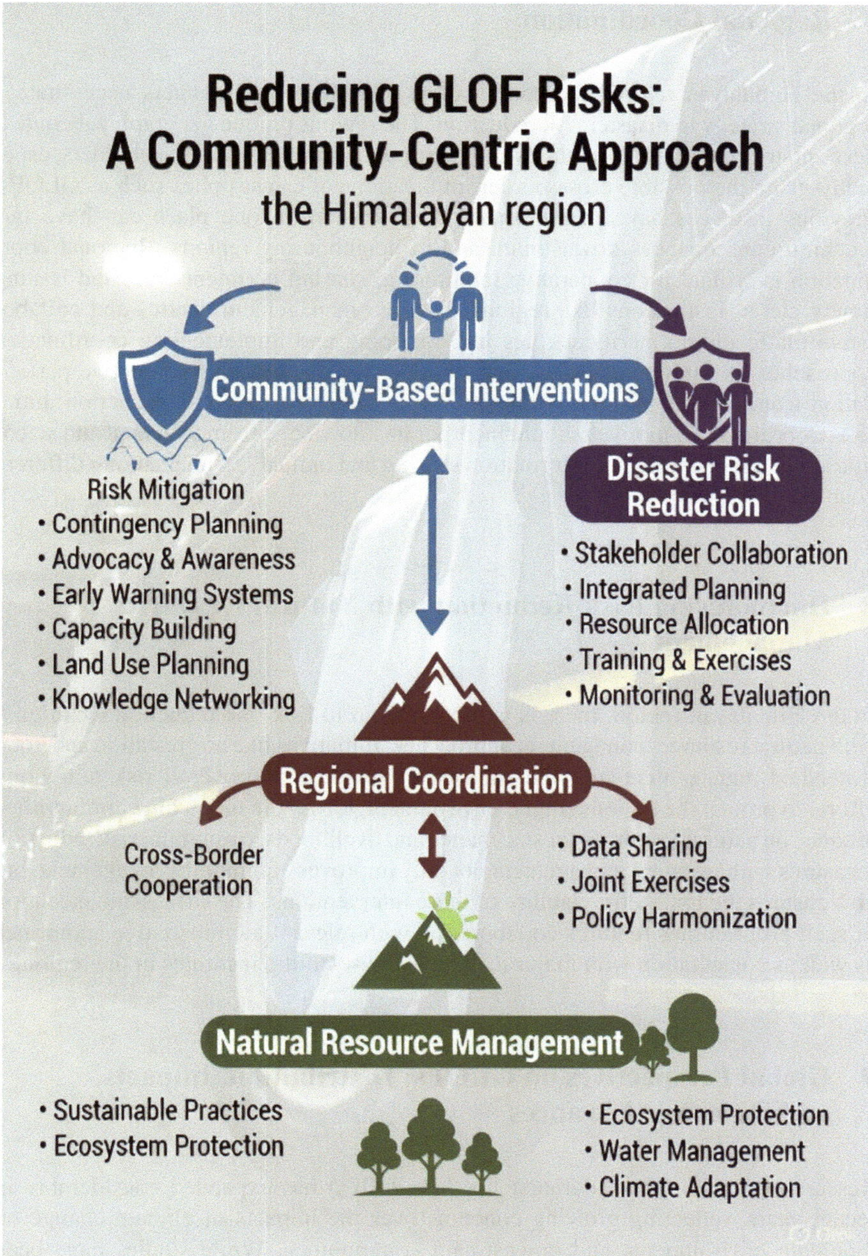

Fig. 4 Community-centric approach to reducing Glacial Lake Outburst Flood (GLOF) risks in the Himalayan region, highlighting key non-structural interventions, integration with disaster risk reduction, regional coordination, natural resource management, and multi-stakeholder collaboration for sustainable resilience and climate change adaptation

7 Regional Coordination

In the Himalayan region, the complexities of hazards and dangers necessitate a regional strategy to disaster risk reduction. The regions unique geographical context necessitates collaborative methods and close coordination among stakeholders, especially given the possible cross-border implications of catastrophes such as GLOFs. Previous instances have demonstrated that disasters in one place can have far-reaching consequences downstream and in neighbouring regions. Regional coordination is critical for cooperative monitoring, sharing pertinent data, and issuing timely alerts. Institutions like regional Disaster management centres and collaborative platforms play critical roles in developing and implementing coordination approaches. A multi-stakeholder and multi-sectoral approach, with active participation from various players, is required for a comprehensive risk reduction initiative. Coordinated initiatives should incorporate knowledge from technical and scientific agencies, encouraging information sharing and mutual learning among different countries.

8 Harmonising Risk Reduction with Natural Resource Management

In the Himalayan region, there is an urgent need to link risk reduction techniques with natural resource management approaches. Initiatives like afforestation and good watershed management must be smoothly integrated into overall risk mitigation efforts to protect the regions fragile environment. Given the mountain communities' reliance on natural resources for sustenance and livelihoods, combining risk reduction measures with resource management not only improves community engagement, but also ensures the long-term viability of these interventions. The successful execution of such programmes requires collaboration with relevant administrative authorities as well as consultation with major technical and scientific institutes in the region.

9 Global Perspectives on GLOFs: Distribution, Impacts, and Research Advances

Research on glacial lake outburst floods (GLOFs) has expanded considerably in recent years, reflecting growing concerns over the impacts of climate change on mountain environments and downstream communities. While studies have been conducted across major glacierized regions worldwide, including the Himalaya, Andes, Alps, Alaska, Greenland, Iceland, Scandinavia, and High Mountain Asia, research attention has increasingly focused on regions where GLOFs pose significant threats to human settlements, infrastructure, hydropower development, and ecosystem services. Consequently, the geographic focus of GLOF research is often influenced not only by the frequency of events but also by their potential societal and environmental consequences. Recent studies have documented GLOFs originating from a variety of lake types, including ice-, moraine-, and bedrock-dammed lakes. Ice-dammed lakes are particularly notable for producing recurrent outburst floods in several mountain regions, whereas moraine-dammed lakes often receive greater attention because of their potential to generate large, high-impact events. Advances in satellite remote sensing, geospatial analysis, and field-based investigations have substantially improved the detection, monitoring, and assessment of potentially dangerous glacial lakes worldwide. Alongside improvements in hazard assessment, increasing emphasis has been placed on understanding the human dimensions of GLOFs. Researchers have highlighted the importance of engaging local communities in risk assessment, adaptation planning, and decision-making processes. Community participation, local knowledge integration, and co-production of knowledge are now widely recognised as essential components of effective and equitable GLOF risk management. Ethical research practices founded on trust, transparency, and collaboration are increasingly viewed as critical for developing sustainable adaptation strategies in mountain regions. Significant progress has also been made in the modelling and simulation of GLOF processes. Early assessments relied primarily on empirical approaches and simplified representations of flood propagation. These have gradually evolved into physically based modelling frameworks capable of simulating complex process chains involving dam breach formation, debris entrainment, sediment transport, and downstream flood inundation. More recently, integrated multi-hazard modelling approaches, machine learning techniques, and three-dimensional numerical simulations have emerged as powerful tools for improving hazard assessment and forecasting. Despite these advances, challenges remain in representing complex outburst mechanisms, reducing uncertainties, and translating sophisticated modelling approaches into operational tools for early warning, risk reduction, and climate adaptation planning.

10 Holistic Approach to GLOF Hazard Mitigation and Climate Change Adaptation

A comprehensive approach to GLOF hazard reduction and climate change adaptation entails combining various fields, perspectives, and approaches to meet the complex and interlinked issues associated with these disasters (Ahmed et al., 2026). This strategy recognises that GLOF dangers and climate change impacts are interrelated and necessitate a comprehensive and collaborative response. Several critical factors are involved in a comprehensive approach to GLOF hazard mitigation and climate change adaptation. To begin, it involves collaborating with a wide range of stakeholders and sectors, particularly scientists, politicians, community people, and private-sector stakeholders, to ensure that diverse viewpoints and interests are taken into account in decision-making processes.

Secondly, a holistic approach entails tackling the underlying causes of GLOF hazards and the effects of climate change, such as greenhouse gas emissions, land use practices, and infrastructure development. Encouraging sustainable land use practices, decreasing emissions through renewable energy and energy efficiency measures, and encouraging green infrastructure that can assist mitigate the effects of GLOF dangers are all examples of ways to do so.

Thirdly, a holistic strategy involves establishing resilience to GLOF hazards and the effects of climate change through a variety of measures such as early warning systems, emergency response planning, protective infrastructure, and community-based adaptation measures. This necessitates working with communities to understand their specific needs and goals, as well as devising context-specific and culturally relevant responses.

Fourth, a holistic approach promotes collaboration and partnerships among various stakeholders to capitalize on resources and expertise, as well as foster the exchange of knowledge and collaboration. Developing partnerships between the public and private sectors, encouraging international collaboration and knowledge sharing, and collaborating with civil society organisations and community groups to promote local ownership and participation are all examples of this.

Finally, a holistic approach involves recognizing and addressing the social and environmental injustices that often underlie GLOF hazards and climate change impacts. This necessitates promoting social equity and justice, ensuring that vulnerable communities are not disproportionately affected by GLOF hazards and the effects of climate change, and recognising the importance of indigenous and traditional knowledge and practices in promoting resilience and adaptation.

A comprehensive and coordinated action plan for GLOF hazard reduction and climate change adaptation is required, incorporating diverse viewpoints, techniques, and disciplines. Communities may successfully tackle the risks linked to these hazards and build resilience to climate change by addressing the root causes of GLOF hazards and climate change impacts, building resilience, promoting collaboration and partnerships, and recognising and addressing social and environmental injustices. Recognising the interconnection of various environmental and social systems is also part of a comprehensive strategy to GLOF hazard reduction and climate change adaptation. This includes thinking about how GLOF threats and climate change affect ecosystems, water supplies, human settlements, and socioeconomic development. It also entails acknowledging that diverse groups of people, such as women, children, and marginalised communities, are typically affected differently by GLOF threats and climate change. To put this comprehensive strategy into action, it is critical to have a long-term view that considers the potential implications of climate change over numerous generations. This necessitates supporting sustainable development that balances economic, social, and environmental goals while emphasising the necessity of protecting natural resources for future generations.

11 Management and Policy Response

Effective management of glacier hazards in the Himalayas requires a multifaceted approach that combines scientific research, policy-making, and community engagement. Remote sensing technologies have proven invaluable in hazard assessment, allowing for the monitoring of glacial lakes and the early detection of potential threats. Satellite imagery has been particularly effective in remote regions, such as Bhutan, where in-situ monitoring is challenging. However, despite the advances in technology, there are significant gaps in glacier research, particularly in regions like Sikkim in the Eastern Himalayas, where rugged terrain and limited human resources hinder comprehensive studies. This lack of data is compounded by the absence of glacier protection laws and glacial lake management plans, leaving communities vulnerable to climate-induced hazards.

Policymakers across the Hindu Kush-Himalayan (HKH) region face the daunting task of developing and implementing adaptation strategies tshat are both effective and coordinated. While some adaptation measures have been taken, such as lowering water levels in dangerous glacial lakes, these efforts are often incremental and not well integrated into broader development plans. Moreover, the increasing variability in climate patterns poses additional challenges for governments in the HKH region, making it difficult to predict and prepare for future risks.

12 Recommendations and Future Directions

To address the growing threat of glacial hazards in the Himalayas, it is imperative to adopt a comprehensive disaster risk management approach. Key recommendations include:

- *Strengthening Early Warning Systems:* Developing and implementing robust early warning systems that can provide timely alerts to vulnerable communities, allowing them to evacuate or take other protective measures.
- *Land Use Zoning and Infrastructure Planning:* Enforcing strict land use zoning regulations to prevent the development of infrastructure in high-risk flood zones, particularly in areas like Nyalam, where recent development has increased vulnerability.
- *Community-Based Adaptation:* Building local response capacities through training and education programs that empower communities to respond effectively to GLOF events.
- *Cross-Border Cooperation:* Given the transboundary nature of many glacial lakes, it is essential for countries within the HKH region to cooperate on monitoring and mitigation efforts, sharing data, and coordinating disaster response strategies.

- *Policy Integration:* National and regional governments should integrate glacier hazard management into broader climate change adaptation and disaster risk reduction frameworks. This includes enacting glacier protection laws and developing comprehensive glacial lake management plans.
- *Research and Monitoring:* Continued investment in glaciological research is critical to understanding the evolving nature of glacier hazards. This includes expanding the inventory of glacial lakes, monitoring their stability, and assessing the impacts of climate change on glacier dynamics.

13 Discussion

The study gives a detailed overview of the hazards associated with GLOFs in high alpine regions, with a special emphasis on the Himalayas. It emphasises the complex character of GLOF dangers, which result from both natural processes like glacier melting and human-caused elements like infrastructure development. The discussion focuses on the mechanisms that cause GLOFs, downstream implications, and mitigation techniques, emphasising the importance of proactive actions to address the changing dynamics of glacial lakes in the face of climate change. The study emphasises the significance of hazard mapping as a critical tool for GLOF risk mitigation. Hazard mapping allows for more informed decision-making and targeted interventions by identifying crucial features like glacial lake locations and downstream risks. Remote sensing techniques and hydrodynamic models emerge as useful instruments in this process, allowing for a thorough assessment of risk factors across a variety of high-mountain environments. Furthermore, the study calls for the creation of a strong analytical framework to guide GLOF hazard and risk management. The methodology uses field surveys, remote sensing data, and modelling tools to examine glacial lake dynamics and anticipate probable flood scenarios. By prioritising risk reduction measures based on vulnerability and exposure assessments, the framework offers a systematic method to mitigating the effects of GLOFs on downstream communities and infrastructure.

A community-centric approach appears as a recurring feature in GLOF risk mitigation initiatives. The report emphasises the need of empowering local communities to actively participate in risk reduction and preparedness efforts. Communities can strengthen their resilience to GLOFs and other natural hazards by implementing contingency plans, public awareness campaigns, and capacity building activities. Furthermore, incorporating disaster risk reduction (DRR) into development planning ensures that infrastructure projects and land use decisions include the possible threats posed by GLOFs, enabling long-term and resilient development in high mountain regions. The study also emphasises the importance of regional coordination and collaboration in managing GLOF risks. Given the transboundary nature of GLOF dangers, efficient communication and information exchange among stakeholders are critical for rapid response and mitigation. By integrating risk reduction measures with natural resource management methodologies, the study calls for a comprehensive strategy to GLOF risk mitigation that acknowledges the interdependence of environmental, social, and economic variables in high mountain regions.

14 Conclusion

Addressing the threats posed by GLOFs in high mountain areas necessitates a multifaceted approach. By combining advanced technology like remote sensing and hydrodynamic modelling with community-centric solutions, we can successfully identify risky locations and empower local populations to reduce hazards. Integrating disaster risk reduction into development planning, as well as aligning our policies with natural resource management, are critical steps towards increasing resilience to climate change-related disasters. Moving ahead, proactive efforts must prioritise GLOF risk reduction while encouraging stakeholder engagement to address the transboundary nature of these threats. We can protect both human livelihoods and vulnerable alpine ecosystems by encouraging cross-border collaboration and integrating GLOF mitigation initiatives into larger development goals. These measures have far-reaching consequences that go beyond immediate risk reduction, contributing to the long-term development and environmental conservation in high mountain regions.

Data Availability NA.

Funding No funding was received to carry out this study.

Conflict of Interest Disclosure The authors declare no conflict of interest.

Ethics Approval This study was conducted in accordance with relevant ethical guidelines and regulations. Ethical approval was not required as this research did not involve human or animal subjects.

Permission to Reproduce Material from Other Sources Permission to reproduce material from other sources was not required for this study.

References

Ahmed, R. (2024). Glacial lake outburst flood (GLOF) hazard and risk management strategies: A global overview. *Water Resources Management, 39*, 1–16.

Ahmed, R. (2025). Glacial lake outburst flood (GLOF) hazard and risk management strategies: A global overview: R. Ahmed. *Water Resources Management, 39*(1), 1–16.

Ahmed, R., Ahmad, S. T., Wani, G. F., Mir, R. A., Ahmed, P., & Jain, S. K. (2022a). High resolution 645 inventory and hazard assessment of potentially dangerous glacial lakes in upper Jhelum 646 basin, Kashmir Himalaya, India. *Geocarto International*, 1–32.

Ahmed, R., Ahmad, S. T., Wani, G. F., Mir, R. A., Almazroui, M., Bansal, J. K., & Ahmed, P. (2022b). Glacial lake changes and the identification of potentially dangerous glacial lakes (PDGLs) under

warming climate in the Dibang River basin, eastern Himalaya, India. *Geocarto International*, 1–27.
Ahmed, R., Rawat, M., Wani, G. F., Ahmad, S. T., Ahmed, P., Jain, S. K., et al. (2022c). Glacial Lake outburst flood hazard and risk assessment of Gangabal lake in the upper Jhelum Basin of Kashmir Himalaya using geospatial technology and hydrodynamic modeling. *Remote Sensing, 14*(23), 5957.
Ahmed, R., Rather, A. F., Banerjee, P., Wani, G. F., Saleem, S., Shamim, T., Wani, T. A., Ahmed, P., Almazroui, M., & Mir, R. A. (2025a). Impact of climate variability on the feeding glaciers of potentially dangerous glacial lakes in the Jhelum basin of Kashmir Himalaya, India. *Natural Hazards Review, 26*(2), 04025001.
Ahmed, R., Sam, L., Bhardwaj, A., Kulkarni, A. V., & Yamasaki, S. (2026). Increasing global threat of outburst floods from overlooked small alpine lakes. *Nature Sustainability*, 1–10.
Ahmed, R., Wani, G. F., Ahmad, S. T., Mir, R. A., Almazroui, M., Jain, S. K., & Ahmed, P. (2021a). Spatiotemporal dynamics of glacial lakes (1990–2018) in the Kashmir Himalayas, India using remote sensing and GIS. *Discover Water, 1*, 1–17.
Ahmed, R., Wani, G. F., Ahmad, S. T., Sahana, M., Singh, H., & Ahmed, P. (2021b). A review of glacial lake expansion and associated glacial lake outburst floods in the Himalayan region. *Earth Systems and Environment, 5*(3), 695–708.
Ahmed, R., Wani, G. F., Ahmad, S. T., Mir, R. A., Faisal, A. A., Rather, A. F., & Saeed, S. (2023). Expansion of moraine-dammed glacial lakes and historical GLOF events in Cordillera Blanca Region of Peruvian Andes. *Earth Systems and Environment, 7*(1), 131–150.
Allen, S. K., Rastner, P., Arora, M., Huggel, C., & Stoffel, M. (2016). Lake outburst and debris flow disaster at Kedarnath, June 2013: Hydrometeorological triggering and topographic predisposition. *Landslides, 13*, 1479–1491.
Allen, S. K., Sattar, A., King, O., Zhang, G., Bhattacharya, A., Yao, T., & Bolch, T. (2022). Glacial lake outburst flood hazard under current and future conditions: Worst-case scenarios in a transboundary Himalayan basin. *Natural Hazards and Earth System Sciences, 22*(11), 3765–3785.
Attri, S. D., & Tyagi, A. (2010). *Climate profile of India* (India Meteorology Department). Environment Monitoring and Research Center.
Bajracharya, S. R., Mool, P. K., & Shrestha, B. R. (2008). Global climate change and melting of Himalayan glaciers. *Melting glaciers and rising sea levels: Impacts and implications*, 28–46.
Barua, P., Mitra, A., & Eslamian, S. (2022). Disaster risk reduction: Detecting Himalayan glacial lake outburst floods. In *Disaster risk reduction for resilience: Disaster and social aspects* (pp. 393–403). Springer International Publishing.
Change, C. (1990). *The IPCC scientific assessment. Contribution of working group I to the first assessment report of the intergovernmental panel on Climate Change* (p. 365).
Dasgupta, A. Factors Responsible for Triggering GLOF-Glacial Lake Outburst Flood & Resilient Remedial Measures: A Literature Review (2021).
Emmer, A., Harrison, S., Mergili, M., Allen, S., Frey, H., & Huggel, C. (2020). 70 years of lake evolution and glacial lake outburst floods in the cordillera Blanca (Peru) and implications for the future. *Geomorphology, 365*, 107178.
Frey, H., Haeberli, W., Linsbauer, A., Huggel, C., & Paul, F. (2010). A multi-level strategy for anticipating future glacier lake formation and associated hazard potentials. *Natural Hazards and Earth System Sciences, 10*(2), 339–352.
Frey, H., Huggel, C., Bühler, Y., Buis, D., Burga, M. D., Choquevilca, W., et al. (2016). A robust debris-flow and GLOF risk management strategy for a data-scarce catchment in Santa Teresa, Peru. *Landslides, 13*, 1493–1507.
Frey, H., Huggel, C., Chisolm, R. E., Baer, P., McArdell, B., Cochachin, A., & Portocarrero, C. (2018). Multi-source glacial lake outburst flood hazard assessment and mapping for Huaraz, cordillera Blanca, Peru. *Frontiers in Earth Science, 6*, 210.
Gądek, B. (2014). Climatic sensitivity of the non-glaciated mountains cryosphere (Tatra Mts., Poland and Slovakia). *Global and Planetary Change, 121*, 1–8.

Gupta, A., Maheshwari, R., Guru, N., Sweta, Rao, B. S., Raju, P. V., & Rao, V. V. (2022). Quantitative prioritization of potentially critical glacial lakes in the Indus River basin using satellite derived parameters. *Geocarto International, 37*(25), 7508–7530.

IPCC. (2013). *Climate change 2013: The physical science basis. Contribution of working group I to the fifth assessment report of the intergovernmental panel on climate change.*

Ives, J. D., Shrestha, R. B., & Mool, P. K. (2010). *Formation of glacial lakes in the Hindu Kush-Himalayas and GLOF risk assessment* (pp. 6-–56).

Khan, I., Bahuguna, H., Kainthola, A., Ahmed, R., & Asgher, M. S. (2025). National-scale landslide susceptibility and risk mapping of India using a hybrid data-driven approach. *Scientific Reports.*

Koike, T., & Takenaka, S. (2012). Scenario analysis on risks of glacial lake outburst floods on the Mangde Chhu River, Bhutan. *Global Environmental Research, 16*(1), 41–49.

Kumar, A., & Ram, J. (2005). Anthropogenic disturbances and plant biodiversity in forests of Uttaranchal, central Himalaya. *Biodiversity and Conservation, 14*, 309–331.

Lal, M., Harasawa, H., & Takahashi, K. (2002). Future climate change and its impacts over small Island states. *Climate Research, 19*(3), 179–192.

Mitigation, C. C. (2011). IPCC special report on renewable energy sources and climate change mitigation.

Mohanty, L., & Maiti, S. (2021). Probability of glacial lake outburst flooding in the Himalaya. *Resources, Environment and Sustainability, 5*, 100031.

Mool, P. K., Tao, C., & Bajracharya, S. R. (2005). Monitoring of glaciers and glacial lakes from 1970 s to 2000 in Poiqu Basin, Tibet Autonomous Region, PR China. *Unpublished report prepared by ICIMOD, Kathmandu, Nepal, and CAREERI, Gansu, China.*

Mir, R. A., Ahmed, R., Rather, A. F., Ahmad, S. T., Bhat, I. A., Wani, G. F., & Javaid, S. (2024). A narrative review of geospatial techniques for assessing climate change impacts on cryosphere, geo-environmental hazards and risks in the Himalayan region. *Geo-information for Disaster Monitoring and Management*, 93–121.

Mir, R. A., Jain, S. K., Ahmed, R., & Rather, A. F. (2025). A strategic framework for Glacial Lake Outburst Flood (GLOF) Risk Reduction in the Himalayas under climate change. *Discover Geoscience, 3*(1), 143.

Mir, R. A., Jain, S. K., Lohani, A. K., Ahmed, R., Ahmad, S. T., & Wani, G. F. (2022). Application of mike 11 for one-dimensional GLOF modeling of a rapidly expanding dalung proglacial lake, indus river basin, Western Himalaya. In *Challenges of disasters in Asia: Vulnerability, adaptation and resilience* (pp. 147–161). Springer Nature Singapore.

Nie, Y., Liu, Q., Wang, J., Zhang, Y., Sheng, Y., & Liu, S. (2018). An inventory of historical glacial lake outburst floods in the Himalayas based on remote sensing observations and geomorphological analysis. *Geomorphology, 308*, 91–106.

Rather, A. F., Ahmed, R., Ahmed, P., Bansal, J. K., Banerjee, P., Wani, T. A., ... & Mir, R. A. (2024a). Examining the glacier-glacial lake interactions of potentially dangerous glacial lakes (PDGLs) under changing climate in Shyok catchment of the upper Indus Basin. *Physics and Chemistry of the Earth, Parts A/B/C, 136*, 103686.

Rather, A. F., Ahmed, R., Bansal, J. K., Mir, R. A., Ahmed, P., Malik, I. H., & Varade, D. (2024b). Glacial lake outburst flood risk assessment of a rapidly expanding glacial lake in the Ladakh region of Western Himalaya, using hydrodynamic modeling. *Geomatics, Natural Hazards and Risk, 15*(1), 2413893.

Rather, A. F., Ahmed, R., Jun, C., Bateni, S. M., Ahmed, P., & Mir, R. A. (2024c). Understanding glacial Lake evolution and the associated GLOF Hazard in the Shyok catchment of the upper Indus Basin using geospatial techniques. *Natural Hazards Review, 25*(3), 04024014.

Rawat, M., Ahmed, R., Jain, S. K., Lohani, A. K., Rongali, G., & Tiwari, K. C. (2023). Glacier–glacial lake changes and modeling glacial lake outburst flood in Upper Ganga Basin, India. *Modeling Earth Systems and Environment, 9*(1), 507–526.

Richardson, S. D., & Reynolds, J. M. (2000). An overview of glacial hazards in the Himalayas. *Quaternary International, 65*, 31–47.

Sattar, A., Goswami, A., Kulkarni, A. V., Emmer, A., Haritashya, U. K., Allen, S., et al. (2021). Future glacial Lake outburst flood (GLOF) hazard of the south Lhonak Lake, Sikkim Himalaya. *Geomorphology, 388*, 107783.

Shamim, T., Bhat, M. S., Alam, A., Allaie, A., & Ahsan, S. (2025). Impact of El Nino Southern Oscillation, Indian Ocean Dipole and North Atlantic Oscillation on the drought scenarios in the Upper Indus Basin. *Theoretical and Applied Climatology, 156*(4), 234.

Sharma, A., Thakur, V., Prakash, C., Sharma, A., & Sharma, R. (2024). Deep learning-based glacial lakes extraction and mapping in the Chandra–Bhaga Basin. *Journal of the Indian Society of Remote Sensing, 52*(2), 435–447.

Stoffel, M., & Huggel, C. (2012). Effects of climate change on mass movements in mountain environments. *Progress in Physical Geography, 36*(3), 421–439.

Taylor, C., Robinson, T. R., Dunning, S., Rachel Carr, J., & Westoby, M. (2023). Glacial lake outburst floods threaten millions globally. *Nature Communications, 14*(1), 487.

Trivedi, A., Chauhan, A., & Singh, S. P. (2021). Analyzing glacial lake outburst flood triggers for sustainable disaster risk mitigation: An interpretive structural modelling based approach. *Management of Environmental Quality: An International Journal, 32*(6), 1284–1297.

Veh, G., Korup, O., von Specht, S., Roessner, S., & Walz, A. (2019). Unchanged frequency of moraine-dammed glacial lake outburst floods in the Himalaya. *Nature Climate Change, 9*(5), 379–383.

Veh, G., Lützow, N., Kharlamova, V., Petrakov, D., Hugonnet, R., & Korup, O. (2022). Trends, breaks, and biases in the frequency of reported glacier lake outburst floods. *Earth's Future, 10*(3), e2021EF002426.

Wang, J., Chen, F., Zhang, M., & Yu, B. (2022). NAU-Net: A new deep learning framework in glacial lake detection. *IEEE Geoscience and Remote Sensing Letters, 19*, 1–5.

Evaluating Conditioning Factors Influencing Terrain Stability in Himalayan Intermountain Regions

Bilquis Shah

Abstract The Himalayan region has challenging topography and intricate geological formations, resulting in widespread unstable slopes and frequent tectonic activity that create conditions conducive to the occurrence of landslides. The Kashmir valley, located in the Himalayan-susceptible intermountain zones, includes various landslide-prone areas. The likelihood, magnitude, intensity, and frequency of landslides are directly affected by several factors, such as geological, hydrological, and anthropogenic factors. Despite the region having a notable record of devastating landslides, no specific study has been undertaken to explore the connection between conditioning factors and the occurrence of landslides. The current study filters specific conditioning factors and analyzes their impact on landslides using a frequency ratio method. The history of landslides in Kashmir valley was reconstructed (1990–2020), and the concentration of past landslides in each indicator was analyzed. A map of landslide susceptibility was created and classified into various categories, depending on the slip movement potential. The results were validated using a relative landslide density index. The study shows that approximately 52.3% of the valley falls under medium to very high susceptibility zones, with 25.5% high and 6.8% very high. Key conditioning factors such as elevation, geomorphology, distance to roads, and rivers were assigned significantly higher weights. Only 15% of the area lies in the very low susceptibility class. These results are crucial for the development of risk assessments and mitigation efforts in this region. Incorporating these findings into the planning process will allow for the identification of high-risk regions and support the development of specific plans to reduce the impact of landslides.

Keywords Frequency ratio · Slip movements · Risk assessment · Relative landslide density

---

B. Shah (✉)
Department of Geography and Disaster Management, University of Kashmir, Srinagar, Jammu and Kashmir, India
e-mail: shahbilquisgeo@gmail.com

R. Ahmed et al. (eds.), *Climate Change, Water Resources, and Natural Hazards in the Himalayas*, Advances in Geographical and Environmental Sciences,
https://doi.org/10.1007/978-981-92-0835-7_9

1 Introduction

Landslides, a geological phenomenon, damage infrastructure such as roads, bridges, and residences, but also endanger human lives. Therefore, it is critical to use mapping tools to detect landslide-prone locations. This mapping procedure is crucial for systematically identifying geological hazards and allows for a preventive approach to analyze and manage the risks associated with landslides (Sarkar and Kanungo, 2004). Landslide susceptibility maps effectively evaluate and illustrate the geographical probability of incidence of landslides within a within a certain geographical area. These maps are created by analyzing the association between the positions of previous landslides and a range of indicators that can trigger and intensify such incidents (Corominas et al., 2003; Nadim et al., 2006).

The three primary categories of landslide susceptibility methods are: innovative, deterministic, and statistical. The innovative method relies on user opinions, while the deterministic method employs mathematical calculations (Xiao et al., 2020). The most widely used approach for Landslide Susceptibility Mapping (LSM) is the multi-criteria assessment approach used by Carver et al., (1991); Akgun, (2012); Pourghasemi et al., (2012); Feizizadeh et al., (2014); Ahmad, (2015); Hong et al., (2019); Ghiasi, (2021) and Alam et al., (2023). The frequency ratio method is another extensively used statistical methodology that is well-known for delivering reliable results all across the world (Akgun et al., 2008; Mezughi et al., 2011; Ramesh & Anbazhagan, 2015; Qazi et al., 2023). The ongoing technological progress has brought artificial intelligence to the forefront of landslide research, providing extensive knowledge about various aspects of the phenomenon. This involves a detailed examination of moisture patterns, evaluation of slope stability, monitoring of debris flow depth and movement, and comprehensive analysis of landslide-associated risks. The integration of advanced techniques like Fuzzy Logic (Ercanoglu & Gokceoglu, 2004; Ilanloo et al., 2011; Sharma et al., 2013; Zhu et al., 2014; Kumar & Anbalagan, 2015; Wardhana et al., 2019), Artificial Neural Networks (Arora et al., 2004; Ermini et al., 2005; Tsangaratos et al., 2014; Pradhan et al., 2009; Jacinth Jennifer & Saravanan, 2022; Ali et al., 2022), Random Forest (Akar & Güngör, 2015; Catani et al., 2013; Hong et al., 2017; Saha et al. 2022), and Support Vector Machines (Yoa et al., 2008; Pham et al., 2016; Lee et al., 2017; Kumar et al., 2017; Huang & Zhao, 2018) significantly improves the accuracy and effectiveness of these analyses.

Using the above methods several studies have been done in the Himalayas including an investigation of the Kalimpong region in the Himalayas, featuring diverse landslide attributes, including movement type, activity state, material, distribution, style, and failure mechanism. This study used a deep learning algorithm to identify debris slides, rockslides, and rockfalls, constructing two highly accurate landslide susceptibility models. A Similar study conducted in the Garhwal Himalaya of the Rudraprayag district utilized various techniques, including artificial neural networks (ANN), support vector machines (SVM), random forests (RF), and logistic regression, as well as their combinations to assess landslide susceptibility. Among all the ensembles, ANN-RF-LR showed the highest level of fit, predictive ability, and stability (Saha et al. 2020). Nepal's Kulekhani watershed was used as a case study to assess the effectiveness of the GIS-based heuristic and bivariate methods for measuring landslide exposure in the Himalayan region. The results indicate that although the bivariate methods provide similar outcomes, the heuristic method produces significantly different results. Moreover, the agreed area analysis is found to be a reliable approach for evaluating the spatial and quantitative consistency of landslide susceptibility maps (Kayastha et al., 2013).

The selection of conditioning factors that contribute to landslides varies depending on the specific causes of landslides in different regions. This variability is influenced by the relative importance of each parameter in the region under consideration. The most filtered indicators responsible for causing landslides are slope, elevation, soil type, lithology, rainfall, curvature, vegetation cover, soil moisture, groundwater condition, and land use. Several studies have been conducted along the Himalayan region pertaining to landslide susceptibility and have determined how various factors influence the probability of landslides using different methodological approaches. A study in Uttarakhand Himalaya, India employed the Frequency Ratio (FR) model to assess landslide susceptibility using data from 236 locations. The analysis considered six key factors: elevation, slope aspect, curvature, rainfall, land use, and slope angle, which reveal valuable insights for targeted risk management in the region (Pham et al., 2015). Another study focused on creating macro-zonation maps for landslide susceptibility in a 100 sq. km area along the National Highway-21 section Garamaura-Swarghat of Bilaspur District, Himachal Pradesh. This study integrates terrain evaluation maps, including relative relief structure, land use, slope, morphometry, lithology, and groundwater condition maps for landslide susceptibility zonation (Sharma and Mehta, 2012). Similarly, a weighted composite model Analytic Hierarchy Process (AHP) has been used based on the selected parameters classified into five categories: anthropogenic, surface causal, lithological causal, triggering, and protective factors. The results of this study show that lithological factors are primary contributors to landslide initiation in the Gish River Basin, West Bengal (Basu and Pal, 2017).

Kashmir Valley is an intermountain region nested between the Zanskar range and the Pir Panjal range of the Himalayas (Fig. 1). The region, known for its tectonic activity, is prone to frequent landslides, particularly concentrated along the Pir Panjal range and the Greater Himalayas (Shah et al., 2022; Ali et al., 2022; Shah et al., 2023; Shah et al., 2024). This geological activity has caused substantial damage over time, and the situation is further compounded by the recent significant population expansion, resulting in the unchecked development of human settlements and economic infrastructure on vulnerable terrains, such as mountain slopes. Besides, this growing development is taking place in a global context of a change in the frequency and severity of landslides, attributed to environmental degradation and climate change. Therefore there is a need to verify the dominant causes of landslides and identify the most vulnerable sites of landslides in the region which is projected to rise in the near future, posing serious threats to both established and newly developed areas in this tectonically active region.

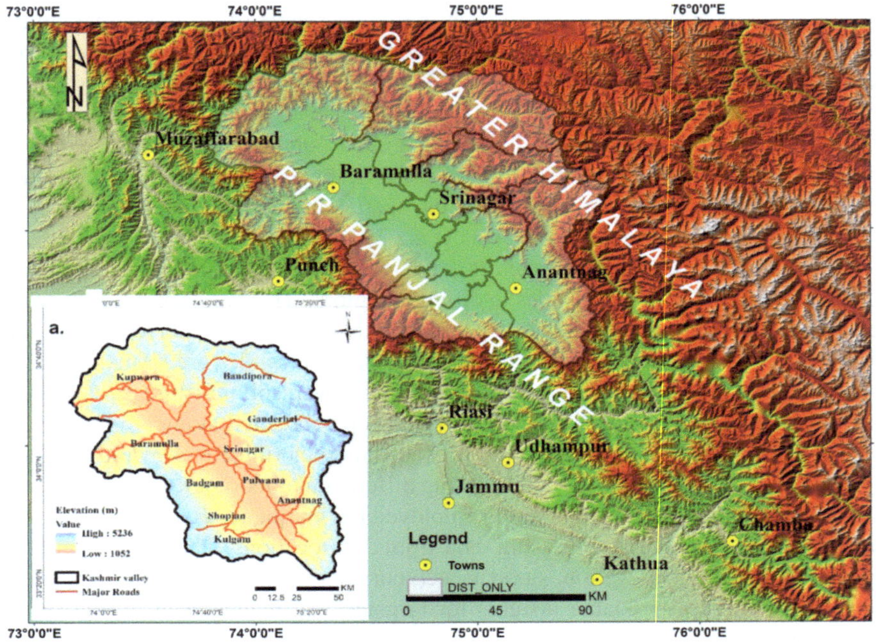

Fig. 1 Location map of Kashmir Valley

However, limited studies have been conducted in this region to establish the relationship between landslide-causative factors and their occurrence. Thus the present study analyzes the frequency ratio of the conditioning factors and landslides in the valley and assesses the most effective causes of landslides in the valley by calculating the prediction rate of each factor. Finally, landslide susceptibility maps are generated for the valley to recognize the potential landslide areas of the region.

2 Methodology

2.1 *Landslide Inventory Mapping*

The landslide inventory of historical events was compiled and used for landslide susceptibility mapping of the study area, including 559 landslide events in the Kashmir Valley (valley) 1990 to 2020, gathered from a variety of secondary sources such as newspaper archival accounts, academic papers, and reports from the Border Road Organisation (BRO). Some images of landslide-prone areas in the valley and along the highway are shown in Figs. 2 and 3.

Fig. 2 Field photographs of the landslides prone areas in the valley: **a** Sonmarg road, **b** Uri Baramulla, **d** Sadhnatop Road, and **e** Slide zone of Mughal road

2.2 Landslide Susceptibility Zonation

2.2.1 Selection of Conditioning Factors of Landslide Activity

Environmental elements significantly impact landslide frequency and severity, necessitating their assessment to map landslide-prone areas (Khan et al., 2019). Several variables impact the choice of the conditioning factors including the availability of data, local terrain conditions, extent of the study area, historical landslip records, geological information, hydrological characteristics, seismic activities, land use patterns, and the expertise and research goals of the field examiners (Das et al., 2023). The study has chosen 12 prominent variables belonging to morphological (slope, elevation, aspect, curvature), geological (lithology, faults, and soil), hydrological (drainage), and land cover (roads, NDVI) groups for the calculation of landslide susceptibility index. The data on lithology, geomorphology, faults, and road

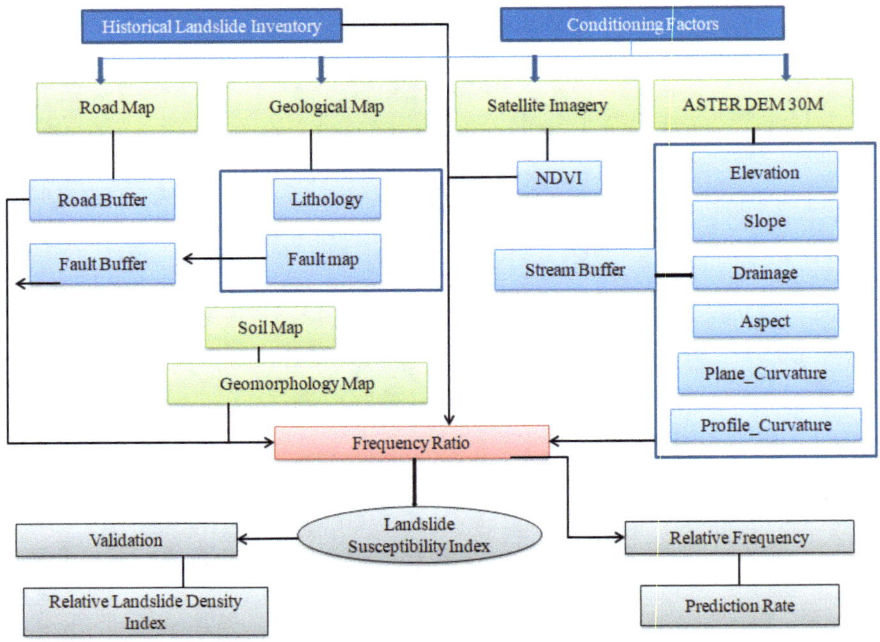

Fig. 3 The landslide susceptibility index framework adopted in this study

networks has been obtained from the Geological Survey of India whereas the soil data has been derived from the National Bureau of Soil Survey and Land Use Planning (NBSS&LUP). The slope, elevation, aspect, and curvature have been calculated using the ASTER DEM (Digital Elevation Model) data. Moreover, the drainage map of the area has been generated from DEM. Figure 3 shows the methodological framework adopted in this study.

The detailed description of the selected variable has been given below;

Slope: The stability of the slope affects the occurrence and severity of landslides, with the greatest magnitude occurring at slopes between 35° and 40° and decreasing on slopes over 40° (Dai FC & Lee CF, 2002). Based on the sharpest downhill or uphill direction, the slope determines how steep or flat the ground is at a certain location.

Elevation: The risk of landslides is substantially impacted by elevation. The ascent involves climbing higher slopes, which increases the force of gravity and the risk of landslides. Risk factors associated with elevation include unique geology, severe weathering, erosion, scant flora, climate, and freeze–thaw cycle of rocks and soil (Kavzoglu et al., 2014; Sheng & Chan, 2017; Shen et al., 2022). Moreover, the likelihood of landslips increases when there is snow or ice present.

Aspect: The direction a slope faces is referred to as its aspect, and is commonly expressed in degrees, ranging from 0° to 360° in a clockwise direction. It indicates the compass orientation of the slope, such as north, east, south, or west (Kavzoglu et al., 2014). Understanding an aspect enables analysis of factors such as rainfall distribution, drying winds, sunshine exposure, and natural processes like erosion responsible for causing landslides (Aniya, 1985; Felicísimo et al., 2013). The aspect measurements were categorized into ten groups: flat slopes (−1°), north-facing slopes (0–22.5°, 337.5–360°), northeast-facing slopes (22.5–67.5°), east-facing slopes (67.5–112.5°), southeast-facing slopes (112.5–157.5°), south-facing slopes (157.5–202.5°), southwest-facing slopes (202.5–247.5°), west-facing slopes (247.5–292.5°), and northwest-facing slopes (292.5–337.5°).

Curvature: Curvature is a measure that describes how the slope of a land surface changes over a specific area and time (Rabby et al., 2020). The velocity of downslope flow is influenced by profile curvature, with convex slopes increasing the water flow and concave slopes slowing it down, possibly leading to erosional and depositional effects as well as landslides. Plan curvature illustrates the dynamics of diverging surface water flow dynamics with negative values and the convergence of flow dynamics with positive values (Kritikos & Davies, 2015).

NDVI: Using satellite or remote sensing data, the normalized difference vegetation index (NDVI) is a valuable statistic that assesses and quantifies the quantity of healthy vegetation present in a specific area (Hong et al., 2017). The normalized difference vegetation index (NDVI), which ranges from −1.0 to 1.0, shows the presence of clouds, water, or snow when it is negative. Values around Zero are frequently observed in areas with exposed rock, little vegetation, or sandy ground. High NDVI values indicate locations with extensive forest cover, whereas moderate NDVI values often correlate to regions with shrubs and grassland vegetation (Ahmad 2015). The NDVI map has been created using data from the Landsat 8 satellite imagery thereby computed by the formula $NDVI = (IR - R)/(IR + R)$ (Pradhan, 2010).

Soil: The composition and texture of the soil are key factors in influencing the stability of the land. Landslides are less likely to occur in sandy soils with bigger particles and adequate drainage, while they are prone to occur in clayey soils with tiny particles and poor drainage, especially during periods of high rainfall (Sidle et al., 1985; Sassa & Canuti, 2008). The soil type is shown in Table 1 which is based on the USDA (United States Department of Agriculture) classification system.

Lithology: Lithology, the study of various rock types and their physical features, has a direct impact on an area's susceptibility to landslides due to its impact on the geological stability and characteristics of surface and subsurface materials (Kamp et al., 2008). The detailed classification of lithology pertaining to study area is shown in Table 2.

Geomorphology: In order to understand the natural processes and landforms that contribute to slope instability, geomorphology is crucial to landslides (Van Westen et al., 2003). The major landforms present in the study area are shown in Table 3.

Table 1 Soil classification of the study area

Soil_sub group	Code	Texture
Aeric haplaquepts	AH	Deep, imperfectly drained, calcareous, fine-loamy soils on gentle slopes with loamy surface, slight erosion and moderately shallow ground water
Aquic hapludalfs	AqH	Deep, imperfectly drained, fine-silty soils on very gentle slopes with loamy surface, slight erosion and moderately deep ground water
Dystric eutrochrepts	DE	Deep, well drained, fine-loamy soils on very gentle slopes with loamy surface and moderate erosion
Fluventic eutrochrepts	FE	Medium deep, moderately well drained, Mesic, fine-loamy soils on steep slopes with loamy surface and severe erosion
Lithic cryorthents	LC	Dominantly rocky landscape; associated with: Very shallow, excessively drained, fragmental soils on very steep slopes with loamy surface,very severe erosion and strong stoniness
Lithic Udorthents	Lu	Shallow, somewhat excessively drained Mesic loamy-skeletal soils on steep slopes with loamy surface, severe erosion and strong stoniness
Typic cryorthents	TC	Medium deep, excessively drained fine-loamy soils on very steep slopes with loamy surface, severe erosion and moderate stoniness
Typic dystrochrepts	TD	Deep, excessively drained, Mesic, fine-loamy soils on steep slopes with loamy surface, severe erosion and slight stoniness
Typic eutrochrepts	TE	Deep,well drained, calcareous, fine-loamy soils on nearly level slopes with loamy surface and very slight erosion
Typic fluvequents	TF	Deep, imperfectly drained, calcareous, coarse-loamy soils on very gentle slopes with loamy surface and moderate flooding
Typic haplaquepts	TH	Deep, poorly drained calcareous, fine-loamy soils on very gentle slopes with loamy surface and moderately shallow ground water
Typic hapludalfs	THs	Deep, well drained, fine-silty soils on very gentle slopes with loamy surface and slight erosion
Typic udifluvents	TU	Deep, moderately well drained, fine-loamy soils on very gentle slopes with loam surface and slight erosion
Typic udorthents	TUs	Medium deep, somewhat excessively drained, Mesic, calcareous, fine-loamy soils on steep slopes with loamy surface, very severe erosion and moderate stoniness
Lithic hapludolls	LH	Very shallow, excessively drained loamy soils on steep slopes with loamy surface and severe erosion associated with

Roads: The proximity of landslides to roadways has been identified as a significant factor that enhances landslide risk. The construction and usage of roads in high terrain can reduce the stability of slopes by causing soil compaction from traffic vibrations, disrupting drainage patterns, preventing water penetration, and destabilizing slopes via excavation (Ayalew & Yamagishi, 2005; Pal & Chowdhuri, 2019; Khan et al., 2019).

Table 2 Lithological description of the study area

Class	Name	CODE	Class	Name	Code
1	Alluvium, moraines, Hillwash and scree	AMHS	32	Sandy shale, siltstone and quartz-arenite	SSQ
2	Arenite/greywacke with siltstone and shale	ASS	33	Shale, limestone and sandstone	SLS
3	Carbonaceous Phyllite with marble and quartzite	CQ	34	Calcareous arenite, shale and limestone	CSL
4	Leucocratic to mesocratic biotite granite	LM	35	Schistose quartzite, slate and gritty quartzite	SSG
5	Andesitic and basaltic lava flows	AB	36	Shale and limestone	SL
6	Shale, sandstone and siltstone	SSS	37	Leuco to mesocratic biotite granite	LM
7	Diamictite, arenite with Phyllite and ash beds	DAPA	38	Pebbly slate, conglomerate with tuff and ash beds	PC
8	Sandstone, Phyllite, schist	SPS	39	Unmapped	UN
9	Slate and carbonaceous Phyllite	SCP	40	Carbonaceous Phyllite, schist, Lst and dolomite	CSLD
10	Limestone and dolomite	LD	41	Quartz vein/reef	Q
11	Granite and Granitoids	GG	42	Phyllite, quartzite, limestone and lava flows	PQLL
12	Arenaceous slate, Phyllite, carbonaceous Phyllite	APC	43	Vescicular basalt, andesite, agglomerate and Chert	VAAC
13	Phyllite, quartzite, Diamictite and lava flows	PQDL	44	Dolerite and gabbro dykes	DG
14	Slate with thin bands of siltstone and limestone	SSL	45	Shale, quartzite, sandstone and limestone	SQSL
15	Arenite, slate, Phyllite with granite and Granitoids	ASPGG	46	Shale, slate, arenite and limestone	SSAL
16	Orthoquartzite with shale bands	OSB	47	Carbonaceous shale, slate, quartzite and limestone	CSQL
17	Splintery shale with Nummulitic limestone	SML	48	Variegated shale, quartz-arenite and conglomerate	VQAC
18	Sandstone, shale, siltstone with limestone	SSSL	49	Arenite and greywacke	AG
19	Golden Brown silt	GB	50	Siltstone, arenite with shale and limestone bands	SAL
20	Clay, Sandy clay, conglomerate with lignite	CSCL	51	Basic flows, Serpentinite and Sedimentaries	BSS
21	Micaceous siltstone and arenite with limestone	MSAL	52	Siltstone and shale	SS

(continued)

Table 2 (continued)

Class	Name	CODE	Class	Name	Code
22	Siltstone, arenite with shale and limestone	SASL	53	Conglomerate, arenite, Quartzwacke and lithic tuff	CAQL
23	Calcareous shale and crystalline limestone	CSCL	54	Bluish limestone	BL
24	Limestone, sandstone, shale and calcareous bands	LSSC	55	Basic/intermediate Volcanics, agglomerate and tuff	BAT
25	Metabasic (amphibolite)	M	56	Leucocratic granite with Felspar porphyries	LF
26	Basal gravel, calcareous loamy clay and silt	BCS	57	Carbonaceous Phyllite schist with graphite	CSG
27	Quartzite, Sandy shale and Biomicritic limestone	QSB	59	Sandstone, Phyllite, schist, quartzite, and conglomerate	SPSQC
28	Shale and limestone with calcareous bands	SLC	60	Basic Volcanics	BV
29	Dolerite/meta-dolerite	D	61	Limestone, arenite with shale layers	LAS
30	Limestone, sandstone and shale	LSS	62	Shale, slate, limestone and quartz-arenite	SSLQ
31	Arenite, shale with plant fossils	ASP	63	Phyllite, slate, carbonaceous shale, limestone	PSCL
			64	Grey argillite, quartzite, limestone and dolomite	GQLD

#Self devised codes

Faults: The occurrence and severity of co-seismic landslides are primarily determined by the location and characteristics of the fault lines. The characteristics of fault lines, including their type, slip rate, and geological features, significantly influence their susceptibility to landslides triggered by seismic activity (Kamp et al., 2008).

Rivers: The impact of proximity to drainage networks and related factors, such as gully erosion and stream-mediated undercutting, on instability due to lateral support removal is a major area of focus in landslide studies, including river undercutting (Dai & Lee, 2002). In addition, the drainage system may induce erosion or saturate the hillside (Gökceoglu & Aksoy, 1996).

2.2.2 Frequency Ratio Method

A GIS-based quantitative technique called the frequency ratio was used to evaluate the susceptibility of landslides using spatial data. The Frequency Ratio (FR), according to Bonham-Carter, represents the likelihood that a certain characteristic

Table 3 Geomorphology of the study area

Class	Name	Code	Class	Name	Code
1	Ridge	R	17	Alluvial Fan	AF
2	Highly Dissected Hills and valleys	HD	18	Older flood plain	OF
3	Piedmont alluvial plain	PA	19	Piedmont slope	PS
4	Strike ridge	SR	20	Low Dissected Hills and valleys	LD
5	Younger alluvial plain	YA	21	Valley	V
6	Lake	L	22	Lacustrine terrain	LT
7	Swamp	S	23	Terrace	T
8	Marsh	M	24	Older alluvial plain	OA
9	Landslide	L	25	Plain	P
10	Moderately Dissected Hills and valleys	MD	26	Lateral Bar	LB
11	Channel Bar	CB	27	WatBod—Lake	La
12	Channel Island	CI	28	Dam and reservoir	DR
13	WatBod—Pond	WP	29	Valley glacier	VG
14	Glacial terrain	GT	30	Point Bar	PB
15	Active flood plain	AF	31	Younger deltaic plain	YD
16	WatBod—River	WR	32	Colluvial Fan	CF

#self devised codes

will develop. The underlying assumption of this model is that the circumstances surrounding landslides in the future will be quite similar to those seen in past incidents (Bonham-Carter, 1994; Lee & Pradhan, 2007). Moreover, by examining the relationship between previous landslide locations and other landslide-related elements, the frequency ratio technique can aid in predicting future landslides (Solaimani et al., 2013).

A region with a value of 1 indicates that landslides occurred roughly as predicted throughout the entire area, acting as the area's baseline or average. A greater correlation and likelihood of landslides occurring are indicated by values greater than 1. On the other hand, when the value is lower than 1, it indicates a weaker relationship and a decreased chance of landslides.

Consequently, the frequency ratio for each group of characteristics was ascertained by examining the frequency with which these factors were linked to landslip incidents. Equation (1) provides a clear definition of the frequency ratio, which is used to quantify the impact of each component on landslide occurrence by evaluating the relationship between each element and the incidence of landslides.

$$\mathrm{FR} = \frac{\left(\mathrm{NLSpix}/\sum_{\mathrm{i}=1}^{\mathrm{N}}\mathrm{NLSpix}\right) * 100}{\left(\mathrm{NCpix}/\sum_{\mathrm{i}=1}^{\mathrm{N}}\mathrm{NCpix}\right) * 100} \tag{1}$$

The number of landslide pixels in a factors class is represented by NLSpix, while the total number of landslide pixels in the entire region is represented by NLSpix. The number of pixels in a factors class is represented by NCpix and the total number of pixel classes in the entire area is represented by NCpix.

Finally, a landslide susceptibility map has been created based on the Landslide Susceptibility Index (LSI), which has been calculated by summing the FR values of the parameters using Eq. (2) (Lee & Talib, 2005).

$$\mathrm{LSI} = \sum \mathrm{FR} \tag{2}$$

Prediction Rate

In the next part of the model, for every factor class, Frequency Ratios (FRs) were normalized using Eq. (3) into Relative Frequency (RF) values within a probability range of 0 to 1.

$$\mathrm{RF} = \frac{\mathrm{FR_{IJ}}}{\sum_{\mathrm{I}=1}^{\mathrm{M}} \mathrm{FR_{IJ}}} \tag{3}$$

Despite this normalization, the RF continued to treat each conditioning element equally. Prediction Rate (PR) was established to get over this restriction and take into account the interactions between independent variables. By recognizing the differing significance of distinct parameters in forecasting landslip susceptibility, this modification attempted to improve the model (Acharya & Lee, 2019). PR entailed rating every conditioning element according to how well it performed using the dataset as Eq. 4:

$$PR = \frac{(RF_{max} - RF_{\min})}{(RF_{max} - RF_{min})} \tag{4}$$

The frequency ratio and Prediction Rate of Causative factors of landslides in the valley and the NH-44 is given Table 4.

Validation

The accuracy of the landslide susceptibility map for various sensitivity classes was validated using the Relative Landslide Density Index (R-index). The R-index is a precise quantitative measure used to evaluate the correlation between reported landslides and the spatial distribution susceptibility. The R-index can be obtained using Eq. (5), where the total number of pixels in a specific susceptibility class is represented by Ni and the number of landslides in that class is represented by ni. An increased R-index value signifies a more noticeable concentration of landslides in locations anticipated to be more vulnerable (Shahabi & Hashim, 2015).

Table 4 Frequency ratio of factors to landslide occurrences in the valley

Factors	Classes	% of class count	% of landslide count	F.R	R.F	$RF_{(MAX)}$—$RF_{(MIN)}$	P.R
Distance to faults	1	4.535894814	2.813299233	0.62023	0.159620856	0.107	1.07
	2	1.398971015	1.023017903	0.731265	0.188196351		
	3	1.953102296	1.534526854	0.785687	0.202202343		
	4	1.788611436	1.278772379	0.714953	0.183998349		
	5	90.32342044	93.35038363	1.033512	0.265982102		
	Sum	100	100	3.885647			
Distance to roads	1	23.23523764	81.58567775	3.511291	0.534026597	0.529	5.29
	2	5.574938542	7.928388747	1.422148	0.216292237		
	3	5.625709955	3.836317136	0.681926	0.103713011		
	4	4.458058333	4.092071611	0.917904	0.139602617		
	5	61.10605553	2.557544757	0.041854	0.006365537		
	Sum	100	100	6.575124			
Slope (degrees)	1	23.85159042	10.23017903	0.42891	0.08116718	0.194	1.94
	2	17.82397939	21.73913043	1.219656	0.23080864		
	3	20.4227268	23.52941176	1.152119	0.218027817		
	4	24.91159624	25.57544757	1.026648	0.194283649		
	5	12.99010715	18.9258312	1.456942	0.275712714		
	Sum	100	100	5.284275			
Plan_curvature	1	3.25559241	3.324808184	1.021261	0.189041477	0.1	1
	2	15.73247207	17.13554987	1.089184	0.201614429		
	3	59.24626061	50.63938619	0.854727	0.158215133		
	4	17.59055081	24.55242967	1.395774	0.258366124		

(continued)

Table 4 (continued)

Factors	Classes	% of class count	% of landslide count	F.R	R.F	$RF_{(MAX)}$—$RF_{(MIN)}$	P.R
	5	4.175124098	4.347826087	1.041365	0.192762837		
	Sum	100	100	5.40231			
Profile _curvature	1	3.348584286	2.813299233	0.840146	0.165124864	0.144	1.44
	2	13.27738496	12.53196931	0.943858	0.185508817		
	3	60.47035874	51.91815857	0.858572	0.168746399		
	4	18.21532043	28.64450128	1.57255	0.309073827		
	5	4.688351585	4.092071611	0.872817	0.171546093		
	Sum	100	100	5.087943			
Aspect	1	0.542494297	0	0	0	0.269	2.69
	2	6.934683554	4.347826087	0.626968	0.070918556		
	3	14.06253457	7.16112532	0.509234	0.057601268		
	4	13.33985886	9.462915601	0.709371	0.080239479		
	5	11.44630127	17.39130435	1.519382	0.17186259		
	6	11.26032864	26.85421995	2.384852	0.26975894		
	7	11.88265177	19.43734015	1.635775	0.185028162		
	8	12.27644939	8.95140665	0.729153	0.082477005		
	9	11.93001013	3.836317136	0.321569	0.036373749		
	10	6.324687522	2.557544757	0.404375	0.045740251		
	Sum	100	100	8.840679			
Distance _rivers	1	12.69787926	37.08439898	2.920519	0.449964571	0.39	3.9
	2	15.77812175	29.66751918	1.880295	0.289697133		
	3	6.33894641	3.324808184	0.524505	0.080810498		

(continued)

Table 4 (continued)

Factors	Classes	% of class count	% of landslide count	F.R	R.F	$RF_{(MAX)}$—$RF_{(MIN)}$	P.R
	4	14.06641859	11.25319693	0.800004	0.123256733		
	5	51.11863399	18.67007673	0.36523	0.056271065		
	Sum	100	100	6.490553			
Geomorphology	1	1.206603361	0	0	0	0.45	4.5
	2	49.39971679	65.2173913	1.320198	0.028879881		
	3	14.19516696	14.83375959	1.044987	0.022859523		
	4	16.15229276	2.046035806	0.126672	0.002770993		
	5	6.698653039	4.603580563	0.68724	0.015033659		
	6	0.059674339	0	0	0		
	7	0.328652676	0	0	0		
	8	0.941293177	0	0	0		
	9	0.002470389	0	0	0		
	10	5.081157635	3.069053708	0.604007	0.013212903		
	11	0.022586416	0	0	0		
	12	0.010004542	0	0	0		
	13	0.002588027	0	0	0		
	14	0.268513135	0.511508951	1.904968	0.04167198		
	15	0.047776901	0.255754476	5.353099	0.117101295		
	16	1.337951072	4.603580563	3.440769	0.075268272		
	17	0.888383	0	0	0		
	18	0.298745994	3.069053708	10.27312	0.22472885		
	19	0.06777529	0	0	0		

(continued)

Table 4 (continued)

Factors	Classes	% of class count	% of landslide count	F.R	R.F	$RF_{(MAX)}$—$RF_{(MIN)}$	P.R
	20	0.370071801	0	0	0		
	21	0.178386703	0	0	0		
	22	1.513193643	0	0	0		
	23	0.085420928	1.79028133	20.95835	0.458472644		
	24	0.39451903	0	0	0		
	25	0.070598592	0	0	0		
	26	0.002470389	0	0	0		
	27	0.314974634	0	0	0		
	28	0.002588027	0	0	0		
	29	0.040007473	0	0	0		
	30	0.005176054	0	0	0		
	31	0.010116832	0	0	0		
	32	0.002470389	0	0	0		
	Sum	100	100	45.7134			
Soil sub_group	1	11.31077574	14.57800512	0	0		
	2	8.754885766	1.534526854	0.114222	0.009199848		
	3	25.04699856	20.97186701	0.837301	0.067439223		
	4	15.48101015	17.64705882	1.139916	0.091813003		
	5	6.156800784	10.4859335	1.703146	0.137177584		
	6	11.64867428	13.29923274	1.141695	0.091956247		
	7	3.238392693	0	0	0		
	8	1.231064636	3.836317136	3.11626	0.250994843		

(continued)

Table 4 (continued)

Factors	Classes	% of class count	% of landslide count	F.R	R.F	$RF_{(MAX)}$—$RF_{(MIN)}$	P.R
	9	1.324502955	0.255754476	0.193095	0.015552545		
	10	4.227517096	1.023017903	0.24199	0.019490771		
	11	5.747869778	3.836317136	0.667433	0.053757459		
	12	1.349783421	0	0	0		
	13	0.529696944	0	0	0		
	14	3.843485066	12.53196931	3.260574	0.262618477		
	15	0.108542122	0	0	0		
	Sum	100	100	12.41563			
Lithology	1	22.39256976	32.48081841	1.450518	0.023648368	0.27	2.7
	2	5.896993997	9.207161125	1.561331	0.025455006		
	3	8.736176151	2.301790281	0.263478	0.004295587		
	4	0.997508665	2.301790281	2.307539	0.03762073		
	5	16.90605793	11.50895141	0.680759	0.011098686		
	6	1.286296681	3.58056266	2.783621	0.045382485		
	7	2.726076415	6.138107417	2.251627	0.036709173		
	8	0.047257681	0	0	0		
	9	0.062806775	0	0	0		
	10	0.007426513	0	0	0		
	11	0.140916344	0	0	0		
	12	1.621833713	0.255754476	0.157695	0.002570958		
	13	1.282280902	0.255754476	0.199453	0.003251758		
	14	0.680449568	0.511508951	0.751722	0.012255624		

(continued)

Table 4 (continued)

Factors	Classes	% of class count	% of landslide count	F.R	R.F	$RF_{(MAX)}$—$RF_{(MIN)}$	P.R
	15	0.00247372	0	0	0		
	16	0.455967551	0	0	0		
	17	0.203723119	0.255754476	1.255402	0.020467324		
	18	0.007656751	0	0	0		
	19	10.12318856	1.79028133	0.17685	0.002883249		
	20	5.077282586	0.767263427	0.151117	0.00246372		
	21	0.713196903	2.046035806	2.868823	0.046771568		
	22	2.325944247	4.092071611	1.759316	0.028682835		
	23	0.99175807	0.767263427	0.77364	0.012612957		
	24	1.95340851	1.023017903	0.523709	0.008538239		
	25	0.054909078	0.511508951	9.315563	0.15187533		
	26	1.087419267	0.767263427	0.705582	0.011503384		
	27	0.238483698	4.092071611	17.15871	0.27974522		
	28	0.789657325	0	0	0		
	29	0.039825814	0	0	0		
	30	6.003583467	13.29923274	2.215216	0.036115546		
	31	0.120773199	0.255754476	2.117643	0.034524772		
	32	0.283481834	0	0	0		
	33	0.115124337	0	0	0		
	34	0.190995779	0	0	0		
	35	0.40236494	0	0	0		
	36	0.307046422	0	0	0		

(continued)

Table 4 (continued)

Factors	Classes	% of class count	% of landslide count	F.R	R.F	$RF_{(MAX)}$—$RF_{(MIN)}$	P.R
	37	0.537868015	0	0	0		
	38	0.007538955	0	0	0		
	39	0.548341165	0	0	0		
	40	0.173315645	0	0	0		
	41	0.007308717	0	0	0		
	42	0.676412372	0	0	0		
	43	0.831491028	0	0	0		
	44	0.00247372	0	0	0		
	45	0.570620704	0	0	0		
	46	0.565234206	0	0	0		
	47	0.034996171	0	0	0		
	48	0.090247928	0.255754476	2.83391	0.046202359		
	49	0.449156791	1.023017903	2.277641	0.037133288		
	50	0.327077125	0	0	0		
	51	0.03027897	0	0	0		
	52	0.364884341	0.255754476	0.700919	0.011427367		
	53	0.301279764	0	0	0		
	54	0.015318856	0	0	0		
	55	0.020143145	0	0	0		
	56	0.002591516	0	0	0		
	57	0.040173848	0	0	0		
	58	0.010018029	0	0	0		

(continued)

Table 4 (continued)

Factors	Classes	% of class count	% of landslide count	F.R	R.F	$RF_{(MAX)}$—$RF_{(MIN)}$	P.R
	59	0.050432823	0	0	0		
	60	0.062453387	0.255754476	4.095126	0.066764468		
	61	0.002591516	0	0	0		
	62	0.004834997	0	0	0		
	Sum	100	100	61.3369			
Elevation	1	9.518198369	1.278772379	0.13435	0.023986996	0.362	3.62
	2	18.60669889	12.02046036	0.646029	0.115342443		
	3	10.62551003	18.41432225	1.733029	0.3094164		
	4	9.349091612	20.20460358	2.16113	0.385849813		
	5	51.9005011	48.08184143	0.926423	0.165404346		
	Sum	100	100	5.600962			
NDVI	1	16.42483527	10.4859335	0.638419	0.127021711	0.263	2.63
	2	14.3257057	24.04092072	1.678167	0.333892722		
	3	22.60822726	33.24808184	1.470619	0.292598413		
	4	29.24838664	26.342711	0.900655	0.179196876		
	5	17.39284514	5.882352941	0.338205	0.067290279		
	Sum	100	100	5.026065			

$$\mathrm{R} = \left(\frac{\mathrm{ni}}{\mathrm{Ni}}\right) / \sum \left(\frac{\mathrm{ni}}{\mathrm{Ni}}\right) * 100 \tag{5}$$

3 Results

3.1 Landslide Conditioning Factors of the Kashmir Valley

Various topographic factors, including elevation, plan curvature, profile curvature, aspect, drainage, and slope, were derived from ASTER DEM. Elevation, ranging from 1060 to 5338 m, has been classified into five classes: 1090–1900 m, 1900–2500 m, 2500–3100 m, 3100–3700 m, and 3700–5236 m. The landslide frequency (Fig. 5a) depicts that the highest occurrence of landslides is found in class five at the highest elevation, whereas the lowest landslide frequency is found at lower elevations.

The slope was measured in degrees, ranging from the highest value of 67° to the lowest value, which was close to zero, as obtained from the DEM. The slope map is categorized into five classes: 0–5°, 5–15°, 15–25°, 25–35°, and 35–65°. Given the significance of the slope as a contributing factor to landslide occurrence, it has been observed that landslides are prevalent across all classified slopes in the valley. Specifically, the slope range of 25–35° exhibited the highest concentration of landslides, as illustrated in Fig. 5b. Higher slope values show a correspondingly higher concentration of landslides, and the tendency for steeper slopes to have a larger concentration of landslides highlights the significance of the slope gradient as a major initiator of landslides in the valley.

Profile curvature within the valley ranges from −5.9 to 5, classified into five categories: −5.59—(−0.65), −0.65—(−0.19), −0.19—0.14, 0.14–0.55 and 0.55–5. The landslide frequency was highest in the range of −0.19 to 0.14 and lowest in the range of −5.59 to −0.65 (Fig. 5c). Regarding plan curvature, values span from −5.9 to 5.6, further divided into five classes: −5.9 to (−0.58), −0.58 to (−0.16), −0.16 to 0.15, 0.15 to 0.56, and 0.56 to 5.6. The landslide frequency in the valley is most prominent in the range of −0.16–0.15 and least prominent in the range of −5.9–(−0.58) (Fig. 5d). The high concentration of landslides in the medium class of both profile and plan curvatures underscores the significance of curvature in landslide occurrence. Considering both the plan and profile curvatures, an examination of the valley curvature shows several noteworthy patterns. The profile curvature of the valley varies from concave to convex, and sections where the curvature is between slightly concave and virtually flat have the highest landslide frequency. The plan curvature, however, shows comparable patterns, with the concentration of landslides being the largest in regions where the terrain has a small concavity. The consistent occurrence of landslides in these medium-curvature classes indicates a significant relationship between the valley curvature and susceptibility to landslides.

The five classes of the Normalized Difference Vegetation Index (NDVI) in the valley are −0.37–0.069, 0.069–0.190, 0.190–0.293, 0.293–0.391, and 0.391–0.63. The NDVI varies from −0.37 to 0.3. The medium class, which had a 0.19–0.29 landslip frequency, had the least amount of vegetation cover. The class with the highest NDVI value represents true plant cover with the lowest frequency of landslides, ranging from 0.31 to 0.63. From the frequency analysis of the NDVI, it has been found that landslides are more common in places that are devoid of vegetation. This pattern highlights the negative relationship between NDVI levels and landslide occurrence (Fig. 5e).

Lithology plays an important role in landslide occurrence, and in the valley, various types of lithology prevail. The primary lithologies associated with landslides are Alluvium, Moraines, Hillwash and Scree (AMHS), Carbonaceous Phyllite with Marble and Quartzite (CQ), Andesitic and Basaltic Lava Flows (AB), Diamictite, Arenite with Phyllite and Ash Beds (DAPA), Phyllite, Quartzite, Diamictite and Lava Flows (PQDL), Splintery Shale with Nodular Limestone (SNL), Calcareous Shale and Crystalline Limestone (CSCL), Siltstone, Arenite with Shale and Limestone (SASL), Limestone, Sandstone, Shale and Calcareous Bands (LSSC), Boulder Conglomerate and Sandstone (BCS), Limestone, Sandstone, and Shale (LSS), Variegated Shale, Quartz-Arenite and Conglomerate (VQAC), and Siltstone and Shale (SS). Among all lithological types, the most significant landslide occurrences were AMHS and LSS, which exhibited the highest landslide frequency in the valley (Fig. 5f). Landslide frequency analysis of the lithology shows that the susceptibility of landslides is significantly influenced by the geological composition of the surrounding area.

The major landforms in the valley contributing to landslides include Highly Dissected Hills and Valleys (HD), Piedmont Alluvial Plain (PA), Strike Ridge (SR), Younger Alluvial Plain (YA), Moderately Dissected Hills and Valleys (MD), Glacial Terrain (GT), Active Flood Plain (AF), WatBod-River (WR), Older Flood Plain (OF), and Terrace (T). The highest landslide frequency was observed in the Highly Dissected Hills and Valleys (HD), surpassing other landforms in the region, followed by the PA and WR. The analysis of landslide frequency in geomorphology has proven to be significant for understanding the causation of landslides (Fig. 5g).

Soil subtypes in the valley that contribute to landslides include Typic Eutrochrepts (TE), Lithic Cryorthents (LC), Typic Udifluvents (TU), Typic Cryorthents (TC), Lithic Udorthents (LU), Aeric Haplaquepts (AH), Aquic Hapludalfs (AqH), Typic Hapludalfs (THs), Fluventic Eutrochrepts (FE), and Typic Dystrochrepts (TD). The highest landslide frequency was observed in the LC soil subtype, followed by the TU, LU, and TD. Conversely, the landslide frequency was minimal in the TE, THs, and AqH (Fig. 5h).

The aspect layer for the region was derived from ASTER DEM, covering a range from 0° to 360°, and segmented into 10 directional categories. The slope aspect was classified as flat, north (N), northeast (NE), east (E), south (S), southeast (SE), southwest (SW), west (W), northwest (NW), or north. Analysis of landslide frequency revealed that the highest occurrences were observed in the southeast (SE), southwest (SW), and south (S) directions, with the east (E) and west (W) directions showing notable frequencies. In contrast, the north (N) aspect exhibited the lowest frequency of landslides (Fig. 5i).

A fault map was utilized to generate a fault buffer map for the valley to analyze the distance to the fault parameter, incorporating five buffer classes ranging from 0-100 m, 100 to 300 m, 300 to 600 m, 600 to 1200 m, and above 1200 m. Notably, the highest landslide frequency was observed in the above 1200-m buffer zone. It is noteworthy that the influence of faults in triggering landslides is relatively low in the valley, as historical landslide occurrences have primarily been along transportation routes away from the fault lines (Fig. 5j).

In the case of the distance-to-roads factor, a road map was employed to create road buffers for the valley, divided into five categories spanning 0-100 m, 100–300 m, 300–600 m, 600–900 m, and above 900 m. The analysis revealed that the 100-m buffer zone exhibited the highest landslide frequency compared to the other buffer zones (Fig. 5k). Examination of the distance to the road layer indicates that landslides are predominantly concentrated along roads, indicating a significant impact on landslide occurrences in the region.

Furthermore, the distance to rivers is a crucial conditioning factor for landslide susceptibility. The drainage map, derived from the ASTER DEM, was used to create a buffer map for the valley, segmented into five zones: 0–100 m, 100–300 m, 300–600 m, 600–900 m, and above 900 m. The analysis revealed that the highest concentration of landslides was found in the 0-100 m buffer zone, followed by the 100–300 m zone. Proximity to water bodies emerged as a significant factor in the valley, as most landslides occurred in close proximity to rivers (Figs. 4 and 5).

3.2 Landslide Susceptibility Index

After preparing the conditioning factors, the frequency ratio, relative frequency, and prediction rate were calculated, as listed in Tables 3 for the valley. Using the Eq. (4), the landslide susceptibility index was calculated for the study area. The generated layers were further rectified using low-pass and majority filters to enhance the quality of the susceptible layers (Fig. 6a). The susceptibility to landslides was divided into five classes: very low, low, medium, high, and very high. The results of the LSI of the valley show that approximately 15% of the total area lies in the very low susceptibility zone, displaying a minimal predisposition to landslides. Topographical and geological characteristics contributed to the low likelihood of landslide occurrence in these areas. Following the low susceptibility class, approximately 32% of the total area has low susceptibility to landslides. The results show that a large portion comes under a low-susceptibility area compared with other susceptibility classes. The medium susceptibility zone, constituting approximately 20% of the total area, denotes regions with a moderate level of vulnerability to landslides. The landslide causative factors in these areas may contribute to a higher potential for landslides compared with low susceptibility zones. Furthermore, the high and very-high-susceptibility zones

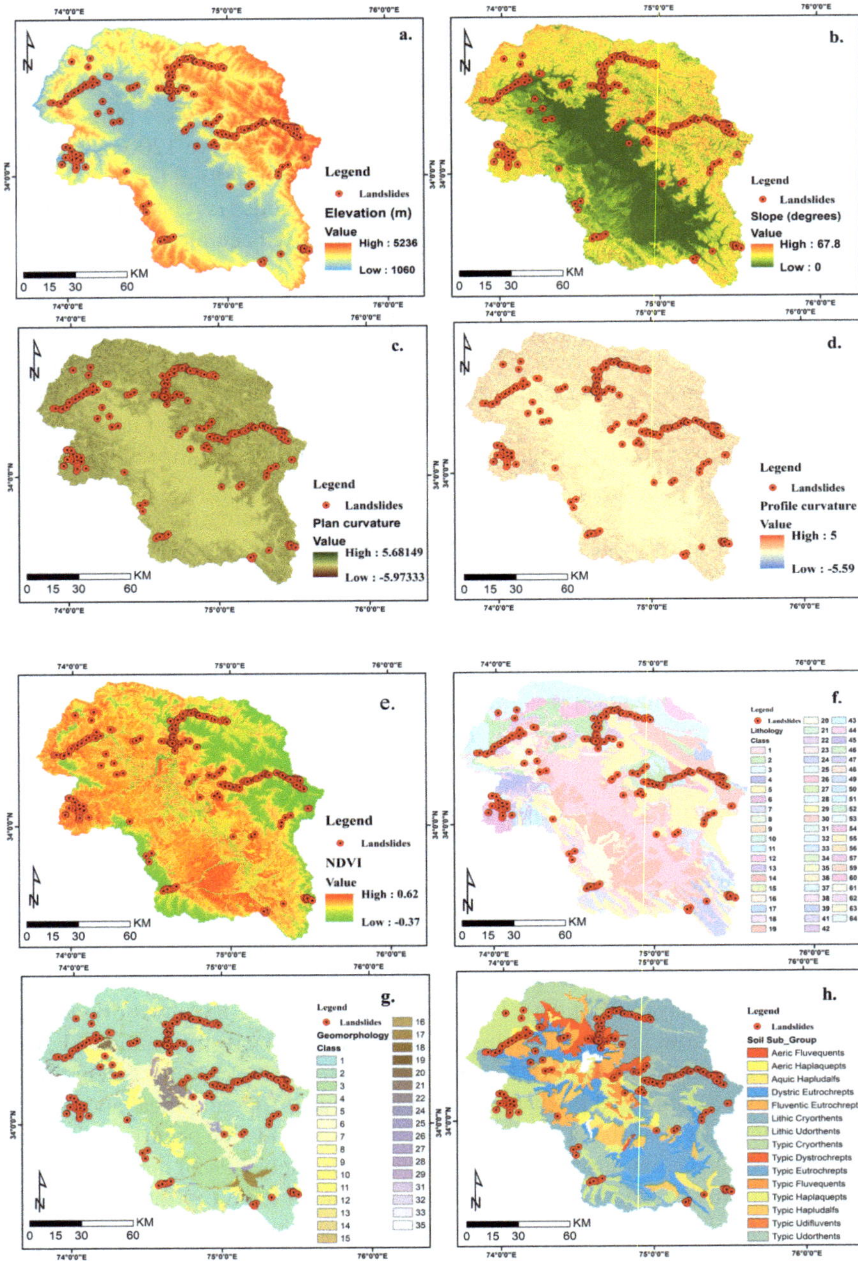

Fig. 4 Spatial distribution of landslides overlapped on the landslide causative factors of the valley

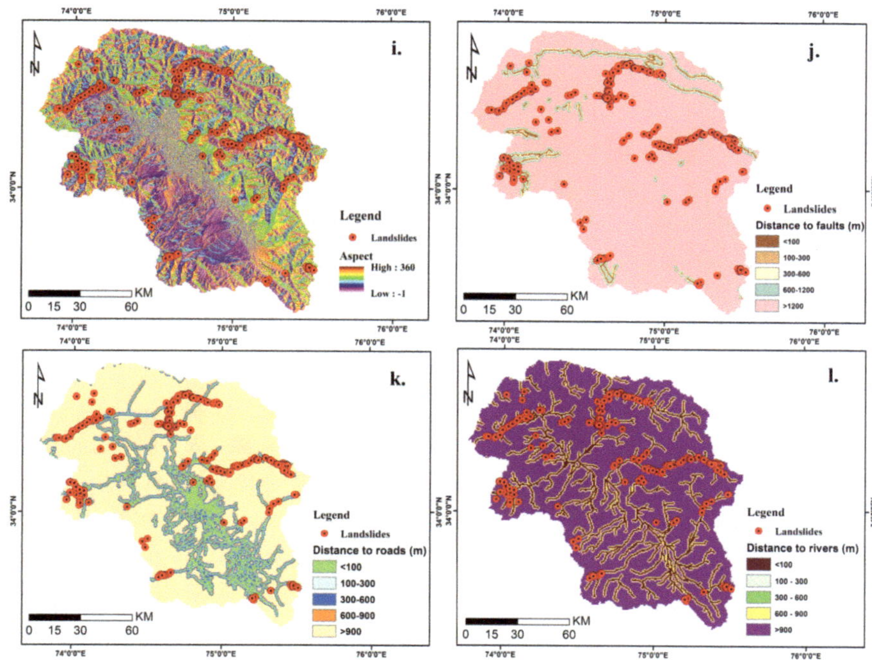

Fig. 4 (continued)

comprise 25.5% and 6.8%, respectively. The landslide-prone areas in these two zones represent the most critical areas for landslides in the valley (Fig. 6a).

The accuracy of the susceptibility maps was validated by the historical land slide inventory of the region and the relative landslide density index using Eq. 4. The accuracy of the maps can be determined from Fig. 7. In the valley, the relative landslide density index was 1.27% in the low susceptibility zone, 5% in the very low susceptibility zone, 28% in the medium susceptibility zone, 49% in the high susceptibility zone, and 16.1% in the very high susceptibility zone. The findings of the accuracy assessments showed that susceptibility levels and landslide prevalence were correlated, with higher susceptibility zones showing a higher occurrence of landslides in the valley.

To determine the prediction rates, the relative frequencies of the different conditioning variables were used, as listed in Tables 4. This statistical approach aims to identify the elements that cause landslides and determine how important they are in causing them. After calculating the prediction rates for each susceptibility indicator in the valley, it was found that some elements were extremely significant in determining the frequency of landslides. Conditioning factors such as elevation, geomorphology, distance to roads, and distance to rivers have been allocated weights that are considerably higher than others. However, slope, lithology, and the normalized difference vegetation index (NDVI) showed a moderate to high prediction rate, indicating their

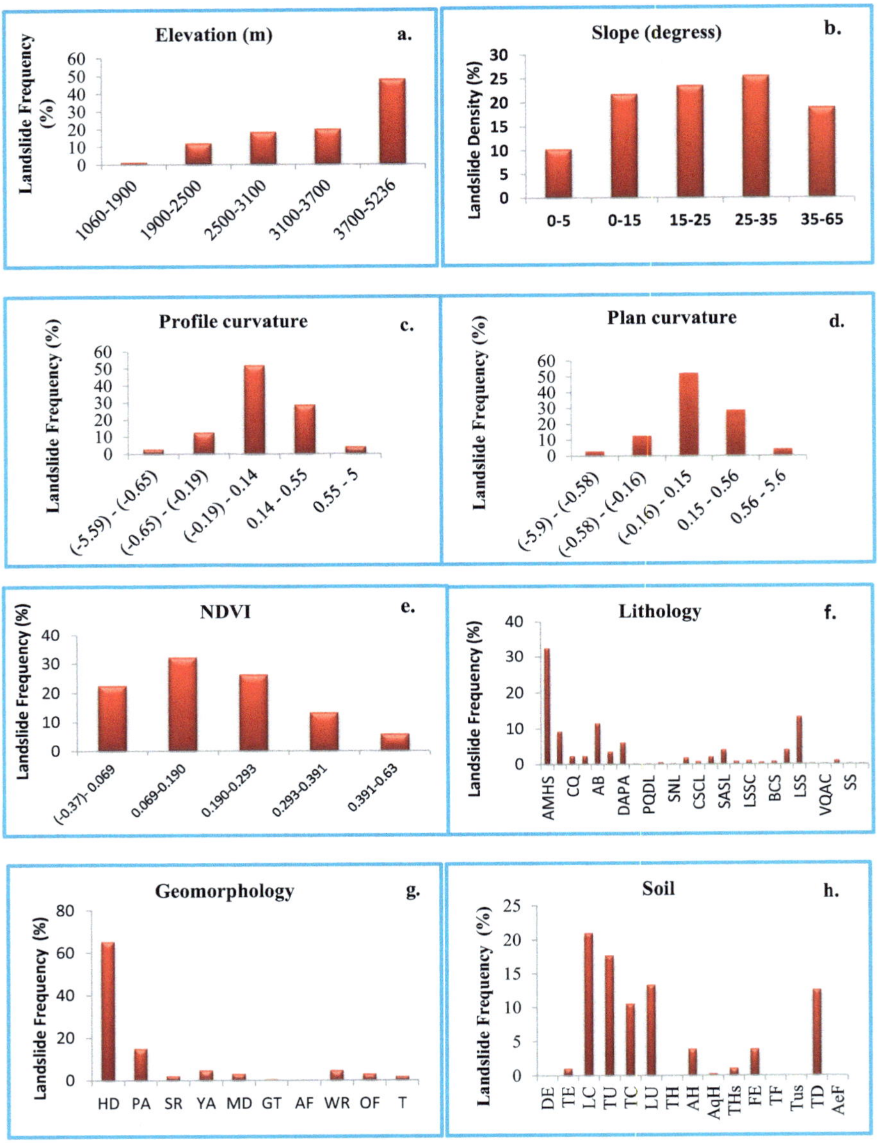

Fig. 5 Relationship of landslides occurrences with the causative factors: **a** elevation, **b** slope, **c** profile curvature, **d** plan curvature, **e** NDVI, **f** lithology, **g** geomorphology, **h** soil, **i** aspect, **j** distance to faults, **k** distance to roads, and **l** distance to rivers

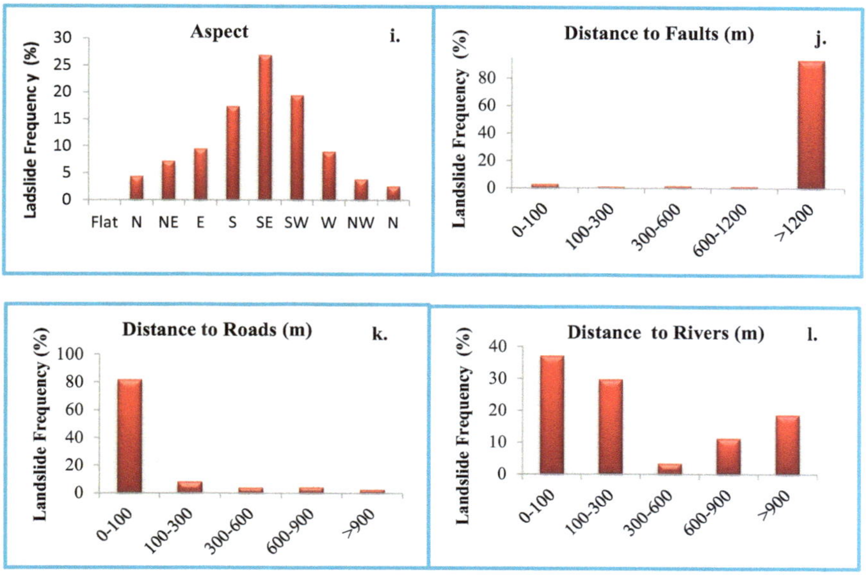

Fig. 5 (continued)

significant impact on the probability of landslides. Thus, these selected factors are important determinants that influence the vulnerability of the region to landslides (Fig. 1)

The results show that proximity to roads in both regions is the most significant element and carries the highest weight. This is because the track record of landslides occurs more frequently when they are closer to roads. Moreover NDVI is a strong predictor of susceptibility, which refers to low vegetation cover. The high prediction rate of the NDVI demonstrates how well it can detect landside-prone sections on highways.

In addition geomorphology is one of the most important conditioning factors. This element is important for risk assessment because of its significant impact on landslide susceptibility. Similar effects on landslip causation were shown by three equally important components at the locations under study: slope, lithology, and elevation. This balance of influences highlights how several elements interact to create the intricate dynamics that lead to landslide incidents.

4 Conclusion

A significant number of landslide-prone areas have been identified by the evaluation of the Landslide Susceptibility Index (LSI) results along the highways that crosses the Pir Panjal range and Greater Himalaya in the valley. The findings reveal that

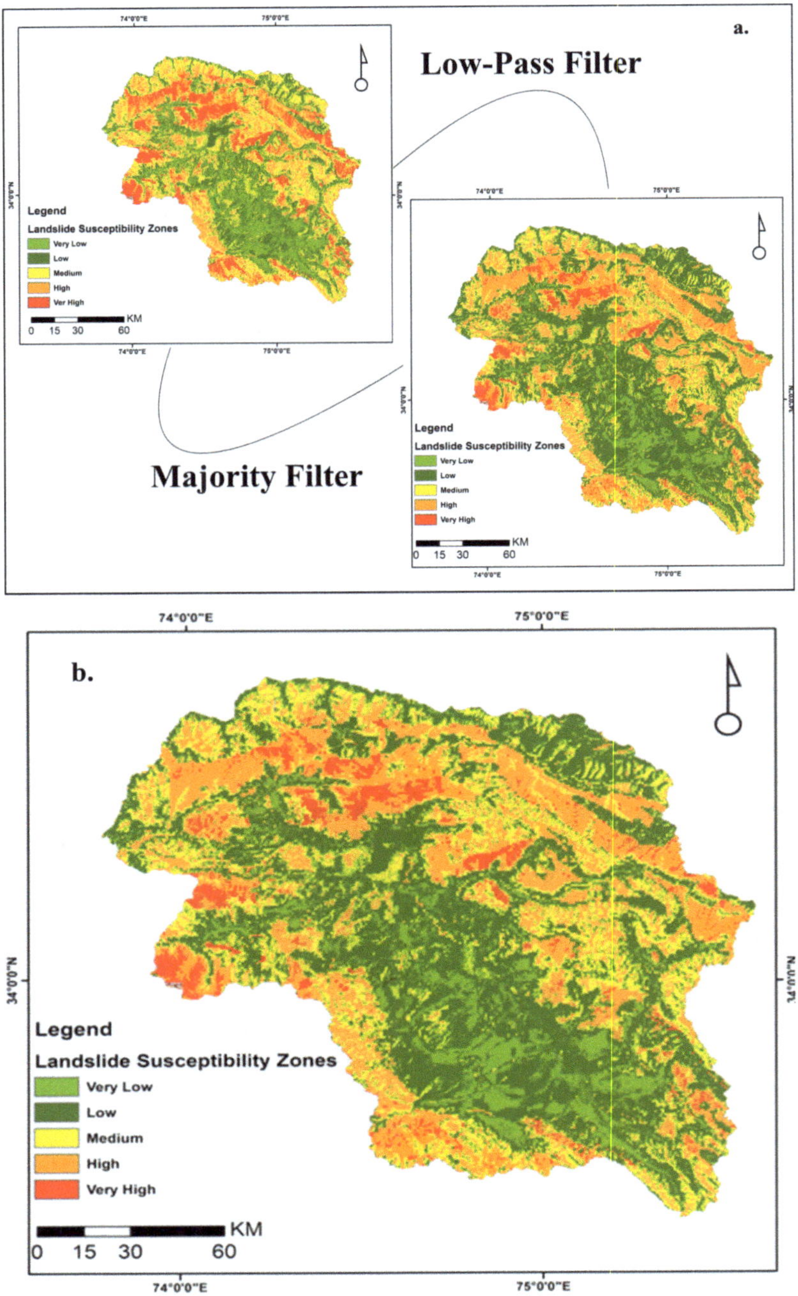

Fig. 6 Image enhancement by applying the filters **a** and **b** the landslide susceptibility map of the valley

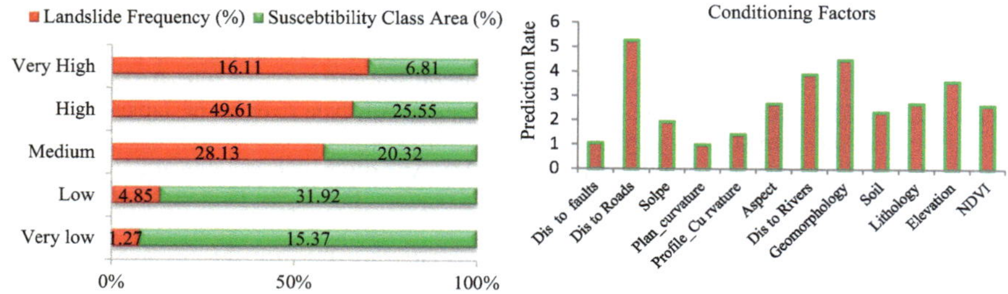

Fig. 7 Relative landslide density index of the valley and prediction rate of the conditioning factors of the valley

a sizable portion is susceptible to landslides, particularly when the valley floor is excluded from the scenario. When determining landslide susceptibility, the prediction rate emphasizes the importance of providing conditioning elements with proper weights and analyzing the relative importance of these variables in causing landslides. Identification of these important factors makes landslip risk assessments more accurate and hence mitigation options can be prioritized.

Given its proximity to both the Pir Panjal and Greater Himalayan ranges, there is an inherent risk of landslides. Therefore, the identified locations falling under the high and very high landslide susceptibility categories require more emphasis and strategic planning to mitigate the potential consequences of future landslides, thereby promoting increased resilience in the region.

References

Acharya, T. D., & Lee, D. H. (2019). Landslide susceptibility mapping using relative frequency and predictor rate along Araniko highway. *KSCE Journal of Civil Engineering, 23*, 763–776.

Ahmad, B. (2015). Landslide susceptibility mapping using frequency ratio model in the Himalayan region. *Arabian Journal of Geosciences, 8*(11), 9507–9521.

Akar, Ö., & Güngör, O. (2015). Integrating multiple texture methods and NDVI to the random Forest classification algorithm to detect tea and hazelnut plantation areas in Northeast Turkey. *International Journal of Remote Sensing, 36*(2), 442–464.

Akgun, A. (2012). A comparison of landslide susceptibility maps produced by logistic regression, multi-criteria decision, and likelihood ratio methods: *A case study at İzmir, Turkey. Landslides, 9*(1), 93–106. https://doi.org/10.1007/s10346-011-0283-7

Akgün, A., Dag, S., & Bulut, F. (2008). Landslide susceptibility mapping for a landslide-prone area (Findikli, NE of Turkey) by likelihood-frequency ratio and weighted linear combination models. *Environmental Geology, 54*(6), 1127. https://doi.org/10.1007/s00254-007-0892-2

Alam, A., Ahmed, B., Sammonds, P., & Kamal, A. M. (2023). Applying rainfall threshold estimates and frequency ratio model for landslide hazard assessment in the coastal mountain setting of South Asia. *Natural Hazards Research, 3*(3), 531–545.

Ali, N., Alam, A., Bhat, M. S., & Shah, B. (2022). Using historical data for developing a hazard and disaster profile of the Kashmir valley for the period 1900–2020. *Natural Hazards, 114*(2), 1609–1646.

Ali, S. A., Parvin, F., Pham, Q. B., Khedher, K. M., Dehbozorgi, M., Rabby, Y. W., et al. (2022). An ensemble random forest tree with SVM, ANN, NBT, and LMT for landslide susceptibility mapping in the Rangit River watershed, India. *Natural Hazards, 113*(3), 1601–1633.

Aniya, M. (1985). Landslide-susceptibility mapping in the Amahata river basin, Japan. *Annals of the Association of American Geographers, 75*(1), 102–114.

Arora, M. K., Das Gupta, A. S., & Gupta, R. P. (2004). An artificial neural network approach for landslide hazard zonation in the Bhagirathi (ganga) valley, Himalayas. *International Journal of Remote Sensing, 25*(3), 559–572.

Ayalew, L., & Yamagishi, H. (2005). The application of GIS-based logistic regression for landslide susceptibility mapping in the Kakuda-Yahiko Mountains, Central Japan. *Geomorphology, 65*(1–2), 15–31. https://doi.org/10.1016/j.geomorph.2004.06.010

Basu, T., & Pal, S. (2017). Identification of landslide susceptibility zones using logistic regression and GIS techniques in Darjeeling Himalaya, India. *Geoenvironmental Disasters, 4*(1), 1–17.

Bera, A., Gupta, S., & Das, S. (2021). GIS-based landslide susceptibility mapping using machine learning algorithms in the Himalayan region. *Natural Hazards, 108*(2), 1821–1848.

Bonham-Carter, G. (1994). *Geographic information systems for geoscientists: Modelling with GIS*. Elsevier.

Carver, S. J. (1991). Integrating multi-criteria evaluation with geographical information systems. *International Journal of Geographical Information Systems, 5*(3), 321–339. https://doi.org/10.1080/02693799108927858

Catani, F., Lagomarsino, D., Segoni, S., & Tofani, V. (2013). Landslide susceptibility estimation by random forests technique: Sensitivity and scaling issues. *Natural Hazards and Earth System Sciences, 13*(11), 2815–2831.

Corominas, J., Moya, J., Ledesma, A., Lloret, A., & Gili, J. A. (2003). Prediction of ground displacements and velocities from groundwater level changes at the Vallcebre landslide (Eastern Pyrenees, Spain). *Landslides, 1*(2), 83–96.

Dai, F. C., & Lee, C. F. (2002). Landslide characteristics and slope instability modeling using GIS, Lantau Island, Hong Kong. *Geomorphology, 42*(3–4), 213–228.

Ercanoglu, M., & Gokceoglu, C. (2004). Use of fuzzy relations to produce landslide susceptibility map of a landslide prone area West Black Sea region, Turkey. *Engineering Geology, 75*(3–4), 229–250.

Ermini, L., Catani, F., & Casagli, N. (2005). Artificial neural networks applied to landslide susceptibility assessment. *Geomorphology, 66*(1–4), 327–343.

Feizizadeh, B., Blaschke, T., Nazmfar, H., & Rezaei-Moghaddam, K. (2014). Landslide susceptibility mapping for the Urmia Lake Basin, Iran: A multi-criteria evaluation approach using GIS. *Journal of Mountain Science, 11*(4), 1059–1073.

Felicísimo, Á. M., Cuartero, A., Remondo, J., & Quirós, E. (2013). Mapping landslide susceptibility with logistic regression, multiple adaptive regression splines, classification and regression trees, and maximum entropy methods: A comparative study. *Landslides, 10*, 175–189.

Ghiasi, V. (2021). Landslide susceptibility mapping using machine learning algorithms and GIS techniques. *Natural Hazards, 106*(2), 1809–1835.

Gökceoglu, C., & Aksoy, H. Ü. S. E. Y. İ. N. (1996). Landslide susceptibility mapping of the slopes in the residual soils of the Mengen region (Turkey) by deterministic stability analyses and image processing techniques. *Engineering Geology, 44*(1–4), 147–161.

Kamp, U., Growley, B. J., Khattak, G. A., & Owen, L. A. (2008). GIS-based landslide susceptibility mapping for the 2005 Kashmir earthquake region. *Geomorphology, 101*(4), 631–642.

Kavzoglu, T., Sahin, E. K., & Colkesen, I. (2014). Landslide susceptibility mapping using GIS-based multi-criteria decision analysis, support vector machines, and logistic regression. *Landslides, 11*, 425–439.

Khan, H., Shafique, M., Khan, M. A., Bacha, M. A., Shah, S. U., & Calligaris, C. (2019). Landslide susceptibility assessment using frequency ratio, a case study of northern Pakistan. *The Egyptian Journal of Remote Sensing and Space Science, 22*(1), 11–24.

Kritikos, T., & Davies, T. (2015). Assessment of rainfall-generated shallow landslide/debris-flow susceptibility and runout using a GIS-based approach: Application to western southern Alps of New Zealand. *Landslides, 12*, 1051–1075.

Kumar, D., Thakur, M., Dubey, C. S., & Shukla, D. P. (2017). Landslide susceptibility mapping & prediction using support vector machine for Mandakini River basin, Garhwal Himalaya, India. *Geomorphology, 295*, 115–125.

Kumar, R., & Anbalagan, R. (2015). Landslide susceptibility zonation in part of Tehri reservoir region using frequency ratio, fuzzy logic and GIS. *Journal of Earth System Science, 124*, 431–448.

Hong, H., Chen, W., Xu, C., Youssef, A. M., Pradhan, B., & Tien Bui, D. (2017). Rainfall-induced landslide susceptibility assessment at the Chongren area (China) using frequency ratio, certainty factor, and index of entropy. *Geocarto International, 32*(2), 139–154.

Hong, H., Liu, J., Zhu, A. X., Shahabi, H., & Chen, W. (2019). A novel hybrid integration model for landslide susceptibility mapping. *Science of the Total Environment, 686*, 136–150. https://doi.org/10.1016/j.scitotenv.2019.05.437

Huang, Y., & Zhao, L. (2018). Review on landslide susceptibility mapping using support vector machines. *Catena, 165*, 520–529.

Ilanloo, M. (2011). A comparative study of fuzzy logic approach for landslide susceptibility mapping using GIS: An experience of Karaj dam basin in Iran. *Procedia-Social and Behavioral Sciences, 19*, 668–676.

Jacinth Jennifer, J., & Saravanan, S. (2022). Artificial neural network and sensitivity analysis in the landslide susceptibility mapping of Idukki district, India. *Geocarto International, 37*(19), 5693–5715.

Kayastha, P., Dhital, M. R., & De Smedt, F. (2013). Evaluation of the consistency of landslide susceptibility mapping: *A case study in Nepal. Environmental Earth Sciences, 70*(2), 647–661.

Lee, S., Hong, S. M., & Jung, H. S. (2017). A support vector machine for landslide susceptibility mapping in Gangwon Province, Korea. *Sustainability, 9*(1), 48.

Lee, C. F., & Talib, J. A. (2005). Probabilistic landslide susceptibility and factor effect analysis. *Environmental Geology, 47*(7), 982–990. https://doi.org/10.1007/s00254-005-1228-z

Lee, S., & Pradhan, B. (2007). Landslide hazard mapping at Selangor, Malaysia using frequency ratio and logistic regression models. *Landslides, 4*(1), 33–41.

Lee, S. (2002). Application of likelihood ratio and logistic regression models to landslide susceptibility mapping using GIS. *Environmental Geology, 42*(6), 655–661. https://doi.org/10.1007/s00254-002-0570-3

Lee, C. F., & Talib, J. A. (2005). Probabilistic landslide susceptibility and factor effect analysis. *Environmental Geology, 47*(7), 982–990. https://doi.org/10.1007/s00254-005-1228-z

Li, Y., Chen, W., Hong, H., et al. (2017). Spatial prediction of landslide susceptibility using machine learning models. *Catena, 151*, 147–160.

Mezughi, T. H., Akhir, J. M., Rafek, A. G., & Abdullah, I. (2011). Landslide susceptibility assessment using frequency ratio model applied to the Pahang region, Malaysia. *Journal of Geographic Information System, 3*(3), 206–213.

Nadim, F., Kjekstad, O., Peduzzi, P., Herold, C., & Jaedicke, C. (2006). Global landslide and avalanche hotspots. *Landslides, 3*(2), 159–173. https://doi.org/10.1007/s10346-006-0036-1

Pal, S. C., & Chowdhuri, I. (2019). GIS-based spatial prediction of landslide susceptibility using frequency ratio model of Lachung River basin, North Sikkim, India. *SN Applied Sciences, 1*, 1–25.

Pham, B. T., Prakash, I., Singh, S. K., et al. (2015). Landslide susceptibility assessment using support vector machines and GIS. *Environmental Earth Sciences, 73*(12), 8477–8495.

Pham, B. T., Bui, D., Prakash, I., & Dholakia, M. (2016). Evaluation of predictive ability of support vector machines and naive Bayes trees methods for spatial prediction of landslides in Uttarakhand state (India) using GIS. *Journal of Geomatics, 10*(1), 71–79.

Pourghasemi, H. R., Pradhan, B., Gokceoglu, C., & Mohammadi, M. (2012). Application of fuzzy logic and GIS for landslide susceptibility mapping in Iran. *Natural Hazards, 65*(2), 1639–1661.

Pradhan, B. (2010). Landslide susceptibility mapping of a catchment area using frequency ratio, fuzzy logic and multivariate logistic regression approaches. *Journal of the Indian Society of Remote Sensing, 38*, 301–320.

Pradhan, B., Kim, H., Lee, S., & Buchroithner, M. F. (2019). GIS-based landslide susceptibility mapping using machine learning techniques. *Geosciences, 9*(7), 314.

Pradhan, B., Lee, S., & Buchroithner, M. F. (2009). Use of geospatial data and fuzzy algebraic operators to landslide-hazard mapping. *Applied Geomatics, 1*, 3–15.

Qazi, N., Bhat, M. S., Ahmad, B., & Shah, A. A. (2023). Landslide susceptibility mapping using ensemble machine learning techniques in the Kashmir Himalaya. *Natural Hazards, 118*(1), 45–69.

Rabby, Y. W., Ishtiaque, A., & Rahman, M. S. (2020). Evaluating the effects of digital elevation models in landslide susceptibility mapping in Rangamati District, Bangladesh. *Remote Sensing, 12*(17), 2718.

Ramesh, M. V., & Anbazhagan, S. (2015). Landslide susceptibility zonation using frequency ratio and GIS for Nilgiris District, India. *Arabian Journal of Geosciences, 8*(8), 5795–5811.

Sassa, K., & Canuti, P. (Eds.). (2008). *Landslides-disaster risk reduction.* Springer Science & Business Media.

Saha, A. K., Arora, M. K., Gupta, R. P., Virdi, M. L., & Csaplovics, E. (2020). GIS-based statistical modelling for landslide susceptibility assessment in the Himalaya. *Natural Hazards, 101*(2), 525–548.

Saha, S., Saha, A., Hembram, T. K., Mandal, K., Sarkar, R., & Bhardwaj, D. (2022). Prediction of spatial landslide susceptibility applying the novel ensembles of CNN, GLM and random forest in the Indian Himalayan region. *Stochastic Environmental Research and Risk Assessment, 36*(10), 3597–3616.

Sarkar, S., & Kanungo, D. P. (2004). An integrated approach for landslide susceptibility mapping using remote sensing and GIS. *Photogrammetric Engineering & Remote Sensing, 70*(5), 617–625.

Shahabi, H., & Hashim, M. (2015). Landslide susceptibility mapping using GIS-based statistical models and remote sensing data in tropical environment. *Scientific Reports, 5*(1), 9899.

Shah, B., Alam, A., Bhat, M. S., Ahsan, S., Ali, N., & Sheikh, H. A. (2023). Extreme precipitation events and landslide activity in the Kashmir Himalaya. *Bulletin of Engineering Geology and the Environment, 82*(8), 328.

Shah, B., Bhat, M. S., Alam, A., Malik, U. F., Ali, N., & Sheikh, H. A. (2024). Establishing the landslide-triggering rainfall thresholds for the Kashmir Himalaya. *Natural Hazards, 120*(2), 1319–1341.

Shah, B., Sultan Bhat, M., Alam, A., Sheikh, H. A., & Ali, N. (2022). Developing landslide hazard scenario using the historical events for the Kashmir Himalaya. *Natural Hazards, 114*(3), 3763-3785.

Sharma, L. P., & Mehta, B. S. (2012). Landslide hazard zonation using GIS and remote sensing in mountainous terrain. *Natural Hazards, 60*(3), 1267–1288.

Sharma, L. P., Patel, N., Ghose, M. K., & Debnath, P. (2013). Synergistic application of fuzzy logic and geo-informatics for landslide vulnerability zonation—A case study in Sikkim Himalayas, India. *Applied Geomatics, 5*, 271–284.

Shen, H., Huang, F., Fan, X., Shahabi, H., Shirzadi, A., Wang, D., et al. (2022). Improving the performance of artificial intelligence models using the rotation forest technique for landslide susceptibility mapping. *International Journal of Environmental Science and Technology*, 1–16.

Sheng, T., & Chen, Q. (2017). An altitude based landslide and debris flow detection method for a single mountain remote sensing image. In *Image and graphics: 9th international conference, ICIG 2017* (pp. 601–610). Springer International Publishing.

Sidle, R. C., Pearce, A. J., & O'Loughlin, C. L. (1985). *Hillslope stability and land use.* American geophysical union.

Solaimani, K., Mousavi, S. Z., & Kavian, A. (2013). Landslide susceptibility mapping based on frequency ratio and logistic regression models. *Arabian Journal of Geosciences, 6*, 2557–2569.

Tsangaratos, P., & Benardos, A. (2014). Estimating landslide susceptibility through a artificial neural network classifier. *Natural Hazards, 74*, 1489–1516.

Van Westen, C. J., Rengers, N., & Soeters, R. (2003). Use of geomorphological information in indirect landslide susceptibility assessment. *Natural Hazards, 30*, 399–419.

Wardhana, M. D., Sofwan, A., & Setiawan, I. (2019). Fuzzy logic method Design for Landslide Vulnerability. In *E3S web of conferences* (Vol. 125, p. 03004). EDP Sciences.

Xiao, T., Segoni, S., Chen, L., Yin, K., & Casagli, N. (2020). A step beyond landslide susceptibility maps: A simple method to investigate and explain the different outcomes obtained by different approaches. *Landslides, 17*, 627–640.

Yao, X., Tham, L. G., & Dai, F. C. (2008). Landslide susceptibility mapping based on support vector machine: A case study on natural slopes of Hong Kong, China. *Geomorphology, 101*(4), 572–582.

Zhu, A. X., Wang, R., Qiao, J., Qin, C. Z., Chen, Y., Liu, J., et al. (2014). An expert knowledge-based approach to landslide susceptibility mapping using GIS and fuzzy logic. *Geomorphology, 214*, 128–138.

Reimagining Tunnel Safety in the Himalayas: A Human-Centric Framework for Seismic Risk, Resilience, and Sustainability

Shima Al-Balushi and Abdullah Ansari

Abstract Large-scale infrastructure developments demand significant financial investments, yet they can suffer extensive damage within moments when exposed to intense seismic activity. To address this, national disaster management agencies prioritize the development of resilient and sustainable infrastructure to mitigate the adverse impacts of earthquakes. This study focuses on conducting a comprehensive seismic microzonation of the Jammu Region (JR), situated in the northwestern Himalayan belt. The approach emphasizes seismic hazard assessment, with particular attention to site-specific characterization through geophysical field investigations. The seismic risk associated with the 345 km long Udhampur-Srinagar-Baramulla Rail Link (USBRL) is evaluated, particularly concerning its operational capacity following seismic events. The outcome in terms of alignment assessments, offer valuable input for reviewing both existing and future tunneling initiatives. This work contributes to the advancement of earthquake-resilient and sustainable infrastructure planning in tectonically sensitive regions such as the Himalayas.

Keywords Seismic microzonation · Tunnel vulnerability · Resilience · USBRL project · Seismic hazard index (SHI)

1 Introduction

Enhancing infrastructure resilience is essential for sustainable national development, especially in regions prone to natural hazards like earthquakes (Ansari et al., 2025a, b). The northwestern Himalayan region of Jammu and Kashmir (J&K) is particularly vulnerable, having experienced major seismic events such as the 2005 Kashmir

S. Al-Balushi (✉) · A. Ansari
Earthquake Monitoring Center, Sultan Qaboos University, Muscat, Oman

A. Ansari
Department of Civil Engineering, Inha University, Incheon, Republic of Korea

Department of Smart City Engineering, Inha University, Incheon, Republic of Korea

R. Ahmed et al. (eds.), *Climate Change, Water Resources, and Natural Hazards in the Himalayas*, Advances in Geographical and Environmental Sciences,
https://doi.org/10.1007/978-981-92-0835-7_10

earthquake, the 2019 Mirpur earthquake, and the 2022 Afghanistan earthquake. According to India's seismic microzonation standards (IS-1893 2016), Jammu lies in Zone IV, indicating a high seismic hazard with a peak ground acceleration (PGA) of 0.24 g. The region is geographically divided into two major sub-regions: Jammu Region (JR) and Kashmir Valley (KV), with JR spanning over 26,000 km^2 of varying terrain (Fig. 1). Recent years have seen significant infrastructure developments in JR, including railways, tunnels, and bridges, aimed at improving regional connectivity and economic integration. A flagship initiative is the Udhampur–Srinagar–Baramulla Rail Link (USBRL), a 345 km project under construction by the Indian government. The USBRL project serves as a vital link connecting the KV with mainland India. By introducing a railway-based mode of transportation, it will provide the people of the valley with a comfortable and efficient means of travel for trade, business, and other purposes, ultimately boosting the economy of Jammu and Kashmir. An integral highlight of this initiative includes the Chenab Bridge and the Anji Khad Bridge, both remarkable feats of engineering. The successful completion of these mega projects not only benefits the region but also reinforces the nation's position on the global stage, showcasing leadership and excellence in the construction industry.

Although various seismic hazard studies have been conducted in the greater Himalayan context, there is an absence of focused research dealing with the seismic vulnerability and risk assessment of tunnel infrastructure in JR. Past studies by

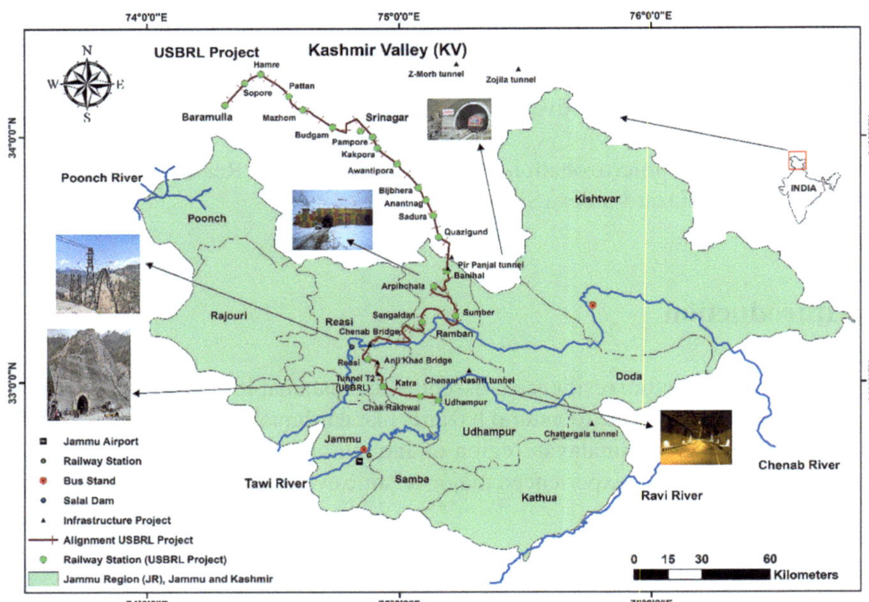

Fig. 1 Location map of the study area in the Jammu region, highlighting the alignment of the Udhampur-Srinagar-Baramulla Rail Link (USBRL) project across Jammu and Kashmir. The map illustrates key infrastructure components including major tunnels and railway

Mahajan et al. (2010), Rafi et al. (2012), and Yousuf and Bukhari (2020) dealt more with urban planning and did not really venture into tunnel-related project-level understanding. Portal stability and the safety of tunnel lining need serious attention, given the complex geology and frequent landslides in the region, particularly against seismic loads. Damage to tunnel portals has been prominently observed in past earthquakes in countries such as China, Taiwan, Italy, and Japan. Such damage has often triggered secondary hazards, including earthquake-induced landslides, ground failures, and slope instabilities, posing significant challenges during post-event recovery phases.

To bridge this gap, the detailed seismic microzonation of JR using data from 242 geophysical test sites is focused in this chapter. Depending on the parameters of the shear wave velocity and predominant frequency, JR was divided into low, moderate, and high hazard zones through the application of the Seismic Hazard Indexing (SHI) model. Tunnel vulnerability and post-earthquake serviceability of the USBRL project were assessed through a Semi-Quantitative Seismic Risk Assessment (SQ-SRA). Recommendation for the future planning, design, and construction of tunnels or other infrastructure projects is also briefly discussed. The study furnishes the risk matrices and serviceability maps that are required for the design of a resilient and sustainable tunnel infrastructure in this high seismic risk zone of the Himalayas.

2 Seismotectonic Prospect of the Jammu Region

The Jammu Region (JR) lies in the NW Himalayas within the Lesser Himalayan Zone, bounded by the Main Central Thrust (MCT) to the north and the Main Boundary Thrust (MBT) to the south. The area's geology mainly consists of sedimentary rocks thrust southward into the Siwalik Molasse following the Reasi Thrust. This complex tectonic activity continually releases energy, causing seismic events. Studies have identified numerous seismic sources within 350 km of Jammu, with several capable of generating significant ground acceleration. The geology features soft alluvial soils in the south and exposed rock formations, including weathered dolomite and Murre formations, in the north making it a key site for detailed seismic risk assessments of tunneling projects. The highly jointed and weathered Himalayan rock formations in this region increase susceptibility to earthquake-induced landslide hazards. Frequent seismic events, driven by both local and regional activities, create a persistent and alarming situation, underscoring the urgent need to address the sustainability and resilience aspects of Jammu and Kashmir.

3 Methodological Framework

The study was conducted in three integrated phases. Initially, geological and seismotectonic data were collected, and an updated earthquake catalog was developed. This catalog supported seismic hazard assessments using deterministic and probabilistic approaches, which estimate expected ground motion levels and account for uncertainties in earthquake occurrence (Cornell, 1968). In the second phase, geophysical survey at 242 locations conducted which led detailed site classification across JR. The final phase produced a microzonation map dividing JR into low, moderate, and high-risk zones using a Seismic Hazard Index (SHI). Circular tunnel models were analyzed to assess damage potential, and empirical relationships were developed between PGA and tunnel vulnerability. Using data from the USBRL project, the study applied a Semi-Quantitative Seismic Risk Assessment (SQ-SRA) to generate risk matrices and evaluate post-earthquake serviceability (Azad et al., 2022; Ansari et al., 2024).

4 Advanced Seismic Microzonation of Jammu Region (JR))

4.1 Seismic Risk Profiling

An updated earthquake catalog of 3114 events from 1552 to 2021 within 350 km of Jammu was compiled (Ansari et al., 2024). Using GMPEs (Ordaz et al., 2014), PGA at bedrock was computed from 87 seismic sources. Based on these values, JR was divided into five seismic zones with varying hazard levels.

4.2 Site Conditions and Ground Response Analysis

This study represented site response in JR by analyzing resonance frequency (f_0) and shear wave velocity (V_s), with the help of MHVSR and MSOR methods (Nakamura, 1989; Park et al., 1999). A change in f_0 indicates a change in geological conditions: low-frequency sites occur in sedimentary zones, and high-frequency sites occur in rock zones. Hence, the sites were classified into different zones (Class A to Class E) concerning frequency range, H/V curve shape, and soil properties, showing the pattern of amplification with a stiff subgrade. Hard Rock Zones such as Kishtwar having with $V_{s30} > 825$ m/s fall in Class A. The equivalent linear response analysis of the ground indicated the range of surface PGA values between 0.13 and 0.55 g, which helped in microzonation.

4.3 *Seismic Risk Classification*

This process classifies a region into various seismic risk zones, acting as vital inputs for safe urban planning and infrastructure design, such as for railways, tunnels, and bridges. Within the Jammu Region, four major seismic parameters were considered: peak ground acceleration (PGA) at bedrock, shear wave velocity at 30 meters depth (V_{s30}), amplification ratio (AR), and resonance frequency (RF). These parameters were prepared in the form of maps of the region and studied through Analytic Hierarchy Process (AHP), which assigns the weights depending on their relative importance with respect to seismic behavior. The analysis leads to the division of the seismic hazard map of JR into three risk zones: high (Zone A), moderate (Zone B), and low (Zone C). Almost 30% of the region is listed as high-risk zones, implying potential locations that could undergo strong ground shaking and consequent damage in the event of an earthquake in the future. The districts of Ramban, Reasi, Poonch, and Rajouri in JR exhibited significant seismic potential. The proximity of the USBRL project to the Main Boundary Thrust (MBT) and Main Central Thrust (MCT) presents a persistent challenge for engineers and planners in the tunnelling industry, as even small to moderate magnitude earthquakes could occur near the tunnel alignments. Additionally, earthquake activity associated with the Kishtwar Window poses further challenges for projects located in Doda and Kishtwar. Collectively, these areas fall within Zone A, representing a region of elevated seismic risk.

4.4 *Tunnel Resilience Under Seismic Loading*

Seismic vulnerability assessment depends on understanding seismic hazard, exposure, and structural susceptibility (Ansari et al., 2024). This study uses damage indices from Argyroudis and Pitilakis (2012) to establish an empirical link between tunnel damage and ground shaking intensity (PGA) via regression (see Table 1). Regression results, with $R^2 > 0.92$, confirm a strong correlation between DI and PGA.

5 Integrated Seismic Risk Analysis for the USBRL Project

5.1 *Engineering Resilience Across the Himalayas*

Before the USBRL project, National Highway 1A (NH1A) served as the only route for crossing the KV. Anyone traveling between the valley and the mainland had no option but to use this highway. However, NH1A, the main link to Srinagar, was challenging for frequent landslides that caused severe traffic disruptions. During the rainy season, rockfalls, debris flows, road cracking, and failures in traffic management often resulted in serious vehicle accidents. At times, the administrative authorities

Table 1 Quantifying the link between damage index and PGA

Seismic microzonation zone	Lining thickness	Overburden depth (m)		Empirical relationship
Zone A	300 mm	Shallow tunnel	5	Ln (DI) = 0.859ln (PGA) + 1.133
		Deep tunnel	200	Ln (DI) = 0.710ln (PGA) + 0.327
	600 mm	Shallow tunnel	5	Ln (DI) = 0.772ln (PGA) + 1.02
		Deep tunnel	200	Ln (DI) = 0.639ln (PGA) + 0.294
Zone B	300 mm	Shallow tunnel	5	Ln (DI) = 0.816ln (PGA) + 1.077
		Deep tunnel	200	Ln (DI) = 0.675ln (PGA) + 0.310
	600 mm	Shallow tunnel	5	Ln (DI) = 0.734ln (PGA) + 0.969
		Deep tunnel	200	Ln (DI) = 0.602ln (PGA) + 0.280
Zone C	300 mm	Shallow tunnel	5	Ln (DI) = 0.687ln (PGA) + 0.907
		Deep tunnel	200	Ln (DI) = 0.818ln (PGA) + 0.216
	600 mm	Shallow tunnel	5	Ln (DI) = 0.618ln (PGA) + 0.816
		Deep tunnel	200	Ln (DI) = 0.722ln (PGA) + 0.116

were forced to close the highway for several days or even weeks, leaving people cut off from their workplaces and daily activities. The USBRL project has emerged as a true lifeline in overcoming these challenges.

The USBRL Project addresses mentioned traffic challenges in Jammu and Kashmir by establishing a 345 km railway line connecting KV to the Indian Railways network (Rao, 2022). The alignment of the railway and adjacent active tectonic faults are illustrated in Fig. 1. Key structures include the Chenab and Anji Bridges in Reasi. The Chenab Bridge, a major engineering achievement, is designed to endure seismic, fire, impact, and blast loads considering the weak Himalayan rock geology. The Anji Khad Bridge, India's first cable-stayed railway bridge, connects Tunnel T2 (Fig. 2) and Tunnel T3. Most tunnels are near major tectonic faults including MBT, MCT, PT, and BT. Tunnel T49, extending 12.75 km, is India's longest tunnel, while Tunnel T80 (Pir Panjal Tunnel), with a maximum overburden of 1300 m, serves as the gateway to Kashmir Valley.

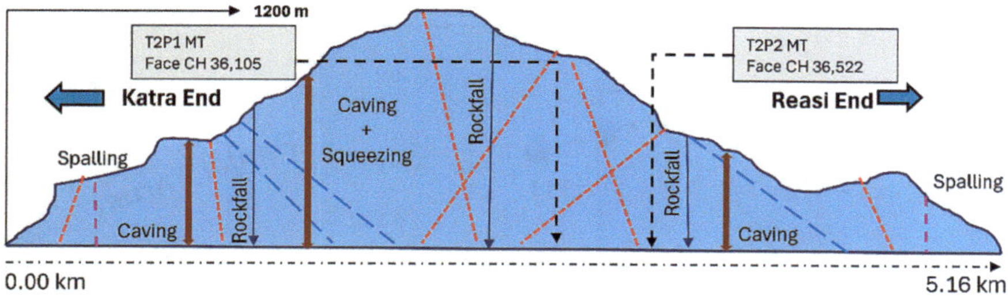

Fig. 2 Geological cross-section along tunnel (T2), illustrating the distribution of failure zones including rockfall, caving, and squeezing conditions

5.2 *Post-Seismic Serviceability Assessment of the USBRL Project*

Seismic fragility functions are valuable for estimating the probability of damage for a specified set of hazard inputs and material properties. Ansari et al. (2024) applied the fragility function derivation approach proposed by Argyroudis and Pitilakis (2012) to develop functions tailored for USBRL tunnels. Tunnel T14 in Phase P21 showed the highest probability (0.44) of extensive damage due to proximity to tectonic zones. Tunnels T12 and T13 in the same phase exhibited lower risk, maintaining Phase P21's accessibility with minor repairs. Likewise, tunnels in Phase P1 (e.g., T1 and T2) are susceptible due to their closeness to active faults.

Semi-Quantitative Seismic Risk Assessment (SQ-SRA) method defines thresholds for acceptable or unacceptable risk and categorizes system performance accordingly (Eskesen et al., 2004). According to SQ-SRA, USBRL serviceability is categorized into A (accessible), B (inaccessible), and C (accessible with moderate repair), distribution of these categories along the Udhampur–Qazigund route is 33%, 56%, and 11%, respectively. Figure 3 visualizes this serviceability map for phase 1. Phase 1 is further divided into three sub-phases: P11, P12, and P13. Phase P1, covering Udhampur to Reasi, showed mainly minor damage (DS_1), including seepage and small rockfalls. Both P11 and P13 are classified as accessible. Sub-phase P13, located adjacent to Reasi, includes tunnels T6, T10, and T11. Accessibility in this sub-phase enables train movement into the Qazigund area. Phase P2 includes moderate damage (DS_2), which can be mitigated with repairs. Phase P3, however, is the least functional, exhibiting extensive (DS_3) and collapse-level (DS_4) damage to tunnel linings and portals. Large cracks and instability in Phase P3 demanded major rehabilitation efforts and threaten project continuity. However, tunnel locations in Phase 3, covering the Qazigund to Sangaldan section, exhibited a high probability of damage, with most tunnels experiencing extensive damage scenarios. Consequently, Phase 3 falls into the inaccessible category, posing significant challenges for resolving traffic disruptions during post-seismic recovery.

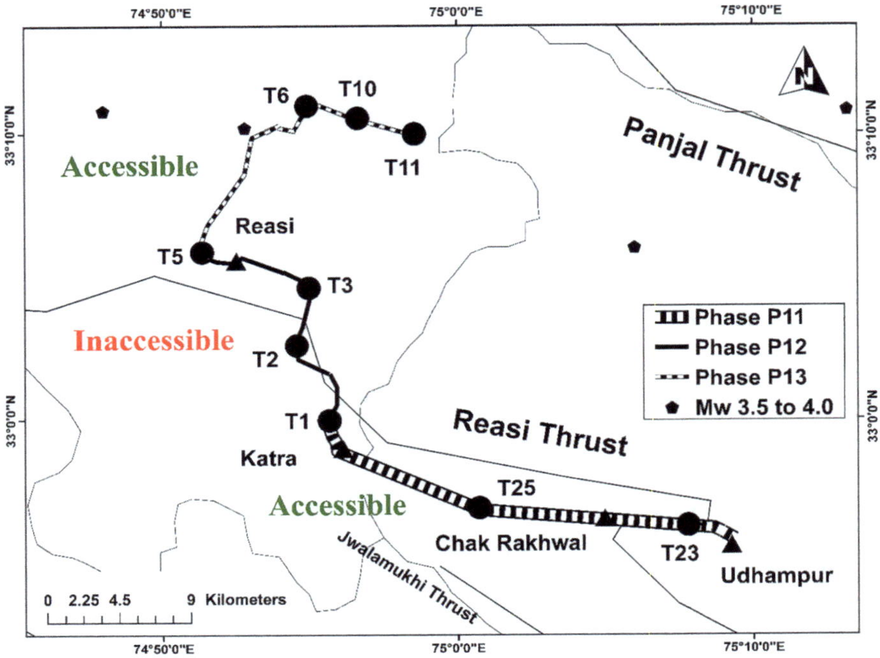

Fig. 3 Serviceability map of USBRL project (Phase 1)

6 Discussions and Conclusions

A Semi-Quantitative Seismic Risk Assessment (SQ-SRA) conducted for the Udhampur–Srinagar–Baramulla Rail Link (USBRL) evaluated both the seismic vulnerability and post-event serviceability of tunnels along the alignment. The findings revealed distinct variations in tunnel performance, recovery time, traffic management, and secondary hazard exposure across the three phases. In Phase 1 (Udhampur to Reasi), most tunnels are expected to remain accessible after seismic events, enabling rapid resumption of train operations with minimal recovery time and negligible traffic congestion. Phase 2 (Reasi to Sangaldan) would remain accessible but require moderate repairs, leading to partial operational delays, potential traffic congestion at diversion points, and extended but manageable recovery periods. In contrast, Phase 3 (Sangaldan to Qazigund) is projected to become inaccessible due to severe seismic damage compounded by landslide-prone slopes and highly weathered rock formations. This scenario would cause prolonged recovery timelines, significant traffic congestion along alternate routes, and heightened exposure to secondary hazards such as slope failures and debris flows, severely impacting post-seismic transport continuity.

In the Himalayan context, many tunnels in Phase 3 are located near active fault zones, and their presence in geologically unstable terrain increases the likelihood of deformation, squeezing, and structural failure during strong seismic events. The study's outputs provide essential support for enhancing tunnel safety and resilience. These findings help raise public awareness of seismic hazards, guide the development of emergency response and tunnel serviceability strategies, and inform cost–benefit and risk analyses for future infrastructure projects. They also assist engineers, consultants, and policymakers in planning and designing sustainable, resilient tunnel systems capable of withstanding seismic impacts in the Himalayan region.

7 Future Recommendations

The analysis of tunnel safety in the Himalayas highlights varying levels of risk, with Udhampur, Reasi, and Ramban showing high safety requirements due to complex geology and seismic vulnerability, while Kathua, Doda, and Rajouri require basic mitigation measures. Table 2 outlines the tunnel safety levels, design-phase safety recommendations, and construction-phase safety recommendations for the 10 districts in JR. Design phase recommendations focus on site-specific geotechnical investigations, seismic-resistant lining, and hazard mapping, whereas construction phase measures emphasize continuous monitoring, temporary support, and emergency preparedness. Implementing these measures ensures resilient and sustainable tunnel infrastructure, enhancing transport reliability and reducing disaster risk. This aligns with UN SDG 9 by promoting resilient infrastructure and sustainable industrialization and SDG 11 by supporting safe, sustainable, and disaster-resilient communities through improved mobility and connectivity.

Table 2 Tunnel safety recommendations

District	Tunnel safety level	Design phase safety recommendations	Construction phase safety recommendations
Jammu	Moderate	Conduct detailed geotechnical investigations; include seismic design criteria; ensure proper lining and support systems	Monitor excavation stability; implement real-time slope and ground monitoring; enforce strict safety protocols
Kathua	Low	Perform detailed site-specific hazard analysis; consider rock reinforcement; design for potential landslides and floods	Stabilize excavation faces; implement temporary supports; maintain emergency evacuation routes
Samba	Moderate	Design tunnel lining for moderate seismic loads; include drainage and waterproofing measures	Use controlled blasting and mechanical excavation; continuously monitor rock deformation; enforce PPE usage
Udhampur	High	Incorporate high seismic loading considerations; advanced ground support systems; detailed rock mass characterization	Implement rigorous monitoring of ground movement; adjust excavation methods based on real-time data; emergency preparedness drills
Reasi	High	Include seismic-resistant design; consider slope stability and potential rockfalls; optimize alignment to avoid weak zone	Continuous monitoring of rock and soil behavior; deploy additional support where needed; maintain clear safety communication channels
Ramban	High	Seismic-resistant lining; extensive hazard mapping; consider hydrological impacts; robust ventilation design	Continuous face inspection; use of flexible support systems; controlled blasting; ensure trained personnel for emergency situations
Kishtwar	Moderate	Include seismic and landslide mitigation; detailed geological mapping; optimized tunnel alignment to avoid faults	Monitor deformation and vibrations; provide temporary supports; ensure proper material handling and safety training
Doda	Low	Conduct detailed rock mass evaluation; implement basic seismic design; consider drainage and slope stability	Controlled excavation; implement temporary support and shoring; emergency response planning
Poonch	Moderate	Design for moderate seismic effects; consider slope stabilization; ensure proper drainage design	Monitor tunnel face stability; use temporary support systems; maintain strict construction safety measures
Rajouri	Low	Basic seismic considerations; geotechnical survey; avoid weak rock zones in alignment planning	Provide temporary support; monitor excavation; ensure proper lighting, ventilation, and emergency protocols

References

Ansari, A., Alluqmani, A. E., Huang, Z., Lee, J. H., Özden, G., Fissha, Y., et al. (2025a). Fuzzy synthetic approach for seismic risk assessment of bridges with insights from the 2023 Kahramanmaras earthquake in Turkiye. *Scientific Reports, 15*(1), 13418.

Ansari, A., Lee, J. H., & Alluqmani, A. E. (2025b). Assessing seismic impact and fragility of circular tunnels in north Gyeongsang, South Korea: Enhancing post-seismic road route functionality. *Structure and Infrastructure Engineering*, 1–15.

Ansari, A., Seshagiri Rao, K., & Jain, A. K. (2024). Seismic microzonation of the himalayan region considering site characterization: Application toward seismic risk assessment for sustainable tunneling projects. *Natural Hazards Review, 25*(1), 04023052.

Argyroudis, S. A., & Pitilakis, K. D. (2012). Seismic fragility curves of shallow tunnels in alluvial deposits. *Soil Dynamics and Earthquake Engineering, 35*, 1–12. https://doi.org/10.1016/j.soildyn.2011.11.004

Azad, M. A., Naithani, A. K., & Singh, S. K. (2022). Application of the NATM methodology for the excavation of rail tunnel in difficult geological conditions: A case from Garhwal Himalaya, Uttarakhand. *Journal of the Geological Society of India, 98*(11), 1553–1559.

Cornell, C. A. (1968). Engineering seismic risk analysis. *Bulletin of the Seismological Society of America, 58*(5), 1583–1606.

Eskesen, S. D., Tengborg, P., Kampmann, J., & Veicherts, T. H. (2004). Guidelines for tunnelling risk management: International tunnelling association, working group no. 2. *Tunnelling and Underground Space Technology, 19*(3), 217–237.

Mahajan, A. K., Thakur, V. C., Sharma, M. L., & Chauhan, M. (2010). Probabilistic seismic hazard map of NW Himalaya and its adjoining area, India. *Natural Hazards, 53*, 443–457.

Nakamura, Y. (1989). A method for dynamic characteristics estimation of subsurface using microtremor on the ground surface. *Quarterly Reports of Railway Technical Research Institute, 30*(1).

Ordaz, M., Cardona, O., Salgado-Gálvez, M. A., Bernal, G., Singh, K., & Zuloaga, D. (2014). Probabilistic seismic hazard assessment at global level. *International Journal of Disaster Risk Reduction, 10*, 419–427.

Park, C. B., Miller, R. D., & Xia, J. (1999). Multichannel analysis of surface waves. *Geophysics, 64*(3), 800–808.

Pitilakis, K., Riga, E., & Anastasiadis, A. (2012). Design spectra and amplification factors for Eurocode 8. *Bulletin of Earthquake Engineering, 10*(5), 1377–1400.

Rafi, Z., Lindholm, C., Bungum, H., Laghari, A., & Ahmed, N. (2012). Probabilistic seismic hazard of Pakistan, Azad-Jammu and Kashmir. *Natural Hazards, 61*, 1317–1354.

Rao, K. S. (2022). The art of Tunnelling in challenging Himalayan ground conditions. In *8th IGS Ferroco Terzaghi oration 2022, IIT*.

Yousuf, M., & Bukhari, K. (2020). Hazard estimation of Kashmir Basin, NW Himalaya using probabilistic seismic hazard assessment. *Acta Geophysica, 68*, 1295–1316.

Climate Impacts on Forests, Ecosystems, and Biodiversity

Forest Sector Vulnerability Assessments and Adaptation to Climate Change in the Kashmir Himalayas

Talib Bashir Bhat, Shah Murtaza Mushtaq, Akhlaq Amin Wani, Asif Ali Gatoo, Mir Muskan Un Nisa, and Shahid Saleem

Abstract The Indian Himalayan Region (IHR) is highly vulnerable to climate change and its variability leading to the degradation of forest ecosystems. Thus, this study employs an indicator-based approach to assess the forest vulnerability at both block and range levels in the Sindh Forest Division of Jammu and Kashmir, India. Total fourteen indicators were selected from socioeconomic, biophysical and ecologic domain. Systematic sampling belt-transaction methods were used to lay 105 sample plots at various locations. Weights were assigned using the pair wise comparison method (PCM), standardizing indicator values, computing the composite vulnerability index (CVI) along with its visual representation. The Composite Vulnerability Indices developed in this study are based on a set of 14 common indicators. The findings revealed that the Akhal I block had the highest Composite Vulnerability Indices of 0.67 (1st rank), while Harran II had a value of 0.65 (2nd rank), indicating that these blocks were more among the most vulnerable. Conversely, the Chattergul forest block had the lowest Composite Vulnerability Indices of 0.39, making it as the least vulnerable with a (15th rank). The Harran-Shallabugh range had the highest Composite Vulnerability Indicesof 0.62 (1st rank), followed by the Manasbal range with a value of 0.44 (2nd rank) and least vulnerable was Sindh range (0.43). The major drivers for all relative to the case study were high forest fire frequency, highest population density, low aspect profile, a high proportion of BPL households, a greater percentage of area under open forest, the highest drainage density, percent area under slope greater than 30°, low literacy rate, invasive species prevalence, and low values of Important Vital Index and Shannon diversity Index.

T. B. Bhat (✉) · S. M. Mushtaq · A. A. Wani · A. A. Gatoo
Division of Natural Resource Management, Faculty of Forestry, Sher-E-Kashmir University of Agricultural Sciences and Technology of Kashmir, Srinagar, Jammu and Kashmir, India

M. Muskan Un Nisa
Division of Forest Products and Utilization, Faculty of Forestry, Sher-E-Kashmir University of Agricultural Sciences and Technology of Kashmir, Srinagar, Jammu and Kashmir, India

S. Saleem
Department of Geography and Disaster Management, School of Earth and Environmental Sciences, University of Kashmir, Srinagar, India

R. Ahmed et al. (eds.), *Climate Change, Water Resources, and Natural Hazards in the Himalayas*, Advances in Geographical and Environmental Sciences,
https://doi.org/10.1007/978-981-92-0835-7_11

Keywords Climate change · Vulnerability · Indian himalayan region · Adaptation · Indicator

1 Introduction

Climate Change is one of the most dangerous and pressing issues facing environmental and development agendas (Scholes, 2016; Oppenheimer et al., 2014; Masson-Delmotte, 2026). Its effects vary over time, as some ecosystems, economies and societies are more vulnerable to its consequences than others (Marí-Dell'Olmo et al., 2022). For instance, the Himalayan region is currently confronted with various environmental and socioeconomic challenges due to a troubling temperature increase of 0.15–0.65 degrees per decade (Ahmed et al., 2021; Kraaijenbrink et al., 2017). Forest ecosystems, in particular, remain a critical component of the global biogeochemical cycle and substantially affect the world's climate (Lu et al., 2026). These ecosystems will also act as carbon reservoirs and assist in controlling and capturing carbon dioxide content in the environment, thereby influencing climate patterns (González De Andrés, 2019; Pugh et al., 2019). Forests are home to over 80% of terrestrial plants and animals and almost 60% of the world's tribal people (WWF, 2017; Bradford et al., 2019). Climate change poses significant threats to forest ecosystems and forest-dependent communities, necessitating comprehensive vulnerability assessments and adaptive management strategies (Anjos & De Toledo, 2018; Chettri et al., 2023). Forest vulnerability to climate change is influenced by various factors, including exposure to climatic stressors, sensitivity of forest ecosystems, and adaptive capacity (Ordóñez & Duinker, 2014). Key elements contributing to forest vulnerability include drought, heat stress, wind, susceptibility to insects and diseases, and sensitivity of young trees and specific tree species (Marchin et al., 2021; López et al., 2022). Additionally, forest conversion, fragmentation, and disturbance exacerbate the effects of climate change on forest ecosystems (Roshani et al., 2022). Vulnerability has been defined in many ways for diverse purposes, there is an assessment of the metrics in a unified and consolidated form for its evaluation (Li et al., 2014; Chauhan et al., 2020). Vulnerability is a theoretical notion that describes a non-observable and non-measurable condition of a system (Hinkel, 2011). A common approach involves the use of vulnerability indicators and indices. For instance, the construction of a forest ecosystem vulnerability index using exposure, sensitivity, and adaptive capacity components (Roshani et al., 2024). Similarly, vulnerability indicators such as productivity, timber and carbon stocks, biodiversity, disturbances, and silvicultural flexibility (Seidl et al., 2010). Indicators like biological richness, disturbance index, canopy cover, and slope to assess inherent vulnerability of forests (Sharma et al., 2015; Sharma et al., 2017). Interestingly, some studies propose more comprehensive frameworks for vulnerability assessment as defined by the Intergovernmental Panel on Climate Change a standardized procedure based on exposure, hazard magnitude, susceptibility, and lack of adaptive capacity, (Lecina-Diaz et al.,

2020). Assessing the vulnerability of forests to climate change is a key component in developing successful adaptive strategies (Kumar et al., 2018). Although nevertheless, several hurdles often impede the practical application of vulnerability evaluations and adaptive measures (Brandt et al., 2014; Stanturf, 2015; Maier & Winkel, 2017). A more comprehensive approach to evaluating vulnerability requires considering exposure, sensitivity, and resilience in conjunction (Li et al., 2018). These factors enhance the understanding to assess vulnerability of forest ecosystems to climate change (Dong et al., 2020; Faulkner et al., 2020).

The forests of Jammu and Kashmir (J&K) are indeed acknowledged for their rich biodiversity including temperate forests, alpine zones, and rangelands, contribute to its significant plant diversity and biomass production, making it an important region for biodiversity conservation and sustainable resource management (Tali et al., 2019; Ridwan et al., 2023; Wani & Pant, 2023). Between 2001 and 2023, forests in Jammu and Kashmir emitted 86.2 ktCO_2e/year, and removed −4.50 MtCO_2e/year. This represents a net carbon sink of −4.42 MtCO_2e/year (www.globalforestwatch.org). Disparities in ecological exposure, susceptibility, economic sectors, and vulnerability of forests are the main factors responsible for sensitivity towards forest ecosystems (Ghosh & Ghosal, 2020; Dimri et al., 2021). Interestingly, vulnerability patterns can vary across species and regions, with some studies finding that adaptation to site conditions may prevail over species-specific determinism in forest responses to climate change (Dorado-Liñán et al., 2019). Various approaches and analytical frameworks have been utilized in these evaluations to examine the intricate relationships between factors that influence climate vulnerability throughout the nation (Gopalakrishnan et al., 2011; Chaturvedi et al., 2011; Hosonuma et al., 2012; Kumar et al., 2019). Currently, limited literature exists that focuses on community dynamics, changes that have occurred over time, and how these changes have evolved and are viewed in the Kashmir Himalayas. Therefore, the present study aimed to assess vulnerability and examine the impacts of climate change in the Sindh Forest Division of Ganderbal, Jammu, and Kashmir, India.

2 Materials and Methods

2.1 Study Area

The Union Territory of Jammu and Kashmir is located in the northwestern Himalayan region, covering an area of 42,241 km^2, with diverse land use and forested landscapes (Saleem et al., 2021; Mushtaq et al., 2024; Saleem et al., 2024). This study was conducted in the Sindh Forest Division of the Jammu and Kashmir Union Territory (Fig. 1), situated in the southwestern part of the Kashmir Valley under the Central Forest Circle. Geographically, the area lies between 34°07′00″ and 34°28′00″ North Latitude and 74°24′00″ and 75°26′00″ East Longitude, covering approximately 37,901 hectares. The elevation ranges from 1828.8 m above mean sea level (AMSL)

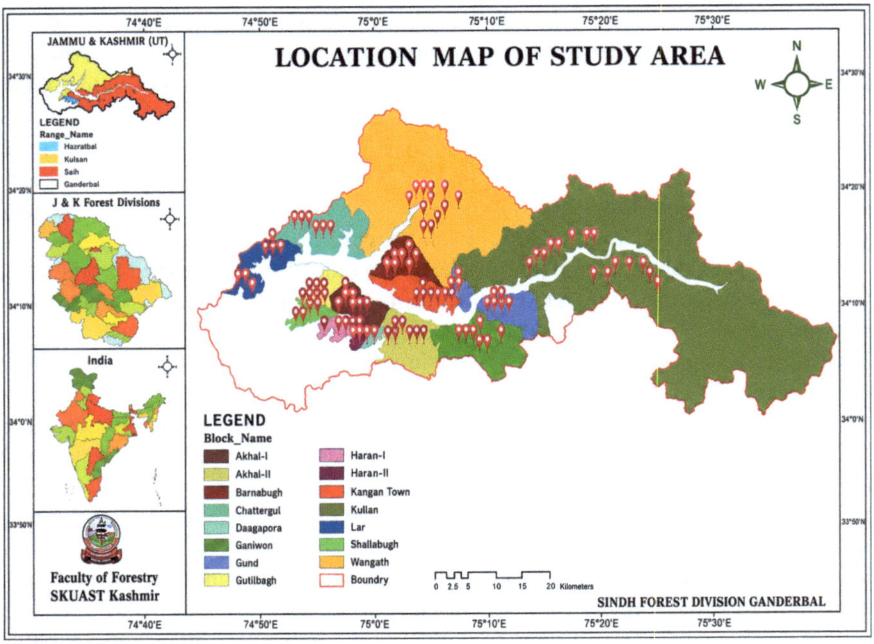

Fig. 1 Location map of study area map, ground truth, and distribution of sample

near Harran village to 5248 m AMSL at Harmukh peak. The terrain is predominantly mountainous in the Sindh and Manasbal ranges, while relatively flat in the Harran range, reflecting the complex topography of the Kashmir Himalayas (Saleem et al., 2024; Ahmed et al., 2025). The southern and northern aspects are almost equally distributed, with the Manasbal range largely southern-facing and the Sindh range predominantly northern-facing. The Sindh Forest Division is administratively divided into three ranges—Sindh (21,807 ha), Manasbal (14,856 ha), and Harran Shallbugh (1117 ha)—further subdivided into fifteen blocks including Akhal I & II, Ganiwon, Gund, Kullan, Kangan Town, Lar, Chattergul, Wangat, Branabugh, Gultibagh, Harran I & II, Dangapora, and Shallbugh.

2.2 Data Acquisition and Preparation

The evaluation of forest vulnerability utilized an indicator-based approach using fourteen indicators: population density, literacy rate, BPL households, % area under slope >30 degrees, aspect, elevation, road density, drainage density, % area under open forest, canopy cover, occurrence of fire, IVI, Shannon index, and preponderance of invasive species (Fig. 2). Population density and literacy rates were sourced from the 2011 Jammu and Kashmir District Census (Census of India, 2011). The total

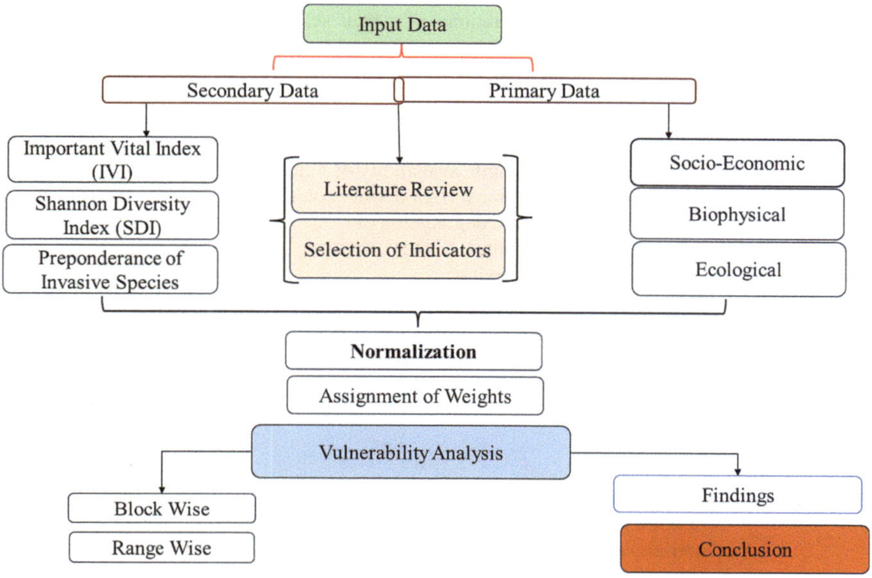

Fig. 2 Methodological flow chart

number of BPL households per block was obtained from the district census books of Jammu and Kashmir and the food cooperation database. Slope, aspect, and elevation layers were derived from a 30 m resolution shuttle radar topography mission DEM (SRTM-DEM), which was obtained from the USGS Earth Explorer for the study area. The drainage density used in the study was generated from ASTER-DEM 30 m data using the ArcGIS hydrological tool. A drainage proximity map was created with a layer of the river network at a distance of 1–5 km. This layer was utilized to determine the varying degrees of forest vulnerability with respect to distance from the drainage network. The canopy cover layer was obtained from the FSI database. Data regarding fire incidence over the past three years were obtained from the MODIS-FIRMS ArcGIS (Table 1).

2.3 Preponderance of Invasive Species and Important Vital Index

The preponderance of invasive species was estimated using the area proportion of the forest floor covered by invasive species in the sampling plot (Sharma et al., 2013). The random quadrant method was employed to calculate the importance value index (IVI) to examine various community characteristics of the tree species (Misra, 1968). Quadrates of a specific size (20 m × 20 m) were randomly placed at each chosen location. Within each quadrant, tree species were categorized and the number of

Table 1 List of indicators, components, rationale for their selection sources for the all-division block-level and range-level analysis

Indicator	Component	Rationale for selection	Sources	Weight
Population density	Total population, area of the region (km^2), urban vs. rural distribution	Population density determines the extent of dependency and per capita availability of finite resources. High density could lead to degradation of resources, further increasing sensitivity	Jammu and Kashmir district census book	0.075
Literacy rate	Number of literate individuals, Total population, gender-based literacy rates	Educated individuals and societies have better preparedness and response to disasters	Jammu & Kashmir district census book	0.03
BPL households	Total number of households, households below the poverty line	A higher percentage of BPL households indicates lesser adaptive capacity	Jammu & Kashmir food coorporation database	0.045
Percentage area under slope >30 degree	Area, slope (steep and gentle)	Steep topographical feature implies lack of availability of flat difficulty in access; likely to be adversely affected during floods, landslide, cloudburst, etc. and increases sensitivity and land	Shuttle radar Toporaphy Misson DEM (SRTM)	0.087
Aspect	Slope direction	Aspect strongly affects the temperature conditions and soils thus significant affect in forest growth increases vulnerability assessment		0.037
Elevation	Elevation of the surface	Ground elevation helps in identifying the vulnerable zones. Areas with high elevation receives ample amount of sunshine and encourages the growth of vegetation		0.05
Road density	Distance from road	Forest situated along the road network are more prone to disturbance than the forest farther from road network	ArcGIS: LISS-IV 5.8 m	0.02
Drainage density	Distance from drainage	The forests near to water bodies are more vulnerable to inundation	ArcGIS; ASTER DEM 30 m	0.05

(continued)

Table 1 (continued)

Indicator	Component	Rationale for selection	Sources	Weight
Percentage area under open forest	Forest area mapping, Forest cover classes, Total geographical area	Large tracts of open forests indicate a higher level of forest disturbance and degradation. Forests provide vital environmental services and thus degradation of forests indicate higher sensitivity	FCD (Forest canopy density) mapping	0.18
Canopy cover	Canopy density	Canopy cover indicates the forest health. Decrease in canopy cover raises the exposure and sensitivity of the forests	Forest survey of India	0.06
Occurrence of fire	Number of fire incidents, duration of fire events from last three years area affected,	The role of fire is in hampering regeneration of native species and aiding proliferation of invasive species must be considered seriously. So, more the fire incidences in forest area more will the vulnerability	MODIS-FIRMS	0.09
Important vital index (IVI)	Relative density, relative dominance, relative frequency of species	Importance value index (IVI) shows the complete or overall picture of ecological importance of the species in community. Higher the IVI, lesser will be vulnerability to forest	IVI = RD + RF + RBA (Curtis, 1959)	0.09
Shannon diversity index (SDI)	Number of species, abundance of each species, calculation of diversity indices	Species diversity differs from species richness in that it takes into account both the numbers of species present and the dominance of species in relation to one another. Higher the species richness, lesser the exposure of community to climatic hazards	$H = \sum (p_i) (\ln p_i)$. Shannon-Weaver (1963)	0.06
Preponderance of invasive species	Percentage of invasive species, impact on native biodiversity	The presence of invasive species increases sensitivity because: Forest ecosystems become more fragile and less resilient to disturbances	(Sharma et al., 2013)	0.12

individuals per species was tallied to determine various phytosociological aspects. Curtis (1959) was used to compute vegetation density, frequency, and basal area. The relative values for frequency, density, and dominance were established according to (Phillips, 1959). The important value index (IVI) for each site was worked out by using formula as below.

Importance Value Index (IVI) = Relative Density + Relative Frequency + Relative Cover.

Where,

$$\text{Relative density} = \frac{\text{Number of individuals of the species}}{\text{Number of individuals of all the species}} \times 100$$

$$\text{Relative frequency} = \frac{\text{Number of occurrences of the species}}{\text{Number of occurrences of all the species}} \times 100$$

$$\text{Relative basal area} = \frac{\text{Total basal area of the species}}{\text{Total basal area of all the species}} \times 100$$

2.4 *Shannon Diversity Index*

The Shannon diversity index (H), is a prominent index for describing species variety within a population. Species diversity varies with species richness because it considers both the number of species present and their dominance over one another. The greater the species diversity, the less vulnerable the community is to climate change.

The Shannon-Weiner diversity index (Shannon & Weaver, 1963) was calculated as given below.

$$\mathrm{H} = -\sum_{\mathrm{i}=1}^{\mathrm{s}} \mathrm{p}_i \log \mathrm{p}_i$$

where

$p_i = n_i/N$

H = Shannon Weiner index

n_i = total number of individuals of one species

N = total number of individuals of all species

S = Numbers of species encountered

Log = natural logarithm

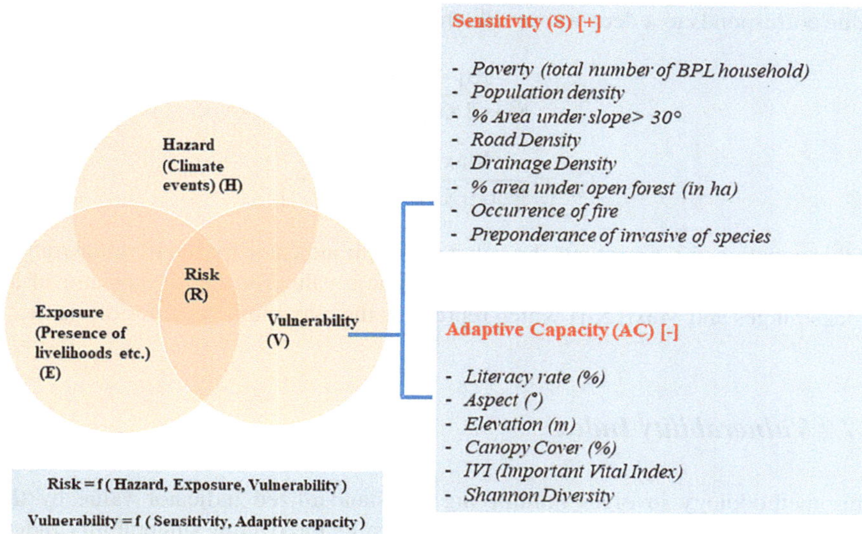

Fig. 3 Framework for vulnerability assessment

2.5 *Framework for Vulnerability Assessment*

Hazard, exposure, and vulnerability interact and result in risk within the total climatic and non-climatic physical and socio-political environment, (IPCC, 2014) (Fig. 3). The main aim of this framework is to investigate the effects of climate risk on the design of adaptation solutions. The proposed objectives were achieved by following the twelve stages outlined by Sharma et al. (2018). The evaluation process follows a series of ordered steps: conceptualizing vulnerability, selecting appropriate indicators, standardizing indicator values, and computing the composite vulnerability index (CVI) along with its visual representation (IHCAP, 2018). The Pair wise Comparison Method (PCM) was used to assign weights to the chosen indicators (Saaty et al., 2008). In the study area, 105 sample plots (Gimaret-Carpentier et al., 1998) were evenly placed, required as per the area ratio of block locations, followed by systematic sampling using the belt transaction method (Kerr et al., 2002).

2.6 *Normalization*

Normalization is based on a linear maximum and minimum scaling procedure that considers the functional link between indications and vulnerability. Equation (1) has been used to positively connect indicators, that is, those where vulnerability rises as the indicator's value rises. Equation (2) has been employed to establish an inverse relationship between indicators, specifically in cases where an indicator's increasing value corresponds to a decrease in vulnerability.

$$X_{ij}^{p} = \frac{\text{Xij} - \text{Mini}\{\text{Xij}\}}{\text{Maxi}\{\text{Xij}\} - \text{Mini}\{\text{Xij}\}} \quad (1)$$

$$X_{ij}^{N} = \frac{\text{Maxi}\{\text{Xij}\} - \text{Xij}}{\text{Maxi}\{\text{Xij}\} - \text{Mini}\{\text{Xij}\}} \quad (2)$$

In this equation, Xij represents the value of the jth indicator for the ith block/range. That is Mini{Xij} which represents the minimum value for the jth indicator of all blocks/ranges and Maxi{Xij} which represents the maximum value for the same.

2.7 Vulnerability Index

This methodology involves multiplying the standardized indicator value by the weight employed and then adding it. To determine the overall vulnerability index, the following equation was used to combine the normalized indicator values with their respective weights:

$$V1 = \sum_{j} \left(x_{ij} \times w_{j}\right)$$

where,

VI is the Vulnerability Index
xij are normalized values and
wij are the assigned weights

3 Results

Understanding the vulnerability of forests to climate change in the Kashmir Himalaya is crucial for addressing the complex factors affecting forest ecosystems and developing effective adaptation strategies. The Himalayan region, including Kashmir, is highly vulnerable to climate change impacts due to its unique biogeography and sensitive social-ecological systems (Pandey & Bardsley, 2015; Upgupta et al., 2015). Vulnerability assessments in this region reveal that climate change affects the social, economic, political, and environmental dimensions, shaping how climate risks are experienced and responded to by communities (Malik & Ford, 2024). These assessments help identify priority areas for adaptation planning, such as blocks and ranges with high vulnerability indices, enabling policymakers and forest managers to allocate resources effectively and implement targeted interventions.

The assessment of socio-economic, biophysical, and ecological indicators revealed significant variations across the Sindh Forest division. Harran II had the highest population density at 2220 individuals per square kilometer, followed by Harran I (1,387) and Dagapora (1140). Gund exhibited the lowest population density at 90 persons per square kilometer. Lar recorded the highest literacy rate (73.77%), while Dagapora had the lowest (38.4%). Kangan Town led in open forest area coverage at 67%, while Akhal I recorded the lowest at 3.9%. In terms of canopy cover, Wangat had the highest at 71.98%, and Harran II had the lowest (28.61%). Kullan and Ganiwon had the highest elevation at 2400 m, whereas Harran II had the lowest at 1459 m. The Harran Shallabugh Range had the highest population density at 1208 individuals per square kilometer, followed by the Manasbal Range (298) and the Sindh Range (241). In terms of literacy, the Manasbal Range recorded the highest rate (57.77%), while the Harran Shallabugh Range had the lowest (42.94%). Open forest coverage was highest in the Harran Shallabugh Range (57.25%), followed by the Manasbal Range (26.4%) and the Sindh Range (19.5%). Elevation also varied significantly, with the Manasbal Range averaging 2213 m, while the Harran Shallabugh Range had the lowest altitude at 1528 m. Kangan Town had the highest number of Below Poverty Line (BPL) households (1892), whereas Akhal II had the lowest (419). Among ranges, Sindh had the highest number of BPL households (6834), followed by Manasbal (5759), and Harran Shallabugh had the lowest (3564).The highest percentage of area under a mean slope greater than 30° was recorded in the Manasbal Range (63.2%) and the Akhal II forest block (75%). The Harran Shallabugh Range had the lowest percentage of steep slopes, with its blocks (Harran, Harran II, Shallabugh, and Dagapora) being predominantly flat. Maximum aspect values (315° north) were found in the Gund, Kullan, and Gutlibagh forest blocks, as well as the Sindh Forest range. Conversely, the Harran Shallabugh Range had no aspect values due to its flat terrain. The Sindh Range had the highest drainage density (1.84 sq. km), while the Manasbal Range recorded the lowest (1.37 sq. km). Among forest blocks, Akhal II had the highest drainage density (2.65 sq. km), and Harran II had the lowest (1.1 sq. km). Road density was highest in the Manasbal Range (0.75 sq. km), followed by the Sindh Range, while the Harran Shallabugh Range had the lowest (0.32 sq. km). The Chattergul forest block had the maximum road density (0.89 sq. km), and the Dagapora forest block recorded the lowest (0.21 sq. km). The Kullan forest block had the highest canopy density (68.75%), while Harran II had the lowest (28.61%). Among ranges, the Manasbal Range recorded the largest canopy cover (57.77%), followed by the Sindh Range (52.13%), with the Harran Shallabugh Range being the lowest (33.92%). Importance Value Index (IVI) was highest in Akhal II (186.65) and lowest in Kangan Town (126.37). The Sindh Range had the maximum IVI (196), while the Harran Shallabugh Range had the lowest (117). Shannon-Wiener Index (SWI) values were highest in the Chattergul block (0.81) and Sindh Range (0.86), while the lowest values were recorded in Dagapora (0.35) and Harran Shallabugh Range (0.47) A floristric analysis revealed that the prominat broad leaved species at lower altitudes including *Aesculus indica, Ailanthas altissimo, Celtisaustralis, Morusalba, Prunes armeniaca, Poplus deletoidas, Populus alba Robinia pseudoacacia,, Salix alba, Salix fragilis. While conferious species such as Abies pindrow,*

Cedrus deodara Pinus wallichiana Picea simithiana and *Taxus contorta* dominate at the higher elevations. With respect to preponderance of invasive species the Gund forest block had the highest average density of invasive species (756 ± 37 m^2/ha), while Kangan Town had the lowest (667 ± 33 m^2/ha). Among ranges, the Harran Shallabugh Range recorded the maximum density of invasive species (781 ± 36 m^2/ha), followed by the Manasbal Range. Prominent invasive species recorded such as Anthemis *cotula, Arctium lappa, Conyza canadensis, Datura stramonium, Impatiens glandulifera, Leucanthemum vulgare and Stipa sibirica.* The Sindh Range had the highest annual fire incidence (19 hectares), followed by the Manasbal Range (12 hectares) and the Harran Shallabugh Range (2 hectares) (Tables 2 and 3).

This comprehensive assessment highlights the variability in socio-economic, biophysical, and ecological indicators across the Kashmir Himalaya. The findings underscore the importance of targeted conservation and management strategies, particularly in areas with high vulnerability indices. By addressing these vulnerabilities, policymakers and forest managers can enhance the resilience of forest ecosystems and the communities that depend on them.

3.1 Block-Level and Range-Wise Composite Vulnerability Index

Composite Vulnerability Indices (CVIs) are categorized as the most vulnerable forest blocks and ranges, along with drivers for current climate-related threats. It evaluated 15 blocks and three ranges within the division by using cross-common socio-economic, biophysical, and ecological parameters. In the analysis that follows, the Akhal I Block was defined as the most vulnerable block with the highest composite vulnerability index of 0.67, while the Chattergul Forest Block had the lowest composite vulnerability index (0.39 (Table 4 and Fig. 4)). The vulnerability factors identified in these blocks included the highest percentage of area falling under open forest, the highest human population density, the highest BPL households, the highest aspect profile, and the highest road density. Secondary factors affecting the study area were the relative density of invasive species, high drainage density, steep mean slopes, and low species diversity. Comparatively, a higher average composition vulnerability index was found in the Harran-Shallabugh Range (0.62) than in the Manasbal Range (0.44), and the lowest vulnerability index (0.43) was for the Sindh Range (Table 5 and Fig. 5). In terms of the proportion areas, the Sindh range impacted (57.9%), followed he most by the Manasbal Range (39.19%), while the Harran-Shallabugh Range portion was only (2.94%). Significant levels of vulnerability in these ranges can be attributed to low literacy levels, high BPL household ratios, low elevation, high drainage density, low canopy cover, and frequent fire. Secondary factors include high human population density, increased road density, large human-impacted open forest zones, and invasive species. It is also pertinent to mention that the vulnerability indices are comparative; therefore, the ranking does

Table 2 Sub indicators values of blocks of Sindh forest division

Blocks	Socio-economic indicators				Biophysical					Ecological					
	Total area in Sq km	Pop density/ Sq Km	Litercy rate (%)	Total No. BPL House Hold	Percentage area under slope > 30°	Aspect	Elevation (m)	Drainage density:km/ Sq Km	Road density/ Sq km	Percentage area under open forest in ha	Canopy COVER (%)	Important vital index	Shannon Wenier index	Pro. of invasive spp.(Average density per ha,	Forest fire incidence from last three years in ha.
Akhal I	26.91	164	60.1	685	42	45	2135	1.66	0.26.	3.9	50.1	164.34	0.68	554	8
Akhal II	24.34	153	54.18	419	75	135	2234	2.65	0.54	6.47	44.18	186.65	0.72	543	1
Ganiwon	36.63	248	48.97	1494	71	225	2400	1.55	0.34	10.5	58.97	176	0.75	434	0
Gund	69.84	90	56.85	1204	69	315	1950	1.44	0.55	6.69	66.85	157.11	0.68	756	8
Kullan	55.32	160	48.75	1140	66	315	2400	2.54	0.75	23	68.75	137.21	0.64	667	2
Kangan town	1.85	636	43.94	1892	50	45	1894	1.25	0.45	67	23.94	126.37	0.68	434	0
Lar	25.53	143	73.77	643	54	225	2189	1.68	0.75	26	43.77	141.31	0.74	543	1
Chattergul	71.24	472	54.91	1639	66	0	2267	2.38	0.89	17	54.91	144.042	0.81	868	5
Wangat	39.22	201	50.98	1463	65	135	2398	1.078	0.73	27	71.98	152.22	0.7	842	3
Branabugh	21.51	220	55.1	512	74	45	2234	1.32	0.84	18	55.1	170.34	0.76	324	1
Gultibagh	13.44	455	54.11	1502	57	135	1979	2.43	0.56	9	64.11	150.5	0.4	675	2
Harran I	1.33	1387	40.61	755	0	0	1500	1.24	0.34	63	36.61	144.82	0.45	644	1
Harran II	1.01	2220	40.61	682	0	0	1459	1.1	0.43	54	28.61	170.18	0.66	485	0
Shallabugh	3.54	988	52.08	485	0	0	1589	1.81	0.33	66	32.08	160.18	0.63	1139	1
Dagapora	5.29	1140	38.4	1642	0	0	1566	1.62	0.21	46	38.4	145.47	0.35	856	0

Table 3 Sub indicators values of ranges of Sindh forest division

Ranges	Socio-economic indicators				Biophysical					Ecological					
	Total area in Sq km	Pop Density Sq. Km	Litercy Rate (%)	Total BPL House Hold	Percentage area under slope > 30°	Aspect	Elevation(m)	Drainage density:km/ Sq Km	Road density/ Sq km	Percentage area under open forest in ha	Canopy cover (%)	IVI	SWI	Pro. of invasive spp. (Average density per ha	Forest fire incidence from last three years in ha.
Sind range	214.89	241.85	52.13	6834	62.6	315	2168	1.84	0.48	19.59	55.13	196	0.86	546.66	19
Manasbal range	170.94	298	57.77	5759	63.2	45	2213	1.37	0.75	26.4	57.77	167	0.77	650.66	12
Harran shallabugh	11.17	1208	42.94	3564	0	0	1528	1.44	O.32	57.25	33.92	117	0.41	781	2

Table 4 Vulnerability index values of the main indicators, composite vulnerability index values and corresponding ranks of blocks of Sindh forest division

Blocks	Vulnerability index values of the main indicators			Composite vulnerability index	Rank
	Socio Economic	Biophysical	Ecological		
Akhal I	0.79	0.55	0.66	0.67	1
Harran II	0.45	0.97	0.51	0.65	2
Harran I	0.41	0.93	0.52	0.62	3
Dagapora	0.34	0.9	0.54	0.59	4
Branabugh	0.83	0.3	0.63	0.59	5
Shallabugh	0.56	0.86	0.29	0.57	6
Akhal II	0.88	0.17	0.64	0.56	7
Lar	0.73	0.32	0.6	0.55	8
Kangan Town	0.45	0.59	0.53	0.52	9
Ganiwon	0.66	0.28	0.64	0.52	10
Gund	0.73	0.32	0.45	0.5	11
Wangat	0.67	0.39	0.41	0.49	12
Gultibagh	0.55	0.24	0.63	0.47	13
Kullan	0.76	0.07	0.54	0.46	14
Chattergul	0.51	0.23	0.45	0.39	15

not mean that the Chattergul Forest Block or Harran-Shallabugh Range is not vulnerable in an absolute sense. These rankings are constrained by the indicators used in this assessment, mostly using socioeconomic and livelihood indices and selected biophysical and ecological conditions.

3.2 *Quartile-Based Categorization of Vulnerability of Forest Blocks*

Using vulnerability ranking, the 15 forest blocks of the Sindh Forest Division were divided into four quartiles. The central points of the blocks correspond to the three quartiles, with each quartile representing 25% of the data.

(a) Quartile I (top 25%, Most Vulnerable Blocks).

Akhal I, Harran II and Harran I with vulnerability indices ranging from 0.62 to 0.67. Main drivers have been observed to influence vulnerability include high proportion of open forests, high population density, highest number of BPL households, high road density, frequent forest fire and invasive alien species.

(b) Quartile II (Upper Middle 25% Vulnerable Blocks):

Dagapora, Branabugh,Shallabugh, Akhal II and Lar with VI of between 0.55 and 0.59 Key drivers are population pressure, High percentage bpl households, high mean slope, the presence of undesirable species, and forest fire.

(c) Quartile III (Lower Middle 25% Vulnerable Blocks):

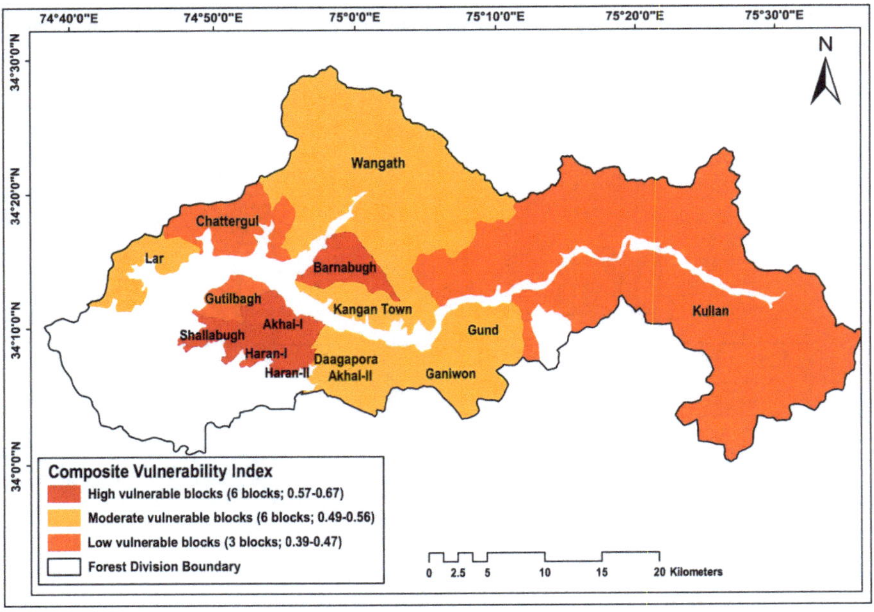

Fig. 4 Block level composite vulnerability map of Sindh forest division

Table 5 Vulnerability index values of the main indicators, composite vulnerability index values and corresponding ranks of blocks of Sindh forest division

Blocks	Vulnerability index values of the main indicators			Composite vulnerability index	Rank
	Socioeconomic	Bio-physical	Ecological		
Harran Shallabugh	0.5	0.97	0.41	0.62	1
Manasbal Range	0.56	0.35	0.41	0.44	2
Sindh Range	0.57	0.079	0.65	0.43	3

Kangan Town, Ganiwon, Gund, Wangat, and Gutlibagh had vulnerability indices of 0.47, 0.52. The main drivers include forest fires, high road density, high proportion of open forests, reduced tree cover, and high population.

(d) Quartile IV (Bottom 25% Least Vulnerable Blocks):

Kullan and Chattergul, vulnerability indices, vary from 0.39 to 0.46. Key drivers include a large percentage of BPL households, high population density, frequent forest fires, and high aspect profile

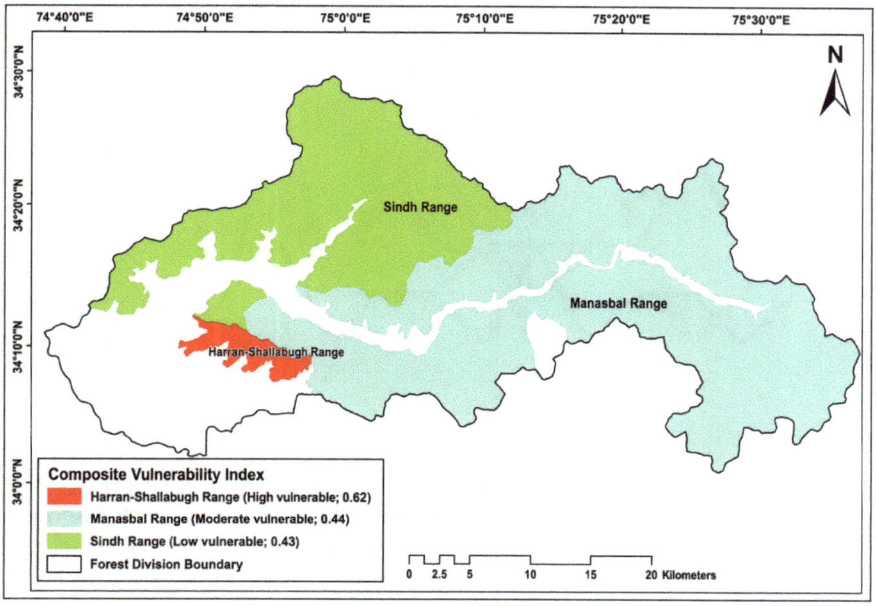

Fig. 5 Range level composite vulnerability map of Sindh forest division

3.3 Frequency Distribution of Vulnerability Drivers

The quartile distribution of drivers is shown in (Fig. 6). Quartile blocks were the most vulnerable and had a relatively large open forest area, high population density, many BPL households, high road density, many forest fires, and invasions by destructive species. In Quartile, the primary agents influencing them were population den Quartile II, households, steep slopes, invasive species, and forest fires. Quartile III blocks are threatened by road density, open forest area, low canopy cover, and fires, while Quartile IV blocks face threats from population pressure, fires, and high aspects. Factors, including high population density and high aspect profiles, showed that response frequency was the highest in two out of the three ranges. The other indicators were recorded at a frequency of one (Fig. 7). Thus, this analysis shows that the proposed adaptation.

4 Discussion

Climate-sensitive industries, such as forestry, agriculture, and fishing, have severely impacted India (Bahl, 2015; Mehraj et al., 2023). Forests are particularly important in areas such as Kashmir because they are integral to the local economy and livelihood (Haq et al., 2020; Pokhriyal et al., 2020). Assessing the level of vulnerability of

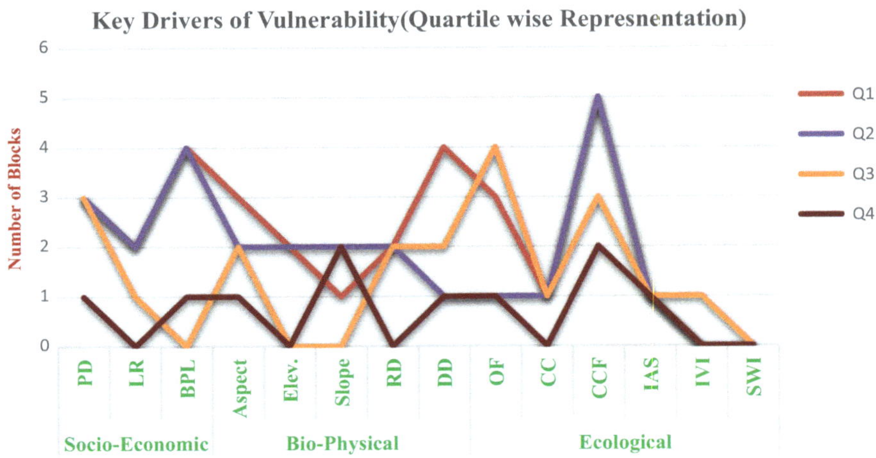

Fig. 6 Key drivers of vulnerability (quartile-wise reoresentation). PD: Population density; LR: Literacy rate; BPL: Below poverty line; RD: Road density; DD: Drainage density; OF: Open forest; CC: Canopy cover; CCF: Occurance of fire; IAS; Inasive alien species; IVI: Important vital index; and SWI: Shannon-weiner-index

Fig. 7 Drivers of vulnerability of ranges PD. Population density; Ast: Aspect; LR: Literacy rate; BPL: Below poverty line; RD: Road density; DD: Drainage density; OF: Open forest; CC: Canopy cover; CCF: Occurance of fire; IAS; Inasive alien species; IVI: Important vital index; and SWI: Shannon-weiner-index

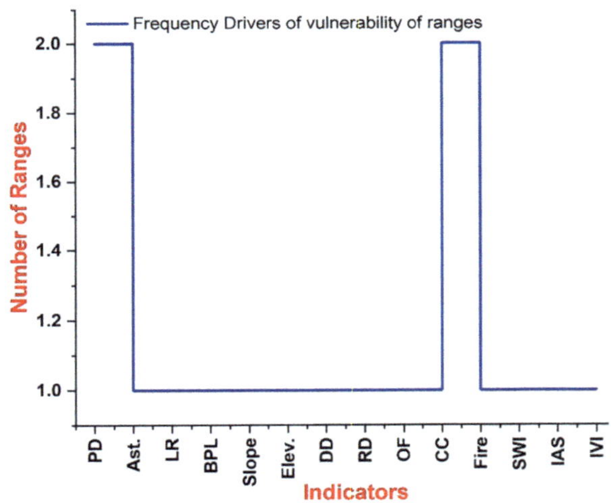

forest ecosystems is important for identifying their sensitivity to climate change and determining the required protection measures. This study offers valuable insights for assessing the level of susceptibility and adaptivity of different forest blocks and ranges in the Sindh Forest Division. This study revealed that the vulnerability, in this case, varied significantly geographically and about different forest blocks. Analysing the data related to the climate vulnerability index, the Akhal I Block was found to be the most sensitive area, with a VI of 0.67. However, the Chattergul forest block depicted lower vulnerability on the same scale, with an index value of 0.39, demonstrating greater resilience of the area. Among the ranges, Harran-Shallabugh had the highest composite vulnerability index of 0.43, while the Sindh range had the lowest index of 0.62. Such evidence suggests that the variability in local environmental conditions and socioeconomic factors directly affects the level of forest vulnerability. Other considerations regarding vulnerability include incidences of fire recurrence, high population density, large BPL population, lowest literacy level, areas of open forestry, and undesirable species. Population density determines the reliance on and availability of limited resources, increasing vulnerability to resource depletion and climate risks (Woodward et al., 1998; Poudel et al., 2020). Individuals with extremely low incomes are considered among the most vulnerable: they possess minimal to no financial resources, and consequently, have the least ability to respond to climate-related threats. (O'Brien & Leichenko, 2000). Other factors that can also contribute to vulnerability include Biophysical and Ecological indices, such as high drainage density and road density, low canopy cover, low IVI, and low SWI. These results agree with those of earlier studies conducted in the Indian Western Himalayas (Nandy et al., 2015; Gupta et al., 2020; Kumar et al., 2021). Pragya et al. (2023) investigated the vulnerability of forests in multiple dimensions. The distribution of forests and their health conditions are influenced by climatic factors such as changes in precipitation and temperature. However, forest conditions are further deteriorated by different disturbances, such as fire, natural disasters, and human intervention towards developmental activities (Yang et al., 2017). Numerous studies have emphasized that the increase in forest vulnerability is due to forest fires, and the level of intensity is influenced by both natural causes and human actions (Lecina-Diaz et al., 2021; Mina et al., 2023). Therefore, Vulnerability is a function of the intersectional relationships among various climatic, ecological, and socioeconomic factors (Thakur et al., 2019; Gupta et al., 2019). In the evaluation of vulnerability using indicator-based approaches, Kumar et al. (2021) revealed that lower-altitude forests were more prone to climate change than upper-altitude forests. Furthermore, over-exploitation of medicinal plants for local use, overgrazing, landslides, and soil erosion have been identified as some of the major challenges affecting forest ecosystems in vulnerable regions (Ganie et al., 2019).

Evaluating and undertaking local adaptation measures is important to minimize the risks of forest vulnerability in changing climate scenarios. The successful construction of adaptive forest conservation and the enhancement of a sustainable community and biological diversity requires the involvement of communities in such practices. Community participation in conservation increases its effectiveness and promotes local livelihoods (Bolte et al., 2009; Hayes & Persha, 2010). Poverty alleviation and improved literacy rates can contribute significantly to forest resilience. Poverty is characterized by a monthly income below Rs. 965 in urban settings and Rs. 781 in rural areas (Census of India, 2011). A higher number of households classified as Below the Poverty Line (BPL) indicate a reduced capacity to adapt (Singh & Kaushal, 2023; Gill-Wiehl et al., 2022). In the Kashmir Himalayas, massive biotic pressures, including overgrazing, floods, drought, landslides, forest fires and deforestation, have greatly influenced forest decline (Altaf et al., 2022). Various studies have indicated that anthropogenic distribution and biotic pressure are among the foremost causes of the poor forest cover in the Indian Himalayas (Chakraborty et al., 2016; Chakraborty et al., 2018). Climate variability is among the most significant factors that lead to changes in land use and land cover in the Kashmir Himalayas, and spatial transformation of land use and land cover has been noted due to rampant lateral urbanization and changing climatic patterns (Ahmed et al., 2023). In addition, studies have provided evidence that climate change impacts the nomadic herders of the western Himalayas of Jammu and Kashmir (Fayaz et al., 2020). Furthermore, studies on invasion pressure and the dependency of communities on forests have pointed to increased pressure on forests. Invasive Alien species are major contributors to forest degradation in the Kashmir Himalayas (Sharma et al., 2013; Mehraj et al., 2017). In the IHR, approximately 297 alien plant species have become part of the natural flora and belong to 65 different families. Most of these species are found in Jammu and Kashmir, and approximately 192 (64.6%) species were recorded from this region. Among these, *Lantana camara, Ageratinaa denophora, Parthenium hysterophorus,* and *Ageratum conyzoides* are invasive species that have invaded IHR, including Kashmir (Lamsal et al., 2018; Pathak et al., 2019; Gokdemir et al., 2021). The presence of invasive species increases sensitivity, because forest ecosystems become more fragile and less resilient to disturbances (*Invasive Species Management*, 2009; Weller et al., 2017; Dyderski & Jagodziński, 2020; Polce et al., 2023). Habitat reduction and fragmentation have consequences on the forest ecosystem's capacity for adaptation, where forests need more protection from external stress factors. Similar to the methods used by Thompson et al. (2009), that revealed the drainage patterns were linked to terrestrial vegetation changes. The results were further supported by other prior research conducted by (Joshi et al., 2002; Chandra & Bhardwaj, 2015) wherein they highlighted that forest degradation destabilises the biological and ecological equilibrium. Kashmir Himalayan forests are significant carbon sinks as they store a substantial quantum of carbon in trees. Forest carbon, an essential climatic indicator, is determined by a complex interaction of other ecological variables, particularly stand structural features. The study provides insights into the role of these natural forests in climate change mitigation and in REDD+/national commitments to offset the carbon Dar & Parthasarathy (2022). Forest fire incidents negatively affect native species regeneration and positively influence the establishment of invasive species, which should not be overlooked. Hence, experience shows that a higher fire occurrence in forested zones increases the level of risk involved (Sarkar et al., 2024). As per "DEE&RS: FOREST FIRE RISK ZONATION & VULNERABILITY ASSESSMENT OF JAMMU & KASHMIR FORESTS—2021 report final forest fire vulnerability maps shows that, out of total area of 27869.87 km^2 of 6646 compartments of forest divisions of Jammu & Kashmir, 4286.21 km^2 fall in the high vulnerable zone and needs immediate attention to minimize the damage, 9490.15 km^2 falls in the medium vulnerable zone, 4636.47 km^2 falls in the low vulnerable zone of forest fire and 9457.04 km^2 fall in unlikely vulnerable zone. It is also observed that out of the total 4354 forest fire incidents occurred in the earlier years in Jammu and Kashmir, 2374 fire incidents has occurred in the highly vulnerable zone, 1592 fire incidents in the medium vulnerable zone and 166 fire incidents in low vulnerable zone.

However, it is surprising that the vulnerability to climate-induced stresses is lower in forests with high levels of species richness and multi-layered canopies. Consequently, canopy cover values in the analysis were inverse; the higher the canopy cover, the healthier a forest is, as higher canopy cover means less forest exposure and sensitivity (Upgupta et al., 2015). Thus, there are important lessons that need to be learned from the need to conserve species wealth in forests (Norris et al., 2011). Other physical characteristics, such as slope steepness, also affect forest growth and management. The slope is one of the paramount predictors of forest dynamics and sustainability across large territories. Management practices that consider these physical parameters might enhance forest resilience and productivity (Devisscher et al., 2019; Martin et al., 2020). Technological advancements, such as remote sensing and GIS, have opened new avenues for assessing forest vulnerability to current and future climate change scenarios (Voight et al., 2019; Gao et al., 2020). A study by Yuh et al. (2022) revealed that multi-temporal satellite data were useful in evaluating the forests, green spaces, and agricultural areas to underpin conservation planning. This shows that there is a need to use both current technology and the traditional approach to solve the issues associated with forest sensitivity. Climate Change Vulnerability Assessments (CCVAs) can use the information in this study as initial data for all subsequent adaptive planning. Vulnerability drivers justify the need for more region-specific adaptation measures and forest management plans for the current and future volatility of climate change. This knowledge can be used effectively by policymakers and forest managers when determining which area is currently stressed by climate change and where particular action should be taken at the site level. Strategies to remove socioeconomic and biophysical weaknesses and cover biodiversity status, invasive species impacts, human pressure on forests, and forest sustainability. Such a policy intervention will contribute to climate change response and ensure the future sustainability of the Sindh Forest Division through the protection of natural resources.

5 Conclusion

The present study investigated the inherent forest vulnerability in Sindh Forest Division of Ganderbal, (J&K). The assessment analyzed population and demographic aspects along with physical features (slope, elevation, aspect), biophysical features (drainage, road density), forest attributes (open forest, canopy cover) and environmental aspects (forest fires, invasive species). An indicator-based approach was employed to monitor the vulnerability of the forest sector to climate change. The analysis revealed that several blocks, namely Akhal I, Harran II, Harran I, Dagapora, Branabugh, and Shallabugh, exhibited high vulnerability to climate change. Whereas moderate vulnerable blocks were Akhal II, Lar, Kangan Town Ganiwon, Gund, and Wangat, while for blocks like Gutlibagh, Kullan and Chattergul with low vulnerability. Similarly, vulnerability differed by range, with Harran-Shallabugh being most vulnerable, Manasbal was moderately vulnerable, and Sindh had relatively low vulnerability. The major drivers are responsible for vulnerability, including a high percentage of open forests, high population density, high road and drainage density, frequency of forest fires, steep slopes, low canopy cover, poor biological opulence, and poor species richness. These findings underscore the need for targeted interventions to enhance forest resilience, such as improving canopy cover, controlling invasive species, and mitigating the impacts of forest fires. Future research should aim to refine vulnerability indices by incorporating advanced geospatial techniques and climate projection models to better inform long-term adaptation strategies.

References

Ahmad Dar, J., & Somaiah, S. (2015). Altitudinal variation of soil organic carbon stocks in temperate forests of Kashmir Himalayas India. *Environmental Monitoring and Assessment, 187*(2). https://doi.org/10.1007/s10661-014-4204-9

Ahmed, R., Ahmed, P., Sahana, M., Singh, H., Wani, G. F., & Ahmad, S. T. (2021). A review of glacial Lake expansion and associated glacial Lake outburst floods in the Himalayan region. *Earth Systems and Environment, 5*(3), 695–708. https://doi.org/10.1007/s41748-021-00230-9

Ahmed, S., Iqbal, Z., & Antahal, P. C. (2023). Impact of climate change on nomadic herders livelihoods evidence from Bakarwal community in the Western Himalayas in India. *Environmental Development, 48*, 100930. https://doi.org/10.1016/j.envdev.2023.100930

Ahmed, R., Rather, A., Banerjee, P., Wani, G., & Saleem, S. (2025). Impact of climate variability on the feeding glaciers of potentially dangerous glacial lakes in the Jhelum Basin of Kashmir Himalaya, India. *Natural Hazards Review.* https://doi.org/10.1061/NHREFO.NHENG-2140

Altaf, A., Haq, S. M., Shabnum, N., & Jan, H. A. (2022). Comparative assessment of phyto diversity in Tangmarg Forest division in Kashmir Himalaya, India. *Acta Ecologica Sinica, 42*(6), 609–615.

Anjos, L. J. S., & De Toledo, P. M. (2018). Measuring resilience and assessing vulnerability of terrestrial ecosystems to climate change in South America. *PLoS One, 13*(3), e0194654. https://doi.org/10.1371/journal.pone.0194654

Bahl, P. N. (2015). Climate change and pulses: Approaches to combat its impact. *Agricultural Research, 4*(2), 103–108. https://doi.org/10.1007/s40003-015-0163-9

Bhat, I. A., & Shyju, P. J. (2014). Prospects of Ecotourism Development in Ganderbal District of Jammu and Kashmir. *Atna Journal of Tourism Studies, 9*(1), 39–61.

Bolte, A., Ammer, C., Löf, M., Madsen, P., Nabuurs, G. J., Schall, P., Spathelf, P., & Rock, J. (2009). Adaptive forest management in Central Europe: Climate change impacts, strategies and integrative concept. *Scandinavian Journal of Forest Research, 24*(6), 473–482. https://doi.org/10.1080/02827580903418224

Bradford, J. B., Lauenroth, W. K., Palmquist, K. A., Campbell, S. B., Maestas, J. D., Schlaepfer, D. R., & Chambers, J. C. (2019). Climate-driven shifts in soil temperature and moisture regimes suggest opportunities to enhance assessments of dryland resilience and resistance. *Frontiers in Ecology and Evolution, 7*. https://doi.org/10.3389/fevo.2019.00358

Brandt, L., Blume-Weaver, R., Swanston, C., Dijak, W. D., Nelson, P., Prasad, A., Oberle, B., Wang, W. J., Walk, J. W., Depuy, J., Westin, S., Tirpak, J. M., Dinkel, G., Schneiderman, J. E., Watts, L., Thompson, F. R., Shannon, P. D., Smith, A. B., Studyvin, C., et al. (2014). *Central hardwoods ecosystem vulnerability assessment and synthesis: A report from the central hardwoods climate change response framework project* (Vol. 124). US of agriculture forest service northern research station. https://doi.org/10.2737/nrs-gtr-124

Census of India. (2011).Goverment of India (GoI): The Registrar General & Census Commissioner

Chakraborty, A., Ghosh, A., Sachdeva, K., & Joshi, P. K. (2016). Characterizing fragmentation trends of the Himalayan forests in the Kumaon region of Uttarakhand, India. *Ecological Informatics, 38*, 95–109. https://doi.org/10.1016/j.ecoinf.2016.12.006

Chakraborty, A., Joshi, P. K., Saha, S., & Sachdeva, K. (2018). Vulnerability of forests in the Himalayan region to climate change impacts and anthropogenic disturbances: A systematic review. *Regional Environmental Change, 18*(6), 1783–1799. https://doi.org/10.1007/s10113-018-1309-7

Chandra, K. K., & Bhardwaj, A. K. (2015). Incidence of forest fire in India and its effect on terrestrial ecosystem dynamics, nutrient and microbial status of soil. *International Journal of Agriculture and Forestry, 5*(2), 69–78.

Chaturvedi, R. K., Jayaraman, G. M., Bala, G., Joshi, N. V., Sukumar, R., & Ravindranath, N. H. (2011). Impact of climate change on Indian forests: A dynamic vegetation modeling approach. *Mitigation Adaptation Strategies for Global Change, 16*(2), 119–142

Chauhan, N., Shukla, R., & Joshi, P. K. (2020). Assessing inherent vulnerability of farming communities across different biogeographical zones in Himachal Pradesh, India. *Environmental Development, 33*, 100506. https://doi.org/10.1016/j.envdev.2020.100506.

Chettri, N., Adhikari, B., Chaudhary, S., & Wangchuk, K. (2023). Changing discourses in the third pole: A systematic review of climate change impact on biodiversity in the Hindu Kush Himalaya. *Ecological Indicators, 155*, 111046. https://doi.org/10.1016/j.ecolind.2023.111046

Cui, G., Chae, Y., Kim, J.-S., Kwak, H., Lee, W.-K., Kim, M., Lim, C.-H., & Choi, S. (2015). Assessing vulnerability of forests to climate change in South Korea. *Journal of Forestry Research, 27*(3), 489–503. https://doi.org/10.1007/s11676-015-0201-2

Curtis, J. T. (1959). *The vegetation of Wisconsin: An ordination of plant communities* (p. 657). University of Wisconsin Press.

Dar, A. A., & Parthasarathy, N. (2022). Patterns and drivers of tree carbon stocks in Kashmir Himalayan forests: Implications for climate change mitigation. *Ecological Processes, 11*(1). https://doi.org/10.1186/s13717-022-00402-z

Devisscher, T., Konijnendijk, C., Nesbitt, L., Lenhart, J., Salbitano, F., & Lwasa, S. (2019). SDG 11: Sustainable cities and communities-Impacts on forests and forest-based livelihoods. In *Chapter 11-SDG 11: Sustainable cities and communities–impacts on forests and forest-based livelihoods* (pp. 349–385). Cambridge University Press.

Devisscher, T., Spies, J., & Griess, V. C. (2021). Time for change: Learning from community forests to enhance the resilience of multi-value forestry in British Columbia, Canada. *Land Use Policy, 103*, 105317. https://doi.org/10.1016/j.landusepol.2021.105317

Dimri, A. P., Stoffel, M., Bolch, T., Mal, S., Tiwari, P., Maharana, P., Ballesteros-Cánovas, J. A., Shukla, A., Pandey, A., Huggel, C., Rohrer, M., Thayyen, R. J., & Allen, S. (2021). Climate change, cryosphere and impacts in the Indian Himalayan region. *Current Science, 120*(5), 774. https://doi.org/10.18520/cs/v120/i5/774-790

Dong, X., Jiang, L., Zeng, S., Guo, R., & Zeng, Y. (2020). Vulnerability of urban water infrastructures to climate change at city level. *Resources Conservation and Recycling, 161*, 104918. https://doi.org/10.1016/j.resconrec.2020.104918

Dorado-Liñán, I., Vacchiano, G., Zang, C., Cañellas, I., Jantsch, M. C., Gea-Izquierdo, G., Fernández-De-Uña, L., Hornstein, D., Levanič, T., Castagneri, D., Martínez-Sancho, E., Zlatanov, T., Piovesan, G., Di Filippo, A., Mellert, K. H., Menzel, A., Ewald, J., & Gutiérrez, E. (2019). Geographical adaptation prevails over species-specific determinism in trees' vulnerability to climate change at Mediterranean rear-edge forests. Global Change Biology, 25(4), 1296–1314. https://doi.org/10.1111/gcb.14544.

Dyderski, M. K., & Jagodziński, A. M. (2020). Impact of invasive tree species on natural regeneration species composition, diversity, and density. *Forests, 11*(4), 456. https://doi.org/10.3390/f11040456

Faulkner, J. P., Murphy, E., & Scott, M. (2020). Developing a holistic 'vulnerability-resilience' model for local and regional development. *European Planning Studies, 28*(12), 2330–2347. https://doi.org/10.1080/09654313.2020.1720612

Fayaz, A., Shafiq, M. U., Singh, H., & Ahmed, P. (2020). Assessment of spatiotemporal changes in land use/land cover of North Kashmir Himalayas from 1992 to 2018. *Modeling Earth Systems and Environment, 6*(2), 1189–1200. https://doi.org/10.1007/s40808-020-00750-9

Forest fire Risk Zonation & Vulnerability Assessment of Jammu & Kashmir Forests ©2021, Department of Ecology, Environment & Remote Sensing, Forest Department J&K.
Ganie, A. H., Tali, B. A., Khuroo, A. A., Reshi, Z. A., & Nawchoo, I. A. (2019). Impact assessment of anthropogenic threats to high-valued medicinal plants of Kashmir Himalaya, India. *Journal for Nature Conservation, 50*, 125715. https://doi.org/10.1016/j.jnc.2019.125715
Gao, F., Anderson, M. C., & Hively, W. D. (2020). Detecting cover crop end-of-season using VENµS and Sentinel-2 satellite imagery. *Remote Sensing, 12*(21), 3524. https://doi.org/10.3390/rs12213524
Ghosh, M., & Ghosal, S. (2020). Climate change vulnerability of rural households in flood-prone areas of Himalayan foothills, West Bengal, India. *Environment, Development and Sustainability, 23*(2), 2570–2595. https://doi.org/10.1007/s10668-020-00687-0
Gill-Wiehl, A., Brown, T., & Smith, K. (2022). The need to prioritize consumption: A difference-in-differences approach to analyze the total effect of India's below-the-poverty-line policies on LPG use. *Energy Policy, 164*, 112915. https://doi.org/10.1016/j.enpol.2022.112915
Gimaret-Carpentier, C., Pélissier, R., Pascal, J. P., & Houllier, F. (1998). Sampling strategies for the assessment of tree species diversity. *Journal of Vegetation Science, 9*(2), 161–172. https://doi.org/10.2307/3237115
Gokdemir, C., Rubin, Y., Li, X., & Xu, H. (2021). A vulnerability assessment method to evaluate the impact of tunnel drainage on terrestrial vegetation under various atmospheric and climatic conditions. *Advances in Water Resources, 147*, 103796.
González De Andrés, E. (2019). Interactions between climate and nutrient cycles on forest response to global change: The role of mixed forests. *Forests, 10*(8), 609. https://doi.org/10.3390/f10080609
Gopalakrishnan, R., Jayaraman, M., Bala, G., & Ravindranath, N. H. (2011). Climate change and Indian forests. *Journal of Current Science, 101*, 348–355.
Gupta, A. K., Negi, M., Nandy, S., Alatalo, J. M., Singh, V., & Pandey, R. (2019). Assessing the vulnerability of socio-environmental systems to climate change along an altitude gradient in the Indian Himalayas. *Indian Journal of Ecology, 106*, 105512.
Gupta, A. K., Negi, M., Nandy, S., Kumar, M., Singh, V., Valente, D., Petrosillo, I., & Pandey, R. (2020). Mapping socio-environmental vulnerability to climate change in different altitude zones in the Indian Himalayas. *Ecological Indicators, 109*, 105787.
Haq SM, Rashid I, Khuroo AA, Malik AH, Hamid M, Dar G H, & Ahmad R (2020) Forest ecosystems of Jammu and Kashmir state (pp 191–208) springer Singapore https://doi.org/10.1007/978-981-32-9174-4-8.
Hayes, T., & Persha, L. (2010). Nesting local forestry initiatives: Revisiting community forest management in a REDD + world. *Forest Policy and Economics, 12*(8), 545–553. https://doi.org/10.1016/j.forpol.2010.07.003
Hinkel, J. (2011). Indicators of vulnerability and adaptive capacity, towards a clarification of the science–policy interface. *Global Environmental Change, 21*(1), 198–208.
Hosonuma, N., Herold, M., Desy, V., De-Fries, R. S., Brockhaus, M., Verchot, L., & Romijn, E. (2012). An assessment of deforestation and forest degradation drivers in developing countries. *Environmental Research Letters, 7*(4), 56–77.
IHCAP. (2018). Climate vulnerability assessment for the Indian Himalayan. *The Region, 58.*
Invasive Species Management. (2009). oxford university pressoxford. https://doi.org/10.1093/oso/9780199216321.001.0001
IPCC. (2014). Technical summary. Climate change 2014: Impacts, adaptation and vulnerability. Part a: Global and sectorial aspect. In C. B. Barros, V. R. Dokken, D. J. Mach, K. J. Mastrandrea, M. D. Bilir, T. E. Chatterjee, M. Ebi, K. L. Estrada, Y. O. Genova, R. C. Girma, B. Kissel, E. S. levy, A. N. MacCracken, S. Mastrandrea, P. R. white, L. L. field, C. B. Barros, & V. R. Mach (Eds.), *Contribution of Working group II to the fifth assessment report of the intergovernmental panel on climate change.*
Italiano, S. S. P., Castellaneta, M., Colangelo, M., Pizarro, M., Borghetti, M., Ripullone, F., & Camarero, J. J. (2023). Assessing forest vulnerability to climate change combining remote

sensing and tree-ring data: Issues needs and avenues. *Forests, 14*(6), 1138. https://doi.org/10.3390/f14061138

Joshi, H. C. (2002). *Assessment of habitat diversity, forest vegetation and human dependence in the buffer zone of Nanda Devi biosphere Reserve of West Himalaya Ph.D.* Thesis submitted to Kumaun University Nainital.

Joshi, P. K., Roy, P. S., Singh, S., Agarwal, S., Yadav, D. (2002). Biome level characterization (BLC) of western India-a geospatial approach. *Tropical Ecology, 43*(1), 213–228.

Kerr G, Mason B, Boswell R, & Pommerening A, (2002). Monitoring the transformation of even-aged stands to continuous cover.Iii, 12. http://hdl.handle.net/10242/38991

Kraaijenbrink, P. D. A., Bierkens, M. F. P., Lutz, A. F., & Immerzeel, W. W. (2017). Impact of a global temperature rise of 1 5 degrees Celsius on Asia's glaciers. *Nature, 549*(7671), 257–260. https://doi.org/10.1038/nature23878

Kumar, M., Kalra, N., Singh, H., Sharma, S., Singh Rawat, P., Kumar Singh, R., Kumar Gupta, A., Kumar, P., & Ravindranath, N. H. (2021). Indicator-based vulnerability assessment of forest ecosystem in the Indian Western Himalayas: An analytical hierarchy process integrated approach. *Ecological Indicators, 125*, 107568. https://doi.org/10.1016/j.ecolind.2021.107568

Kumar, M., Rawat, S., & Singh, H. (2018). Dynamic forest vegetation models for predicting impacts of climate change on forests: An Indian perspective. *Indian Journal of Forestry, 41*, 1–12.

Kumar, M., Singh, S. H., Pandey, P., Singh, M. P., Ravindranath, N. H., & Kalra, N. (2019). Assessing vulnerability of forest ecosystem in the Indian Western Himalayan region using trends of net primary productivity. *Biodiversity and Conservation, 28*, 2163–2182.

Lamsal, P., Aryal, A., Atreya, K., & Kumar, L. (2018). Invasive alien plant species dynamics in the Himalayan region under climate change. *Ambio, 47*(6), 697–710. https://doi.org/10.1007/s13280-018-1017-z

Lecina-Diaz, J., Martínez-Vilalta, J., Vayreda, J., Alvarez, A., & Retana, J. (2021). Assessing the risk of losing forest ecosystem services due to wildfires. *Ecosystems, 24*(7), 1687–1701. https://doi.org/10.1007/s10021-021-00611-1

Lecina-Diaz, J., Retana, J., Alvarez, A., Martínez-Vilalta, J., Banqué, M., Birkmann, J., Vayreda, J., & Feldmeyer, D. (2020). Characterizing forest vulnerability and risk to climate-change hazards. *Frontiers in Ecology and the Environment, 19*(2), 126–133. https://doi.org/10.1002/fee.2278

Li, Y., Wang, W., Pan, J., Lin, E., Ju, H., Hu, W., Li, K., Xiong, W., & Berry, P. (2014). Integrated assessment of China's agricultural vulnerability to climate change: A multi-indicator approach climatic change, *128*(3–4), 355–366. https://doi.org/10.1007/s10584-014-1165-5

Li, D., Zhang, Y., Liu, L., Li, S., & Wu, S. (2018). Vulnerability of the global terrestrial ecosystems to climate change. *Global Change Biology, 24*(9), 4095–4106. https://doi.org/10.1111/gcb.14327

López, R., Ramírez-Valiente, J. A., & Pita, P. (2022). How plants cope with heatwaves in a drier environment. *Flora, 295*, 152148. https://doi.org/10.1016/j.flora.2022.152148

Lu, X., Zhu, X., Chen, W., Gao, Y., Hamani, A. K. M., Tobiloba, L. H., Abbas, N., Bai, J., Qiu, Z., Pan, J., Maqsood, N. (2026). Nitrogen biogeochemical cycling in forest ecosystems with the globalization of nitrogen deposition. *Nitrogen Cycling, 2*(1).

Maier, C., & Winkel, G. (2017). Implementing nature conservation through integrated forest management: A street-level bureaucracy perspective on the German public forest sector. *Forest Policy and Economics, 82*, 14–29. https://doi.org/10.1016/j.forpol.2016.12.015

Malik, I. H., & Ford, J. D. (2024). Monitoring climate change vulnerability in the Himalayas. *Ambio, 54*(1). https://doi.org/10.1007/s13280-024-02066-9

Marchin, R. M., Ossola, A., Backes, D., Ellsworth, D. S., Leishman, M. R., & Tjoelker, M. G. (2021). Extreme heat increases stomatal conductance and drought-induced mortality risk in vulnerable plant species. *Global Change Biology, 28*(3), 1133–1146. https://doi.org/10.1111/gcb.15976

Marí-Dell'Olmo, M., Sheehan, M., Villalbí, J. R., Barón-Miras, L. E., Ventayol, I., Montalvo, T., Ariza, C., Gómez-Gutiérrez, A., Mercuriali, L., Borrell, C., & Oliveras, L. (2022). Climate change and health in urban areas with a Mediterranean

Martin, M., Girona, M. M., Morin, H., & Zang, R. (2020). Driving factors of conifer regeneration dynamics in eastern Canadian boreal old-growth forests. *Plos One 15*(7), e0230221. https://doi.org/10.1371/journal.pone.0230221

Masson-Delmotte, V. (2026). Scenarios, projections, and climate-related risks. *Comptes Rendus. Géoscience, 358*(G1), 309–329.

Mehraj S, Ashraf S, Rather AMUD, Bisati IA, Malik HA, Pandit AH, & Ali MT (2023) Climate change:- general overview and implications on agriculture and allied sectors (pp 97–110) springer. https://doi.org/10.1007/978-3-031-21079-2-7

Mehraj, G., Khuroo, A. A., Friedman, C. R., Qureshi, S., Rashid, I., & Muzafar, I. (2017). Patterns of alien plant diversity in the urban landscapes of global biodiversity hotspots: A case study from the Himalayas. *Biodiversity and Conservation, 27*(5), 1055–1072. https://doi.org/10.1007/s10531-017-1478-6

Mina, U., Farswan, S., & Dimri, A. P. (2023). Forest fires and climate attributes interact in Central Himalayas: An overview and assessment fire ecology, *19*(1). https://doi.org/10.1186/s42408-023-00177-4

Misra, R. (1968). *Ecology workbook* (pp. 131–133). Oxford and IBH Publishing Company.

Mushtaq, S., Saleem, S., Ahmed, R., Tass, M. S., Rather, J. A., & Rather, G. M. (2024). Spatio-temporal analysis of land use land cover changes in South Kashmir region of North-Western Himalayas using Landsat data. *Discover Geoscience, 2*(1), 37.

Nandy, S., Singh, C., Das, K. K., Kingma, N. C., & Kushwaha, S. P. S. (2015). Environmental vulnerability assessment of eco-development zone of great Himalayan National Park Himachal Pradesh India. *Ecological Indicators 57*, 182–195. https://doi.org/10.1016/j.ecolind.2015.04.024

Norris, C., Hobson, P., & Ibisch, P. L. (2011). Microclimate and vegetation function as indicators of forest thermodynamic efficiency. *Journal of Applied Ecology, 49*(3), 562–570. https://doi.org/10.1111/j.1365-2664.2011.02084.x

O'Brien, K. L., & Leichenko, R. M. (2000). Double exposure: Assessing the impacts of climate change within the context of economic globalization. *Global Environmental Change, 10*(3), 221–232.

Oppenheimer, M., Campos, M., Warren, R., Birkmann, J., Luber, G., O'Neill, B., & Takahashi, K. (2014). IPCC-WGII-AR5–19. Emergent risks and key vulnerabilities. In *Climate change 2014: Impacts, adaptation, and vulnerability Part A: global and sectoral aspects contribution of working Group II to the fifth assessment report of the intergovernmental panel on climate change* (pp. 1039–1099)

Ordóñez, C., & Duinker, P. N. (2014). Assessing the vulnerability of urban forests to climate change. *Environmental Reviews, 22*(3), 311–321. https://doi.org/10.1139/er-2013-0078

Pandey, R., & Bardsley, D. K. (2015). Social-ecological vulnerability to climate change in the Nepali Himalaya. *Applied Geography, 64*, 74–86. https://doi.org/10.1016/j.apgeog.2015.09.008

Pathak, R., Negi, V. S., Rawal, R. S., & Bhatt, I. D. (2019). Alien plant invasion in the Indian Himalayan region: State of knowledge and research priorities. *Biodiversity and Conservation, 28*(12), 3073–3102. https://doi.org/10.1007/s10531-019-01829-1.

Phillips, E. A. (1959). *Methods of vegetation study.* Henry Holt.

Pokhriyal, P., Sahana, M., Kumar, M., Areendran, G., Rehman, S., Sajjad, H., Pandey, R., & Raj, K. (2020). Assessing forest cover vulnerability in Uttarakhand, India using analytical hierarchy process. *Modeling Earth Systems and Environment*, *6*(2), 821–831. https://doi.org/10.1007/s40808-019-00710-y.

Polce, C., Zulian, G., Deriu, I., Cardoso, A. C., Tsiamis, K., Maes, J., Vigiak, O., & Gervasini, E. (2023). Invasive alien species of policy concerns show widespread patterns of invasion and potential pressure across European ecosystems. *Scientific Reports, 13*(1). https://doi.org/10.1038/s41598-023-32993-8

Poudel, S., Mishra, B., Funakawa, S., & Shinjo, H. (2020). Understanding households' livelihood vulnerability to climate change in the Lamjung district of Nepal. *Environment, Development and Sustainability, 22*(8), 8159–8182. https://doi.org/10.1007/s10668-019-00566-3

Pragya, P., Sahu, N., Tripathi, D. K., Verma, N. K., Bhadwal, S., Majid, S. I., Avtar, R., Kumar, M., & Tiwari, A. (2023). Integrated spatial analysis of forest fire susceptibility in the Indian Western Himalayas (IWH) using remote sensing and GIS-based fuzzy AHP approach remote sensing, *15*(19), 4701. https://doi.org/10.3390/rs15194701

Pugh, T. A. M., Arneth, A., Kautz, M., Poulter, B., & Smith, B. (2019). Important role of forest disturbances in the global biomass turnover and carbon sinks. *Nature Geoscience, 12*(9), 730–735. https://doi.org/10.1038/s41561-019-0427-2

Ridwan, Q., Siddiqui, S., Pant, S., Alamri, S., Wani, Z. A., Shah, A. A., & Hanief, M. (2023). Indigenous knowledge and perception of local people towards biodiversity conservation in Rajouri District of Jammu and Kashmir, India. *Sustainability, 15*(4), 3198. https://doi.org/10.3390/su15043198

Romshoo SA, Rashid I, Dar GH, & Altaf S (2020) Jammu and Kashmir state: An overview (pp 129–166) springer Singapore https://doi.org/10.1007/978-981-32-9174-4-6.

Roshani, R., Kumar, P., Sahana, M., Ahmed, R., Rehman, S., Rahaman, M. H., Masroor, M., & Sajjad, H. (2022). Forest vulnerability to climate change: A review for future research framework. *Forests, 13*(6), 917. https://doi.org/10.3390/f13060917

Roshani, R., Sajjad, H., Rahaman, M. H., Masroor, M., Sharma, Y., Sharma, A., & Saha, T. K. (2024). Vulnerability assessment of forest ecosystem based on exposure, sensitivity and adaptive capacity in the Valmiki Tiger Reserve, India: A geospatial analysis. *Ecological Informatics, 80*, 102494. https://doi.org/10.1016/j.ecoinf.2024.102494

Saaty, T. L. (2008). Relative measurement and its generalization in decision making why pairwise comparisons are central in mathematics for the measurement of intangible factors the analytic hierarchy/network process. *RACSAM-Revista de la Real Academia de Ciencias Exactas, Físicas y Naturales. Serie A, Matemáticas, Serie, 102*(2), 251–318.

Saleem, S., Ahmed Hajam, F., & Rather, J. A. (2021). Spatio-temporal analysis of land use land cover changes in Sind catchment of the Kashmir Valley, India. *Geo-Eye, 10*(2), 35–42.

Saleem, S., Rather, J. A., Ahmed, S., Mushtaq, S., Ahmed, R., & Malik, I. H. (2024). Above ground biomass estimation for alpine grasslands of Kashmir Himalayas using remote sensing and field data. *Rangeland Ecology & Management, 96*, 117–127.

Sarkar, M. S., Majhi, B. K., Pathak, B., Biswas, T., Mahapatra, S., Kumar, D., Bhatt, I. D., Kuniyal, J. C., & Nautiyal, S. (2024). Ensembling machine learning models to identify forest fire-susceptible zones in Northeast India. *Ecological Informatics, 81*, 102598. https://doi.org/10.1016/j.ecoinf.2024.102598

Scholes, R. J. (2016). Climate change and ecosystem services. *WIREs Climate Change, 7*(4), 537–550. https://doi.org/10.1002/wcc.404

Seidl, R., Rammer, W., & Lexer, M. J. (2010). Climate change vulnerability of sustainable forest management in the eastern Alps. *Climatic Change, 106*(2), 225–254. https://doi.org/10.1007/s10584-010-9899-1

Shannon, C., & Weaver, W. (1963). *The mathematical theory of communication.* University of Illinois.

Sharma, J., Chaturvedi, R. K., Bala, G., & Ravindranath, N. H. (2013). Challenges in vulnerability assessment of forests under climate change. *Journal of Carbon Management, 4*, 403–411.

Sharma, J., Indu, K., Murthy, E. T., Negi, P., Sushma, S., Dasgupta, S., Barua, A., Bala, G., & Ravindranath, N. H. (2018). Vulnerability and risk assessment: Framework methods and guideline. *Department of Science and Technology (DST), government of India Indian institute of Science.* http://www ihcap in/resources html.

Sharma, J., Jayaraman, M., Chaturvedi, R. K., Ravindranath, N. H., Bala, G., & Upgupta, S. (2017). Vulnerability of forests in India: A National Scale Assessment. *Environmental Management, 60*(3), 544–553. https://doi.org/10.1007/s00267-017-0894-4

Sharma, J., Ravindranath, N. H., Upgupta, S., Bala, G., Kumar, R., & Chaturvedi, R. K. (2015). Assessment of inherent vulnerability of forests at landscape level: A case study from Western Ghats in India. *Mitigation and Adaptation Strategies for Global Change, 22*(1), 29–44. https://doi.org/10.1007/s11027-015-9659-7

Singh, B., & Kaushal, S. L. (2023). Poverty reduction through MGNREGA in Himachal Pradesh. *Productivity, 64*(2), 202–209. https://doi.org/10.32381/prod.2023.64.02.9

Stanturf, J. A. (2015). Future landscapes: Opportunities and challenges. *New Forests, 46*(5–6), 615–644. https://doi.org/10.1007/s11056-015-9500-x

Tali, B. A., Khuroo, A. A., Ganie, A. H., & Nawchoo, I. A. (2019). Diversity, distribution and traditional uses of medicinal plants in Jammu and Kashmir (J&K) state of Indian Himalayas. *Journal of Herbal Medicine*, 17, 100280–18. https://doi.org/10.1016/j.hermed.2019.100280

Thakur, S., Negi, V. S., Pathak, R., Dhyani, R., Durgapal, K., & Rawal, R. S. (2019). Indicator based integrated vulnerability assessment of community forests in Indian west Himalaya. *Forest Ecology and Management, 457*, 117674. https://doi.org/10.1016/j.foreco.2019.117674.

Thompson, I., Mackey, B., McNulty, S., & Mosseler, A. (2009). Forest resilience, biodiversity and climate change in secretariat of the convention on biological diversity Montreal. *Technical, 43*, 1–67.

Upgupta, S., Sharma, J., Jayaraman, M., Kumar, V., & Ravindranath, N. H. (2015). Climate change impact and vulnerability assessment of forests in the Indian Western Himalayan region: A case study of Himachal Pradesh, India. *Climate Risk Management, 10*, 63–76. https://doi.org/10.1016/j.crm.2015.08.002

Voight, C., Hernandez-Aguilar, K., Garcia, C., & Gutierrez, S. (2019). Predictive modeling of future Forest cover change patterns in southern Belize remote sensing, *11*(7), 823. https://doi.org/10.3390/rs11070823

Wani, Z. A., & Pant, S. (2023). Status of biodiversity in a protected area of Kashmir Himalaya: Gulmarg wildlife sanctuary. *Nordic Journal of Botany, 2023*(7). https://doi.org/10.1111/njb.03982

Weller, S. G., Lorence, D. H., Clark, M., Tangalin, N., Wood, K., Sakai, A. K., Kishida, W., & Flynn, T. (2017). The effects of introduced ungulates on native and alien plant species in an Island ecosystem: Implications for change in a diverse Mesic forest in the Hawaiian islands. *Forest Ecology and Management, 409*, 518–526. https://doi.org/10.1016/j.foreco.2017.11.023

Woodward, A., Weinstein, P., & Hales, S. (1998). Climate change and human health in the Asia Pacific region: Who will be most vulnerable? *Climate Research, 11*(1), 31–38. https://doi.org/10.3354/cr011031

WPSFD (Working Plan Sindh Forest Division). (2000). *J & K Forest department Sindh Forest Division Ganderbal Kashmir India*. Ministry of Environment.

WWF. 2017. Forests with value for people and the planet: the way forward for FSC. https://medium.com/@WWF/forests-with-value-for-people-and-the-planet-the-way-forward-for-fsc-fa9114967210.

www.globalforestwatch.org

Yang, W., Wang, Y., Webb, A. A., Li, Z., Tian, X., Han, Z., Wang, S., & Yu, P. (2017). Influence of climatic and geographic factors on the spatial distribution of Qinghai spruce forests in the dryland Qilian Mountains of Northwest China. *Science of the Total Environment, 612*, 1007–1017. https://doi.org/10.1016/j.scitotenv.2017.08.180

Yuh, Y. G., Tracz, W., Matthews, H. D., & Turner, S. E. (2022). Application of machine learning approaches for land cover monitoring in northern Cameroon. *Ecological Informatics, 74*, 101955. https://doi.org/10.1016/j.ecoinf.2022.101955

Habitat Suitability Zone Analysis Using Geoinformatics for Elephants to Mitigate Human-Elephant Conflict in Mayurjharna Elephant Reserve, West Bengal, India

Lopamudra Ghosh and Nirupam Acharyya

Abstract The Mayurjharna Elephant Reserve, located in eastern India, spans parts of Paschim Medinipur, Jhargram, and Bankura districts in West Bengal. This research focuses on identifying habitat suitability zones for elephants in the reserve using the Analytic Hierarchy Process (AHP) method. Parameters such as the Soil Adjusted Vegetation Index (SAVI), water availability (NDWI), road network, agricultural land, settlements, elevation, slope, and aspect are used to create thematic raster layers, weighted based on their significance. Additionally, a Conflict Risk Zone map, created from Forest Office survey data, presents high, medium, and low conflict zones between humans and elephants. The study also includes a Forest Fire Risk Zone map, created using the AHP method with factors such as road network, settlements, land use and land cover (LULC), elevation, slope, and aspect. These criteria were used to generate thematic raster layers and assign weighted significance. The analysis reveals that 53.865 km^2 of the reserve is highly suitable for elephants, while 29.409 km^2 is of very low suitability. Furthermore, 158 villages fall under high conflict risk zones, and 64 villages are in low conflict risk zones. Approximately 11.5 km^2 of the reserve is classified as high-risk for forest fires, while 194.35 km^2 is in the very low-risk category. The study also explores the causes of increasing human-elephant conflicts. Strategies for elephant conservation and solutions to mitigate human-elephant conflicts are proposed to ensure the coexistence and safety of both species.

L. Ghosh (✉)
Department of Remote Sensing and Geoinformatics, Birla Institute of Technology, Mesra, Ranchi, Jharkhand, India
e-mail: lopamudraghosh0101@gmail.com

N. Acharyya
Department of Remote Sensing and GIS, Vidyasagar University, Rangamati, Midnapore, West Bengal, India
e-mail: nirupam@mail.vidyasagar.ac.in

R. Ahmed et al. (eds.), *Climate Change, Water Resources, and Natural Hazards in the Himalayas*, Advances in Geographical and Environmental Sciences,
https://doi.org/10.1007/978-981-92-0835-7_12

Keywords Mayurjharna Elephant Reserve · Habitat suitability · AHP · Human-elephant conflict · Geoinformatics · Forest fire risk

1 Introduction

Forest is an important part of our environment. Forest helps to maintain ecological balance. Greenery dense forest is habitat and also home of wildlife. They get food, water and shelter from their habitat, they take rest in the forest core area. Food, water and shelter are very well related to landuse. Elephants, the wild herbivore mammals completely depend on forest for food, water and shelter (Mandal & Das Chatterjee, 2021). United States Department of Agriculture (USDA) states that Habitat is a combination of food, water, shelter, and space arranged to meet the needs of wildlife. When more than one larger wildlife habitats join, it creates a connection between wildlife habitats, this is called Wildlife Corridor. Wildlife Corridor, alternatively known as Green Corridors are areas that helps migratory animals to move safe and securely from one habitat to another habitat. Corridors also helps to solve fragmentation of forests, as with animals, wildlife corridors offer immense benefits to the various species of plants. They help plants grow and thrive in a healthy protected environment. Corridors prevents animal vehicle collisions and minimize human-animal interactions (Acharyya & Bandyopadhyay, 2017). Elephants are the largest mammals in the world. Three living species of elephants are currently recognised: the, the African forest elephant (*Loxodonta cyclotis*), the African bush elephant (*Loxodonta africana*) and the Asian elephant (*Elephas maximus*). Elephants are scattered throughout Sub-saharan Africa, South-east Asia, South Asia, and are found in different habitats, including Savannahs, forests, deserts, and marshes. Elephants are live in tropical evergreen forests, grasslands, moist deciduous forests, dry thorn forests, semi-evergreen forests, dry deciduous forests and also lives in scrublands cultivated and secondary forests (Ram, 2014).

The Asian elephant (Elephas maximus), often referred to as the Asiatic elephant, is the sole surviving member of the genus Elephas. Its range extends across the Indian subcontinent and parts of Southeast Asia. Populations occur in countries including India and Nepal, extend southward to Sumatra, eastward to Borneo, and are also present in Sri Lanka. In 1986, the Asian elephant was classified as Endangered on the IUCN Red List following evidence of a population decline of at least 50% over three generations, equivalent to roughly 60–75 years. The species faces multiple threats, primarily habitat loss, degradation, and fragmentation, which have intensified interactions and conflicts between humans and elephants. Illegal hunting remains another serious concern, as elephants are targeted for ivory as well as for products derived from their skin, meat, and other body parts. In recent years, the use of elephant skin in certain traditional medicinal practices has further increased exploitation pressure on the species.

Conservation of wildlife and their habitats has become one of the most pressing environmental challenges of the present time. Animal habitats, particularly those of elephants, are increasingly threatened by unsustainable human activities and changing natural conditions (Sanare et al., 2015). Human–wildlife conflict has intensified across the country, especially in transitional zones where human settlements overlap with wildlife habitats. This conflict has escalated largely due to habitat destruction, which reduces the availability of food and resources within forests. As a consequence, wildlife species are compelled to move into nearby villages and agricultural lands, leading to frequent confrontations between people and animals (Badhe & Jaybhaye, 2021). Forest fires represent a major environmental concern, particularly in Mediterranean regions where such events occur regularly. These areas require comprehensive assessment frameworks that consider multiple contributing factors and the broader impacts of fire at regional scales. The preparation of forest fire risk zonation maps can help reduce fire incidence by identifying vulnerable areas. Effective fire risk evaluation must account for both ignition probability and potential fire spread. Estimating how fire may propagate under specific burning conditions within different forest stands is essential. The spread of forest fires threatens vegetation cover, wildlife habitats, and human safety. Early detection plays a crucial role in minimizing fire-related damage (Erten et al., 2004). Geospatial technologies combined with the Analytical Hierarchy Process method have proven effective for assessing forest fire risk (Lamat et al., 2021). A significant research gap exists in the delineation of elephant habitat suitability zones, as only limited studies have addressed this aspect. In the case of the Mayurjharna Elephant Reserve, research on elephant habitat assessment is particularly scarce, with only one documented study available.

Mayurjharna Elephant Reserve, which does not have the status of a fully protected forest, has experienced substantial deforestation. This ongoing habitat degradation has forced elephants to move progressively toward nearby villages, resulting in a steady decline in their forest-based habitat. Several settlements are located within or adjacent to forest boundaries, increasing the likelihood of encounters between elephants and residents. Such interactions frequently lead to conflict, often resulting in casualties among both humans and elephants. In north-eastern India, human–elephant conflict has intensified in recent decades due to expanding economic activities, population growth, settlement expansion, and forest clearance. Over the past few decades, extensive forest areas have been converted into military installations and commercial tea plantations, further fragmenting elephant habitats (Naha et al., 2019). After finding the habitat suitability zone of the Elephants in Mayurjharna Elephant Reserve forest through Remote Sensing and GIS, taking into consideration of nine parameters then the high, medium and low zones of habitat suitability are marked in this forest. In this case the actual suitable living place of the Elephant must be found for their peace living there. Remote Sensing and GIS is very useful to understand Human-Elephant conflict. RS & GIS also helps to conserve natural resources (Kanga et al., 2017). After finding the conflicting zones of the human and the Elephant and also its high, medium and low conflicted zones through Remote Sensing and GIS, particularly taking into consideration of high conflicted zones, some steps must be considered to reduce such conflicts between humans and the Elephant as far as possible, as a result the incident of death of either human and the

Elephant will be reduced to the positively. Conservation of habitat is very important for conservation of wildlife. Habitat Suitability analysis is play very important role to conserve wildlife like elephants and also RS and GIS plays very important role to conserve wildlife (Sharma et al., 2019). Remote Sensing and GIS helps scientist and forest officers efficiently to conserve wildlife. Aerial, satellite and UAV remote sensing technology can support scientists and conservationists to identify, track, monitor, and access wildlife in the most remote areas around the world, including marine habitats.

2 Research Objectives

1. To find out habitat suitability zone for Elephants in Mayurjharna Elephant Reserve
2. To assess the reason behind Human-Elephant conflict in Mayurjharna Elephant Reserve
3. To map Human-Elephant Conflict Risk Zones
4. To delineate Forest Fire Risk Zone of Mayurjharna Elephant Reserve.

This main focus of the study is to address the complex interaction of ecological, anthropogenic, and environmental factors to improve peaceful coexistence of human and elephants and make sure the lifelong existence of elephants in the Mayurjharna Elephant Reserve.

3 About the Study Area

The Mayurjharna Elephant Reserve is located at the confluence of the districts of Jhargram, Purulia, Bankura of West Bengal State, India. The study area is situated between 86°24′48.38″ E to 86°53′09.77″ E Longitude and 22°33′38.72″ N to 22°54′33.58″ N Latitude. The area of this elephant reserve is 414.06 km^2. and adjoining area is 1436 km^2, this area is also declared as 'ZONE OF INFLUENCE' (Mandal & Das Chatterjee, 2021). The present study covers 906.61 km^2 (Fig. 1). The reserve is declared on 24th October 2002 by the Government of West Bengal. Forest areas of three divisions Belpahari, Bhulaveda, and Banspahari forest areas of Jhargram division, Fulkusma, Motgoda, Ranibandh, Jhilimili forest areas of Bankura division and Yamuna forests of Kangsabati Soil Conservation Division II come under this Elephant reserve Zone. Mayurjharna Elephant Reserve situated in a border of Jharkhand and West Bengal state. Dalma Wildlife Sanctuary of Jharkhand State and Mayurjharna Elephant Reserve of West Bengal are connected by vast tract of forest which was again regenerated through Joint Forest Management in late 80s. That made a corridor between Dalma Wildlife Sanctuary and Mayurjharna Elephant Reserve, which is helps elephants to migrate Jharkhand to West Bengal (Ministry of

Environment, Forest and Climate Change. (n.d.). *Brief note on vulnerability of forest area to erosion*).

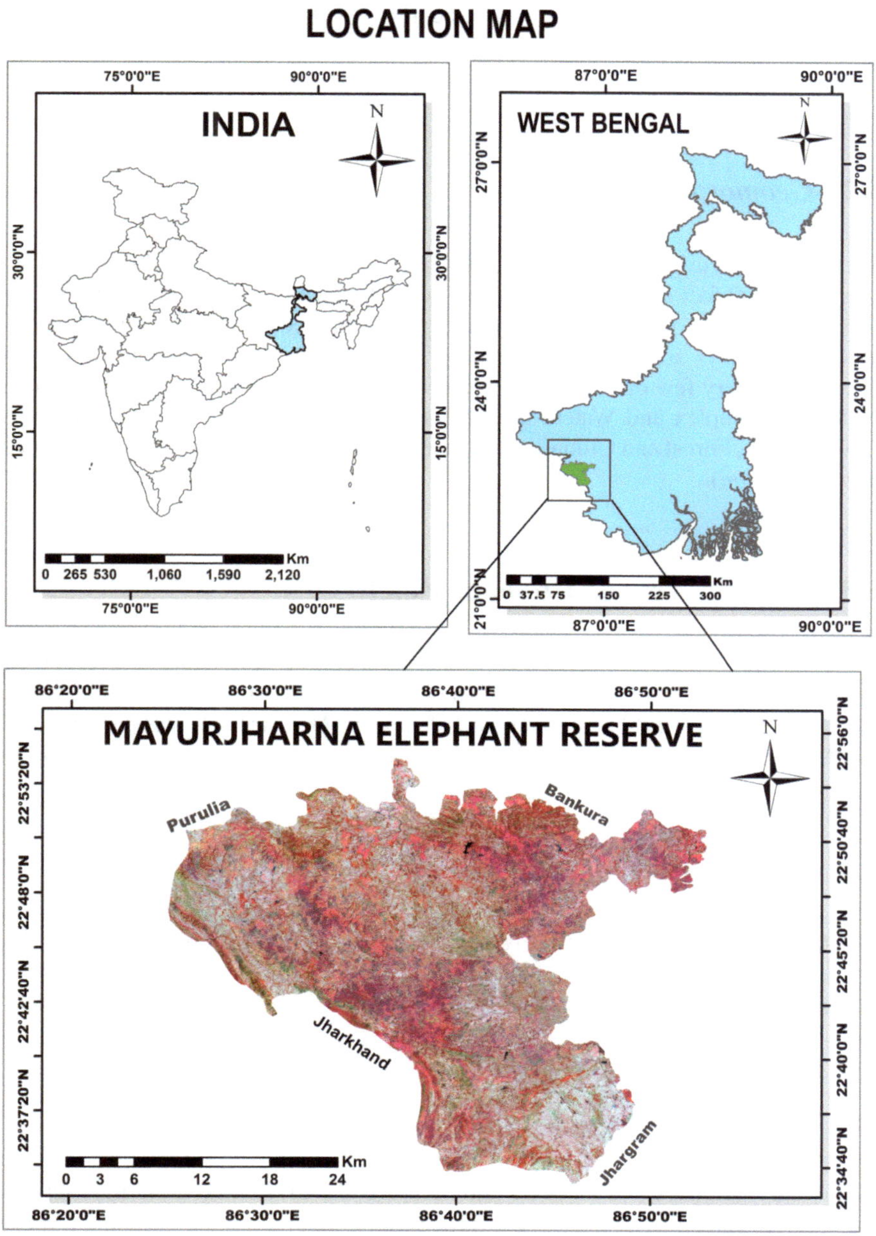

Fig. 1 Showing location of study area

3.1 Geology and Rocks

The Elephant Reserve shows rocks of archean era in the geological formation along with gondowana system. These reserves consist of DALMA GROUP, KUILAPAL GRANITE, SINGHBHUM GROUP (Ministry of Environment, Forest and Climate Change. (n.d.). *Brief note on vulnerability of forest area to erosion*).

3.2 Geomorphology

The entire Mayurjharna Elephant Reserve is situated on low-rise hillocks. They rise 800–1000 feet above the Mean Sea Level. Conical hills are very common and mainly composed of porphyritic or dome gneiss. In this area Moderately Dissected Hills and Valleys are found, lesser site of Highly Dissected Hills and Valleys are found here, Very few of Low Dissected Hills and Valleys are found here. Pediment Pediplain Complex and Waterbodies are seen in the rest of the area (Ministry of Environment, Forest and Climate Change. (n.d.). *Brief note on vulnerability of forest area to erosion*).

3.3 River System and Drainage Pattern

Mayurjharna Elephant Reserve support many seasonal streams and only a few perennial ones, eminent among them are Domohani, Tarafeni, Mayurjharna, Aamjharna, Kuilapal, Dulung, Kuldiha Yamuna, Yujardhara etc. (Ministry of Environment, Forest and Climate Change. (n.d.). *Brief note on vulnerability of forest area to erosion*).

3.4 Climatic Conditions

The Mayurjharna Elephant Reserve has a tropical climate with high humidity. The hot and dry months of May and June may reach temperatures as high as 46 °C, while the cold evenings of December and January can see temperatures as low as 4 °C.

3.5 Wildlife

There are various types of wild animals are found in Mayurjharna Elephant Reserve. Fishing Cat (*Felis Vivenrina*), Small Indian Civet (*Vivericula Indica*), Rhesus macaque (*Macaca mulata*), Jungle cat (*Felischaus*), Peacock (*Pavo cristatus*),

Jackle (*Canis auris)*, Indian Fox (*Vulpes bengalensis)*, Common palm civet (*Paradoxurus hermaphroditus)* and Asian Elephants (*Elephantus maximus*) are seen in Mammals class. Indian Phython (*Phython molurus)*, Cobra (*Naja naja)*, Monitor Lizard (*Varanus bengalensis)* are seen in Reptiles and Amphibians class (Ministry of Environment, Forest and Climate Change. (n.d.). *Brief note on vulnerability of forest area to erosion*).

4 Data Sources

For this study data has been collected from various sources. Mostly from Open Access Hub—Copernicus, Geological Survey of India and BHUVAN etc. The data and their sources are given in Table 1.

Table 1 Data used and their sources

Maps	Data	Source
Geological map	Geology 2M	Geological Survey of India
Geomorphological map	Geomorphology 250K	Geological Survey of India
Drainage density map	Cartosat-1	BHUVAN
Road network map	Google earth imagery	Google Earth Pro Image
NDVI map	Sentinel-2 MS L2A	Open Access Hub—Copernicus
SAVI map	Sentinel-2 MS L2A	Open Access Hub—Copernicus
NDWI map	Sentinel-2 MS L2A	Open Access Hub—Copernicus
LULC map	Sentinel-2 MS L2A	Open Access Hub—Copernicus
Aspect map	Cartosat-1	BHUVAN
Digital elevation model	Cartosat-1	BHUVAN
Slope map	Cartosat-1	BHUVAN
Rainfall map	Gridded rainfall (0.25×0.25) NetCDF file	Indian Meteorological Department
Soil map	WB soil map	National Bureau of Soil Survey and Land Use Planning
Human-elephant conflict zones map	India_village_cencus_2001WB	Socio Economic Data and Applications Center (SEDAC)

5 Methodology

5.1 Habitat Suitability Zone for Elephants

Sentinel-2 MS L2A satellite image downloaded from ESA Copernicus Open Access Hub. After downloading the image, the band composition and subsetting the study area have also been done. The criteria map for habitat suitability analysis like NDVI, SAVI, and NDWI map has been prepared from sentinel-2 image in ArcGIS 10.8 platform. The Land Use and Land Cover map has been prepared from Sentinel-2 image in ERDAS Imagine software. Road network delineated from Google Earth Pro. web platform and Digital Elevation Model, Aspect and Slope map created from CARTOSAT-1 data which has been downloaded from NRSC BHUVAN web portal.

After that did pairwise comparison between criterion of Habitat suitability Zone map (Tables 2 and 3). Then reclassify all the criterion maps with equal interval and Weighted Overlay Analysis method is applied in ArcGIS 10.8 software to create Habitat Suitability Zone Map of Elephants (see Fig. 2 and Table 3).

The consistency of the pairwise comparison matrix was evaluated to ensure the reliability of the assigned weights. The weighted sum values and corresponding consistency measures for each criterion are presented in Table 4. Based on these calculations, the principal eigenvalue (λ) was derived and used to compute the Consistency Index (CI) and Consistency Ratio (CR). The obtained CR value falls within the acceptable threshold, indicating that the judgments made in the pairwise comparison are consistent and suitable for further analysis (Table 4).

The normalized weights and their corresponding influence on habitat suitability are presented in Table 5. Each criterion was assigned suitability classes based on its relative importance and reclassified into different intensity levels. These classes were used to standardize all thematic layers for further analysis. The final weighted values derived from this classification formed the basis for the Weighted Overlay Analysis in the habitat suitability assessment (Table 5).

5.2 Human-Elephant Conflict Risk Zone

The Human-Elephant conflict related data has been collected from Forest Department. The derived data was analyzed to identify which areas have high, moderate, low conflict risk zone, and created a Human-Elephant Risk Zone Map. 10.8 software (Fig. 3).

Table 2 Pairwise comparison of the evaluation criteria

Criteria	SAVI	NDWI	Agricultural land	Distance from settlements	Distance from roads	Forest fire	Elevation	Slope	Aspect
SAVI	1	3	4	5	6	7	8	9	9
NDWI	1/2	1	3	4	5	6	7	8	8
Agricultural land	1/3	1/2	1	2	4	5	6	7	8
Distance from settlement	1/4	1/4	1/2	1	3	4	5	6	7
Distance from roads	1/6	1/5	1/4	1/2	1	3	4	6	6
Forest fire	1/7	1/6	1/6	1/5	1/2	1	2	4	5
Elevation	1/7	1/7	1/6	1/4	1/3	1/2	1	3	4
Slope	1/9	1/8	1/7	1/6	1/5	1/4	1/3	1	2
Aspect	1/9	1/8	1/7	1/5	1/5	1/3	1/2	1/2	1

Table 3 Standardized pairwise comparison matrix and their weights

Criteria	SAVI	NDWI	Agricultural land	Distance from settlements	Distance from roads	Forest fire	Elevation	Slope	Aspect	Weight
SAVI	0.36	0.54	0.42	0.37	0.3	0.26	0.24	0.2	0.18	0.31
NDWI	0.18	0.18	0.32	0.3	0.25	0.22	0.21	0.18	0.16	0.22
Agricultural land	0.12	0.09	0.11	0.15	0.2	0.2	0.18	0.16	0.16	0.15
Distance from settlement	0.1	0.05	0.05	0.07	0.14	0.14	0.15	0.13	0.14	0.11
Distance from roads	0.06	0.04	0.03	0.04	0.05	0.11	0.11	0.13	0.12	0.08
Forest fire	0.05	0.03	0.02	0.02	0.02	0.03	0.06	0.09	0.1	0.05
Elevation	0.05	0.03	0.02	0.02	0.02	0.02	0.03	0.07	0.08	0.04
Slope	0.04	0.02	0.02	0.01	0.01	0.01	0.01	0.02	0.04	0.02
Aspect	0.04	0.02	0.01	0.01	0.01	0.01	0.01	0.01	0.02	0.02
	1	1	1	1	1	1	1	1	1	1

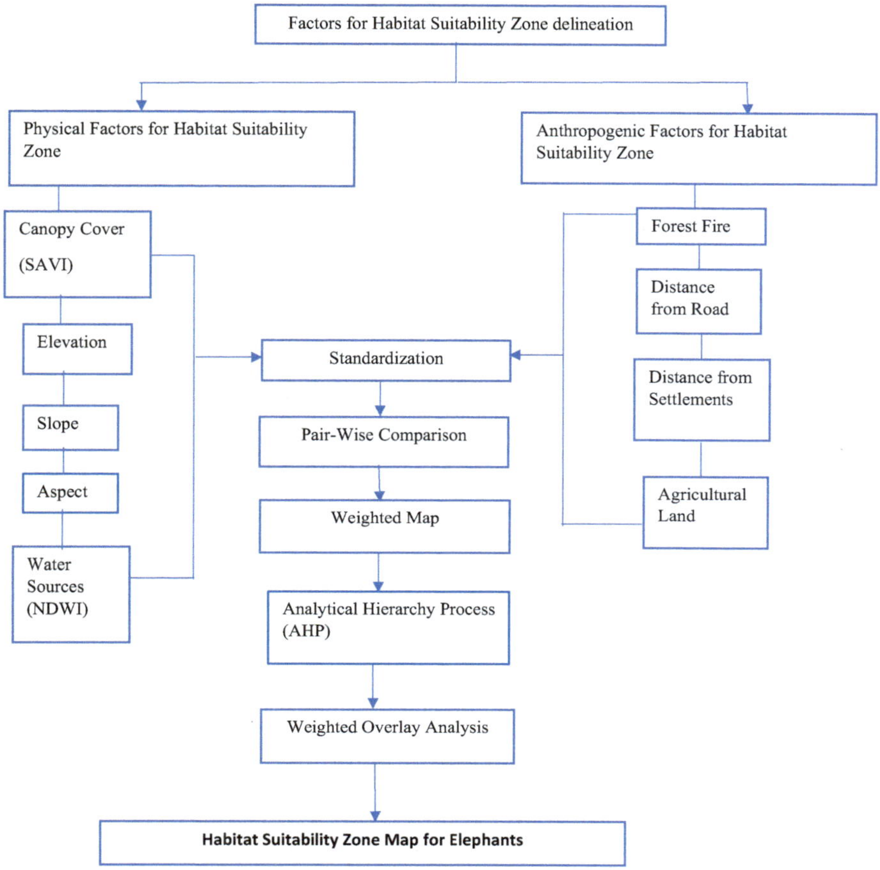

Fig. 2 Flow chart of habitat suitability map of elephants

5.3 *Forest Fire Risk Zone Mapping*

At first Digital Elevation Model, Aspect and Slope map has been prepared from CARTOSAT-1 data. The LULC map has also been prepared with the help of sentinel data in ERDAS Imagine software. The final Road network is digitized from Google Earth Pro. web platform.

The pairwise comparison between different criteria has been conducted to develop a Forest Fire Risk Zone map. Then reclassify all the maps with equal interval method and Weighted Overlay Analysis method is applied in ArcGIS 10.8 software to create Forest Fire Risk Zone Map of Mayurjharna Elephant Reserve (Fig. 4).

Table 4 Determining the consistency ratio

Criteria	Step-1		Step-2	
SAVI	(0.31 * 1) + (0.22 * 3) + (0.15 * 4) + (0.11 * 5) + (0.08 * 6) + (0.05 * 7) + (0.04 * 8) + (0.02 * 9) + (0.02 * 9)	3.63	3.63/0.31	11.7
NDWI	(0.31 * 0.5) + (0.22 * 1) + (0.15 * 3) + (0.11 * 4) + (0.08 * 5) + (0.05 * 6) + (0.04 * 7) + (0.02 * 8) + (0.02 * 8)	2.56	2.56/0.22	11.63
Agricultural land	(0.33 * 0.31) + (0.5 * 0.22) + (1 * 0.15) + (2 * 0.11) + (4 * 0.08) + (5 * 0.05) + (6 * 0.04) + (7 * 0.02) + (8 * 0.02)	1.69	1.69/0.15	11.26
Distance from settlement	(0.25 * 0.31) + (0.25 * 0.22) + (0.5 * 0.15) + (1 * 0.11) + (3 * 0.08) + (4 * 0.05) + (5 * 0.04) + (6 * 0.02) + (7 * 0.02)	1.2	1.2/0.11	10.91
Distance from roads	(0.16 * 0.31) + (0.2 * 0.22) + (0.25 * 0.15) + (0.5 * 0.11) + (0.08 * 1) + (3 * 0.05) + (4 * 0.04) + (6 * 0.02) + (6 * 0.02)	0.79	0.79/0.08	9.87
Forest fire	(0.14 * 0.31) + (0.16 * 0.22) + (0.16 * 0.15) + (0.2 * 0.11) + (0.5 * 0.08) + (1 * 0.05) + (2 * 0.04) + (4 * 0.02) + (5 * 0.02)	0.46	0.46/0.05	9.2
Elevation	(0.14 * 0.31) + (0.14 * 0.22) + (0.16 * 0.15) + (0.25 * 0.11) + (0.33 * 0.08) + (0.5 * 0.05) + (1 * 0.04) + (3 * 0.02) + (4 * 0.02)	0.35	0.35/0.04	8.75
Slope	(0.11 * 0.31) + (0.12 * 0.22) + (0.14 * 0.15) + (0.16 * 0.11) + (0.2 * 0.08) + (0.25 * 0.05) + (0.33 * 0.04) + (1 * 0.02) + (2 * 0.02)	0.17	0.17/0.02	8.5
Aspect	(0.11 * 0.31) + (0.12 * 0.22) + (0.14 * 0.15) + (0.2 * 0.11) + (0.2 * 0.08) + (0.33 * 0.05) + (0.5 * 0.04) + (0.5 * 0.02) + (1 * 0.02)	0.16	0.16/0.02	8

Consistency Ratio is
λ = (11.7 + 11.63 + 11.26 + 10.91 + 9.87 + 9.2 + 8.75 + 8.5 + 8)/9 = 9.98
CI = λ – n/n – 1 = 9.98 – 9/9 – 1 = 0.12
CR = CI/RI = 0.12/1.45 = 0.08

5.4 Consistency Analysis and Criteria Weighting for Forest Fire Risk Mapping

A pairwise comparison of the selected criteria influencing forest fire risk was carried out to determine their relative importance (Table 6). The standardized pairwise comparison matrix along with the derived weights is presented in Table 7. Further,

Table 5 Normalized and final weights of different features of thematic layer for assessment.

Main criteria	Weight	Weighted influence (%)	Suitability criteria	Sub criteria	Intensity of importance
SAVI	0.31	31	Very low	− 0.283–0.300	1
			Low	0.300–0.416	2
			Moderate	0.416–0.513	3
			High	0.513–0.952	4
NDWI	0.22	22	High	< 400 m	4
			Moderate	400–800 m	3
			Low	800–1200 m	2
			Very low	> 1200 m	1
Agricultural land	0.15	15	Very low	< 300 m	1
			Low	300–600 m	2
			Moderate	600–900 m	3
			High	> 900 m	4
Distance from settlement	0.11	11	Very low	< 900 m	1
			Low	900–1800 m	2
			Moderate	1800–2700 m	3
			High	> 2700 m	4
Distance from roads	0.08	8	Very low	< 550 m	1
			Low	550–1100 m	2
			Moderate	1100–1650 m	3
			High	> 1650 m	4
Forest fire	0.05	5	Very low	11.5 km^2	1
			Low	213.16 km^2	2
			Moderate	475.9 km^2	3
			High	194.35 km^2	4
Elevation	0.04	4	Very low	40–107 m	1
			Low	107–152 m	2
			Moderate	152–197 m	3
			High	197–435 m	4
Slope	0.02	2	Very low	0°–7.839003°	1
			Low	7.839003°–17.051142°	2
			Moderate	17.051142°–31.819948°	3
			High	31.819948°–88.467598°	4
Aspect	0.02	2	Very low	Flat, north, northeast	1
			Low	West, northwest, north	2
			Moderate	South, southwest	3
			High	East, southwest	4

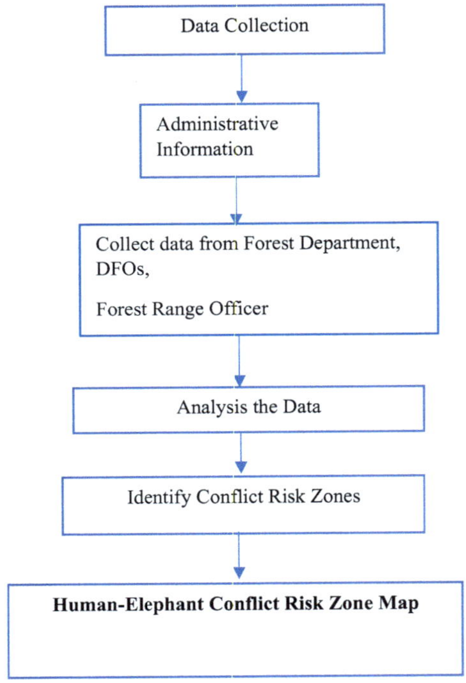

Fig. 3 Flow Chart of human-elephant conflict risk zone map

the consistency of the judgments was evaluated using eigenvalue-based calculations, and the Consistency Index (CI) and Consistency Ratio (CR) were computed, as summarized in Table 8. The obtained CR value indicates acceptable consistency in the decision-making process. Based on these results, the criteria were normalized and classified into different suitability levels, and their final weights and influence on forest fire risk assessment are presented in Table 9.

6 Result and Discussion

6.1 Habitat Suitability Zone for Elephants

To develop a habitat suitability map for elephants, nine criteria were considered. The physical factors included Soil Adjusted Vegetation Index (SAVI), Normalized Difference Water Index (NDWI), Slope, Aspect, and Elevation. Additionally, four anthropogenic factors were incorporated: Forest fire, Distance from settlements, Distance from roads, and Agricultural land.

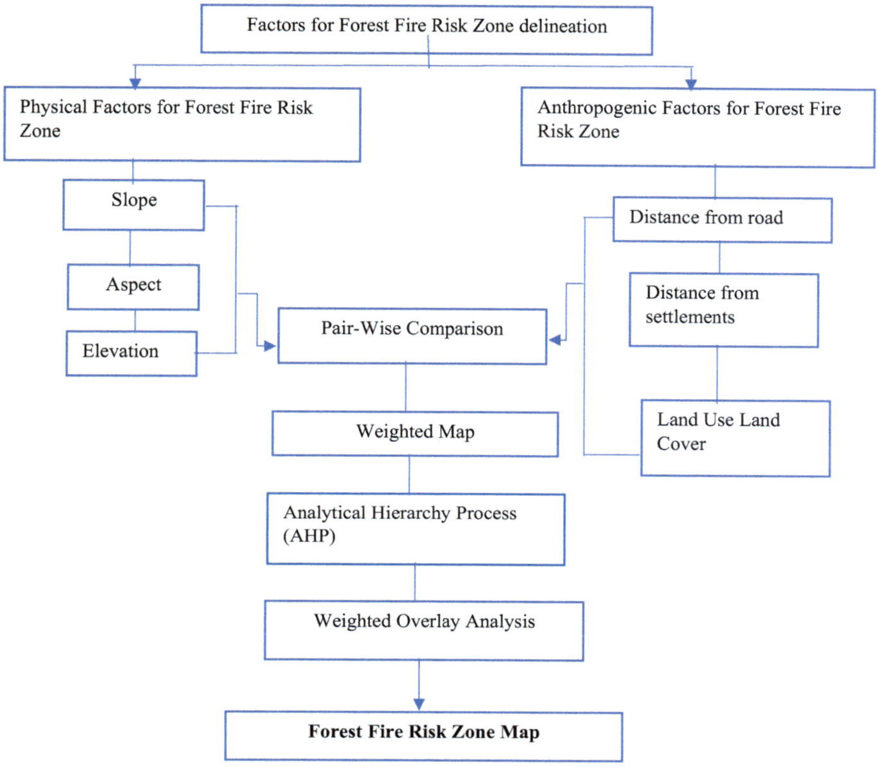

Fig. 4 Flow chart of forest fire risk zone map

Table 6 Pairwise comparison of the evaluation criteria

Criteria	Distance from roads	Distance from settlement	LULC	Aspect	Slope	Elevation
Distance from roads	1	1 / 2	3	4	5	5
Distance from settlement	4	1	5	6	6	7
LULC	1/2	1/3	1	3	3	4
Aspect	1/4	1/5	1/3	1	2	3
Slope	1/5	1/6	1/3	1/2	1	2
Elevation	1/6	1/7	1/4	1/3	1/2	1

Table 7 Standardized pairwise comparison matrix and their weights

Criteria	Distance from roads	Distance from settlement	LULC	Aspect	Slope	Elevation	Weight
Distance from roads	0.16	0.21	0.3	0.27	0.29	0.23	0.24
Distance from settlement	0.65	0.43	0.5	0.4	0.34	0.32	0.44
LULC	0.08	0.14	0.1	0.2	0.17	0.18	0.15
Aspect	0.04	0.09	0.03	0.07	0.11	0.14	0.08
Slope	0.03	0.07	0.03	0.03	0.06	0.09	0.05
Elevation	0.03	0.06	0.03	0.02	0.03	0.04	0.04
	1	1	1	1	1	1	1

Table 8 Determining the consistency ratio

Criteria	Step-1		Step-2	
Distance from roads	(0.24 * 1) + (0.44 * 0.5) + (0.15 * 3) + (0.08 * 4) + (0.05 * 5) + (0.04 * 5)	1.7	1.7/0.24	7.08
Distance from settlement	(0.24 * 4) + (0.44 * 1) + (0.15 * 5) + (0.08 * 6) + (0.05 * 6) + (0.04 * 4)	3.3	3.3/0.44	7.5
LULC	(0.24 * 0.5) + (0.44 * 0.33) + (0.15 * 1) + (0.08 * 3) + (0.05 * 3) + (0.04 * 4)	1.01	1.01/0.15	6.73
Aspect	(0.24 * 0.25) + (0.44 * 0.2) + (0.15 * 0.33) + (0.08 * 1) + (0.05 * 2) + (0.04 * 3)	0.51	0.51/0.08	6.4
Slope	(0.24 * 0.2) + (0.44 * 0.16) + (0.15 * 0.33) + (0.08 * 0.5) + (0.05 * 1) + (0.04 * 2)	0.31	0.31/0.05	6.2
Elevation	(0.24 * 0.16) + (0.44 * 0.14) + (0.15 * 0.25) + (0.08 * 0.33) + (0.05 * 0.5) + (0.04 + 1)	0.2	0.2/0.04	5

Consistency Ratio is
$\lambda = 7.08 + 7.5 + 6.73 + 6.4 + 6.2 + 5/6 = 6.48$
$CI = \lambda - n/n - 1 = 6.48–6/6–1 = 0.09$
$CR = CI/RI = 0.09/1.24 = 0.07$

6.1.1 Soil Adjusted Vegetation Index (SAVI)

SAVI of Mayurjharna Elephant Reserve is divided into four classes, (1) – 0.283–0.300 (2) 0.300–0.416 (3) 0.416–0.513 and (4) 0.516–0.952. – 0.283 to 0.300 range is very low suitable for elephants to stay, 0.300–0.416 range is low suitable, 0.416–0.513 is moderately suitable and 0.516–0.952 range is highly suitable for elephants to stay (Fig. 5).

Table 9 Normalized and final weights of different features of thematic layer for assessment

Main criteria	Weight	Weighted influence (%)	Suitability criteria	Sub criteria	Intensity of importance
Distance from roads	0.24	24	High	< 400	1
			Moderate	400–800	2
			Low	800–1200	3
			Very low	> 1200	4
Distance from settlement	0.44	44	High	< 800	1
			Moderate	800–1600	2
			Low	1600–2400	3
			Very low	> 2400	4
LULC	0.15	15	High	Agricultural land	1
			Moderate	Built-up area	2
			Low	Vegetation	3
			Very low	Waterbodies	4
Aspect	0.08	8	High	South, south-west	1
			Moderate	Flat, north, north-east	2
			Low	East, south-east	3
			Very low	West, north–west	4
Slope	0.05	5	High	31.91772163–88.46759796	1
			Moderate	16.99965609–31.91772162	2
			Low	7.632498649–16.99965608	3
			Very low	0–7.632498648	4
Elevation	0.04	4	High	44.3319931–106.9455829	1
			Moderate	106.945583–151.6695757	2
			Low	151.6695758–196.3935684	3
			Very low	196.3935685–424.4859314	4

6.1.2 Waterbody Buffer

Waterbody buffer derived from NDWI is divided into four buffer zones areas within < 400 m is highly suitable for elephants because this area near to waterbodies. 400–800 m moderately suitable 800–1200 m is low suitable and areas > 1200 m is very low suitable for elephants because this area is far away from waterbodies (Fig. 5).

6.1.3 Agricultural Land

The amount of Agricultural Land is 21% in Mayurjharna Elephant Reserve. In order to meet the demand of foods the amount of agricultural land is gradually increasing.

In order to increase the agricultural land deforestation is required as a result the forest area is gradually decreasing. For that reason, Human-Wildlife conflict is gradually increasing.

Agricultural Land of study area classified into four buffer zones <300-meter area is very low suitable, 300–600 m is low suitable, 600–900 m is moderately suitable and > 900 m is highly suitable for elephants (Fig. 5).

6.1.4 Distance from Settlement

As the population increases day by day, the amount of settlement area is also increasing. With the increasing in settlements, forest cover is gradually decreasing. To accommodate more cottages in tourist spots, trees in the nearby forest are being cut down. Settlement is cover 5% in Mayurjharna Elephant Reserve. Tribal families live in the middle of the forest and near to the forest, many cottages are located in the forest for tourists.

Settlement buffer zone classified into four categories, < 900 m is very low suitable, 900–1800 m is low suitable, 1800–2700 m is moderately suitable, > 2700 m is highly suitable for elephants (Fig. 5).

6.1.5 Distance from Road

Every year many tourists come to Mayurjharna Elephant Reserve to experience its scenic beauty. Dalma hills, dams and forests are very popular tourist spot in this reserve. Footpath leads the tourist into the dense forest. Also State Highway-5 goes through middle of the forest which is connected Purulia, Jhargram, Bankura district.

Forest is divided into two or more ecosystems due to road, for that reason normal life of wildlife was disturbed, many animals die during crossing the road after getting fatal injury, The rail line passing through the forest causes the death of the animals during crossing the rail line after getting fatal injury from running train. High decibel horn disturbed they're well living in a forest.

Road buffer zone of study area divided into four classes, < 550 m is very low suitable, 550–1100 m is low suitable, 1100–1650 m is moderately suitable and > 1650 m is highly suitable for elephants (Fig. 5).

6.1.6 Forest Fire

Global temperature of The Earth due to global warming and that reason facing forest fire regular manner all over the world. Humans are intentionally setting fire in the forest. Many animals die with trees being burned into ashes due to forest fire. Forest ecosystem completely destroyed due to forest fire.

In forest fire map 11.5 km^2 area is highly risk zone, 213.16 km^2 area is moderately risk zone, 475.9 km^2 area is low risk zone, 194.35 km^2 is very low risk zone (Fig. 5).

6.1.7 Elevation

Elevation is an important criterion for elephant's habitat. High elevation is not suitable for elephant's because they are not move easily in highly elevated areas. Also, low elevation is not suitable for elephant's habitat because low elevation areas forest is not available so, medium elevation is most appropriate for elephant's habitat. In this study area, 197–435 m is highly suitable, 152–197 m is moderately suitable, 107–152 m is low suitable, 40–107 m is very low suitable for elephants (Fig. 5).

6.1.8 Slope

Slope is a crucial topographic parameter in studies of human–wildlife conflict and environmental geography. In the study area, it is closely linked with forest distribution and habitat quality. Increasing slope generally corresponds to reduced human disturbance, thereby providing more suitable conditions for elephants (Badhe & Jaybhaye, 2021). Within the area, slopes of 0°–7.839003° are categorized as very low suitability, 7.839003°–17.051142° as low suitability, 17.051142°–31.819948° as moderately suitable, and 31.819948°–88.467598° as highly suitable for elephant habitat (Fig. 5).

6.1.9 Aspect

Aspect is an important terrain parameter for evaluating slope stability and ecological conditions. It significantly influences vegetation density and species composition by controlling solar radiation exposure and moisture availability. In the study region, slopes receiving greater solar radiation tend to experience higher evaporation rates and lower moisture retention (Badhe & Jaybhaye, 2021). Based on aspect classification in the study area, flat, north, and northeast-facing slopes are considered very low suitability for elephants. West, northwest, and north-facing slopes fall under low suitability. South and southwest-facing slopes are categorized as moderately suitable, whereas east and southeast-facing slopes are identified as highly suitable habitats. Eastern slopes receive early morning sunlight, while southern slopes in the Northern Hemisphere receive greater overall solar exposure, promoting better vegetation growth. The resulting availability of food resources and denser canopy cover makes the southern and eastern aspects more favorable for elephants compared to other slope directions (Fig. 5).

Habitat Suitability Map of Elephants was prepared by AHP method and geospatial technique. SAVI, Waterbody buffer, Agricultural Land, Distance from Settlement, Distance from Road, Forest Fire, Elevation, Slope, Aspect is taken for preparing Habitat Suitability Map of Elephants. Pairwise Comparison is done with these nine parameters after that all parameters are converted into raster format and reclassify all parameters. Final Habitat Suitability Map of Elephants is prepared by Weighted Overlay Analysis.

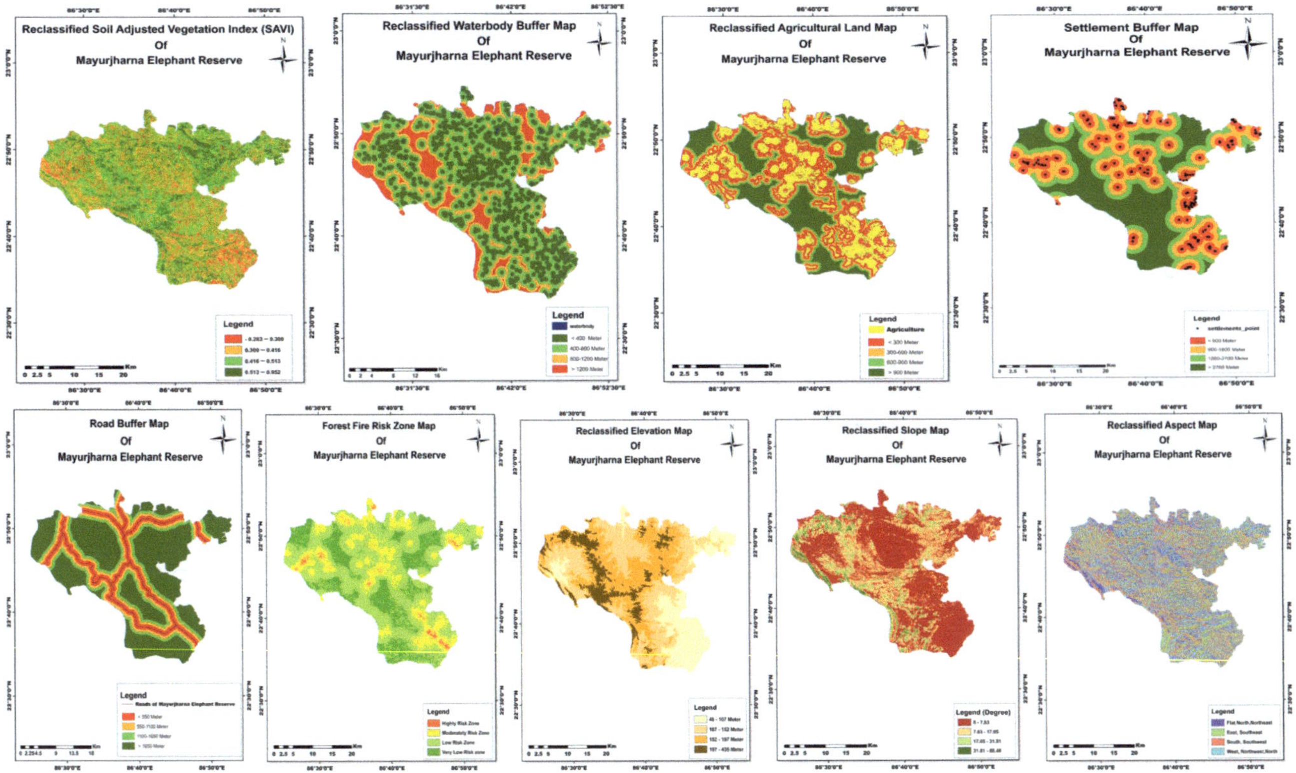

Fig. 5 Parameters for delineate habitat suitability zone for elephants, SAVI, Waterbody buffer, agricultural land, distance from settlement, distance from road, forest fire risk zone, elevation, slope, aspect

Habitat Suitability Map of Elephants is classified into four classes as follows, Very Low Suitable (29.409 km^2), Low Suitable (461.501 km^2), Moderately Suitable (349.525 km^2), and Highly Suitable (53.865 km^2) (Figs. 6 and 7; Table 10).

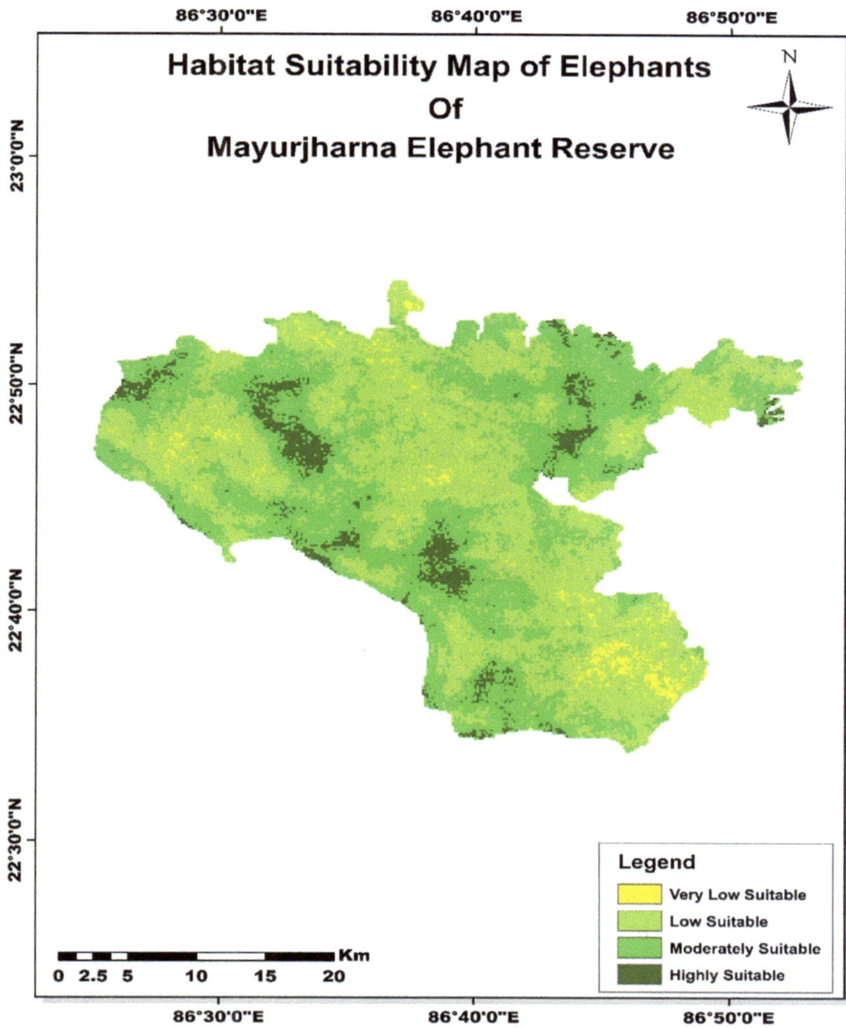

Fig. 6 Showing habitat suitability of elephants in the study area

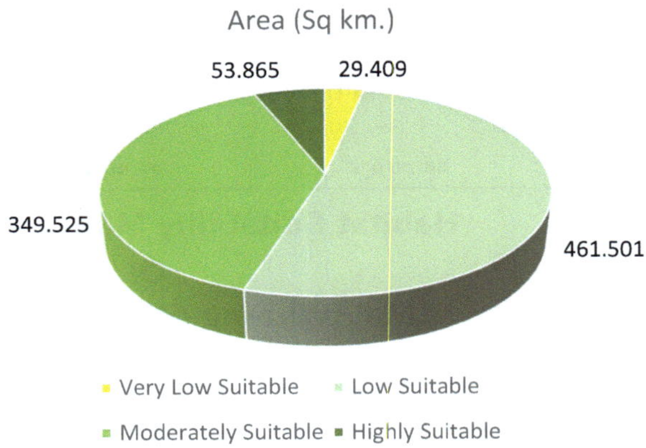

Fig. 7 Pie chart of area (km^2) of habitat suitability

Table 10 Area (km^2) of habitat suitability

Habitat suitability	Area (km^2)
Very low suitable	29.409
Low suitable	461.501
Moderately suitable	349.525
Highly suitable	53.865

6.2 *Human-Elephant Conflict Risk Zone*

Conflict for space is one of the primary factors of conflict between humans and wildlife in general. Given that they need comparatively extensive areas to dwell in, this is particularly true for wild Asian elephants. Elephant existence is seriously threatened by forest destruction from logging, encroachment, slash-and-burn, shifting cultivation, and monoculture tree plantations. Human-elephant conflicts emerge when elephants raid crops of shifting cultivators in fragmented agricultural areas, scattered across a vast landscape interspersed with forests. Another significant area of human-elephant conflict is depredation in human settlements, which can occur on elephant migration routes, in isolated pockets of forest, or in encroachments into elephant habitat. In locations experiencing high and moderate conflict, there is typically a high rate of crop and housing destruction, human casualties, and elephant casualties. Three buildings are damaged for elephants, and 69% of the crops in the Belpahari range are damaged. Seventeen homes in the Bhulaveda range have been damaged by elephants.

6.2.1 Causes of Increasing Human-Elephant Conflict

Deforestation is main and important cause of increasing Human-Elephant Conflict.

Habitual food of Elephants is available in very less quantity in forest, due to this reason Human-Elephant Conflict is increasing.

Day by day human encroaching is increasing into the forest.

Cut down the trees or sets fire into the forest for increasing agricultural land.

When Elephants enter into the villages, local people irritating Elephants, threw stone to the Elephants.

Cut down the trees or sets fire into the forest for increasing Settlements.

Elephants get their food in agricultural land like sugarcane, Watermelon, Potato, Paddy, Jackfruit and other many vegetables and fruits, due to this reason Elephants come frequently into local areas.

In Jungle mahal area Tribal people store intoxicant material like "Heria" in their houses, so Elephants come to the local area smelling such materials like "Heria".

6.2.2 Insurance or Compensation when Elephants Damage Properties and Injuries

Estimated compensation 60 rupees/Decimal is provided for agricultural damage.

Treatment for injuries of each person up to 3 days or 7 days are compensated with rupees 25,000 and 50,000 respectively.

If People die due to Elephant attack, then family gets 5 lakh rupees.

People get compensation based on the extent of damage to their houses.

A total of 440 villages are affected by Human-Elephant Conflict, of which 158 villages in the Mayurjharna Elephant Reserve fall under the High Conflict Risk Zone, 218 villages are categorized as being in the Moderate Conflict Risk Zone, and 64 villages are classified as being in the Low Conflict Risk Zone (Figs. 8 and 9; Table 11).

6.3 Forest Fire Risk Zone Map

To develop a forest fire risk zone map of Mayurjharna Elephant Reserve, six criteria were considered. The physical factors included Slope, Aspect, and Elevation. Additionally, three anthropogenic factors were incorporated: Distance from settlements, Distance from roads, and Land Use and Land Cover (LULC).

6.3.1 Distance from Settlement

Tribal families live in the middle of the forest and near to the forest, many cottages are located in the forest for tourists. Tribal people set fire to the forest intentionally and unintentionally to find Mahul under Mahul tree, to save either their houses from snakes or to clear bushes for agricultural land. Tourists also carelessly throw lighted cigarette butts and set campfire improperly and also leave behind lighted fire woods

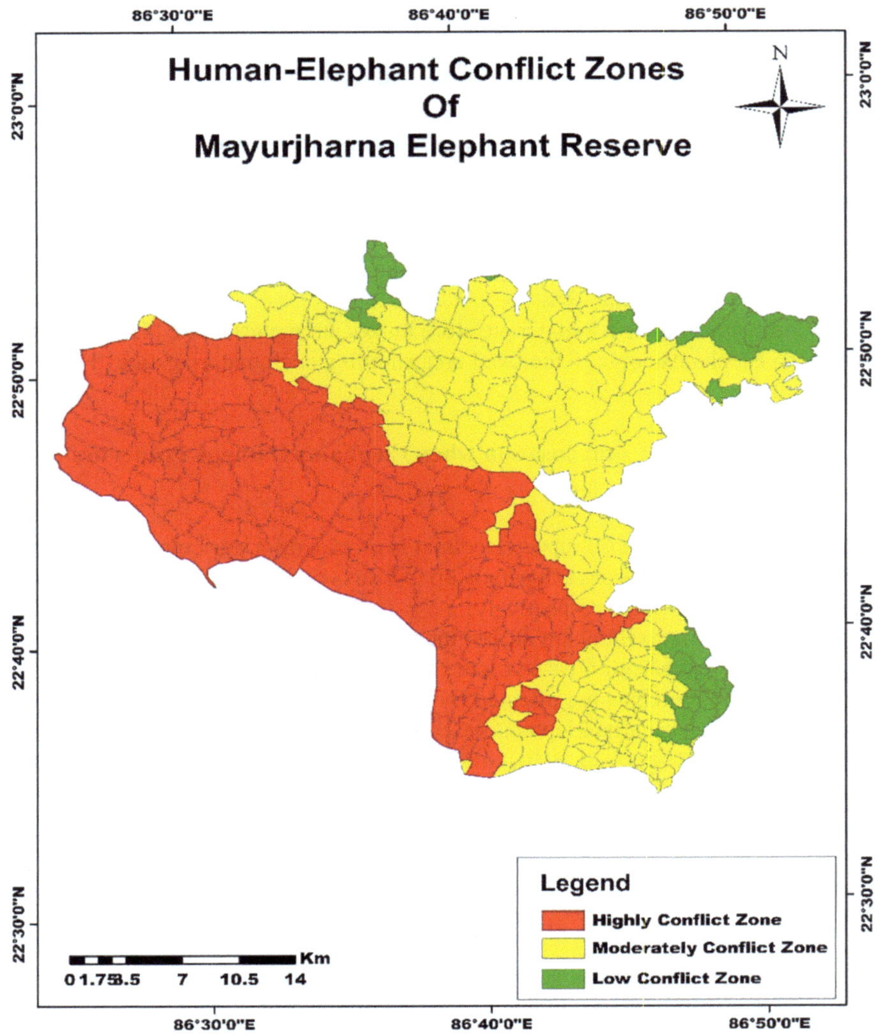

Fig. 8 Showing human-elephant conflict zones in the study area

recklessly as a result the tourist become the cause of forest fire. The forest area, very closed to the human settlement is very prone to forest fire.

Study area Mayurjharna Elephant Reserve are classified into four classes depending on distance from settlement, as follows less than 800 m highly prone zone for Forest fire, 800–1600 m moderately prone zone for Forest fire, 1600–2400 m low prone zone for Forest fire and greater than 2400 m very low prone zone for Forest fire (Fig. 10).

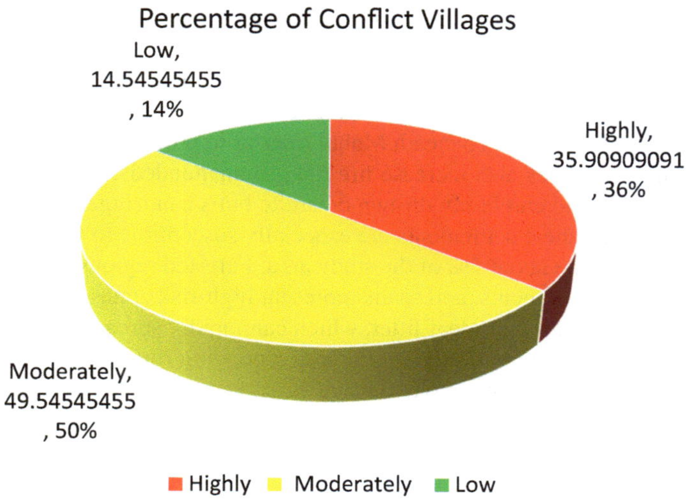

Fig. 9 Pie chart of percentage of conflicted villages

Table 11 Showing Conflicted Villages in percentage (%)

Conflict risk zone	Villages	Total villages	Percentage of conflict villages
Highly	158	440	35.90909091
Moderately	218	440	49.54545455
Low	64	440	14.54545455

6.3.2 Distance from Road

The movement of tourists through these roads can leads accidental forest fires, due to carelessly thrown burning cigarette butts or match sticks, burning tar for road making, camping fire, way side food making by lighting stoves, Bad habit of human is to make fire to the dry leaves as a result forest catches fire severely in winter season. Mayurjharna Elephant Reserve are classified into four classes depending on distance from road, as follows less than 400 m highly prone zone for Forest fire, 400–800 m moderately prone zone for Forest fire, 800–1200 m low prone zone for Forest fire and greater than 1200 m very low prone zone for Forest fire (Fig. 10).

6.3.3 Land Use and Land Cover

In the present study, land use and land cover were reclassified into four categories: forest, agricultural land, waterbodies, and settlements. A substantial proportion of recorded forest fire incidents is associated with slash-and-burn practices carried out for cultivation. Forest patches located adjacent to agricultural fields are particularly vulnerable and fall within high fire-prone zones. Agricultural land accounts for approximately 21 percent of the total study area. In addition, certain traditional practices contribute to fire occurrence. Tribal communities sometimes ignite forest areas to collect Mahul leaves beneath Mahul trees or to protect their dwellings from snakes. Tourist activities also increase fire risk, as unattended campfires, discarded burning firewood, and carelessly thrown cigarette butts can trigger accidental fires. Consequently, areas near settlements are especially susceptible to forest fire events. Settlements constitute about 5% of the study area. Forested regions situated close to agricultural land, settlements, and roads represent high-risk zones, as human activities often involve burning dry leaf litter, which can rapidly spread during the winter season. Forest cover dominates the landscape, occupying roughly 71 percent of the study area. In contrast, forest areas located near waterbodies show relatively low fire susceptibility. Waterbodies account for approximately 3 percent of the total area (Fig. 10).

6.3.4 Aspect

Aspect refers to the directional orientation of a slope, which influences the amount of solar radiation it receives. Slopes facing east are exposed to sunlight during the early morning hours, whereas west-facing slopes receive greater solar input later in the day. In the Northern Hemisphere, south-facing slopes are subjected to comparatively higher solar radiation, resulting in elevated surface temperatures and faster drying of combustible materials. Consequently, southern aspects are considered more susceptible to forest fire occurrence (Mukherjee & Raj, 2014). Seasonal wind patterns also contribute to fire risk. During winter, prevailing winds typically move from the northeast toward the southwest. Slopes oriented toward the northeast therefore become more exposed and are associated with increased fire susceptibility in this season. In contrast, summer winds generally flow from the southwest to the northeast, making southwest-facing slopes comparatively more vulnerable to fire during the summer months (Fig. 10).

6.3.5 Slope

Slope represents the gradient or degree of inclination of the land surface and plays an important role in fire behavior. Fire generally spreads more rapidly upslope than downslope because heat rises and preheats the fuel located above the flame front. Steeper gradients can therefore facilitate faster fire propagation, as fuels on upper slopes tend to become drier and more combustible compared to those on lower slopes. In the present study, the terrain is predominantly undulating, including areas categorized under lower slope classes (Lamat et al., 2021). For analytical purposes, the slope map of the study area was reclassified into four categories: very low (0°–7.63°), low (7.63°–16.99°), moderate (16.99°–31.91°), and high (31.91°–88.46°).

The results indicate that most forest fire occurrences are concentrated within low and moderate slope zones, particularly in areas dominated by forest cover (Fig. 10).

6.3.6 Elevation

Because elevation influences temperature and precipitation, it is considered to be an essential factor in forest fires. Higher elevations are generally less favourable for flames since rainfall tends to increase and temperature tends to decrease. On the other hand, at lower elevations, conditions that are more conducive to fire occurrences are created by greater temperatures and less rainfall.

Elevation is consistently a lead predictor variable of fire return intervals (FRI) (Rogeau & Armstrong, 2017). In the study area elevation is reclassify into four classes as follows High (44.33–106.94 m), Moderate (106.94–151.66 m), Low (151.66–196.39 m), Very Low (196.39–424.48 m) (Fig. 10).

It was discovered that while there were fewer fire events at high elevations in the research area, the majority of forest fire incidences were found to be relatively high in the lower and intermediate elevation ranges. This can be ascribed to the soil's and vegetation's high moisture content, which prevented fire from starting and spreading.

Forest Fire Risk Zone Map was prepared by AHP method and geospatial technique. Distance from Settlement, Distance from Road, Land Cover and Land Use, Aspect, Slope, Elevation are taken for preparing Forest Fire Risk Zone Map. Pairwise Comparison is done with these six parameters after that all parameters are converted into raster format and reclassify all parameters. Final Forest Fire Risk Zone Map is prepared with Weighted Overlay Analysis. Forest Fire Risk Zone Map is classified into four classes as follows Highly Risk Zone (11.5 km^2), Moderately Risk Zone (213.16 km^2), Low Risk Zone (475.9 km^2), Very Low Risk Zone (194.35 km^2) (Figs.11 and 12; Table 12).

7 Management and Suggestions

- Mayurjharna Elephant Reserve is still not active as protected forest, so it must be made active as a protected forest as soon as possible.
- Habitual food like Sugercane, Bamboo grass etc. plants must be planted in the forest for their survival.
- Water reservoirs must be arranged in the forest for their drinking.
- Human encroachment in the forest must be reduced, so that elephants can live peacefully in the forest.
- Human being must be made understood not to cut the trees in the forest.
- Elephant conservation should be associated with local people participation.
- Conservation of elephant corridors characterized by dense vegetation cover and minimal human population density should be prioritized to ensure safe movement and reduce conflict.

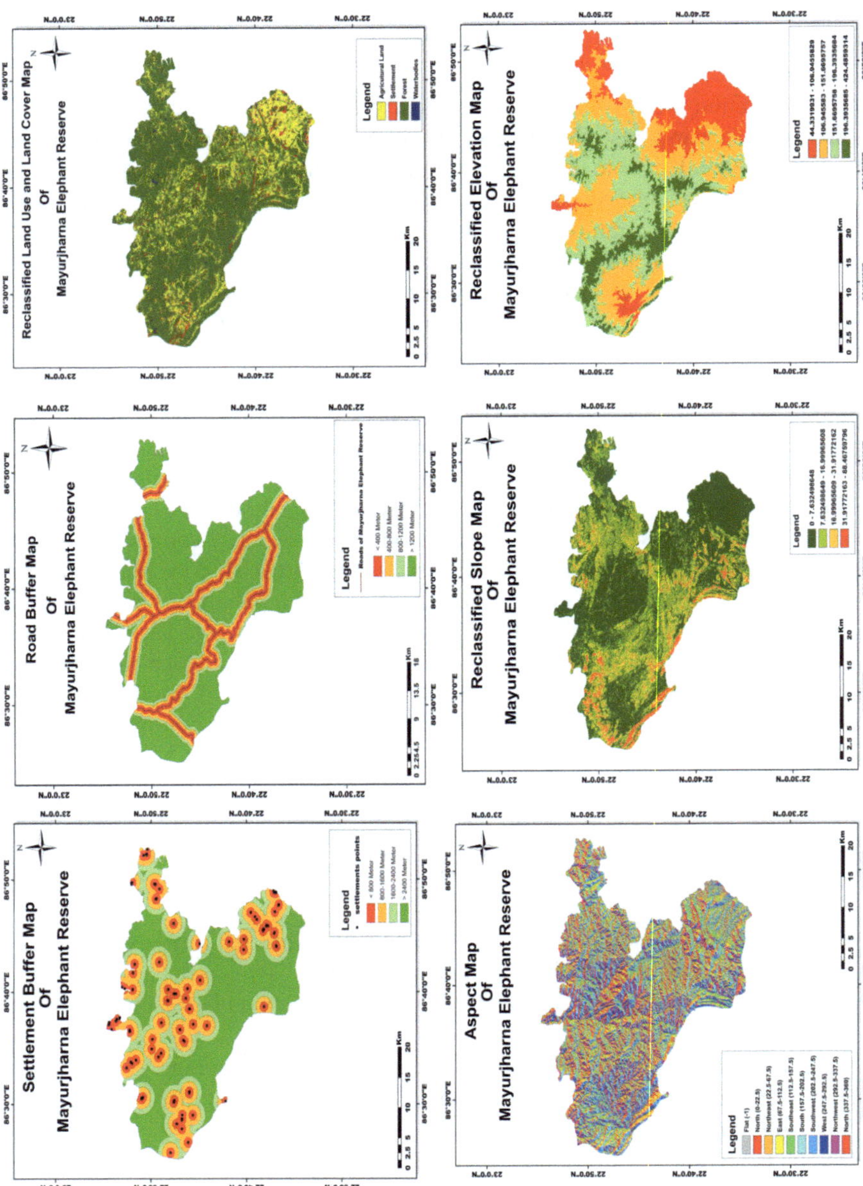

Fig. 10 Parameters for delineate forest fire risk zone, distance from settlement, distance from road, LULC, aspect, slope, elevation

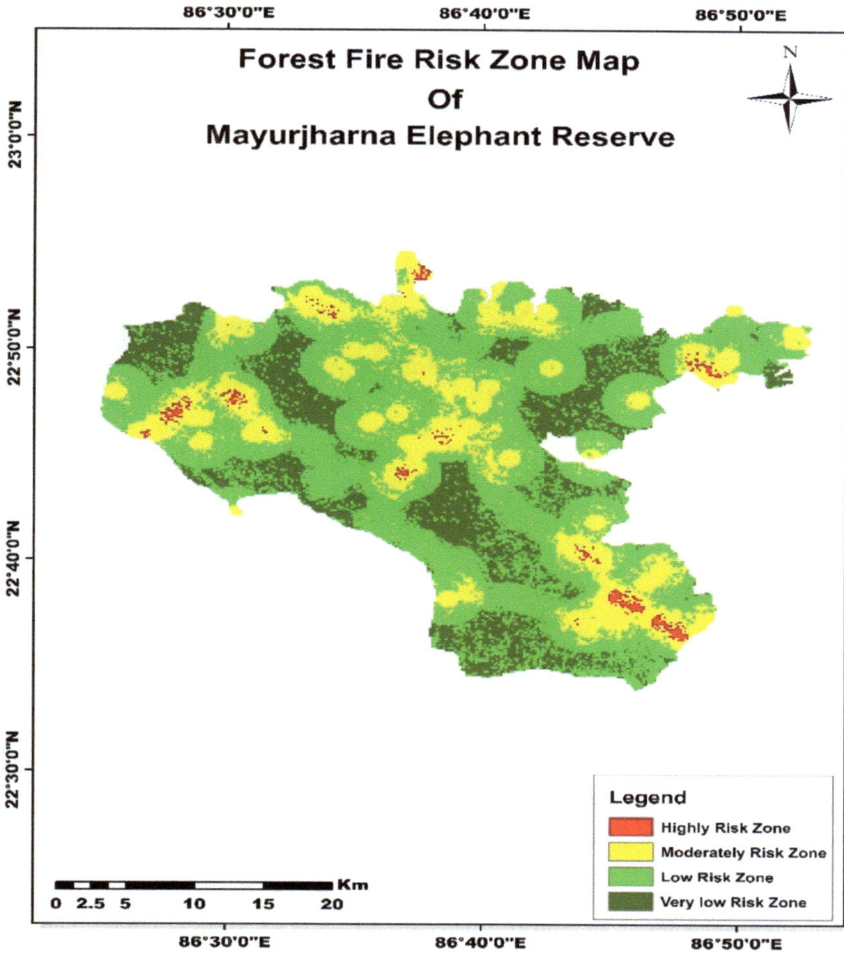

Fig. 11 Showing forest fire risk zone of study area

- The forest departments under Mayurjharna Elephant Reserve have learnt the techniques of elephant management quickly and is managing the present situation quite well.
- People living close to the forest must be made understood about elephants.
- People also need to be educated about the conservation, importance of the elephants, so giants are treated with more sympathy and kindness.
- Speed of the vehicle passing through the middle of the forest road must be limited to safe value.
- Forest Department can introduce a new technology like RTM algorithm or a web-based GIS on recent occurrence of the elephants, their location, numbers, area of activity, time of finding and direction of movement can be uploaded by the users.

Fig. 12 Pie chart of area (km^2) of Forest Fire Risk Zone

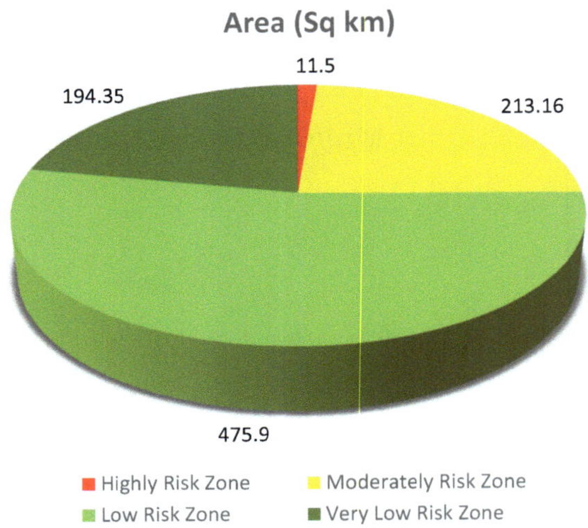

Table 12 Showing area (km^2) of Forest Fire Risk Zone

Risk zone	Area (km^2)
Highly risk zone	11.5
Moderately risk zone	213.16
Low risk zone	475.9
Very low risk zone	194.35

- People must be made understood not to make catch fire in the forest, and Strict punishment must be given to the people violating the forest law.
- Rehabilitation of those villages which are coming under site suitability zone with discussions with the people of those particular villages.
- The Elephant habitat suitability maps should be used by the agents, decision makers, wildlife management's agency and other related organization of interest to aid in wildlife conservation and programmer's awareness.
- Mayurjharna Elephant Reserve is currently under pressure and threats due to cottages, roads, settlements and deforestation, so it is highly recommended that to monitor and manage this landscape in well manner and proper ways.
- The current situation of the elephants of Mayurjharna Elephant Reserve is under high threats therefore its recommended that the further research should be carried out by considering more factors and more techniques for properly conserving those species.

8 Conclusion

Anthropogenic activities, including the illicit felling of trees in the core forest area, pose a significant threat to the habitat of wild animals while also violating existing forest laws and regulations. As a result, wild animals lost their habitats; they are migrating to find their favorable habitats and foods, and gradually the conflict between humans and wildlife is increasing. Mayurjharna Elephant Reserve is still not active as protected forest, due to this reason humans are easily cutting down the trees of the forest and making settlements, cottages for tourists, increasing agricultural land. After calculating their habitat suitability place in this forest being 53.865 km^2 is considered as highly habitat suitable zone for the elephants, in the case of moderate suitable zone area about 349.525 km^2, the presence of the Elephant is found to a some bits of number in this area, again in the low and very low habitat suitable zone respectively about 461.501and 29.409 km^2, the presence of the Elephant is almost nil. Due to the scarcity of sufficient food, elephants come to agricultural land to eat crops, which damages houses, and humans and elephants both get injured. As a result, after the conflicts between villagers and the Elephant the affected villages as high, moderate and low with the numbers calculated as 158, 218, 64 respectively. Rising global temperatures have led to an increased risk of forest fires. Furthermore, in some regions, humans intentionally set fires in forests during the winter season; campfires in the forest area are also another reason for forest fires with high and very low risk of forest firing zones as calculated about 11.5 and 194.35 km^2 respectively.

The adversities as found in this case as follow "Transportation system is very poor from Jhargram to Mayurjharna Elephant Reserve, public bus is very less available, buses have not right time, due to these reasons facing travel problem during field work", "In the Office of the Divisional Forest Officer, Jhargram Division did not give some of the conflict or migration related data due to government policy", "In Bhulaveda Forest Range Office, they cannot give any information related to damage of human life, crops, due to they have not any records of that", "In Block Seed Farm Office, they did not give Temperature data because their own privacy policy".

To protect the forest ecosystem from fire the various factors are being analyzed and categorized into different risk levels to better understand and mitigate the causes of forest fires. To protect our wildlife and environment, Remote Sensing and GIS, Artificial Intelligence (AI) play an important role. It is easy to monitor, track, and assess wildlife with the help of RS & GIS and AI. The elephant habitat map has been delineated as far as possible with the help of remote sensing and GIS techniques, as well as, prepared a map of the human-elephant conflict risk zone and migration route of elephants. Wildlife helps to keep balance in our environment, animals are beneficial to various aspect of human beings, so to save wildlife from extinction we should set conservation goals, including habitat conservation at landscape scales, conserve wildlife corridors, and generate public awareness on conservation issues.

Some Field Photographs (Figs. 13, 14, 15, 16, 17, 18, 19, 20, 21, 22, 23 and 24).

Fig. 13 With range officer at Bhulaveda Range Office

Fig. 14 Conducting survey at Mayurjharna Dam

Fig. 15 Trying to extinguish forest fire with the help of forest officials

Fig. 16 Burned trees due to forest fire

Fig. 17 Agricultural land in middle of the forest

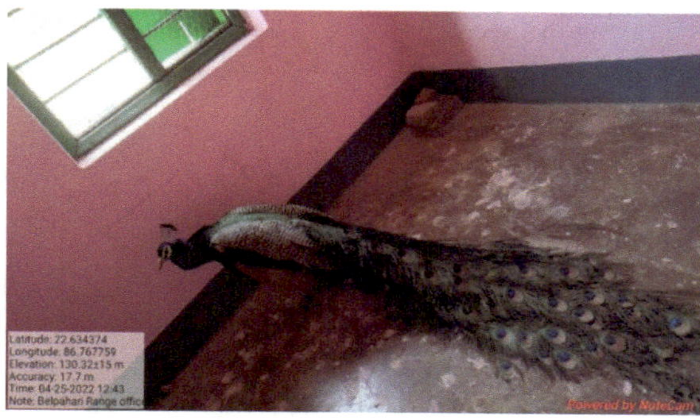

Fig. 18 Injured peacock under treatment in Belpahari Range Office

Fig. 19 Settlements near forest

Fig. 20 Foot print of the elephant

Fig. 21 Entrance corridor to the Mayurjharna Elephant Reserve

Fig. 22 Vegetation cover in Mayurjhara Elephant Reserve Forest

Fig. 23 Forest area at Banspahari, Mayurjhara Elephant Reserve Forest

Fig. 24 Dense Sal (Shorea robusta) Forest in Mayurjhara Elephant Reserve Forest

References

Acharyya, N., & Bandyopadhyay, J. (2017). Remote sensing and GIS for wildlife management. In *National Level Seminar on Defaunation and conservation* (pp. 65–71). Tucson Herpetological Society.

Badhe, Y. P., & Jaybhaye, R. G. (2021). Habitat suitability area analysis for leopard to mitigate human-wildlife conflict in Junnar Forest Division of Pune Forest Circle. *Applied Ecology and Environmental Sciences, 9*(5), 524–542.

Erten, E., Kurgun, V., & Musaoglu, N. (2004, July). Forest fire risk zone mapping from satellite imagery and GIS: A case study. In *XXth Congress of the International Society for Photogrammetry and Remote Sensing*, Istanbul, Turkey (pp. 222–230).

Kanga, S., Pandey, A. C., Shaheen, A., & Singh, S. K. (2017). Elephant habitat suitability using geoinformatics for Palamau Tiger Reserve, Jharkhand (India). *International Journal for Research in Applied Science and Engineering Technology, 5*(6), 493–499.

Lamat, R., Kumar, M., Kundu, A., & Lal, D. (2021). Forest fire risk mapping using analytical hierarchy process (AHP) and earth observation datasets: A case study in the mountainous terrain of Northeast India. *SN Applied Sciences, 3*(4), 1–15.

Mandal, M., & Das Chatterjee, N. (2021). Geospatial approach-based delineation of elephant habitat suitability zones and its consequence in Mayurjharna Elephant Reserve, India. *Environment, Development and Sustainability, 23*(12), 17788–17809.

Naha, D., Sathyakumar, S., Dash, S., Chettri, A., & Rawat, G. S. (2019). Assessment and prediction of spatial patterns of human-elephant conflicts in changing land cover scenarios of a human-dominated landscape in North Bengal. *PLoS One, 14*(2), e0210580.

Ram, A. K. (2014). Geospatial modeling to assess Asian elephant. Elephas maximus.

Sanare, J. E., Ganawa, E. S., & Abdelrahim, A. M. S. (2015). Wildlife habitat suitability analysis at Serengeti National Park (SNP), Tanzania case study *Loxodonta* sp. *Journal of Ecosystem & Ecography, 5*, 164. https://doi.org/10.4172/2157-7625.1000164. Page 2 of 8 *Journal of Ecosystem & Ecography, 5*(3), 1000164. ISSN 2157-7625 JEE, an open access journal.

Sharma, P., Adhikari, H., Tripathi, S., Ram, A. K., & Bhattarai, R. (2019). Habitat suitability modeling of Asian elephant Elephas maximus (Mammalia: Proboscidea: Elephantidae) in Parsa National Park, Nepal and its buffer zone. *Journal of Threatened Taxa, 11*(13), 14643–14654.

Rogeau, M. P., & Armstrong, G. W. (2017). Quantifying the effect of elevation and aspect on fire return intervals in the Canadian Rocky Mountains. *Forest Ecology and Management, 384*, 248–261.

Mukherjee, S., & Raj, K. (2014). Analysis of forest fire of Meghalaya using geospatial tools. In *15th ESRI India User Conference.*

Assessing Soil Quality for Sustainable Land Management Under Different Tillage Systems in the Temperate Himalaya

Shazia Ramzan, Sabia Akhter, Haadiya Saleem, Ifra Ashraf, Pervez Ahmad, and Nageena Nazir

Abstract Soil is a fundamental constituent of our environment. In recent years Soil quality indices (SQI) have been anticipated as an important tool for analyzing soil management practice effects on the quality of soil together with tillage. The suitable management practices could be selected based on soil quality evaluation. The key motive of this study was to develop the SQI and the evaluation of SQI under different tillage systems in the Temperate Himalayan zone. The experimental site was the research station of SKUAST-K, Shalimar, Srinagar. The field was no-tilled for the last 18 years. In this study, composite soil samples were collected from three distinct tillage systems: conventional tillage (CT), reduced tillage (RT), and no-tillage (NT). Three key soil functions of the soil were determined which include soil capacity for root development (RD), nutrient supply (NS) capacity of soil and water storage (WS). The contribution of soil capacity for RD was highest in RT (32.23%) while NS

S. Ramzan
Krishi Vigyan Kendra (KVK), Haran District, Budgam, Jammu and Kashmir, India

S. Akhter
SMS Agronomy, Krishi Vigyan Kendra (KVK), Haran District, Budgam, Jammu and Kashmir, India

H. Saleem · P. Ahmad
Department of Geography and Disaster Management, School of Earth and Environmental Sciences, University of Kashmir, Srinagar, India

H. Saleem (✉)
Indira Gandhi National Open University (IGNOU), New Delhi, Maidan Garhi, New Delhi 110068, India

I. Ashraf
College of Agricultural Engineering and Technology, Sher-e-Kashmir University of Agricultural Sciences and Technology of Kashmir, Shalimar, India

N. Nazir
Division of Agricultural Statistics, Sher-e-Kashmir University of Agricultural Sciences and Technology of Kashmir, Shalimar, Jammu and Kashmir, India
e-mail: nazirnageena@skuastkashmir.ac.in

R. Ahmed et al. (eds.), *Climate Change, Water Resources, and Natural Hazards in the Himalayas*, Advances in Geographical and Environmental Sciences,
https://doi.org/10.1007/978-981-92-0835-7_13

capacity and WS capacity of the soil were found to be maximum in RT (61.14 and 22.83% respectively). Of the 31 soil quality parameters considered for the study, four parameters like water holding capacity (WHC), mean weight diameter (MWD), bulk density (BD), organic carbon (OC), and soil porosity and were recognized as the key indicators of soil quality. Across the treatments RT (0.86) recorded the highest soil quality rating followed by CT (0.74) and NT (0.69), indicating RT is the most apt management technique that should be adopted owing to its potential of conserving/ enhancing the soil quality (SQ).

Keywords Soil quality index · Tillage systems · Scoring · Temperate Himalayas · Principal component analysis

1 Introduction

Soil is the most important natural resource that maintains water and air quality, protracts agricultural productivity and supports human health and habitation (Laishram et al., 2012). The importance of soil quality is sparked by numerous problems like soil contamination, erosion, loss of soil organic matter, intense cultivation, and fertility (Lal, 2015). Soil quality has traditionally been equated with agricultural productivity. According to the Soil Science Society of America (1995), the description of soil quality is: "The capacity of a specific soil to function, within natural or managed ecosystem boundaries, to sustain plant and animal productivity, maintain or enhance water and air quality, and support human health and habitation". This concept is essential to enumerate the effects of human activities on the environment, generally, those associated with the sustainability of agricultural practices (Andrews et al., 2004). For the reason that inappropriate management can lead to harmful changes in soil function, e.g., tillage is the key factor signifying the loss of soil organic matter (Lal, 2015; Rasmussen and Collins, 1991), therefore a prerequisite for the tools and techniques to monitor and assess the SQ were identified.

Maintaining high output without significant environmental or soil deterioration results in superior soil quality (Govaerts et al., 2006). On the other hand, quantification of soil quality is essential for indicating early warning indications of adverse trends, identifying problem areas/regions and preventing soil degradation (Qi et al., 2009). According to the Soil Quality Institute (USDA, 2006), the crucial goal for evaluating soil quality is protecting and restoring water quality, long-term agricultural productivity and habitations of all creatures including humans.

Quality of soil as well as soil management practices sustainability in agro ecosystems can be accessed via soil attributes, that are susceptible to change through management and can be used as indicators (Andrews et al., 2004; Shukla et al., 2006). Indicators used to analyze soil quality must be highly responsive to recognize the effects of management practices, however, they should not be exaggerated by short-term weather patterns. Karlen et al. (1992) recommended some extremely effectual indicators, viz., respiration, microbial biomass, and ergo sterol concentrations, for evaluating the long-term impacts of crop and soil management practices on SQ. Sharma et al. (2008) recommended SQ indicators that include easily oxidisable N ($KMnO_4$ oxidisable-N), microbial biomass carbon (MBC), hydraulic conductivity (HC), DTPA extractable Zn and Cu, mean weight diameter (MWD) of soil aggregates. Similarly, Chun-Juan et al. (2013) suggested EC, NO^{-3}-N, pH, Corg, Cbio, AP, Cd, Hg, DDTs and HCHs as the least data set for evaluating SQ. Moreover, Chaudhury et al. (2005) acknowledged mean weight diameter (MWD) of aggregate, available P, total soil N, and dehydrogenase activity as important indicators for alluvial soils. Li et al. (2013) recognized only AP, SK, SOM, AN, and sand as prospective soil quality indicators through prime component investigation. Since soil quality appraisal is purpose- and site-specific, various researchers in diverse regions may utilize different indicators (Shukla et al., 2006).

A static descriptor known as a "soil quality index" (SQI) is used to examine management-induced changes over time or to compare various management approaches (Sharma et al., 2008). For developing an incorporated soil quality index, a number of systematic advancements have been taken into consideration. Imaz et al. (2010) and Shukla et al. (2006) determined SQI using factor analysis. Doran and Parkin (1994) suggested the use of a simpler multiplicative function to review the soil quality; that considers socioeconomic, geographical, and climatic concerns.

Fernandes et al. (2011) established the soil quality index under four management systems i.e., no-tillage with a double-disk planter (NTd), reduced tillage (RT), no-tillage with a hoe planter (NTh), and conventional tillage (CT), subsequent to the methodology anticipated by Karlen and Stott (1994), where the soil quality indices assorted amid treatments being 0.64, 0.68, 0.86 and 0.79 under NTh, NTd, RT and CT, correspondingly. Sofi et al. build up the soil quality index in diverse cropping systems of the northwest Himalayan zone and established that the inclusion of legumes in the soil of apple orchards resulted in the highest soil quality and cereal-based cropping systems had the lowest soil quality rating.

Most systematic research efforts on evaluating soil quality in temperate climates have been made recently (Nortcliff, 2002; Arshad, 2002; Dalal & Moloney, 2000; Andrews et al., 2002a; Shukla et al., 2006). However, an inadequate number of research studies have assessed soil quality under the Temperate Himalayas. Initially research was mainly focused on studying the physical, chemical and biological properties as separate entities while ignoring the quality indexing together. Our study combines the physical, chemical, and biological features of the soil by generating the MDS by the PCA method to assess the most appropriate SQI indicators. Hence, an effort was made in this paper to assess the tillage effect on soil quality indexing in temperate conditions.

2 Materials and Methods

2.1 Site Characteristics

This experiment was performed in 2013 at the Research Station, located at SKUAST-K, Shalimar between the coordinates (34° 8′ 42″ and 34° 9′ 3″ N latitudes) and (74° 39′ 5″ and 74° 53′ 5.6″ E longitude) in Srinagar, Kashmir.

The study area experiences a temperate and moderate climate which is characterized by cold winters and mild summers having normal annual minimum and maximum temperatures of 6.80 °C and 19.53 °C, respectively, with a normal annual rainfall of 786.2 mm. The average elevation of the study region is 1615 m above sea level.

2.2 Management Practices

No-tillage system was practised for some parts of the study area for the last 15 years. The other tillage practices undertaken on the farm are reduced tillage (RT) and conventional tillage (CT). So, this experiment was conducted on three tillage systems i.e., conventional tillage (CT), reduced tillage (RT), and no-tillage (NT). The engine used was IH Heavy-Offset 790 offset with 85 kW (121 hp) in all the tillage operations. The randomized block design with five replications for each treatment was used. Fertilizer application in the NT treatment was manual and the ploughing mechanism was a double disk both for planting and fertilizing the field. For RT, a chisel equipped with 5 tines was used. The ploughing mechanism was a single disk for both operations. Conventional tillage was performed with a heavy-disc harrow (AR-630, Jandear) composed of six discs followed by passes with a light-disc harrow with 32 discs. The tilth of chisel plow and moldboard plow were 15.5 and 15.6 cm, respectively. Maize was planted as per recommendation and fertilizer was also applied as per package of practices.

2.3 Soil Sampling and Analyses

Both disturbed as well as undisturbed soil samples were taken from the upper layer of soil from 0–20 cm depth. Disturbed samples were subjected to air drying and sieving through 2 mm sieve (except for carbon) and finally stored for analysis. The number of samples was 15 (3 management systems × 5 replicates). Undisturbed soil samples were collected with cores of diameter 100 mm and effective height of 127.3 mm, and 1000 ml capacity for the analyses of mean weight diameter (MWD), bulk density (BD), saturated hydraulic conductivity (Ks), and total porosity. Soil resistance to penetration (Rp) was measured at the field on every sampling point. For soil resistance, three replicas were taken from each site so as to minimize the error.

The analyses of soil parameters in the current study were carried out with standard procedures outlined below:

2.3.1 Soil Physical Properties

Mechanical Analysis

The mechanical analysis of soil samples was carried out in accordance to the International Pipette method given by Piper (1966). No chemical pre-treatments were given during the fractionation of soils into sand, silt and clay in order to minimize the risk of alteration of sensitive minerals. The soil fractions were subjected to chemical dispersion using 1% sodium hexametaphosphate salt solution. For 10 min, each suspension was shaken mechanically. Wet sieving was used to separate clay (less than 0.002 mm), silt (0.02–0.002 mm), and sand (0.2–0.02 mm) separated by the decantation and sedimentation process. The textural classes were established using the textural triangle of the International Society of Soil Science.

Field-Saturated Hydraulic Conductivity (Ks)

Field-saturated hydraulic conductivity was measured using a falling head permeameter (Richard, 1954). It consists of a metallic mould, 100 mm internal diameter, 127.3 mm effective height and 1000 ml capacity. The mould is provided with a detachable extension collar having 60 mm height and 100 mm diameter. The soil ring and specimen with ring are placed on bottom porous stone. Upper plate is placed on the permeameter and fastened securely to the base with 3 nuts to ensure ring seal. Top porous stone and piston was inserted. The bleeding value at the base of apparatus was closed. The soil samples were initially saturated under a specific head condition. The standpipe was filled and the height of water in tube was noted while allowing water to pass through the soil without keeping a constant pressure head. Above steps were repeated several times to establish an average hydraulic conductivity for the sample.

The hydraulic conductivity under falling head method has been made using following equation:

$$K_s = 2.3\left(\frac{aL}{At}\right)\log_{10}\left(\frac{h_0}{h_1}\right)$$

Where, 'a' in the above equation represents the cross-sectional area of stand pipe and L represents the length of the soil column. A and t represent the cross-sectional area of soil column and time interval to head drop, whereas h_0 and h_1 is total head before the test and after test, respectively.

Aggregate Size Distribution

The aggregate size distribution was derived by utilizing the Yoder's apparatus (Yoder, 1936). The apparatus was set to provide a ¾inch stroke at a pace of 29 strokes per minute. Samples that had been air dried and taken from the research region were crushed by hand into tiny crumbs and fractionated through an 8 mm sieve. Following an overnight saturation, the material remaining in a 4 mm sieve was analyzed to determine the aggregate size distribution. The sieve net was submerged in water that was 22 0.5 °C (the GSIL, Kolkata norm). The apparatus was operated for a duration of 30 minutes. The obtained samples were then placed on their respective sieves and subsequently dried in an oven set at 105 °C. Once dried, the samples were weighed, and different size fractions were calculated. To ensure accurate classification and prevent misidentification of coarse primary particles as aggregates, the results were calibrated. This calibration involved dispersing the materials obtained from each sieve using dispersing agents, specifically 1N NaOH. The material was mechanically stirred and then washed back through the same sieve to ensure accurate classification.

After the second sieving, the weight of retained aggregates was subtracted from the total weight of the dispersed material retained after the first sieving. The mean weight diameter (MWD) was computed with the help of procedure as out lined by Van Bavel (1949).

$$\mathrm{MWD} = \sum_{i=1}^{n} X_i\, W_i$$

where,

'X_i' and 'W_i' are the mean diameter (mm) and weight of different size fractions in the sample, respectively. In this case, $X_1 = 5.00$ mm, $X_2 = 2.00$ mm, $X_3 = 1.00$ mm, $X_4 = 0.50$ mm, $X_5 = 0.20$ mm and $X_6 = 0.1$ mm, and Wi values represent the proportion in mass of each size fraction in the sample, expressed as the mass of aggregates in the i-th size fraction (in grams) divided by the total mass of the aggregates in the sample (in grams) on an oven-dry basis.

Bulk Density (BD)

To achieve this, a hammer is typically employed to pound the core into the ground. From each desired depth, four cores are extracted, and the soil's fresh weight and moisture content are determined using the gravimetric method. The dry bulk density is then estimated by establishing the mass-volume relationship based on the average dry weight and volume of the cores.

The core sampler method given by Black (1965) was used to derive the bulk density of the soil. This method involves using a cylindrical core with a sharp edge at the bottom, allowing it to penetrate the soil smoothly. A hammer was used to pound the core into the ground. From each desired depth, four cores were extracted, and the moisture content and fresh weight of the soil were determined using the gravimetric method. Finally, the dry bulk density was estimated through the mass volume relationship.

Porosity (P)

The formula provided by Rattan (2009) was used to calculate the porosity.

$$\text{Porosity } (\%) = 1 - \frac{\text{Bulk density}}{\text{Particle density}} \times 100$$

Resistance to Penetration (RP)

Soil resistance to penetration was measured using a penetrometer (SC-60, Soil Control), with six sub-samples per replication and obtained values in MPa

Water Holding Capacity (WHC)

WHC was obtained by method given by Jackson (1973).

2.3.2 Chemical Properties

Soil Reaction (pH)

The soil pH was measured using a digital glass electrode pH meter in a 1:2.5 soil-to-water suspension, following the method described by Jackson (1973).

Electrical Conductivity (EC) (dSm^{-1})

It was estimated by using solubridge conductivity meter (Jackson, 1973).

Cation Exchange Capacity (CEC)

The determination of CEC was carried out following the procedure outlined by Rhoades (1982).

Available Macro-Nutrients

The determination of various nutrients and elements followed specific methods as described by different researchers. Available nitrogen (N) was measured using the potassium permanganate method developed by Subbiah and Asija (1956). Available phosphorus (P) was extracted using 0.5M $NaHCO_3$ according to the procedure outlined by Olsen et al. (1954). Available potassium (K) was extracted using Neutral Normal Ammonium Acetate, following the method described by Jackson (1973). Available sulfur (S) was determined using the turbidimetric method developed by Chesnin and Yien (1951). Available calcium (Ca) and magnesium (Mg) were analyzed using the Atomic Absorption Spectrophotometer procedure suggested by Hesse in 1972. Available micronutrients (copper (Cu), zinc (Zn), manganese (Mn), iron (Fe), and nickel (Ni)) and toxic elements (lead (Pb) and cadmium (Cd)) were extracted using the DTPA extraction method developed by Lindsay and Norvell (1978) and analyzed using Atomic Absorption Spectrophotometer (AAS).

2.3.3 Biological Indicators

Organic Carbon (OC)

The determination of organic carbon (OC) was carried out using the rapid titration method developed by Walkley and Black (1934). A 10 ml solution of potassium dichromate and concentrated sulphuric acid was used to digest one gram of soil (20 ml). The measurement of the excess potassium dichromate, which was not reduced by the organic matter in the soil, was conducted using diphenylamine as an indicator and a standard ferrous ammonium sulfate solution in the presence of ortho-phosphoric acid.

2.4 Statistical Analyses

In order to examine the impact of different tillage systems on soil quality, an analysis of variance (ANOVA) was conducted. To reduce the number of independent variables and address issues related to multicollinearity, principal component analysis (PCA) with the assistance of SPSS software (Version 10.0) was utilized as a data-reduction tool. This approach was chosen due to its ability to select a minimum data set (MDS) (Andrews et al., 2002a). PCA is a statistical technique that transforms a set of potentially correlated variables into a smaller number of uncorrelated variables, known as principal components (PC). These components are linear combinations of the original variables and account for the maximum variance in the dataset. PCs are determined by calculating the vectors that are closest to the observations in a p-dimensional space while being orthogonal to each other. The main purpose of PCA is to reduce the dimensionality of the dataset and identify underlying variables that explain the observed variability.

The first PC takes into account as much of the data's uncertainty as it can, and each succeeding component takes into account as much of the remaining variability as is probable. In short, the diverse consecutive steps of analysis followed to recognize key indicators and to figure soil quality index (SQI) comprised of the following: (1) to define the goals, (2) to evaluate the relative relevance of different soil indicators using different management techniques, (3) to apply PCA to select a representative MDS, (4) to conduct correspondence analysis to address any misleading grouping among highly weighted variables within each PC, and (5) to rank the MDS indicators based on their performance in terms of soil function and SQI calculation.

3 Results

3.1 Impact of Different Tillage Systems on Soil Properties and Soil Functions

The data on 26 considered soil quality parameters are presented in Tables 1, 2 and 3. Table 1 presents the physical characteristics, i.e., soil resistance, clay, silt, sand, bulk density (BD), porosity, MWD, WHC, and Ks of soil aggregates. Table 2 presents the chemical properties including EC, pH, CEC, available P, N, S, K, transferable Mg and Ca, DTPA-extractable micronutrients, i.e., Zn, Fe, Cu and Mn. Additionally, the biological parameter of SOC is presented in Table 3. Among them nine (9) parameters were drastically influenced by tillage and were WHC, Soil resistance, aggregates in the size range of < 0.1, 0.5, and 2.00 mm, SOC, pH, available Fe and Pb. Lower soil penetration resistance was found in reduced tillage as compared to other tillage systems. Clay, silt as well as sand content was almost similar in all systems but numerically clay was highest in NT, silt and sand in CT. Bulk density was same under all systems but highest in CT, contributing to reduce its SQI, since high bulk densities are indicator of low pore space volume available for storing water. Across the tillage systems, WHC, MWD, 0.5 mm and 2 mm aggregates were found maximum in NT systems, thus having positive contribution to SQI, as the soils with MWD ranging from 1 to 3 mm are categorized as reasonably water stable (Rabbi et al., 2004). Highest values of pH, EC, porosity and CEC were recorded in RT but except pH none of them differ considerably among tillage systems. Nitrogen content was highest in NT while as P, K, S, Zn and Cu was highest in RT, probably due to accumulation of manures and fertilizers. Fe, Mn, Ni and Pb were highest in NT. Tillage effect was significant ($P < 0.05$) for SOM and was highest under RT due to less oxidation of organic matter in NT system.

For soil quality assessment under different tillage systems only those soil properties were selected that influence vital functions of soil. Three soil quality functions i.e., development of root (RD), ability to supply and retain nutrients (SN) and storage

Table 1 Impact of different tillage systems on mean value of soil physical indicators

Physical indicators	NT	RT	CT
Sand (%)	23.12	24.16	24.53
Silt (%)	45.81	44.91	45.95
Clay (%)	31.07	30.93	29.52
BD (mg m^{-3})	1.22	1.24	1.31
Rp (MPa)	2.80	2.11	1.89
Porosity (%)	48.06	51.37	50.95
WHC (%)	48.57	52.52	50.33
Ks (cm h^{-1})	0.79	0.87	0.89
MWD (mm)	1.22	1.30	1.27
< 0.1 mm	25.58	33.75	44.49
0.1 mm	7.05	4.36	9.38
0.2 mm	5.68	5.86	4.93
0.5 mm	17.12	8.10	12.11
1 mm	7.049	6.17	7.91
2 mm	26.99	19.05	16.79
5 mm	12.89	14.22	10.53

BD = Bulk density; RP = Resistance to penetration; WHC = Water holding capacity; Ks = Saturated hydraulic conductivity; MWD = Mean weight diameter

of water (WS) contributed 16.03% to 32.23%, 49.65% to 61.14% and 15.15% to 22.83%, respectively to the SQI (Fig. 1). Tillage had constructive influence only on the function of root development (RD). This may be described by the higher values of Rp observed under NT in comparison to those observed under RT and CT. In addition, present study showed negative influence of tillage on other two functions of soil (i.e., SN and WS). On the other hand, maximum contribution to the general SQI (48.65–59.14%) was for the imperative function of soil i.e., nutrient supply (NS) (Fig. 1). Among the three-tillage systems CT and RT accounts the lowest values for this function. Hasty oxidation of organic matter, loss of nutrients and low CEC under these two tillage systems may have decreased capability to supply and retain nutrients due to changes in soil structure (Fig. 1). The main function of water storage (WS) contributed to 15.15–22.83% of the SQI (Fig. 1). This was least affected soil function under diverse tillage systems. The enhanced water retention capacity observed in soils under NT system can be attributed to factors such as stable soil aggregates, elevated clay content, increased organic matter content, greater porosity, larger MWD, and higher WHC.

Table 2 Impact of different tillage systems on mean value of soil chemical indicators

Chemical indicators	NT	RT	CT
pH	6.44	7.12	6.85
EC	0.15	0.17	0.17
CEC	13.06	13.77	13.24
N	359.30	331.8	316.10
P	25.50	38.11	30.25
K	172.10	220	190
S	13.71	15.04	13.85
Ca	1881	2171	1689
Mg	216.90	224.20	196.70
Zn	0.63	0.67	0.58
Cu	2.22	2.39	1.79
Fe	54.76	35.02	41.18
Mn	18.79	12.28	17.98
Ni	0.38	0.36	0.33
Pb	1.13	0.49	1.44
Cd	0.08	0.07	0.07

Cation exchange capacity (CEC) (c mol (p^+) kg^{-1}), electrical conductivity (EC) (dSm^{-1}), Available nitrogen (N), phosphorus (P), sulfur (S) (kg ha^{-1}), DTPA extractable calcium (Ca), magnesium (Mg), zinc (Zn), copper (Cu), iron (Fe), manganese (Mn), lead (Pb), cadmium (Cd) (mg kg^{-1}) of soil

Table 3 Impact of different tillage systems on mean value of soil biological indicators

Biological indicators	NT	RT	CT
SOC	1.25	1.77	1.21

SOC Walkley and Black soil organic carbon (%)

3.2 *Soil Quality Index (SQI)*

In the current study, the methodology most used for evaluating SQI is proposed by Karlen and Stott (1994). Following the methodology all the 24 soil quality indicators were considered for PCA. According to the criteria established by Rezaei et al. (2006), PCs with eigenvalues greater than or equal to 1, and capable of explaining at least 5% of the variance in the data, were measured. As the best representative of system attributes, the PC with higher eigen values and variables with high factor loading were assessed (Sofi et al., 2016). PC1, PC2, PC3, PC4, PC5, PC6, and PC7 accounted for 12.98%, 9.05%, 8.87%, 7.63%, 6.15%, 5.80%, and 5.50% of the overall unpredictability, respectively, however the collective variation was 56% (Table 4). Only factors with absolute values within 10% of the highest factor loading inside each PC were kept for the minimal dataset (Table 4). In cases where multiple factors

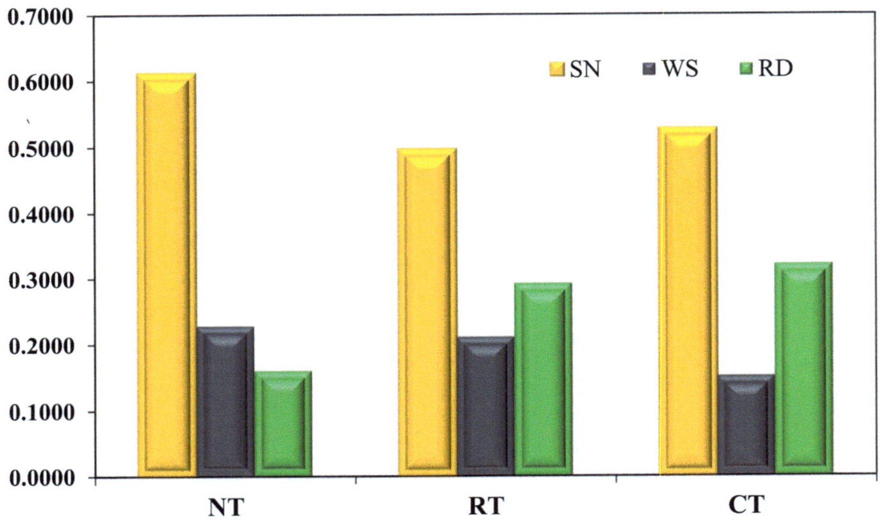

Fig. 1 Percentage contribution of each soil function to the soil quality index assessment for different tillage systems

were retained within a single PC, correlation analysis was conducted to determine whether certain variables could be deemed unnecessary and subsequently excluded from MDS (Andrews et al., 2002b). Only non-significant related values from the set of variables that displayed significant relationships with each other were retained (Mukherjee & Lal, 2014). On the other hand, flexibility criteria (expert opinion) were also taken into consideration in the rare conditions depending upon the significance of the variables. The remaining variables were elucidated from the dataset.

The principal component loading matrix showed that the first PC (PC1) included three variables with high factor loadings, all within 10.0% of the maximum loading. The three indicators were organic carbon (OC), < 0.1 mm aggregate and MWD (Table 4). According to the criteria proposed by Andrews et al. (2002a), all of these variables were found to be strongly correlated, as indicated in Table 5. For addition in MDS, MWD, which has the maximum factor loading were chosen to signify the first PC (Table 4). From the view point of soil fertility organic carbon was considered imperative (Qi et al., 2009) and thus retained. Finally, in PC1, MWD and OC were retained (Table 5). Soil aggregate in the size fraction 5 mm was the only property causal to the second PC (Table 4), signifying that soil aggregates provided unique information in this PC. In PC3, the variables which qualified for next step were BD, Porosity, WHC and Ks. After applying Pearson's correlation analysis, certain variables displayed insignificance, as indicated in Table 5, and were subsequently removed. Only porosity, WHC, and Ks were retained for further analysis. As soil bulk density is a red flag indicator predominantly so it was also retained. In the fourth PC (PC4), only one soil property i.e., sand was retained. In PC5, aggregates in the

Table 4 Performance of SQIs based on factor loadings/eigen vector values in PCA

	Principal component						
	I	II	III	IV	V	VI	VII
Eigen value	4.283	2.988	2.926	2.517	2.029	1.912	1.814
Variability	12.98	9.05	8.87	7.63	6.15	5.80	5.50
Cumulative variability	12.98	22.03	30.90	38.53	44.68	50.47	55.97
Factor loading/Eigen vectors							
Factors							
Clay	0.395	0.135	− 0.366	− 0.257	0.234	**− 0.44**	− 0.199
Silt	− 0.103	− 0.475	− 0.026	− 0.556	− 0.264	0.128	**0.526**
Sand	− 0.145	0.372	0.25	**0.692**	0.109	0.149	− 0.382
BD	− 0.102	− 0.17	**− 0.458**	0.231	0.163	**0.515**	0.081
Porosity	− 0.042	0.27	**0.509**	− 0.369	− 0.221	**− 0.486**	− 0.22
WHC	0.38	0.069	**0.531**	− 0.126	− 0.185	0.142	0.252
Ks	0.006	0.294	**0.506**	− 0.077	0.205	0.278	− 0.419
MWD	**0.783**	0.444	− 0.299	0.07	− 0.109	0.017	0.11
RP	0.184	− 0.189	− 0.204	− 0.272	0.252	**0.504**	0.099
< 0.1 mm	**− 0.711**	0.243	0.002	− 0.434	0.304	0.146	0.037
0.1 mm	− 0.487	− 0.259	0.397	0.378	− 0.018	− 0.015	0.11
0.2 mm	0.248	− 0.177	0.274	− 0.043	**− 0.44**	− 0.087	0.141
0.5 mm	0.289	− 0.655	− 0.201	0.305	− 0.152	− 0.195	− 0.253
1 mm	0.174	0.552	0.354	− 0.057	− 0.268	0.224	0.007
2 mm	0.778	− 0.216	− 0.071	0.267	− 0.093	− 0.191	− 0.064
5 mm	0.418	**0.681**	− 0.334	− 0.107	− 0.023	0.134	0.197
pH	− 0.623	0.204	− 0.328	0.08	− 0.264	− 0.07	− 0.13
EC	− 0.192	0.275	− 0.164	0.093	− 0.36	0.148	0.151
CEC	0.214	0.208	− 0.387	− 0.177	**0.401**	− 0.318	− 0.203
OC	**0.736**	0.058	0.263	0.028	− 0.265	0.26	− 0.021
N	0.352	0.19	0.357	− 0.239	0.298	0.164	− 0.057
P	− 0.083	0.415	− 0.342	0.08	− 0.098	− 0.059	0.286
K	− 0.126	0.229	− 0.353	0.444	− 0.335	− 0.081	0.03
S	0.21	0.136	− 0.207	0.044	0.104	**0.455**	0.135
Ca	0.105	− 0.285	− 0.23	0.061	0.139	0.303	− 0.127
Mg	0.049	− 0.152	0.232	0.378	− 0.07	0.168	0.063
Zn	0.191	− 0.246	0.009	− 0.17	0.101	0.197	− 0.088
Cu	0.276	− 0.074	− 0.155	0.046	− 0.047	0.153	− 0.332
Fe	0.472	− 0.329	0.245	0.079	**0.415**	− 0.056	0.127

(continued)

Table 4 (continued)

	Principal component						
	I	II	III	IV	V	VI	VII
Mn	– 0.023	0.177	0.155	0.278	0.298	– 0.268	**0.512**
Ni	0.115	0.112	0.138	0.317	0.42	– 0.298	**0.483**
Pb	– 0.025	0.079	0.198	0.435	0.249	– 0.029	0.241
Cd	0.165	– 0.08	– 0.195	0.011	– 0.295	– 0.174	0.057

Bold-faced values represent the eigenvectors that fall within 10% of the highest factor

size range of 0.2 mm, CEC and Fe were retained. In PC6 and PC7, the variables that are retained are clay, sulphur and Mn, silt respectively (Table 4). Our least data set included the soil quality indicators that have been recommended by other researchers too (Chaudhury et al. 2005; Doran & Parkin, 1994; Masto et al., 2007; Qi et al., 2009).

Each PC accounts for a specific percentage of the total variation in the dataset. To determine the relative importance of the indicators chosen under a particular PC, the

Table 5 Inter-correlations between highly weighted variables under different PCs

PC1	MWD	< 0.1 mm	OC	
MWD	1	0.533	0.52	
< 0.1 mm	0.533	1	0.582	
OC	0.52	0.582	1	
PC3	BD	Porosity	WHC	Ks
BD	1	0.789S	0.209	0.065
Porosity	0.789S	1	0.244S	0.237
WHC	0.209	0.244S	1	0.218
Ks	0.065	0.237S	0.218	1
PC5	0.2 mm	CEC	Fe	
0.2 mm	1	0.295	0.068	
CEC	0.295	1	0.021	
Fe	0.068	0.021	1	
PC6	Clay	S		
Clay	1	0.202		
S	0.202	1		
PC7	Mn	Ni	Silt	
Mn	1	0.439S	– 0.088	
Ni	0.439S	1	– 0.145	
Silt	– 0.088	– 0.145	1	

*Correlation is significant at the 0.05 level (2-tailed)
**Correlation is significant at the 0.01 level (2-tailed)

percentage of variation explained by that PC is divided by the cumulative percentage of variation explained by all PCs with eigenvalues greater than 1. The resulting subjective factors for PC1, PC2, PC3, PC4, PC5, PC6, and PC7 are 0.23, 0.16, 0.16, 0.14, 0.11, 0.10, and 0.10, respectively. These values represent the proportion of the variation explained by each PC relative to the overall variation explained by all PCs.

Following the identification of the MDS indicators, the linear scoring approach suggested by Andrews et al. (2002b) was used to modify each surveillance of each MDS indicator. Indicators were ranked based on whether higher values were considered "good" or "bad" in terms of soil functions to determine the scores. In case of "more is better" indicators, each observation was divided by the maximum observed value to establish a score, with the numerator being the value and the denominator being the observation. The highest score was assigned a value of 1, resulting in a higher score for better performance. On the other hand, for "less is better indicators", the lowest observed value received a score of 1. In this study, the indicators or parameters included in the MDS were considered "good" for soil quality when they exhibited an increasing trend. Consequently, the "more is better" method was employed, where higher values of these indicators were indicative of improved soil quality. The scores of the indicators in the MDS were transformed using a linear scoring method and multiplied by the weighted factor obtained from the PCA analysis, as described in Eq. (1).

$$\mathbf{SQI} = \sum_{\mathbf{i}=1}^{\mathbf{n}} \mathbf{W_i S_i} \tag{1}$$

where S_i in the above equation represents the score for the subscripted variable, whereas Wi is the weighing factor derived from the PCA. (S); where S is the score obtained from homothetic transformation and coefficients are derived from weighting factors obtained from the results of PCA. The resulting SQI were then normalised in relation to the highest likely SQI, which was calculated by adding the maximum PCA weighting factors for each key indicator. The results revealed that the SQI were 0.69, 0.86 and 0.74 for NT, RT and CT respectively (Fig. 2). It can be seen from the above analyses that the quality status of the soil under RT was at a higher level than at both no-till and conventional till, thus highlighting the role of RT in maintaining soil quality. SQI of 1 indicates ideal soil conditions for both most favorable productions as well as for optimal ecological quality. The value of 0.86 for RT (Fig. 2) thus indicates that some effects of RT system are not constructive for soil quality and need to be assessed.

SQI values for NT treatments were analogous to those reported by Fernandes et al. (2011) who resolute soil health index under diverse management systems, i.e., no-tillage with a double-disk planter (NTd), no-tillage with a hoe planter (NTh), CT and RT, and found the SQIs speckled between treatments being 0.68, 0.64, 0.79 and 0.86 under NTd, NTh, CT and RT, respectively. On the other hand, SQI values reported by Erkossa et al. (2007) under different tillage and management systems are much lesser than the SQI determined in the current study. This might be due

Fig. 2 Soil quality index of different tillage systems

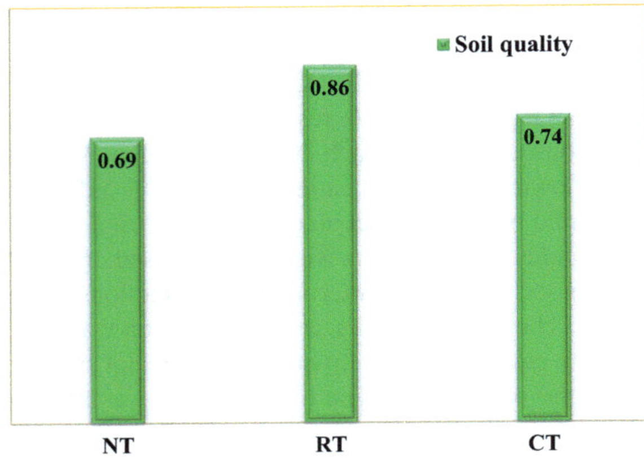

to dissimilar least datasets used in both studies. Sofi et al. (2016) evaluated the SQI under diverse cropping systems in northwest Himalaya region of India. Their findings revealed lower soil quality ratings for cereal-based cropping systems, while the apple-bean cropping system exhibited the highest soil quality rating. This highlighted the importance of incorporating legume crops in orchard floors to sustain soil fertility. In another study, Sharma et al. (2008) evaluated the soil quality index of soil under diverse soil-nutrient management and tillage practices; they observed that SQI under both RT and CT, the sole organic treatment exhibited the greatest improvement in soil quality, with a significant increase of 31.8% compared to the control. Furthermore, the use of conjunctive nutrient treatments resulted in a soil quality improvement ranging from 24.2% to 27.2%. In contrast, the sole inorganic treatment showed a comparatively lower increase of 18.2% in soil quality compared to the control. In a similar study, Melo Filho et al. (2007) applied the same tactic for evaluating the SQI of a soil under natural forest in Bahia (Brazil). They reported a SQI value of 0.46, which is much lesser than those found in our study and indicates a soil of deprived quality for crop production.

3.3 *Contributions of Soil Indicators to SQI*

Figure 3 illustrates the computed common linear scores for each of the indicators under different tillage systems, providing insights into the contribution of each indicator towards soil quality. The assistance of most indicators to soil health had no significant difference in the three tillage systems. The key indicators improving soil quality were OC, MWD, porosity, BD and WHC, with the contribution rate always above 10%. However, the offerings of S varied greatly among different tillage systems. Sulphur (S) is a vital nutrient for the plants. It contributes more to CT than NT and RT representing that S had restricted soil quality level in NT and RT.

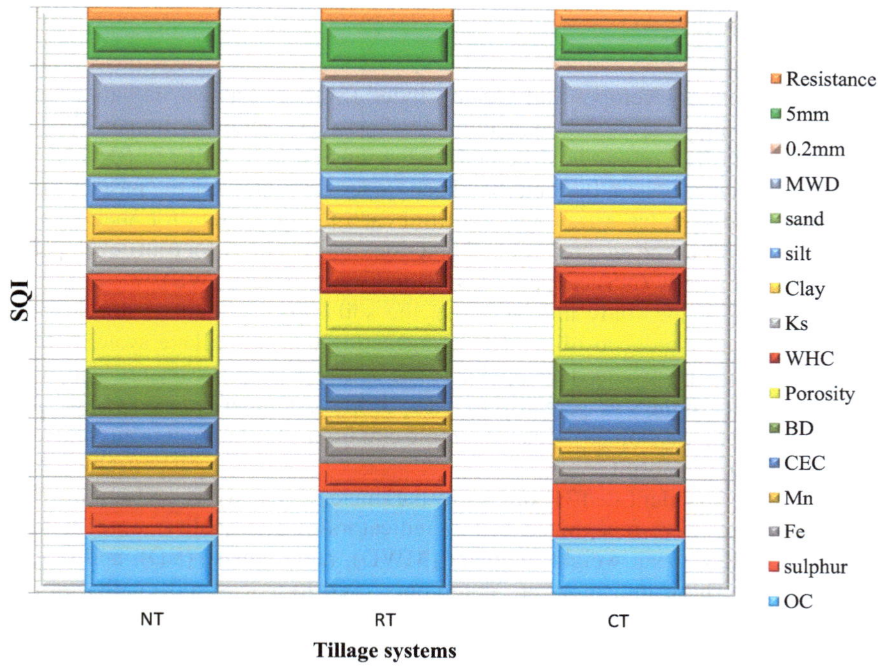

Fig. 3 Contribution of soil quality indicators to soil quality index in various tillage systems

The soil physical indicators of incursion resistance and 0.2 mm aggregates contributed slightest to the soil quality under three tillage systems, signifying a poor soil health status when their values increase in the soil. Amid physical soil quality indicators, the contribution of MWD is maximum because under the circumstances of soil instability caused by different tillage practices, it is a primary factor of soil resistivity to mechanical stresses, together with the impacts of rainfall and surface runoff, and thus to water erosion (Arshad et al., 1996; Canasveras et al., 2010). Moreover, "Ks" is significant for these soils because it affects key soil properties including aeration, water infiltration rate, conductance, porosity, and water transmission, among others. Therefore, "Ks" serves as a reliable indicator for assessing soil quality. Based on the preceding discussion, it can be concluded that the selection of key indicators employed in the present study was appropriate for calculating the soil quality of these soils under various tillage systems.

4 Discussion

The study aims to develop a Soil Quality Index (SQI) and assess its values under three distinct tillage systems: conventional tillage (CT), reduced tillage (RT), and no-tillage (NT). The research was conducted at the research station of SKUAST-K, Shalimar, Srinagar, where the field had been under no-tillage for the past 18 years. Composite soil samples were collected from each tillage system for analysis. The study primarily focused on three key soil functions: soil capacity for root development (RD), nutrient supply (NS) capacity of soil, and water storage (WS).

The results of the study indicated that the soil capacity for root development was highest under reduced tillage (RT), suggesting that this tillage system allows for better root growth and development. The nutrient supply capacity and water storage capacity of the soil were also found to be highest under reduced tillage (RT), indicating that this system provides favorable conditions for nutrient availability and water retention in the soil.

The researchers identified 31 soil quality parameters for analysis, and among them, four parameters were recognized as key indicators of soil quality: water holding capacity (WHC), mean weight diameter (MWD), bulk density (BD), and organic carbon (OC). These indicators are commonly used to assess soil quality as they provide valuable information about soil structure, nutrient availability, and organic matter content.

Based on the analysis of the soil quality parameters, the study calculated the Soil Quality Index (SQI) for each tillage system. The SQI values indicated that reduced tillage (RT) had the highest soil quality rating with a value of 0.86, followed by conventional tillage (CT) with a value of 0.74, and finally, no-tillage (NT) with a value of 0.69. These findings suggest that reduced tillage is the most suitable management technique for soil conservation and enhancement in the Temperate Himalayan zone.

The results of this study have important implications for soil management practices in the region. By assessing the soil quality under different tillage systems, farmers and land managers can make informed decisions about the most appropriate tillage technique to adopt. The findings highlight the benefits of reduced tillage in terms of root development, nutrient supply, and water storage capacity, which are crucial factors for sustainable agriculture in the area.

5 Conclusion

Although the tillage had a noteworthy effect on physical, chemical as well as biological attributes of soils calculated but the physical soil parameters were the most important for sustainable soil management in our study site. Three soil functions calculated i.e., water storage capacity of the soil (WS), soil capacity for root development (RD) and nutrient supply capacity of the soil (NS) were greatly influenced by different tillage systems. Soil quality evaluation seems to be an effective tool to monitor and assess the effects of various tillage systems on the quality of soil. In this study, soil quality evaluation was conducted by selecting indicators through principal component analysis, with emphasis placed on higher factor loadings. To avoid redundancy, interclass correlation was performed. Among 33 soil parameters studied; OC, MWD, BD, soil porosity and WHC were designated as key soil quality indicators. Homothetic transformation was applied to score the selected indicators. According to the results of our study reduced tillage seems to be the most suitable soil management across the different tillage systems for the study area since it combines the best maintenance of soil quality. The methodology employed and the obtained results in this study can be valuable for researchers and land managers, aiding in the identification and calculation of soil quality index (SQI) for different management systems in research farms.

Acknowledgements **Conflicts of Interest** The authors state that they have no conflicts of interest related to this study.

Authors' Contributions All the authors have contributed to the compilation of this study.

Consent to Participate Not applicable.

Consent to Publish Not applicable.

Availability of Data and Materials Not applicable.

References

Andrews, S. S., Karlen, D. L., & Cambardella, C. A. (2004). The soil management assessment framework: A quantitative soil quality evaluation method. *SSSA Journal, 68*, 1945–1962.

Andrews, S. S., Karlen, D. L., & Mitchell, J. P. (2002a). A comparison of soil quality indexing methods for vegetable production systems in northern California. *Agriculture, Ecosystems & Environment, 90*, 25–45.

Andrews, S. S., Mitchell, J. P., Mancinelli, R., Larlen, D. L., Hartz, T. K., Horwarth, W. R., Pettygrove, G. S., Scow, K. M., & Munk, D. S. (2002b). On farm assessment of soil quality in California's Central Valley. *Agronomy Journal, 94*, 12–23.

Arshad, M. A., Lowery, B., & Grossman, B. (1996). Physical tests for monitoring soil quality. In *Methods for assessing soil quality* (Vol. 49, pp. 123–141). SSSA Special Publication, Soil Science Society of America.

Arshad, M. A., & Martin, S. (2002). Identifying critical limits for soil quality indicators in agro ecosystems. *Agriculture, Ecosystems & Environment, 88*, 153–160.

Black, C. A. (1965). *Methods of soil analysis part-I*. American Society of Agronomy.

Canasveras, J. C., Barron, V., Del Campillo, M. C., Torrent, J., & Gomez, J. A. (2010). Estimation of aggregate stability indices in Mediterranean soils by diffuse reflectance spectroscopy. *Geoderma, 158*, 78–84.

Chaudhury, J., Mandal, U. K., Sharma, K. L., Ghosh, H., & Mandal, B. (2005). Assessing soil quality under long-term rice-based cropping system. *Communications in Soil Science and Plant Analysis, 36*, 1141–1161.

Chesnin, L., & Yien, C. H. (1951). Turbidimetric determination of available Sulphur. *Proceedings of SSSAJ., 15*, 149–151.

Chun-Juan, B. I., Zhen-Lou, C., Jun, W., & Dong, Z. (2013). Quantitative assessment of soil health under different planting patterns and soil types. *Pedosphere, 23*, 194–204.
Dalal, R. C., & Moloney, D. (2000). Sustainability indicators of soil health and biodiversity. In P. Hale, a. Petrie, D. Moloney, & P. Sattler (Eds.), *Management for sustainable ecosystems*. Centre for Conservation Biology, the University of Queensland.
Doran, J. W., & Parkin, T. B. (1994). Defining and assessing soil quality. In J. W. Doran, D. C. Coleman, D. F. Bezdicek, & B. A. Stewart (Eds.), *Defining soil quality for a sustainable environment* (pp. 3–21). SSSAJ, Inc..
Erkossa T., Itanna F., Stahr K.Stahr K, 2007. Indexing soil quality: A new paradigm in soil science research. Australian Journal of Soil Research 45: 129–137.
Fernandes, J. C., Gamero, C. A., Lanc, J. G., & Rodrigues Avalos, J. M. M. (2011). Determination of the quality index of a Paleudult under sunflower culture and different management systems. *Soil & Tillage Research, 112*, 167–174.
Govaerts, B., Sayre, K. D., & Deckers, J. (2006). A minimum data set for soil quality assessment of wheat and maize in the highlands of Mexico. *Soil & Tillage Research, 87*, 163–174.
Imaz, M. J., Bescansa, I. V. P., Enrique, I. V. A., Fernandez-Ugalde, O., & Karlen, D. L. (2010). Soil quality indicator response to tillage and residue management on semi-arid Mediterranean cropland. *Soil and Tillage Research, 107*, 17–25.
Jackson, M. L. (1973). *Soil chemical analysis*. Prentice Hall of India Private Limited.
Karlen, D. L., Eash, N. S., & Unger, P. W. (1992). Soil and crop management effects on soil quality indicators. *American Journal of Alternative Agriculture, 7*, 48–55.
Karlen, D. L., & Stott, D. E. (1994). A framework for evaluating physical and chemical indicators of soil quality. In J. W. Doran, D. C. Coleman, D. F. Bezdicek, & B. A. Stewart (Eds.), *Defining soil quality for a sustainable environment* (Vol. 35, pp. 53–72). SSAJ.
Laishram, J., Saxena, R., Maikhur, I. R. K., & Rao, K. S. (2012). Soil quality and soil health: A review. *International Journal of Ecology and Environmental Sciences, 38*(1), 19–37.
Lal, R. (2015). Restoring soil quality to mitigate soil degradation. *Sustainability, 7*, 5875–5895.
Li, P., Zhang, T., Wang, X., & Yu, D. (2013). Development of biological soil quality indicator system for subtropical China. *Soil and Tillage Research, 126*, 112–118.
Lindsay, W. L., & Norvell, W. A. (1978). Development of a DTPA soil test for zinc, iron, manganese and copper. *Soil Science Society of America Journal, 42*, 421–428.
Masto, R. E., Chhonkar, P. K., Dhyan Singh, & Patra, A. K. (2007). Soil quality response to long-term nutrient and crop management on a semi-arid Inceptisol. *Agriculture, Ecosystems & Environment, 118*, 130–142.
Melo Filho, J. F., Souza, A. L. V., & Souza, L. S. (2007). Determinac¸a˜o do ı'ndice de qualidade subsuperficial em um latossolo amarelo coeso dos tabuleiros costeiros, sobfloresta natural. *Revista Brasileira de Ciência do Solo, 31*, 1599–1608. (in Portuguese).
Mukherjee, A., & Lal, R. (2014). Comparison of soil quality index using three methods. *PLoS One, 9*, e105981. www.plosone.org
Nortcliff, S. (2002). Standardization of soil quality attributes. *Agriculture, Ecosystems & Environment, 88*, 161–168. https://doi.org/10.1016/S0167-8809(01)00253-5.
Olsen, S. R., Cole, C. V., Watanabe, F. S., & Dean, L. A. (1954). *Estimation of available phosphorus in soils by extraction with sodium carbonate* (p. 939). U.S. Department of Agriculture Circular.
Piper, C. S. (1966). *Soil and plant analysis* (p. 164). Hans Publishers.
Qi, Y., Darilek, J. L., Huang, B., Zhao, Y., Sun, W., & Gu, Z. (2009). Evaluating soil quality indices in an agricultural region of Jiangsu Province, China. *Geoderma, 149*, 325–334.
Rabbi, S. M. F., Kibria, K. Q., Rahman, A., Islam, M. S., Bhuiyan, M. R., & Ahsan, M. (2004). Aggregate stability of Ganges tidal floodplain soils and its relationship with soil physical and chemical properties. *Bangladesh Journal of Soil Science, 30*, 61–69.
Rasmussen, P. E., & Collins, H. P. (1991). Long-term impacts of tillage, fertilizer and crop residue on soil organic matter in temperate semi-arid regions. *Advances in Agronomy, 45*, 93–134.
Rattan, R. K. (2009). *Fundamentals of soil science* (p. 91). National Agricultural Science Centre Complex.

Rezaei, S. A., Gilkes, R. J., & Andrews, S. S. (2006). A minimum data set for assessing soil quality in rangelands. *Geoderma, 136*, 229–234.

Rhoades, J. D. (1982). Cation exchange capacity. In A. L. Page, R. H. Miller, & D. R. Keeney (Eds.), *Methods of soil analysis: Chemical and microbiological properties. Part-II*. American Society of Agronomy and Soil Science Society of America.

Richard, L. A. (1954). Diagnosis and improvement of saline and alkali soils. USDA. In *Hand book no. 80*. Oxford and IBH Publishing Co.

Sharma, K. L., Grace, K. B. J., Mandal, U. K., Pravin, N., Gajbhiye, K., Srinivas, G. R., Korwar, V., HimaBindu, V., Ramesh, Ramachandran, K., & Yadav, S. K. (2008). Evaluation of long-term soil management practices using key indicators and soil quality indices in a semi-arid tropical Alfisol. *Australian Journal of Soil Research, 46*, 368–377.

Shukla, M. K., Lal, R., & Ebinger, M. (2006). Determining soil quality indicators by factor analysis. *Soil and Tillage Research, 87*, 194–204.

Sofi, J. A., Bhat, A. G., Kirmani, N. A., Wani, J. A., Lone, A. H., Ganie, M. A., & Dar, G. I. H. (2016). Soil quality index as affected by different cropping systems in northwestern Himalayas. *Environmental Monitoring and Assessment, 188*, 161. https://doi.org/10.1007/s10661-016-5154-1

Soil Science Society of America. (1995). *SSSA statement on soil quality* (p. 7). Agron News.

Subbiah, B. V., & Asija, G. L. (1956). A rapid procedure for the estimation of available nitrogen in soils. *Current Science, 25*, 259–260.

USDA. 2006. Soil quality institute. Natural Resources Conservation Service. http://soils.usda.gov/sqi/

Van Bavel, C. H. M. (1949). Mean weight diameter of soil aggregation as a statistical index of aggregation. *Soil Science Society of America Proceedings, 14*, 20–23.

Walkley, A. J., & Black, C. A. (1934). An estimation of the Degtjareff method for determining soil organic matter and a proposed modification of the chromic acid titration method. *Soil Science, 37*, 29–38.

Yoder, R. E. (1936). A direct method of aggregate analysis and study of the physical nature of erosion losses. *Journal of the American Society of Agronomy, 28*, 337–351.

Geospatial Technologies and Earth Observation for Climate and Hazard Studies

Flood Mapping Using Google Earth Engine (GEE): A Short Review

T. V. Vandhana and Girish Gopinath

Abstract Flood Mapping, a core application in disaster management and mitigation, is a useful tool for environmental monitoring, flood risk, and flood resilience. Open satellite archives and cloud computing now enable timely, basin-to-nation-scale inundation products. Google Earth Engine (GEE), a cloud-based Geospatial platform, provides unique capabilities in large-scale and real-time flood analysis, supports flood detection and monitoring using optical (Sentinel-2, Landsat) and synthetic aperture radar (Sentinel-1) data. GEE has lowered the operational and technical barriers to flood mapping by collocating global archives and scalable computation. Simple, transparent baselines—adaptive thresholds and change detection—remain powerful starting points, especially when combined with topography-aware masks and morphological cleanup. Machine learning and deep learning deliver gains in complex settings when supported by representative labels and sound validation. This review chapter delivers various flood mapping methodologies including flood change identification, machine learning and thresholding techniques, multi-sensor data integration, and time-series analysis in GEE and its capacity for integrating diverse satellite and hydrological data and summarize practical preprocessing on GEE (cloud/shadow masking, radiometric calibration, speckle mitigation, orbit filtering), feature construction (NDWI/MNDWI, VV/VH ratios, coherence, textures), and post-processing (topography-aware masks and morphological filters). Validation design is treated explicitly, including stratified sampling, spatial cross-validation, confusion-matrix reporting (Precision, Recall, F1/IoU), and strategies for class imbalance. Case studies from 2019 to 2022 of the GEE application in flood mapping in the diverse geographical contexts are highlighted. Exploring numerous literatures, this paper discusses challenges and limitations such as data quality issues, computational constraints, and points out potential gaps, and covers a broad discussion on technological advancement and interdisciplinary collaboration of Artificial Intelligence for the precision and efficacy of flood mapping. Summarizing this chapter signifies the perspective of GEE in ongoing research and inducing emergency operations, risk assessment, and long-term resilience planning across diverse geographies.

T. V. Vandhana (✉) · G. Gopinath
Department of Climate Variability and Aquatic Ecosystems, Kerala University of Fisheries and Ocean Studies, Kochi, India

R. Ahmed et al. (eds.), *Climate Change, Water Resources, and Natural Hazards in the Himalayas*, Advances in Geographical and Environmental Sciences,
https://doi.org/10.1007/978-981-92-0835-7_14

Keywords Flood mapping · Google Earth Engine (GEE) · Remote sensing · Limitations · Artificial intelligence · Disaster risk reduction

1 Introduction

Floods are among the most destructive natural disasters globally, impacting millions of people each year. They occur when water inundates land that is usually dry, often resulting from heavy rainfall, river overflow, storm surges, or the rapid melting of snow and ice. The frequency and intensity of floods have increased in recent decades, exacerbated by climate change, urbanization, and deforestation, making them a critical area of study in environmental science and engineering. Floods rank as one of the most destructive occurrences of natural disasters, resulting in extensive harm to infrastructure, sources of income, and the natural surroundings. In recent years, the frequency and intensity of floods have been on the rise due to various factors such as climate change, urbanization, and land use changes (Zeng et al., 2016). Flooding can have catastrophic impacts, including loss of life, economic damage, and long-term disruptions to communities. The World Bank surmised that floods alone account for nearly 50% of all weather-related disasters globally, affecting more than 2.3 billion people (Hallegate et al., 2016). The social and economic implications are profound, necessitating more sophisticated methods for predicting, managing, and mitigating floods. Research in flood management has evolved significantly, shifting from reactive to proactive approaches. Early studies focused primarily on historical flood data to understand patterns and predict future events (Smith & Ward, 1998). Flood mapping is an important aspect of disaster management and environmental monitoring, which is essential for risk mitigation and has far-reaching consequences, making it imperative in the decision-making process. It identifies and defines flood-prone areas, especially the lowlands next to rivers, and gives important data on their durations and extents (Kwak et al., 2016). The data is important for sustainable urban planning, protecting lives and property, and disaster risk reduction (Zin et al., 2018). In defining a flood-risk area, it is vital and very suitable to use flood mapping. This includes elevation, distance from land surface to streams, precipitation, and soil texture, comingling to identify the probability of a flood (Azizat & Omar, 2018). However, there have been attempts in recent studies to understand how machine learning algorithms could be used for flood hazard mapping with acceptable accuracy and use of minimal preparation of geo-environmental data (Leggesse et al., 2024). Flood mapping is essential for effective flood risk management, especially in transboundary river basins where floods do not recognize borders (Chini et al., 2017). It aids in creating early warning systems, evacuation planning, and identifying critical facilities and communication channels (Ali et al., 2017). Moreover, sustainability-based flood hazard mapping incorporates economic, social, and environmental factors, providing a more comprehensive perspective for decision-makers (Ahmadisharaf et al., 2017). By integrating community-based approaches, flood hazard mapping can also empower local communities to understand their vulnerabilities and play an active role in managing flood risks (Bhere & Reddy, 2022; Osti et al., 2008). Flooding represents one of the most pervasive and destructive climate-related hazards globally, inflicting substantial environmental and socioeconomic damages annually (Misra et al., 2025). The escalating frequency and severity of these events, driven by climate change and accelerated urbanization, necessitate advanced methodologies for effective flood monitoring and assessment (Wedajo, 2017). Traditional two-dimensional flood mapping techniques, while foundational, often provide limited insights into the complex dynamics of flood events, thus underscoring the imperative for more sophisticated, often three-dimensional, approaches (Jia et al., 2025). Accurate flood mapping can provide crucial insights into the extent and severity of flooding, allowing for targeted and efficient flood management strategies. Hydrodynamic modelling and flood mapping of river basins are essential for understanding the potential impacts of floods and informing decision-making processes (Afzal et al., 2022; Tariq et al., 2021). Flood mapping challenges are faced by urban, rural, and semi-urban areas. In urban centers, the complex infrastructure and high population density require detailed mapping and analysis to identify vulnerable areas and develop targeted mitigation strategies. In rural communities, the dispersed nature of settlements and the dependence on agricultural livelihoods necessitate a comprehensive understanding of flood dynamics and their impacts. Mapping flood hazard zones and assessing flood risks involves integrating various hydrological and geographical data to create informative risk management tools. These tools typically utilize flood characteristics, such as peak flood levels and flood depth, in conjunction with exposure data and the vulnerability of affected elements to generate accurate flood hazard maps (Nguyen et al., 2023). Google Earth Engine (GEE), a satellite imagery tool unveiled at the UN Climate Change Conference (COP 16), provides a plethora of environmental data (Abutu, 2010). GEE offers a stable and scalable infrastructure for accessing, analysing, and visualising many publicly available remote sensing datasets. The application varies in different sectors, from environmental studies to disaster studies to medical fields (Amani et al., 2020; Kumar and Mutanga, 2018). High-tech advancements in cloud computing and algorithmic processes facilitate environmental monitoring through the utilisation of satellite-derived datasets. The integration of artificial intelligence, particularly machine learning and deep learning, with platforms like Google Earth Engine offers unprecedented opportunities to overcome these limitations by enabling rapid, accurate, and scalable flood inundation mapping (Liu et al., 2024) (Bentivoglio et al., 2022). This integration facilitates the analysis of vast geospatial datasets, leading to improved predictive capabilities for flood extent, depth, and duration, which are crucial for disaster preparedness and response (Ahmad et al., 2025). Moreover, the combination of publicly available satellite imagery, geographical information systems (e.g., geodata), and explainable machine learning algorithms results in improved interpretability of flood risk predictions, which is the key to making informed decisions and choosing the way of creating resilient infrastructure (Antzoulatos et al., 2022). This unified synthesis of computation models and enhanced analytics data marks a paradigm changes in prognostic and responsive flood hazard management, as opposed to the historical observational statistics, and moving to prediction and prevention of future disastrous events. Traditionally, the numerical and physical hydraulic models have played a key role in the hydraulic analysis of flood events and the creation of hazard maps (Breznik et al., 2025). This review explores the current state and future directions of integrating Google Earth Engine with advanced flood mapping techniques, highlighting the transformative potential of machine learning and artificial intelligence in enhancing global flood resilience. Specifically, this involves leveraging global land cover maps and crowdsourced geographic data to refine flood extent and depth estimations, thereby enhancing the precision of disaster management and response efforts (Johary et al., 2023). This approach allows for the development of robust, scalable flood mapping solutions that are critical for mitigating the impacts of increasingly frequent and intense flood events worldwide (Atchyuth et al., 2023). The subsequent sections will delve into specific deep learning architectures and their applications within the Google Earth Engine environment, examining their efficacy in various flood scenarios, from pluvial to fluvial inundation. This comprehensive analysis will also address the challenges and opportunities associated with integrating diverse data sources, such as synthetic aperture radar data from Sentinel-1, with optical imagery for improved flood detection and mapping precision, especially in cloud-prone regions or during nighttime events (Atchyuth et al., 2023). Furthermore, the integration of machine learning models like artificial neural networks and random forests with cloud computing platforms holds significant promise for accurate and real-time land use/land cover forecasting essential for effective flood response (Kumar & Shankar, 2024). Google Earth Engine (GEE) has reshaped how analysts build flood maps. Instead of downloading and storing terabytes of scenes, users query curated archives, apply map-reduce operations server-side, and visualize or export results directly. This cloud-native model lowers barriers to entry, shortens iteration cycles, and facilitates transparent, shareable workflows for research and operations alike. The platform's strengths data gravity, scalable computation, and reproducibility align naturally with the demands of flood response, where analysts must move quickly while keeping methods defensible. A world cloud visualisation of this review chapter is provided in Fig. 1.

2 Flood Mapping

Flood mapping is a critical task in disaster management and mitigation, as it provides valuable information about the extent, depth, and duration of flooding events. Moreover, the integration of satellite imagery with Digital Elevation Models enhances the accuracy of flood inundation maps, allowing for better assessment of flood risks and the formulation of effective response strategies in affected regions (Saksena, 2015; Tariq et al., 2021). Furthermore, the utilization of remote sensing techniques, particularly those employing multispectral satellite imagery, has proven effective in delineating flooded areas and assessing the impact of floods, providing a cost-efficient method to gather extensive data throughout all phases of a flood disaster (Sajjad

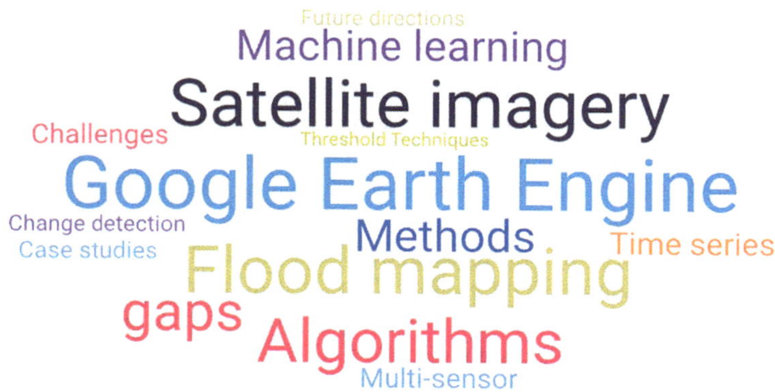

Fig. 1 World-cloud visualisation of overall review

et al., 2023). The accuracy of flood inundation maps can be significantly improved by modeling the spatial distribution of Digital Elevation Model errors (Afzal et al., 2022). Therefore, by applying these advanced modeling techniques, it is possible to obtain more precise flood maps even in regions where high-resolution topographic data is lacking, thereby enhancing flood risk assessments and improving disaster preparedness strategies (Saksena, 2015). Effective flood management requires a comprehensive understanding of flood hazards and their associated risks. Mapping flood hazard zones and assessing flood risks involves integrating various hydrological and geographical data to create informative risk management tools. These tools typically utilize flood characteristics, such as peak flood levels and flood depth, in conjunction with exposure data and the vulnerability of affected elements to generate accurate flood hazard maps (Nguyen et al., 2023). Moreover, the generation of these maps often relies on advanced modeling techniques and geographic information systems to simulate flood scenarios, thereby providing critical insights into the area's most at risk and facilitating informed decision-making for flood risk management strategies (Romali et al., 2018). Additionally, applying hydrodynamic modeling alongside satellite imagery can enhance the accuracy of flood inundation mapping, enabling authorities to assess and predict the extent of flooding events better, which is essential for devising effective mitigation strategies and managing flood-related emergencies more efficiently (Afzal et al., 2022). Flood mapping plays a vital role in disaster preparedness and response, as it delivers essential insights into the spread, severity, and persistence of flood events. By combining hydrologic and hydrodynamic models, authorities can obtain critical data needed to build dependable forecasting systems. Such systems support a wide range of emergency activities, including planning evacuation routes, coordinating relief operations, estimating property damage costs, and providing other valuable decision-making information. Flood mapping can facilitate risk identification and awareness, as well as speed up decision-making before and during a crisis (Hocini et al., 2020). In rural communities, the dispersed nature of settlements and the dependence on agricultural livelihoods necessitate a comprehensive understanding of flood dynamics and their impacts. The accuracy of flood inundation maps can be significantly improved by modeling the spatial distribution of Digital Elevation Model errors (Afzal et al., 2022). Mapping flood hazard zones and assessing flood risks involves integrating various hydrological and geographical data to create informative risk management tools. These tools typically utilize flood characteristics, such as peak flood levels and flood depth, in conjunction with exposure data and the vulnerability of affected elements to generate accurate flood hazard maps (Nguyen et al., 2023). Moreover, the generation of these maps often relies on advanced modeling techniques and geographic information systems to simulate flood scenarios, thereby providing critical insights into the area's most at risk and facilitating informed decision-making for flood risk management strategies (Romali et al., 2018). The sources cited in this research paper provide valuable insights into the importance of flood mapping and the various methods and techniques employed to enhance the accuracy and reliability of these maps (Abdrabo et al., 2020; Haces-Garcia et al., 2023; Meesuk et al., 2017; Trinh & Molkenthin, 2021). Accurate Digital Elevation Model data is essential in increasing the value of flood models because it has detailed information on surface morphology vital in determining areas of hazard. Other methods, including remote sensing methods using synthetic aperture radar images, can also be used to monitor floodwater depths rapidly and at a large scale and in quasi-synchronous terms, which are key criteria in assessing the hazard and representing risk areas. Such developments support the increasing awareness of the potential of artificial intelligence as a revolutionary mechanism of managing flood risk that provides an integrated outlook to aid in informed decision-making (Liu et al., 2024). This is mostly so in the case of AI, machine learning, and deep learning models, the latter two of which could process and analyze large, complex data to better predict floods and estimate risks (Liu et al., 2024). The conventional techniques employed for flood mapping before the emergence of Google Earth Engine (GEE) are shown in Fig. 2.

3 Overview of Google Earth Engine

Google Earth Engine is a state-of-the-art planet-scale cloud-based platform that enables users to process and analyze geospatial data in ways never possible before; therefore, it serves as one aspect to be used as an arsenal for developing flood monitoring systems. GEE is a cloud-based platform that provides access to a huge database of satellite images and environmental data, as well as the computational power necessary to perform large-scale analyses, and is good at managing big data issues related to remote sensing and geospatial analysis. With its cloud-based infrastructure and parallel processing capabilities, users can do sophisticated analysis at scale without the need for significant hardware or coding experience. As a result, Google Earth Engine is becoming increasingly popular for use by researchers and practitioners working in many different areas such as environmental monitoring, urban planning, or disaster response. Each year, the number of academic papers developing research based on Google Earth Engine is growing (Kumar & Mutanga, 2018), reaching over

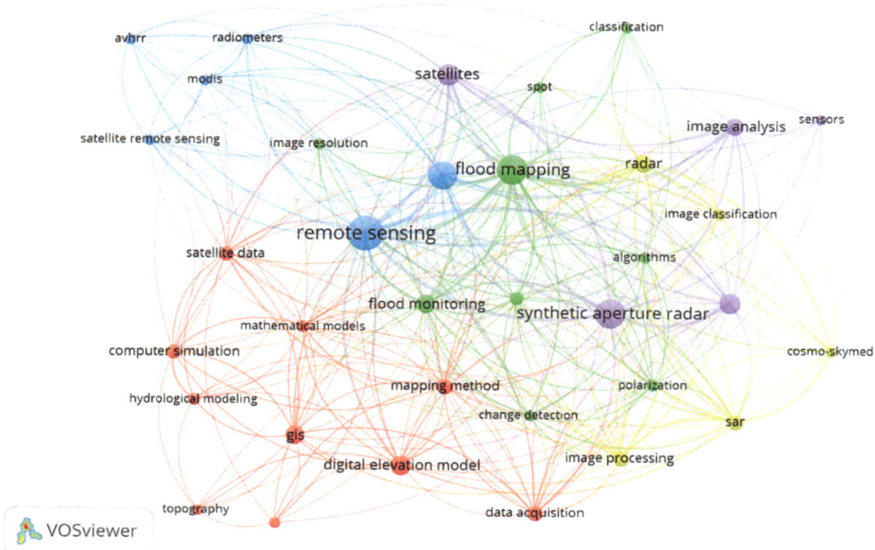

Fig. 2 The conventional techniques employed for flood mapping before the emergence of Google Earth Engine (GEE)

300 journal papers published between 2011 and 2017. Some of the flood mapping using GEE for 2019–2024 case studies are illustrated in Fig. 3. Remote Sensing has been a high-visibility journal for publications centered on Google Earth Engine, as well as Remote Sensing of Environment. The literature also identifies the potential of Google Earth Engine for Flood Monitoring Applications. Its access to very recent, near-real-time satellite images, as well as other environmental data, together with its computational power, makes it a convenient tool for monitoring and analyzing flood events. By leveraging the platform's capabilities, researchers and practitioners can gain valuable insights into the spatial and temporal dynamics of floods, which can inform mitigation and response strategies. Google Earth Engine (GEE) has reshaped how analysts build flood maps. Instead of downloading and storing terabytes of scenes, users query curated archives, apply map-reduce operations server-side, and visualize or export results directly. This cloud-native model lowers barriers to entry, shortens iteration cycles, and facilitates transparent, shareable workflows for research and operations alike. The platform's strengths data gravity, scalable computation, and reproducibility align naturally with the demands of flood response, where analysts must move quickly while keeping methods defensible.

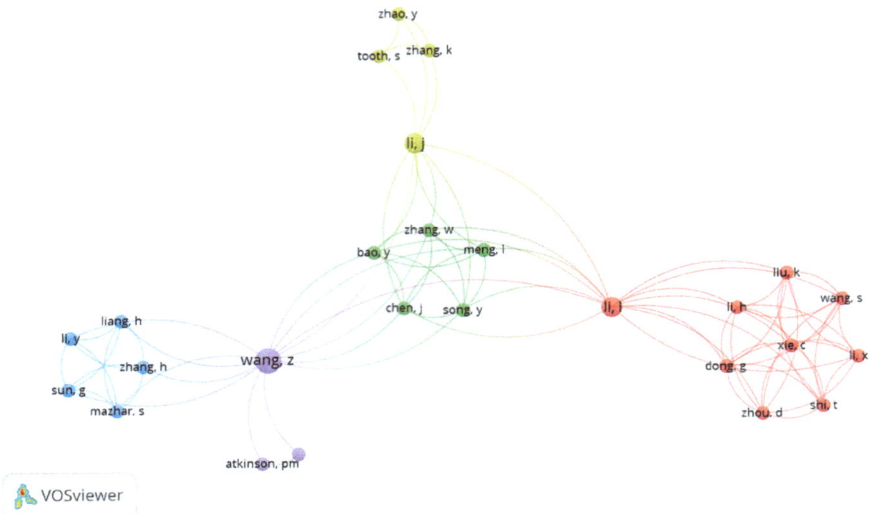

Fig. 3 Case studies of GEE for flood mapping from 2019 to 2024

4 Methodology

The main aim of this chapter is to synthesize recent applications of Google Earth Engine (GEE) for flood detection and mapping. The procedures below underpin the evidence summarized in Fig. 3. A focused search was conducted in Scopus (Title–Abstract–Keywords) to identify studies that use GEE for flood mapping.

Illustrative Boolean Structure Adapted

TITLE-ABS-KEY ((“Google Earth Engine” OR GEE) AND (flood* W/3 (mapping OR detection OR inundation OR delineation)) AND (“Sentinel-1” OR “Sentinel 1” OR “Sentinel-2” OR “Sentinel 2” OR Landsat OR SAR OR NDWI OR MNDWI OR “random forest” OR SVM OR Otsu OR “change detection”)).

Time Window: 2019–2024

Document Types: Journal articles and conference papers; English language.

Fields Exported: Authors, title, year, source, abstract, author keywords, index keywords, DOI, and citations.

For Screening and Eligibility

Two-stage screening was applied to the Scopus export.

Title/abstract screening to remove off-topic records and duplicates.

Full-text assessment against predefined criteria.

Inclusion Criteria

The study uses GEE in any substantive step of flood mapping (preprocessing, feature engineering, classification/segmentation, change detection, or validation).

(ii) Sufficient methodological detail is reported to extract the algorithm family, sensor(s)/dataset(s), and study area.

Exclusion Criteria

Non-GEE studies; editorials/notes without methods; hydrodynamic-only studies with no GEE component.

VOS viewer was used on the Scopus export to build keyword co-occurrence (and optionally co-authorship) networks. Pre-processing included lower-casing, stop word removal, and synonym merging (e.g., "GEE"/ "Google Earth Engine"; "SAR"/ "synthetic aperture radar"). A minimum keyword occurrence threshold (e.g., 3–5) was set to balance coverage and noise. Clustering and overlay visualizations were used to interpret topic groupings that support the narrative. Quantitative tallies (e.g., prevalence of Sentinel-1, Sentinel-2, Landsat-8; use of DEM and LULC/population layers) and algorithm families (thresholding, change detection, RF/SVM/CART, CNN, object-based) were summarized from the coded table and reported study-provided validation where available and noted common sources of bias (SAR backscatter in urban areas, shoreline mixed pixels, DEM errors, cloud contamination).

5 Flood Detection and Mapping Algorithms in GEE

Flooding is a common natural disaster with severe environmental and economic impacts; therefore, accurate and timely detection and mapping are critical to responding to the disaster and mitigating the harm done. Google Earth Engine is a comprehensive platform, powered by the cloud, and containing massive amounts of satellite imagery and geospatial data; thus, it allows advancing the flood detection and mapping algorithms. Accurate detection and mapping are essential for flood response and alleviation efforts. For instance, Google Earth Engine aids flood detection and mapping through numerous algorithms such as thresholding, machine learning, and image classification (Tamiminia et al., 2020). Several thresholding techniques, such as Normalized Difference Water Index and Modified NDWI, have been utilized to detect flooded areas through spectral signatures with water bodies (Tobon-Marin and Canon, 2020). Artificial Intelligence has greatly transformed areas of our lives, and its use in flood mapping is an example of this revolution in technology advancement. Recent progress in machine learning methods has opened possibilities for more precise flood mapping that overcomes the challenges posed by older techniques. The management of flood risks has emerged as an issue, and the merging of AI with technologies like Google Earth Engine presents a hopeful resolution. Machine learning algorithms provide more advanced approaches for detecting and mapping floods by utilizing the computational capability of the Google Earth Engine to process a vast amount of satellite data. For example, research has demonstrated the effectiveness of supervised and unsupervised classification, like Random Forests for classification and K-Means clustering for flood delineation (Shelestov et al., 2017). It also classified image methods, in which land cover classes were predefined, pixels, and pixel classes according to these types of images. Some of the case studies of algorithms used in GEE are shown in Table 1 (Fig. 4).

6 Data Sources and Preprocessing in GEE for Flood Mapping

GEE, recognized as a sophisticated cloud computing platform, serves as a gateway to a heterogeneous array of satellite datasets. A significant remote sensing dataset is accessible for flood delineation (Table 1). GEE has the capacity to utilize software and algorithms to analyze satellite imagery for flood mapping. Cloud cover and terrain defect problems are major issues in satellite imagery. The Google Earth Engine platform offers a wide range of satellite datasets that can be utilized for flood mapping, providing researchers and practitioners with powerful tools for monitoring and assessing flood events. Among the key datasets available in GEE, several stand out as particularly useful for flood monitoring: Landsat, Sentinel-1, and Sentinel-2. Landsat, a joint program between the United States Geological Survey and the National Aeronautics and Space Administration, has a long history of providing high-quality optical imagery with global coverage. These images can be leveraged for flood mapping through the calculation of vegetation indices, such as the Normalized Difference Vegetation Index, which can help distinguish between flooded and non-flooded areas (Morales et al., 2023). Moreover, the availability of Landsat imagery spanning several decades allows for the examination of long-term trends and changes in flood patterns. The Sentinel-1 mission, operated by the European Space Agency, is a constellation of Synthetic Aperture Radar satellites that offer all-weather, day-and-night imaging capabilities. The ability of Sentinel-1 to penetrate cloud cover and capture detailed information on surface water dynamics makes it an asset for flood monitoring, particularly in regions with persistent cloud cover (Gasica et al., 2020). Complementing the Sentinel-1 SAR data, the Sentinel-2 mission provides high-resolution multispectral imagery that can further enhance flood mapping efforts. The combination of these diverse satellite datasets, along with advanced image processing techniques and cloud-based computing platforms like GEE, has significantly improved the ability to monitor and assess flood events across various spatial and temporal scales (Ban et al., 2017; Lo et al., 2021; Mustain et al., 2021). The Sentinel-2 mission, also operated by the European Space Agency, provides high-resolution optical imagery with a wide range of spectral bands, including visible, near-infrared, and short-wave infrared. The combination of high spatial resolution, frequent revisit times, and spectral information from Sentinel-2 can enable detailed mapping of flooded areas, as well as the assessment of flood impacts on vegetation

Table 1 Case studies of algorithms used in flood mapping in GEE

References	Methods	Algorithms used	Dataset used	Study area
Ganjirad and Delavar (2023)	ML	RF and SVM	Rivershapefile, Elevation, Slope, Landuse, Rainfall, plan curvature, Spectral indices (NDVI, MNDWI, NDBI)	United States
Suab et al. (2022)	ML, Object-based classification and Thresholding techniques	ML algorithms, SNIC, MNDWI	Sentinel-1 SAR, Sentinel-2 multispectral, Aerial photos, Historical data	Malaysia
Jiang et al. (2021)	Unsupervised deep learning	Feltz-CNN	SAR	China
Farhadi and Najafzadeh (2021)	ML and RS	RF	SRTM DEM, Landsat 8/ OLI	Iran
Atchyuth et al. (2023)	ML	RF, SVM, CART	Sentinel-1, Sentinel-2	India
Pinheiro et al. (2024)	Thresholding and change detection	Otsu automatic algorithm	SAR	India
Tumaneng et al. (2024)	Change detection, Thresholding approach, Refined techniques	Change detection algorithm	SAR, LULC, Population data	Philippines
Islam and Meng (2022)	Threshold, change detection, Supervised and unsupervised classification	Polarization combination	SAR	United States
Vekaria et al. (2022)	Change detection	Binarisation change detection	Multi-temporal Sentinel-1	India
Priyatna et al. (2023)	Thresholding method	Otsu, statistical sampling	Multi-temporal Sentinel-1	Indonesia
Chen & Zhao (2022)	Threshold method	Topological relationships and digital surface model algorithm	Sentinel-1 SAR	China
Moharrami et al. (2021)	Threshold method	Otsu	Sentinel-1	Iran

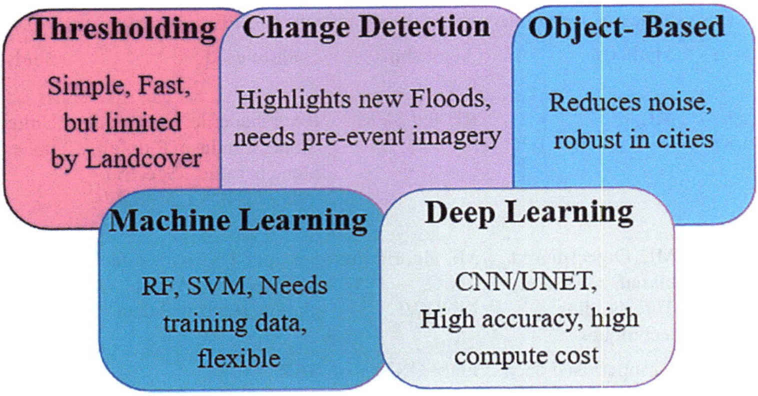

Fig. 4 Comparison of flood detection approaches in GEE

and infrastructure (Tsyganskaya et al., 2018). A recent research study, by Mustain et al. (2021) showcased the use of radar images from the Sentinel 1 mission to map flood inundation in Indonesia's Banten Province precisely. The study highlighted how SAR data could effectively identify changes in surface water levels and offer insights for monitoring floods and responding to emergencies effectively. A different research project focused on utilizing LiDAR data. Remote sensing methods to map flooding in the Pesanggrahan River Watershed in Jakarta. The study conducted by Riyanto and colleagues in 2019 highlighted how quality digital elevation models from LiDAR data were used to enhance the precision of flood predictions and mapping processes. This showcases the significance of sensing technologies in managing floods effectively (Fig. 5).

6.1 Optical Sensors: Sentinel-2 and Landsat

Optical imagery is intuitive for flood delineation because water exhibits distinctive spectral behavior. Sentinel-2 provides 10–60 m multispectral data with frequent revisit, while Landsat offers multi-decadal continuity at 30 m. On GEE, analysts typically work with surface reflectance collections that include per-pixel quality flags for clouds and shadows. Effective workflows start by filtering by date, region, and cloudiness; applying cloud and shadow masks; and computing water-sensitive indices such as the Normalized Difference Water Index (NDWI) and its modified variant (MNDWI). Temporal compositing—median or percentile composites over short windows—helps mitigate residual clouds and stabilize thresholds.

Common Optical Pitfalls and Remedies. Thin clouds and cloud edges can leak through masks. Include shadow masking and consider a short temporal window

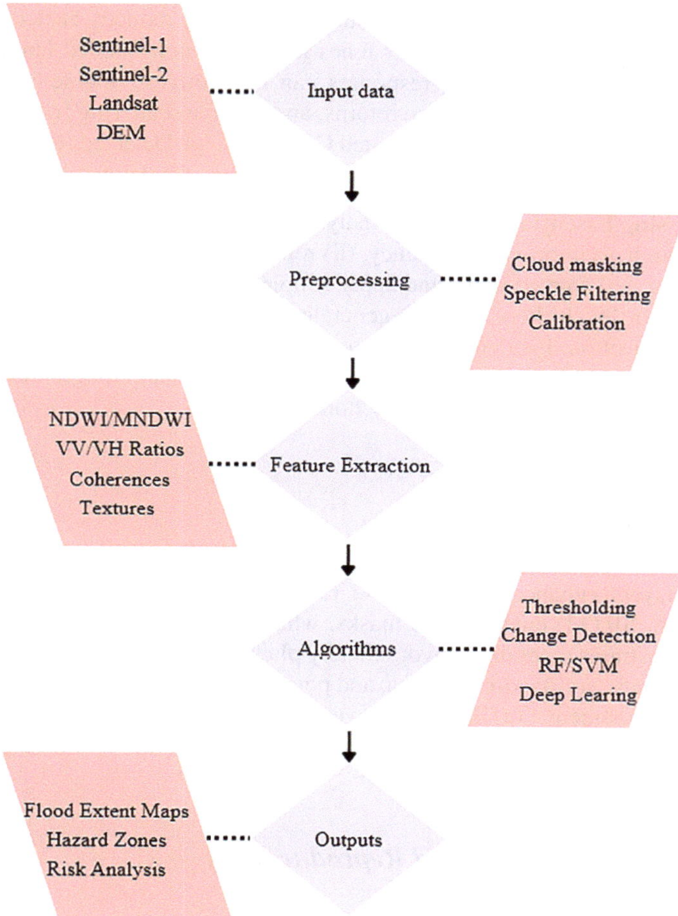

Fig. 5 General workflow of flood mapping using GEE

before/after the event to fill gaps. Urban surfaces with bright roofs or specular reflections may resemble water in single dates; using short-term change and proximity to known water bodies reduces false positives. Vegetation over inundated ground reduces spectral contrast; SAR or multisensor fusion can help.

6.2 *SAR: Sentinel-1*

SAR is indispensable during storms because it is unaffected by clouds and sunlight. Sentinel-1 in Interferometric Wide (IW) mode provides dual polarization (VV and

VH) at ~10 m ground resolution. Inundation typically reduces backscatter over smooth open water, producing darker tones relative to non-flooded land. In vegetated or urban terrain, however, responses can be complex: double-bounce from flooded urban canyons may increase returns, and volume scattering from vegetation can obscure changes. GEE hosts calibrated Ground Range Detected (GRD) products that can be converted to decibels and composited over user-defined windows.

Pre-processing Essentials for SAR. Analysts commonly (i) filter by orbit pass and incidence angle to improve consistency, (ii) remove border noise and extreme incidence angles, (iii) convert to dB and apply simple speckle suppression via temporal compositing or focal filters, and (iv) generate pre-event and post-event composites for change detection. Incidence-angle normalization and terrain-masking (layover/shadow) are advisable in rugged terrain. Ratio or difference images between pre- and post-event composites are effective flood indicators.

6.3 Ancillary Data

Ancillary layers enhance detection and reduce false positives. Digital elevation models (e.g., SRTM) enable slope masks, while Height Above Nearest Drainage (HAND) constrains mapping to hydraulically plausible areas. River networks, water occurrence maps, soil/land-cover data, and population grids support post-processing and impact assessment. On GEE, these layers are readily combined with imagery and used to inform thresholds, refine masks, and prioritize outputs for decision-makers.

6.4 Data Management and Reproducibility

Reproducible analyses explicitly list collection IDs, filters, and parameters. Versioning code, recording threshold values, and exporting parameter files or metadata alongside results allow others to replicate and audit outputs. GEE Apps and shared code links provide lightweight dissemination channels for collaborators and responders.

7 Methods for Flood Detection

Methods were grouped into five families. Each can be implemented in GEE with modest effort, and many benefit from combining optical and SAR.

7.1 Spectral and Index Thresholding (Optical) and Backscatter Thresholds (SAR)

Thresholding is the simplest approach: compute a water-sensitive feature and apply a threshold to separate water from land. In optical imagery, NDWI and MNDWI emphasize water by contrasting green/SWIR or green/NIR reflectance. Fixed thresholds (e.g., NDWI > T) are fast but brittle; adaptive thresholds such as Otsu's method estimate a scene-specific value by maximizing between-class variance. For SAR, thresholds can be applied directly to dB values or to the VV/VH ratio; lower backscatter often indicates open water.

Strengths. Simple, transparent, fast, and easy to audit.

Limitations. Sensitive to illumination, seasonal variation, and land cover. Vegetation and urban areas challenge fixed thresholds.

7.2 Change Detection

Change detection isolates event-related signals by comparing pre- and post-event imagery. In SAR, log-ratio or difference of dB composites is robust to calibration and terrain effects. In optical data, differencing NDWI/MNDWI or using short-term composites highlights newly inundated areas even under partial cloud cover. Thresholds for change images can again be set via Otsu or quantiles. Incorporating proximity to rivers or water occurrence maps helps distinguish floods from permanent water.

Strengths: Reduces confusion from persistent bright/dark surfaces.

Limitations: Requires suitable pre-event imagery and careful choice of windows.

7.3 Object-Based Approaches

Object-based image analysis segments imagery into homogeneous regions and classifies the resulting objects rather than individual pixels. This reduces salt-and-pepper noise and allows richer features (texture, shape, context). On GEE, Simple Non-Iterative Clustering (SNIC) provides efficient segmentation. Object-based methods can be paired with thresholding or machine learning for robust performance, especially in heterogeneous urban landscapes.

7.4 Supervised Machine Learning

Random Forests (RF) and Support Vector Machines (SVM) are widely used for flood mapping. Features include spectral bands, indices, SAR polarizations and ratios, textures (e.g., GLCM), and ancillary layers (HAND, slope, distance to rivers). Properly tuned models can generalize across moderate variations in sensor conditions and land cover. However, label quality and representativeness are critical. Sampling should be stratified, cover the full range of conditions, and reserve spatially independent areas for validation.

Strengths: Leverages multi-source features and non-linear decision boundaries; retains explainability (feature importance).

Limitations: Requires labeled data and careful validation to avoid optimistic estimates.

7.5 Deep Learning

Convolutional neural networks—especially encoder–decoder architectures such as UNet—have advanced water segmentation accuracy by modeling spatial context and multi-scale features. In GEE, deep learning is typically performed outside the platform (e.g., in Python with TensorFlow or PyTorch) using tiles exported from GEE, then predictions are re-ingested. Transfer learning and data augmentation help when labels are limited. Emerging transformer-based models show promise for large-area generalization, though operational complexity is higher and inference latency matters during disasters.

Strengths: State-of-the-art accuracy, especially in complex scenes.

Limitations: Training data demands, compute cost, and potential brittleness across domains.

7.6 Specialized Cues: Coherence, Ratios, and Multi-sensor Fusion

Where available, SAR coherence from single-look complex (SLC) pairs detects changes in surface structure, useful under vegetation and in urban areas. VV/VH ratios reduce incidence-angle and terrain effects. Fusing optical and SAR mitigates each sensor's weaknesses: optical for open water delineation, SAR for cloud-covered or vegetated areas. Simple decision rules (e.g., union of confident detections) often outperform more complex fusion schemes during time-critical operations.

8 Sampling Strategies

Validation begins with representative samples of flooded and non-flooded pixels or objects. Stratified sampling balances classes and ensures coverage across land-cover types, elevations, and distances from rivers. When labels are derived from manual interpretation, a double-blind protocol reduces bias. Crowdsourced or opportunistic data (e.g., road passability, social media geolocation) can supplement imagery-based labels with caution.

8.1 Spatial Cross-Validation

Random splits inflate accuracy because neighboring pixels share characteristics (spatial autocorrelation). Spatial cross-validation—holding out entire tiles or geographic zones—provides a more realistic assessment of generalization. K-fold spatial CV rotates the held-out regions; leave-one-event-out tests temporal transfer.

8.2 Metrics and Reporting

For binary maps, report Precision, Recall, F1-score (or Intersection-over-Union), and overall accuracy alongside class prevalence. Receiver-operating curves and precision–recall curves are informative when models produce probabilities. For polygon products, area-based agreements and hit/false-alarm rates complement pixel metrics. Always include a confusion matrix and disclose any post-processing that affects counts (e.g., size filters).

8.3 Uncertainty and Sensitivity

Thresholds, compositing windows, and masks influence outcomes. Simple sensitivity analyses—varying thresholds within plausible ranges, expanding/contracting morphological kernels, or toggling HAND/slope constraints—reveal robustness. Ensembles (e.g., multiple thresholds or models) provide uncertainty bounds useful for decision-makers (Table 2).

Table 2 Common datasets used for flood mapping in GEE

Satellite datasets	Resolution (m)
Sentinel-1 SAR	10
Sentinel-2	10/20
Landsat 7 ETM+	30
Landsat 8 OLI/TIRS	30

9 Operationalization on GEE

9.1 Workflow Design

Operational mapping emphasizes robustness and speed. Begin with a baseline RF or change-detection pipeline that runs quickly over the full area of interest. Design tiling to respect memory and task quotas; prefer server-side reducers and avoid large client-side arrays. Parameterize thresholds, windows, and masks so they can be tuned per event without rewriting code.

9.2 Exports and Quotas

Batch exports to Drive or Cloud Storage should be planned with realistic max Pixels and scale settings. When producing polygons, dissolve small patches and snap to grid to reduce vector complexity. For sustained operations, promote stable assets (e.g., a floodplain mask) and cache intermediate composites.

9.3 Dissemination and Reproducibility

GEE Apps provide a lightweight way to share interactive maps with responders. Include a legend, version/date, brief method text, and a download link. Share code links with pinned parameters and a changelog so collaborators can reproduce results. When possible, publish training data footprints and validation points with appropriate privacy safeguards.

9.4 Ethics and Communication

Map users may treat outputs as authoritative; communicate uncertainty clearly. Avoid over-confidence in areas with known limitations (dense vegetation, urban canyons, steep terrain). Document data licensing and acknowledge sources.

10 Validation and Accuracy Assessment

Remote sensing and geospatial technologies are now widely used in flood mapping and disaster management efforts. GEE is a platform that is being utilized as a tool for creating and examining flood maps. It is essential to validate these flood maps to guarantee their precision and dependability. Recent research has delved into approaches for validating the flood maps produced through Google Earth Engine. These techniques frequently include utilizing real-world data like observations on flood extent or measurements on water levels to evaluate how well the flood mapping algorithms perform (Nevo et al., 2022; Arora et al., 2021). For example, one study utilized a structure-from-motion technique to extract flood watermarks from remote sensing data, which were then used to validate the flood maps generated in Google Earth Engine (Meesuk et al., 2017; Schnebele et al., 2014). Other researchers have emphasized the importance of incorporating additional data sources, such as urban topography, drainage networks, and river channel layouts, to enhance the accuracy of the flood models and their corresponding maps. (Meesuk et al., 2017) Additionally, the use of machine learning techniques, such as Long Short-Term Memory networks and Linear models, has been explored for improving the performance of flood forecasting and mapping systems. (Nevo et al., 2022). Furthermore, the literature highlights the need to consider not just the accuracy of the flood maps, but also the risk assessment and mitigation strategies. Fusion of satellite imagery and social media data has been proposed to enhance the timeliness and accuracy of disaster response and decision-making (Gialampoukidis et al., 2021). The accuracy of classification models in geospatial applications is a critical consideration, as it directly impacts the reliability and trustworthiness of the derived information. In the context of Google Earth Engine, a popular cloud-based geospatial analysis platform, researchers have employed various accuracy metrics to assess the performance of their models. (Tiecke et al., 2017). One of the most used accuracy metrics is the confusion matrix, which provides a comprehensive overview of a model's performance by comparing the predicted classes with the true classes (Hossin & Sulaiman, 2015). The confusion matrix allows for the calculation of various derived metrics, such as precision, recall, and the popular F1-score, which balance the trade-offs between false positives and false negatives (Raschka, 2014). The Kappa coefficient is another important evaluation measure, often used to assess how well predicted classes align with the actual ones while accounting for agreement that could occur by chance. This measure is especially valuable in multi-class classification tasks, where relying solely on accuracy might overlook key aspects of performance. In the case of binary classification, tools like the receiver operating characteristic (ROC) curve and the area under the curve (AUC) are commonly applied. These methods highlight the balance between the true positive rate and the false positive rate, providing a clearer understanding of the model's ability to distinguish between different classes.

11 Challenges and Limitations

Mapping floods accurately is crucial in areas with gentle slopes where precise elevation data is vital for predicting flooding risks effectively. Creating maps of flash flood hazards faces challenges due to insufficient information about small rivers and how they flood. In the realm of data analysis and algorithm development lie hurdles that experts and professionals encounter to uncover valuable insights from intricate datasets and create efficient computing solutions. One key obstacle involves data resolution. The level of detail and precision in the data can greatly affect the dependability and authenticity of analytical discoveries (Labrinidis & Jagadish, 2012; Kumar & Khosla, 2018). With the increasing amount and complexity of data, in the domain of "Big Data," it is becoming more crucial to handle and analyze high-dimensional data as it grows (Fan et al., 2014; Labrinidis & Jagadish, 2012). A key hurdle faced is the resolution of data. The diverse resolutions of satellite imagery and geospatial data on the platform can result in discrepancies and inaccuracies in the end outcomes. To address this issue effectively, researchers need to choose and prepare the data to maintain a standard of quality and detail. Moreover, dealing with datasets in the Google Earth Engine platform can present an obstacle due to computational constraints. While the platform's cloud-based system enables processing of large-scale calculations, there are still restrictions concerning computing capacity and memory usage. Researchers need to plan out their algorithms and processes to make sure they are using resources efficiently and steering clear of any obstacles, as pointed out by (Amani et al. 2020; Gorelick et al., 2017). Accuracy is another hurdle in algorithm design challenges. Researchers must pay attention to how the algorithms perform, particularly when applied in the Google Earth Engine platform, as it greatly impacts the outcome quality. Evaluating performance and tweaking it as needed is crucial for ensuring accurate and dependable results.

12 Future Directions

Floods pose a significant threat to infrastructure, human lives, and the environment in many parts of the world. To mitigate the impact of floods and effectively plan for future scenarios, it is crucial to have reliable and accurate flood susceptibility mapping models (Rehana et al., 2020). Flood control plays a critical role in urban security management, especially in rapidly urbanizing regions experiencing the effects of climate change (Zhou et al., 2022). By combining flood mapping with climate forecasts and projections, scientists and decision makers can anticipate and get ready for the consequences of flooding, which may worsen due to the evolving global climate conditions. The growing accessibility of satellite imagery, coupled with powerful cloud-based computational platforms such as Google Earth Engine, has further democratized the development of advanced flood mapping algorithms, allowing for real-time analysis across diverse geographical scales. This has

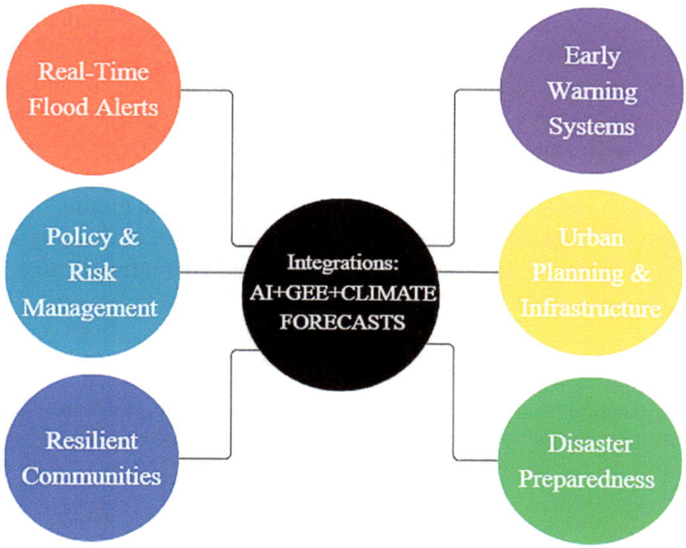

Fig. 6 Future directions of flood mapping with GEE

significantly propelled the application of deep learning methods, which, compared to traditional approaches, offer enhanced accuracy and speed for flood mapping by leveraging inductive biases to better process the spatial characteristics of flooding events. As both aspects are merged, it results in hydrological models that facilitate more precise predictions about flood patterns, scale, and severity levels. This data can be utilized to create flood prevention plans, like improving flood structures and early warning systems while implementing land use regulations to reduce flooding consequences (Afzal et al., 2022; Jha & Afreen, 2020; Zhong et al., 2014). It is essential to integrate flood mapping and climate forecasts into models for risk management and mitigation. Mapping floods accurately can offer information on the scale and impact of flooding events to guide strategies for managing floods more effectively and efficiently (Fig. 6).

13 Conclusions

The Google Earth Engine is now widely recognized as a geospatial analysis tool that is transforming how experts and professionals address flood mapping and monitoring. Historically speaking, evaluating flood risks, and creating maps used to be based on fixed Digital Elevation Models sourced from outlets, like sensing and geographic information systems. The rising occurrence of rainfalls and sea level elevation, along with environmental obstacles, has brought attention to the importance of advanced flood monitoring systems. A notable benefit of GEE is its capability to utilize Earth observation data, like satellite images and sensor networks, to produce flood maps instantly. Google has created a flood prediction system to give timely flood alerts to authorities and the public for river floods in major rivers with gauges in place. The system uses machine learning models for predicting water levels and flooding patterns while also sending out alerts promptly for flood management strategies. Additionally, the combination of GEE with technologies, like GIS and satellite image analysis, has allowed researchers and city planners to carry out thorough assessments of flood risks. The evaluations consider more than the extent of flooding in a region; they also factor in the susceptibility of individuals and vital infrastructure, as well as economic progress, in the impacted zones.

Acknowledgements The authors express their gratitude to the Hon'ble Vice-Chancellor of Kerala University of Fisheries and Ocean Studies (KUFOS), the Dean of the Faculty of Ocean Science and Technology, and the Head of the Department of Climate Variability and Aquatic Ecosystems at KUFOS for their permission and support in providing the facilities required to conduct this study.

References

Abdrabo, K. I., Kantoush, S. A., Saber, M., Sumi, T., Habiba, O. M., Elleithy, D., & Elboshy, B. (2020). Integrated methodology for urban flood risk mapping at the microscale in ungauged regions: A case study of Hurghada, Egypt. *Remote Sensing, 12*(21), 3548.

Abutu, A. (2010). Google Earth Engine launched in Cancun. *SciDev.net-Agriculture.*

Afzal, M. A., Ali, S., Nazeer, A., Khan, M. I., Waqas, M. M., Aslam, R. A., Cheema, M. J. M., Nadeem, M., Saddique, N., Muzammil, M., & Shah, A. N. (2022). Flood inundation modeling by integrating HEC–RAS and satellite imagery: A case study of the Indus River Basin. *Water, 14*(19), 2984.

Ahmad, I., Farooq, R., Ashraf, M., Waseem, M., & Shangguan, D. (2025). Improving flood hazard susceptibility assessment by integrating hydrodynamic modeling with remote sensing and ensemble machine learning. *Natural Hazards*, 1–30.

Ahmadisharaf, E., Kalyanapu, A. J., & Chung, E. S. (2017). Sustainability-based flood hazard mapping of the Swannanoa River watershed. *Sustainability, 9*(10), 1735.

Ali, K., Bajracharyar, R. M., & Raut, N. (2017). Advances and challenges in flash flood risk assessment: A review. *Journal of Geography & Natural Disasters, 7*(2), 1–6.

Amani, M., Ghorbanian, A., Ahmadi, S. A., Kakooei, M., Moghimi, A., Mirmazloumi, S. M., Moghaddam, S. M. H., Mahdavi, S., Ghahremanloo, M., Parsian, S., Wu, Q., & Brisco, B. (2020). Google Earth Engine cloud computing platform for remote sensing big data applications: A comprehensive review. *IEEE Journal of Selected Topics in Applied Earth Observations and Remote Sensing, 13*, 5326–5350.

Antzoulatos, G., Kouloglou, I. O., Bakratsas, M., Moumtzidou, A., Gialampoukidis, I., Karakostas, A., ... & Kompatsiaris, I. (2022). Flood hazard and risk mapping by applying an explainable machine learning framework using satellite imagery and GIS data. *Sustainability, 14*(6), 3251.

Arora, A., Arabameri, A., Pandey, M., Siddiqui, M. A., Shukla, U. K., Bui, D. T., Mishra, V. N., & Bhardwaj, A. (2021). Optimization of state-of-the-art fuzzy-metaheuristic ANFIS-based machine learning models for flood susceptibility prediction mapping in the Middle Ganga Plain, India. *Science of the Total Environment, 750*, 141565.

Atchyuth, B. A. S., Swain, R., & Das, P. (2023). Near real-time flood inundation and hazard mapping of Baitarani River Basin using Google Earth Engine and SAR imagery. *Environmental Monitoring and Assessment, 195*(11), 1331.

Azizat, N., & Omar, W. M. S. W. (2018). Assessment of three flood hazard mapping methods: A case study of Perlis. In *E3S web of conferences* (Vol. 34, p. 02028). EDP Sciences.
Ban, H. J., Kwon, Y. J., Shin, H., Ryu, H. S., & Hong, S. (2017). Flood monitoring using satellite-based RGB composite imagery and refractive index retrieval in visible and near-infrared bands. *Remote Sensing, 9*(4), 313.
Bentivoglio, R., Isufi, E., Jonkman, S. N., & Taormina, R. (2022). Deep learning methods for flood mapping: A review of existing applications and future research directions. *Hydrology and Earth System Sciences Discussions, 2022*, 1–50.
Bhere, S., & Reddy, M. J. (2022). *Multivariate flood risk assessment of the unplanned semi-urban region by incorporating flood hazard, vulnerability, and exposure.*
Breznik, J., Oštir, K., & Rak, G. (2025). The potential of Sentinel-1 imagery for flood event detection: a satellite vs. hydraulic model comparison. *Journal of Hydrology, 651*, 132587.
Chen, Z., & Zhao, S. (2022). Automatic monitoring of surface water dynamics using Sentinel-1 and Sentinel-2 data with Google Earth Engine. *International Journal of Applied Earth Observation and Geoinformation, 113*, 103010.
Chini, M., Hostache, R., Giustarini, L., & Matgen, P. (2017). A hierarchical split-based approach for parametric thresholding of SAR images: Flood inundation as a test case. *IEEE Transactions on Geoscience and Remote Sensing, 55*(12), 6975–6988.
Fan, J., Han, F., & Liu, H. (2014). Challenges of big data analysis. *National Science Review, 1*(2), 293–314.
Farhadi, H., & Najafzadeh, M. (2021). Flood risk mapping by remote sensing data and random forest technique. *Water, 13*(21), 3115.
Ganjirad, M., & Delavar, M. R. (2023). Flood risk mapping using random forest and support vector machine. *ISPRS Annals of the Photogrammetry, Remote Sensing and Spatial Information Sciences, 10*, 201–208.
Gasica, T. A., Bioresita, F., & Murtiyoso, A. (2020). Identification of temporary surface water using sentinel-1 Sar data, case study: Sentani flash flooding, Indonesia. *The International Archives of the Photogrammetry, Remote Sensing and Spatial Information Sciences, 43*, 55–59.
Gialampoukidis, I., Andreadis, S., Vrochidis, S., & Kompatsiaris, I. (2021, July). Multimodal data fusion of social media and satellite images for emergency response and decision-making. In *In 2021 IEEE International geoscience and remote sensing symposium IGARSS* (pp. 228–231). IEEE.
Gorelick, N., Hancher, M., Dixon, M., Ilyushchenko, S., Thau, D., & Moore, R. (2017). Google Earth Engine: Planetary-scale geospatial analysis for everyone. *Remote Sensing of Environment, 202*, 18–27.
Haces-Garcia, F., Maslennikova, N., Glennie, C. L., Rifai, H. S., Hoskere, V., & Ekhtari, N. (2023). *Deep learning hydrodynamic forecasting for flooded region assessment in near-real-time (DL hydro-FRAN).* arXiv preprint https://arxiv.org/abs/2305.12052
Hallegatte, S., Bangalore, M., & Vogt-Schilb, A. (2016). Assessing socioeconomic resilience to floods in 90 countries. *World Bank Policy Research Working Paper, 7663.*
Hocini, N., Payrastre, O., Bourgin, F., Gaume, E., Davy, P., Lague, D., Poinsignon, L., & Pons, F. (2020). Performance of automated flood inundation mapping methods in a context of flash floods: A comparison of three methods based either on the height above nearest drainage (HAND) concept, or on 1D/2D shallow water equations. *Hydrology and Earth System Sciences Discussions, 2020*, 1–23.
Hossin, M., & Sulaiman, M. N. (2015). A review on evaluation metrics for data classification evaluations. *International Journal of Data Mining & Knowledge Management Process, 5*(2), 1.
Islam, M. T., & Meng, Q. (2022). An exploratory study of Sentinel-1 SAR for rapid urban flood mapping on Google Earth Engine. *International Journal of Applied Earth Observation and Geoinformation, 113*, 103002.
Jha, M. K., & Afreen, S. (2020). Flooding urban landscapes: Analysis using combined hydrodynamic and hydrologic modeling approaches. *Water, 12*(7), 1986.

Jia, W., Liang, B., Lu, Y., Khan, M. A., & Zheng, L. (2025, June). A comprehensive survey on deep learning solutions for 3D flood mapping. In *Pacific-Asia conference on knowledge discovery and data mining* (pp. 21–38). Springer Nature Singapore.

Jiang, X., Liang, S., He, X., Ziegler, A. D., Lin, P., Pan, M., Wang, D., Zou, J., Hao, D., Mao, G., Zeng, Y., Yin, J., Feng, L., Miao, C., Wood, E. F., & Zeng, Z. (2021). Rapid and large-scale mapping of flood inundation via integrating spaceborne synthetic aperture radar imagery with unsupervised deep learning. *ISPRS Journal of Photogrammetry and Remote Sensing, 178*, 36–50.

Johary, R., Révillion, C., Catry, T., Alexandre, C., Mouquet, P., Rakotoniaina, S., Pennober, G., & Rakotondraompiana, S. (2023). Detection of large-scale floods using Google Earth Engine and Google Colab. *Remote Sensing, 15*(22), 5368.

Kumar, D., & Shankar, A. (2024). Novel remote sensing technologies for natural hazard management. *Environmental Science and Pollution Research, 31*(41), 53743–53748.

Kumar, L., & Mutanga, O. (2018). Google earth engine applications since inception: Usage, trends, and potential. *Remote Sensing, 10*(10), 1509.

Kumar, V., & Khosla, C. (2018, January). Data cleaning-a thorough analysis and survey on unstructured data. In *In 2018 8th International conference on cloud computing, Data Science & Engineering (Confluence)* (pp. 305–309). IEEE.

Kwak, Y., Park, J., Arifuzzaman, B., Iwami, Y., Amirul, M., & Kondoh, A. (2016). Rapid exposure assessment of Nationwide River flood for disaster risk reduction. *The International Archives of the Photogrammetry, Remote Sensing and Spatial Information Sciences, 41*, 1357–1362.

Labrinidis, A., & Jagadish, H. V. (2012). Challenges and opportunities with big data. *Proceedings of the VLDB Endowment, 5*(12), 2032–2033.

Leggesse, E. S., Derseh, W. A., Zimale, F. A., Tilahun, S. A., & Meshesha, M. A. (2024). Urban flash flood hazard mapping using machine learning, Bahir Dar, Ethiopia. *Journal of Hydroinformatics, 26*(9), 2124–2145.

Liu, Z., Coleman, N., Patrascu, F. I., Yin, K., Li, X., & Mostafavi, A. (2024). Artificial intelligence for flood risk management: A comprehensive state-of-the-art review and future directions. *International Journal of Disaster Risk Reduction, 117*, 105110.

Lo, S. W., Wu, J. H., Chang, J. Y., Tseng, C. H., Lin, M. W., & Lin, F. P. (2021). Deep sensing of urban waterlogging. *IEEE Access, 9*, 127185–127203.

Meesuk, V., Vojinovic, Z., & Mynett, A. E. (2017). Extracting inundation patterns from flood watermarks with remote sensing SfM technique to enhance urban flood simulation: The case of Ayutthaya, Thailand. *Computers, Environment and Urban Systems, 64*, 239–253.

Misra, A., White, K., Nsutezo, S. F., Straka, W., III, & Lavista, J. (2025). Mapping global floods with 10 years of satellite radar data. *Nature Communications, 16*(1), 5762.

Moharrami, M., Javanbakht, M., & Attarchi, S. (2021). Automatic flood detection using Sentinel-1 images on the Google Earth Engine. *Environmental Monitoring and Assessment, 193*(5), 248.

Morales, C., Díaz, A. S. P., Dionisio, D., Guarnieri, L., Marchi, G., Maniatis, D., & Mollicone, D. (2023). Earth map: A novel tool for fast performance of advanced land monitoring and climate assessment. *Journal of Remote Sensing, 3*, 0003.

Mustain, P. N., Mardjono, A., Moe, I. R., Agustia, E., Aribawa, T. M., Prana, A., & Yanbing, B. (2021, July). Flood observation map in Banten Province using radar images, case study: Cidanau, Ciujung, Cidurian River basin. In *IOP conference series: Earth and environmental science* (Vol. 794, No. 1, p. 012067). IOP Publishing.

Nevo, S., Morin, E., Gerzi Rosenthal, A., Metzger, A., Barshai, C., Weitzner, D., et al. (2022). Flood forecasting with machine learning models in an operational framework. *Hydrology and Earth System Sciences, 26*(15), 4013–4032.

Nguyen, T. H., Ricci, S., Piacentini, A., Fatras, C., Kettig, P., Blanchet, G., Luque, S. P., & Baillarin, S. (2023). Assimilation of SAR-derived flood extent observations for improving fluvial flood forecast–A proof-of-concept. In *IOP conference series: Earth and environmental science* (Vol. 1136, No. 1, p. 012018). IOP Publishing.

Osti, R., Tanaka, S., & Tokioka, T. (2008). Flood hazard mapping in developing countries: Problems and prospects. *Disaster Prevention and Management: An International Journal, 17*(1), 104–113.

Pinheiro, G., Raj, A., Minz, S., Choudhury, T., & Um, J. S. (2024). Inundation extend mapping for multi-temporal SAR using automatic thresholding and change detection: A case study on Kosi river of India. *Spatial Information Research, 32*(3), 311–325.

Priyatna, M., Khomarudin, M. R., Wijaya, S. K., Yulianto, F., Nugroho, G., Afgatiani, P. M., Rarasati, A., & Hussein, M. A. (2023). Rapid flood mapping using statistical sampling threshold based on Sentinel-1 imagery in the Barito watershed, South Kalimantan Province, Indonesia. *Journal of Engineering & Technological Sciences, 55*(1), 98–107.

Raschka, S. (2014). *An overview of general performance metrics of binary classifier systems.* arXiv preprint https://arxiv.org/abs/1410.5330

Rehana, S., Rajulapati, C. R., Ghosh, S., Karmakar, S., & Mujumdar, P. (2020). Uncertainty quantification in water resource systems modeling: Case studies from India. *Water, 12*(6), 1793.

Romali, N. S., Yusop, Z., & Ismail, A. Z. (2018, March). Hydrological modelling using HEC-HMS for flood risk assessment of Segamat town, Malaysia. In *IOP conference series: Materials science and engineering* (Vol. 318, No. 1, p. 012029). IOP Publishing.

Sajjad, A., Lu, J., Aslam, R. W., & Ahmad, M. (2023). Flood disaster mapping using geospatial techniques: a case study of the 2022 Pakistan floods. *Environmental Sciences Proceedings, 25*(1), 78.

Sajjad, A., Lu, J., Chen, X., Chisenga, C., Saleem, N., & Hassan, H. (2020). Operational monitoring and damage assessment of riverine flood-2014 in the lower Chenab plain, Punjab, Pakistan, using remote sensing and GIS techniques. *Remote Sensing, 12*(4), 714.

Saksena, S. (2015). Investigating the role of DEM resolution and accuracy on flood inundation mapping. In World environmental and water resources congress 2015 (pp. 2236-2243).

Schnebele, E., Cervone, G., Kumar, S., & Waters, N. (2014). Real time estimation of the Calgary floods using limited remote sensing data. *Water, 6*(2), 381–398.

Shelestov, A., Kolotii, A., Skakun, S., Baruth, B., Lozano, R. L., & Yailymov, B. (2017). Biophysical parameters mapping within the SPOT-5 take 5 initiative. *European Journal of Remote Sensing, 50*(1), 300–309.

Smith, K., & Ward, R. (1998). *Floods: Physical processes and human impacts* (pp. xii–382).

Suab, S. A., Supe, H., Avtar, R., Dambul, R., & Chen, X. (2022, July). Detection and mapping of May 2021 flood in Beaufort, Sabah using Sentinel-1 SAR and Sentinel-2 multispectral in Google Earth Engine. In *IOP conference series: Earth and environmental science* (Vol. 1064, No. 1, p. 012003). IOP Publishing.

Tamiminia, H., Salehi, B., Mahdianpari, M., Quackenbush, L., Adeli, S., & Brisco, B. (2020). Google Earth Engine for geo-big data applications: A meta-analysis and systematic review. *ISPRS Journal of Photogrammetry and Remote Sensing, 164*, 152–170.

Tariq, A., Shu, H., Kuriqi, A., Siddiqui, S., Gagnon, A. S., Lu, L., ... & Pham, Q. B. (2021). Characterization of the 2014 indus river flood using hydraulic simulations and satellite images. *Remote Sensing, 13*(11), 2053.

Tiecke, T. G., Liu, X., Zhang, A., Gros, A., Li, N., Yetman, G., Kilic, T., Murray, S., Blankespoor, B., Prydz, E. B., & Dang, H. A. H. (2017). *Mapping the world population one building at a time.* arXiv preprint https://arxiv.org/abs/1712.05839

Tobón-Marín, A., & Canon Barriga, J. (2020). Analysis of changes in rivers planforms using Google Earth Engine. *International Journal of Remote Sensing, 41*(22), 8654–8681.

Trinh, M. X., & Molkenthin, F. (2021). Flood hazard mapping for data-scarce and ungauged coastal river basins using advanced hydrodynamic models, high temporal-spatial resolution remote sensing precipitation data, and satellite imageries. *Natural Hazards, 109*(1), 441–469.

Tsyganskaya, V., Martinis, S., Marzahn, P., & Ludwig, R. (2018). Detection of temporary flooded vegetation using Sentinel-1 time series data. *Remote Sensing, 10*(8), 1286.

Tumaneng, R. D., Morico, K. E. R., & Principe, J. A. (2024). REMOTE SENSING AND GOOGLE EARTH ENGINE FOR RAPID FLOOD MAPPING AND DAMAGE ASSESSMENT: A CASE

OF TYPHOON GONI (ROLLY) AND VAMCO (ULYSSES). *The International Archives of the Photogrammetry, Remote Sensing and Spatial Information Sciences, 48*, 445–452.

Vekaria, D., Chander, S., Singh, R. P., & Dixit, S. (2022). A change detection approach to flood inundation mapping using multi-temporal Sentinel-1 SAR images, the Brahmaputra River, Assam (India): 2015–2020. *Journal of Earth System Science, 132*(1), 3.

Wedajo, G. K. (2017). LiDAR DEM data for flood mapping and assessment; opportunities and challenges: A review. *Journal of Remote Sensing & GIS, 6*(04).

Zeng, Q., Chen, H., Xu, C. Y., Jie, M. X., & Hou, Y. K. (2016). Feasibility and uncertainty of using conceptual rainfall-runoff models in design flood estimation. *Hydrology Research, 47*(4), 701–717.

Zhong, G., Liu, S., Han, C., & Huang, W. (2014). Urban flood maping for Jiaxing city based on hydrodynamic modeling and GIS analysis. *Journal of Coastal Research, 68*, 168–175.

Zhou, R., Zheng, H., Liu, Y., Xie, G., & Wan, W. (2022). Flood impacts on urban road connectivity in southern China. *Scientific Reports, 12*(1), 16866.

Zin, W. W., Kawasaki, A., Takeuchi, W., San, Z. M. L. T., Htun, K. Z., Aye, T. H., & Win, S. (2018). Flood hazard assessment of Bagoriver Basin, Myanmar. *Journal of Disaster Research, 13*(1), 14–21.

Cloud and Cloud Shadow Detection and Removal Using Landsat-8 Data and Generation of Cloud-Free Image in Beas River Basin of Himachal Pradesh

Gopinadh Rongali, Ashok K. Keshari, Ashvani K. Gosain, and Poonam Vishwas

Abstract Six successful Landsat missions have launched since 1972, contributing to nearly 50 years of earth observation to track the dynamics of land cover and change. On February 11, 2013, the Landsat Data Continuity Mission (LDCM), also known as Landsat-8, successfully launched, fulfilling its goal of collecting Earth imagery under an open data policy. Landsat-8 has two push-broom sensors. The first instrument, the Operational Land Imager (OLI), covers 185 km with 30 m of spatial resolution across nine shortwave spectral bands except for a 15-m panchromatic band. Two thermal bands—band 10 and band 11—were imaged by the Thermal Infrared Sensor (TIRS) spanning 185 km with a 100-m resolution. However, clouds and their shadows make it hard to see the ground surface and limit change pattern investigation. This study shows a new way to detect and remove clouds and cloud shadows using Landsat-8 image data (WRS2: Path/Row = 147/38, obtained on April 24, 2015). This approach employs six of the nine bands for alteration to compute the strength of clouds and cloud shadows, which are then eliminated. The spectral data from the different bands is used in this way. The validation indicates that the cloud shadow and polluted cloud pixels are accurately detected, achieving overall accuracies of 97 and 96%, respectively. Nonetheless, the efficacy of this approach is limited in the presence of thick clouds and cloud shadows. This technique should be improved to account for the weather and make it simpler to identify changes in the terrain.

Keywords Cloud · Cloud shadow · Landsat-8 · OLI · TIRS · LDCM · WRS2

G. Rongali (✉)
National Center for Medium Range Weather Forecasting (NCMRWF), MoES, Noida, India

A. K. Keshari · A. K. Gosain
Department of Civil Engineering, Indian Institute of Technology Delhi, New Delhi, India

P. Vishwas
Centre for the Study of Regional Development, School of Social Sciences, Jawaharlal Nehru University, New Delhi, India

R. Ahmed et al. (eds.), *Climate Change, Water Resources, and Natural Hazards in the Himalayas*, Advances in Geographical and Environmental Sciences,
https://doi.org/10.1007/978-981-92-0835-7_15

1 Introduction

Cloud is one of the challenging problems in snow cover mapping using remote sensing data. Clouds are generally present in the lower atmosphere of the earth in almost all the seasons around. Many open source data is available, but acquisition of the completely atmospheric disturbance free optical satellite imagery has been a challenging task. Cloud cover and corresponding shadow contamination are the major problems that are faced while using the optical remote sensing imagery for various applications Manually interpreted cloud and cloud shadow masks are the primary data source for the development and validation of cloud and cloud shadow identification algorithms (Foga et al., 2017; Hughes and Hayes, 2014; Irish et al., 2006; Li et al., 2022; Qiu et al., 2017; Zhu and Woodcock, 2012). Historically, the Automated Cloud Cover Assessment (ACCA) was used to determine cloud cover percentage in Landsat TM and ETM+ imagery (Irish, 2000; Irish et al., 2006). Utilizing many spectral filters, ACCA effectively estimates a cloud cover score for each image; nevertheless, it lacks precision in pinpointing the positions and borders of clouds (Zhu and Woodcock, 2012). Zhu and Woodcock (2012) developed an F-mask (Function of mask) technique that identifies clouds using a scene-based threshold. This technique may produce a cloud probability layer and is appropriate for the Landsat series 4–8 satellite data. To calculate cloud masks, users may modify the cloud probability threshold. The thermal band is a crucial input in the F-mask technique, since it can detect the "cold" characteristics of clouds (Zhu et al., 2015). However, owing to significant elevational fluctuations, the temperature for clear sky pixels might also differ significantly, which may result in commission and omission mistakes in cloud identification in mountainous places. To mitigate this problem, Qiu et al. (2017) introduced a Mountainous F-mask (MF-mask) technique that normalises the thermal band using Digital Elevation Models (DEMs) grounded on a straightforward linear temperature-elevation model.

Jin et al. (2013) detected clouds by integrating Landsat blue and thermal bands from images taken on two separate dates. Jin et al. (2013) identified cloud shadows by differencing the SWIR and thermal bands from two Landsat images and used the geometric correlation between clouds and their respective shadows to minimise false-positive errors. Goodwin et al. (2013) used LTS from TM/ETM+ for cloud detection. This approach outperforms F-mask in constructing superior cloud masks across Queensland by using the lowest and median values of the blue band as a reference (Zhu & Woodcock, 2012). Nonetheless, it has yet to be evaluated in habitats characterised by varying soils, plant cover, buildings, or regions with snow and ice cover (Goodwin et al., 2013). An algorithm named T-mask (multi-temporal mask) has been specifically developed for the automatic masking of clouds and snow in the monitoring of land cover change for LTS (Zhu & Woodcock, 2014). Hagolle et al. (2010) developed the multi-temporal cloud detection (MTCD) method, which employs time series observations to identify abrupt increases in reflectance in the blue band on a pixel-by-pixel basis, and evaluated the linear correlation of pixel neighbourhoods derived from sequentially acquired image pairs. Hughes and Hayes

(2014) investigated the use of spatial information as an input to a neural network classifier for the identification and classification of clouds in Landsat pictures. Zhou et al. (2016) used a conventional threshold to generate a rudimentary cloud mask and then applied the support vector machine (SVM) classifier to identify clouds in Landsat-8 imagery. While these investigations highlight the efficacy of machine-learning methods in cloud detection, most necessitate a specific understanding of cloud or surface conditions within the images (as training data) and often struggle to identify clouds under certain unique circumstances (Huang et al., 2010). Furthermore, cloud shadow in Landsat imagery may be identified via machine-learning methodologies (Chai et al., 2019; Hughes and Hayes, 2014; Potapov et al., 2011). This strategy, however, significantly depends on the training dataset and exhibits considerable omission or commission mistakes (Hughes & Hayes, 2014).

Zhu and Woodcock (2014) further detected cloud shadows for LTS by picture differencing. While few cloud shadow identification methodologies use multi-temporal Landsat data, these techniques provide superior outcomes compared to those reliant on a single-date Landsat image, particularly for shadows cast by thin clouds (Zhu & Woodcock, 2014). To mitigate misunderstanding between luminous surfaces and haze/cloud, Chen et al. (2015) introduced an iterative haze-optimized transformation (IHOT) to enhance haze/cloud recognition in Landsat images using a matched clear image. By including an iterative regression process into the HOT (Zhang et al., 2002), the reflectance disparity between hazy and clear images and the reflectance of both may facilitate the elimination of land surface information. Zhu and Woodcock (2012) and Zhu et al. (2015) computed cloud shadows by integrating prior shape-similarity matching with cloud-height estimating techniques, conceptualizing the cloud as a three-dimensional entity. This approach is excellent in most scenarios; however, it may struggle to accurately detect cloud shadows in regions with considerable topographic variation or terrain shadows. Qiu et al. (2017) used a double-projection approach to determine cloud shadow morphology and eliminated terrain shadows using a topographic correction model supported by DEMs. Additionally, using heights inferred from nearby clouds, this updated F-mask technique can estimate a cloud's height with greater accuracy. This enhances the ability of Landsat images to identify cloud shadows. This is a constraint in the remote sensing data. A resolution to this challenge is offered by the Landsat OLI data with the inclusion of a cloud detection band, referred known as the cirrus band. In our research, the strength of clouds and cloud shadows was determined by modifying six of the nine spectral bands, after which the clouds and cloud shadows were eliminated from the remote sensing satellite images. The technique use the spectral data of clouds. Landsat OLI data includes an additional cloud detection band, referred to as the cirrus band, for identifying clouds (Kong et al., 2013; Wang et al., 2019). In this study, we have transformed the nine spectral bands into six bands in order to determine the cloud and shadow intensities. The technique makes use of cloud-based spectral data.

Numerous research have focused on techniques for detecting and eliminating clouds. For instance, using a forward modelling method with a suitable radiative transfer model that accounts for specific atmospheric conditions and spectral bands, the impacts of clouds may be calculated and then removed from the observed data

(Berk et al., 1989, 1998). To implement this approach, a comprehensive dataset is obtained that records the atmospheric conditions of the remote sensing data at certain times and locations. From the nine spectral bands, two approaches employ six bands for modification to compute cloud intensity and cloud shadows, which are then eliminated. The cloud may be seen in a two-dimensional spectral space made up of visible bands (OLI2 and OIL4). The OLI 9 (cirrus) band detects just cloud and is unaffected by the kind of surface cover. The two developed approaches may be utilised to measure cloud effectively. Although cloud removal was difficult, the two strategies outlined in this study were successful in removing the thin cloud for OLI data.

2 Study Area

Figure 1 illustrates that the research region for this study encompasses the Beas River basin extending to the Pandoh dam. The Beas River basin is situated between 31° N and 32° N latitude and 77° E and 78° E longitude. It is an essential element of the Indus River system. It begins at 3900 meters in the Rothang Pass and runs mostly in a north-south direction until it arrives to Larji. At Larji, it executes a right turn and proceeds south-west, maintaining that trajectory for the remainder of the route. The Beas River extends 116 km from Pandoh upstream. The tributaries are the glacier-fed Parbati and Sainj Khad Rivers. The catchment area of the Beas River up to the Pandoh dam is 5383 km^2. The catchment region mostly consists of steep inclines including exposed rock formations. Elevated summits are seen to the east and north of the river valley. The Parbati sub-altitude catchment extends from 846 m at the Pandoh dam to 6500 m at the northeastern edge. In winter, a substantial portion of the River is obscured by snow. The primary source of water for the Beas River throughout the summer is snowmelt. The primary purpose of the Pandoh dam is to redirect water from the Beas River to the Satluj River for electricity generation at the Dehar power station located on the right bank of the Satluj River.

3 Data Used

The spectral bands of the Landsat-8 OLI sensor used to collect data for the Beas River basin are used in this investigation. As of April 24, 2015, detection and removal of clouds and cloud shadows using data from the Landsat-8 satellite. Table 1 enumerates the reflective spectral band characteristics for OLI and ETM+. Software such as ERDAS IMAGINE 9.2, ArcGIS 10.3, and ENVI 4.8 were used to analyse Landsat-8 data.

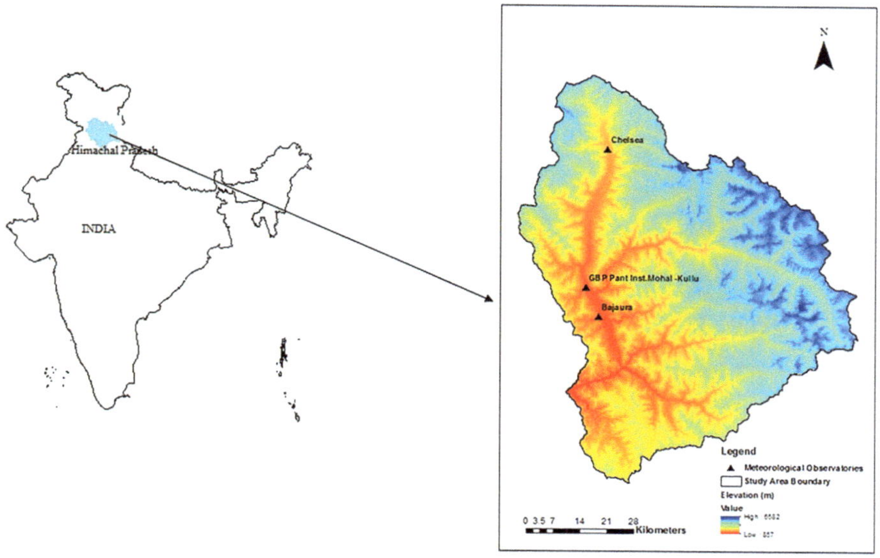

Fig. 1 Locations of the hydro-meteorological stations are shown on the ASTER-DEM

Table 1 Characteristics of OLI and ETM+ Reflective Spectral Bands (Barsi et al., 2014)

OLI spectral bands					ETM+ spectral bands				
Band number	Band name	Spectral range (nm)	GSD (m)	Bit	Band number	Band name	Spectral range (nm)	GSD (m)	Bit
1	Blue	433–453	30	16	1	Blue	450–515	30	8
2	Blue	450–515	30	16	2	Green	525–605	30	8
3	Green	525–600	30	16	3	Red	630–690	30	8
4	Red	630–680	30	16	4	NIR	775–900	30	8
5	NIR	845–885	30	16	5	SWIR	1550–1750	30	8
6	SWIR 2	1560–1660	30	16	6	SWIR	2090–2350	30	8
7	SWIR 3	2100–2300	30	16	7	Pan	520–900	15	8
8	Pan	500–680	15	16					
9	Cirrus	1360–1390	30	16					

3.1 *Landsat-8 OLI Band Characters*

A specific scene (WRS2: Path/Row = 147/38) has been chosen to illustrate a typical surface cover while also including various meteorological conditions, ranging from clear skies to thin and dense cloud cover. The approaches compute the intensity of

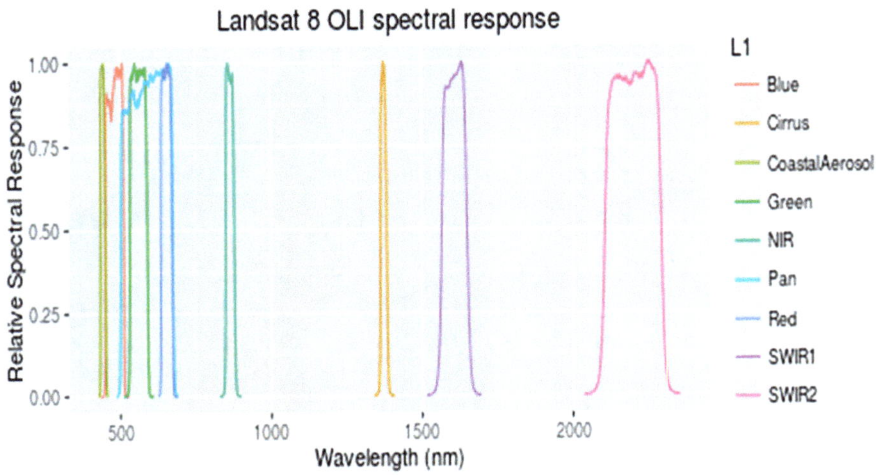

Fig. 2 Relative spectral response curves for each OLI sensors (Barsi et al., 2014)

clouds and cloud shadows by using six bands for transformation from nine spectral bands, while the remaining three bands are used for their exclusion from sequential Landsat OLI pictures. The 16-bit pixel value utilised in the Landsat-8 OLI Level-1 data products (Qcalmax = 65,535) allows for the detection of negative changes in the reflectance and radiance of the land. The Landsat-8 OLI's band and spectral properties were used to identify and remove clouds.

Landsat-8 OLI band characteristics are shown in Table 1. It delineates the prelaunch spectral range, ground sample distance (GSD), and bit depth of digital numbers (DN) for the OLI and ETM+ sensors. The Landsat OLI sensors' relative spectral response (RSR) profiles from prelaunch are seen in Fig. 2. The Landsat-7 sensors have the same spectral range as the OLI bands. It was created with the addition of two additional bands, 1 and 9, for monitoring high and thin clouds, as well as water quality.

4 Methodology

Land cover type can be monitored by using reflectance calculated from satellite image data. Two basic steps are used to calculate at-sensor spectral reflectance. Finding the spectral radiance at the sensor and converting it to the reflectance at the top of the atmosphere (TOA) come first.

4.1 At-Sensor Spectral Radiance Conversion

The Landsat OLI and TIRS band data are transformed into TOA spectral radiance using the radiance rescaling settings specified in the metadata file as follows:

$$L_\lambda = \left(\frac{LMAX_\lambda - LMIN_\lambda}{Q_{calmax} - Q_{calmin}}\right) X \ (Q_{cal} Q_{calmin}) + LMIN_\lambda \tag{1}$$

Or

$$L_\lambda = G_{rescale} X \ Q_{cal} + B_{rescale} \tag{2}$$

where

$$G_{rescale} = \frac{LMAX_\lambda - LMIN_\lambda}{Q_{calmax} - Q_{calmin}} \tag{3}$$

$$B_{rescale} = LMIN_\lambda - \frac{LMAX_\lambda - LMIN_\lambda}{Q_{calmax} - Q_{calmin}} X \ Q_{calmin} \tag{4}$$

where,

$L\lambda$ is top of atmospheric spectral radiance in watts/(m^2 × srad × μm).
$LMAX_\lambda$ is at-sensor spectral radiance scaled to Qcalmax in watts/(m^2 × srad × μm).
$LMIN_\lambda$is at-sensor spectral radiance scaled to Qcalmmin in watts/(m^2 × srad × μm).
Q_{cal} is standard product pixel values (DN) that have been quantized and calibrated.
Q_{calmax} is highest possible pixel value that corresponds to LMAXλ (DN).
Q_{calmin} is the equivalent minimum quantized and calibrated pixel value to LMINλ (DN).
$G_{rescale}$ is gains for spectral rescaling in watts/(m^2 × srad × μm)/DN.
$B_{rescale}$ is bias in spectral rescaling in watts/ (m^2 × srad × μm).

4.2 Conversion to TOA Reflectance

Converting DN value to TOA reflectance for the data of OLI is done by using the Eq. 5 which is different from the method used for other Landsat series data.

$$\rho_\lambda^{,} = \frac{M_\lambda Q_{cal} + A_\lambda}{(\cos\theta_s)} \tag{5}$$

where,

$\rho_\lambda^{,}$ is TOA planetary reflectance.

M_λ is a multiplicative rescaling factor derived from the information for each band.
A_λ is from the metadata, a band-specific additive rescaling factor.
Q_{cal} is standard product pixel values (DN) that have been quantized and calibrated.
θ_s is Solar zenith angle (degrees), which is determined by dividing the metadata's 90-sun elevation angle.

4.3 Cloud Spectral Reflectance Characteristics

In satellite images, it's difficult to determine the difference between high, thin, and thick clouds. Clouds, as well as their shadows, might obstruct measurements. The introduction of the new band in the Landsat-8 OLI sensors has simplified cloud measurement compared to older Landsat sensors. This band detects light in the electromagnetic spectrum when clouds are most prominent. The OLI 9 (cirrus band) parameter displays characteristics similar to those of MODIS band 26 (1.36–1.39 m), which is advantageous for revealing atmospheric conditions.

In the present study, the clouds have been detected based on the spectral reflectance characteristics of clouds. The cloud type can be identified by using the shape of the reflectance spectrum. The difference in the spectral reflectance properties of the cloud free zone and the cloud region were analyzed in this case study for the removal of the cloud and its shadow. Figure 3 displays a colour composite picture (a) made using Landsat OLI image data from April 24, 2015 (7, 4, 3 bands) and a spatial profile curve (b) generated using Eq. (5). The circles, regular triangles, inverted triangles, pentagons, and squares (filled with red) in Fig. 3a denote cloud shadow, cloud, soil, water, and vegetation, respectively. The dense cloud area is shown in the white segment of the false colour composite image, while the shadow zone is depicted in the adjacent black piece. Clouds, their shadow regions, and the other three forms of surface cover have the following spectral characteristics:

1. In all nine reflecting bands, the thick cloud zone's spectral reflectance value is much higher than that of the other cover types.
2. The spectral reflectance values for bands 4, 5, 7, and 9 are uniform and markedly diminished in aquatic environments and areas of cloud shadow.
3. Analysing OLI 9's spectral reflectance values in relation to other bands across all surface cover classifications. The mean value of five distinct feature categories is 5721, whereas the pixel value of the dense cloud in OLI 9 (Cirrus band) is 8380. The mean value of five distinct feature types is 1.9742, but the TOA reflectance of the thick cloud is 2.7157.

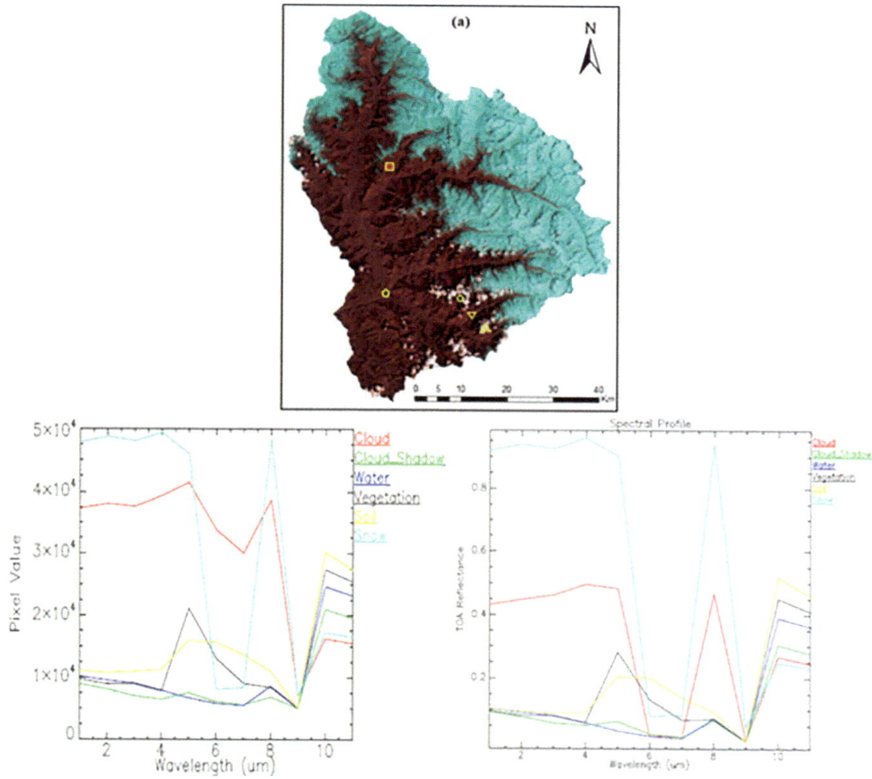

Fig. 3 On April 24, 2015, a Landsat-8 colour composite image derived from Landsat OLI data (**a**) illustrates the spectral characteristics of five land cover categories using pixel values (left) and TOA reflectance (right)

4.4 Clouds Recognition and Detection

Based on the above analysis, the algorithm for the cloud detection has been developed for the Landsat-8 OLI sensors. The following criteria has been used to measure the cloud using a single band.

$$OLI_9 \geq 5154\text{, or reflectance value} \tag{6}$$

$$\rho_9^{TOA} \geq 1.827 \tag{7}$$

where and stand for the reflectance value in the remote sensing image, and 9 stands for the OLI cirrus band.

The algorithm to detect cloud using more bands can be expressed as (Kong et al., 2013):

$$OLI_c = 3.07\ X\ OLI_2 - 2.77\ X\ OLI_3 - 2.64\ X\ OLI_4 - 2.53\ X\ OLI_5 - 2.74\ X\ OLI_6 - 2.61\ X\ OLI_7 \quad (8)$$

$$OLI_c \geq 23671 \quad (9)$$

where c is a cloud component.

4.5 Method for Removing Clouds

The scatterplot of DN values for each spectral band, together with the matching OLI cloud, showed a linear correlation between the two variables. The cloud was removed by altering the pixel values of the cloud area according to Eq. 6 (Kong et al., 2013):

$$OLI_{decloud} = \begin{bmatrix} OLI_2 \\ OLI_3 \\ OLI_4 \\ OLI_5 \\ OLI_6 \\ OLI_7 \end{bmatrix} - \begin{bmatrix} 1.88 \\ 0.89 \\ 1.02 \\ 0.85 \\ 1.40 \\ 0.71 \end{bmatrix} X\ OLI_{cloud} \quad (10)$$

OLIcloud represents cloud pixel values as defined by Eqs. (6) or (8), depending on the context, while OLIdecloud signifies the image data post-cloud removal. The subscripts ranging from 2 to 7 correspond to the OLI band names.

5 Results and Discussion

5.1 Cloud Detection Performance

The suggested methodology for cloud detection with Landsat-8 OLI data was evaluated in the Beas River basin (Path/Row 147/38), encompassing various surface conditions including snow, vegetation, bare soil, and water bodies. The findings (Fig. 4) indicate that both the single-band (cirrus band—OLI 9) and multi-band thresholding methods were successful in identifying cloud-contaminated pixels. The cirrus band demonstrated significant sensitivity to both optically thick and thin clouds owing to its spectral placement in the 1.36–1.39 μm range, characterized by little surface reflectance and elevated cloud reflectance. The algorithm effectively differentiated cloud pixels from surface characteristics, exhibiting minimal misunderstanding with bright targets like snow and urban locations. The reflectance criteria ($\rho_9 \geq 1.827$) and the DN threshold ($DN \geq 5154$) yielded uniform findings throughout the study area.

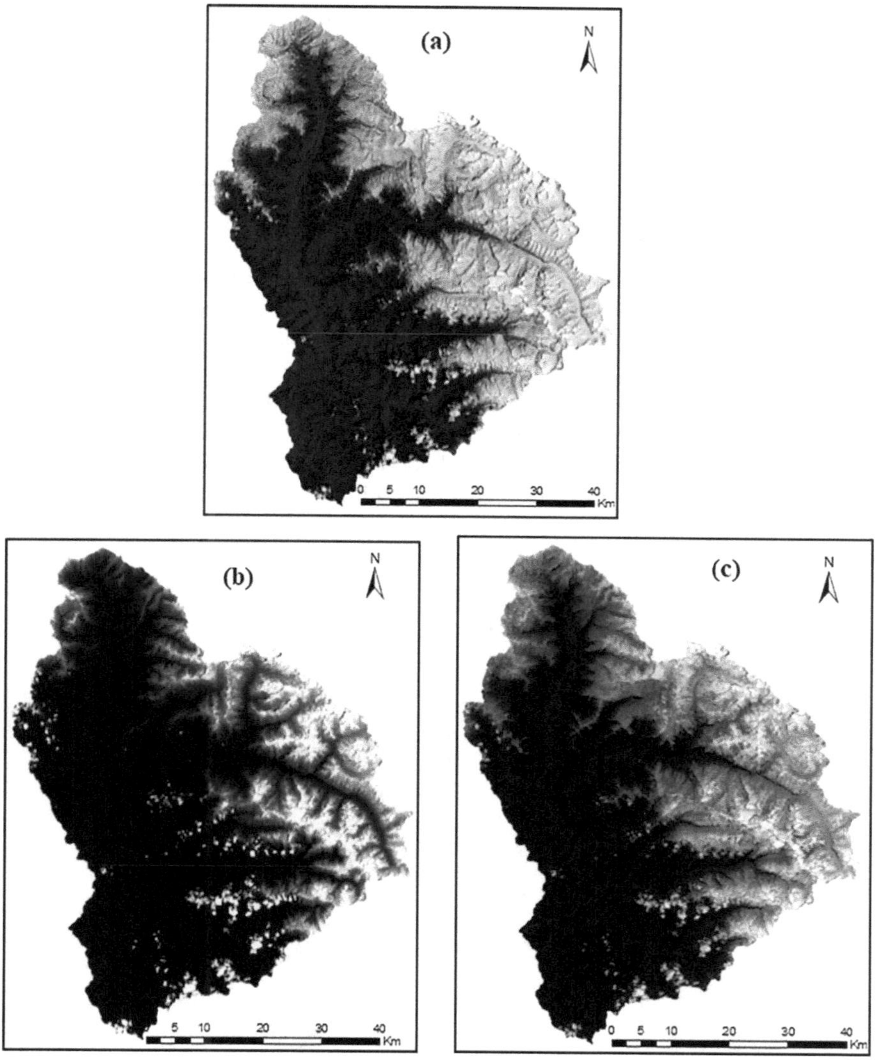

Fig. 4 False color composite (FCC) image (**a**), Cirrus band (B9) image (**b**) and image generated by using Eq. 8 (**c**)

Quantitative validation demonstrated that cloud pixels were recognized with an overall accuracy of around 97%, whilst cloud-shadow detection reached an accuracy of 96%. The results correspond with the findings of Zhu and Woodcock (2014) and Qiu et al. (2017), who illustrated that the incorporation of cirrus and temperature data improves the reliability of cloud recognition systems, particularly in hilly areas.

5.2 Cloud Removal and Image Reconstruction

Upon identification, pixels impacted by clouds and shadows were rectified utilizing a radiometric transformation method (Eq. 10). The regression-based correction successfully replaced polluted pixels with estimated surface reflectance values obtained from adjacent clear pixels and multi-band correlations. Figure 5 demonstrates that the corrected composite image exhibits much enhanced continuity in land-surface features, including vegetation cover, water bodies, and snow-covered regions. The approach proved especially efficacious in eliminating thin and semi-transparent clouds, where the spectral contrast between the clouds and land surface was moderate. The elimination of dense clouds proved more difficult, with lingering haze and localized brightness fluctuations evident in specific high-altitude areas.

The resultant cloud-free mosaic markedly enhances the use of Landsat-8 data for purposes including land-cover mapping, hydrological modeling, and snow-cover research in the Beas River basin. The methodology offers a significant advancement in the practical creation of cloud-free images utilizing moderate-resolution optical datasets.

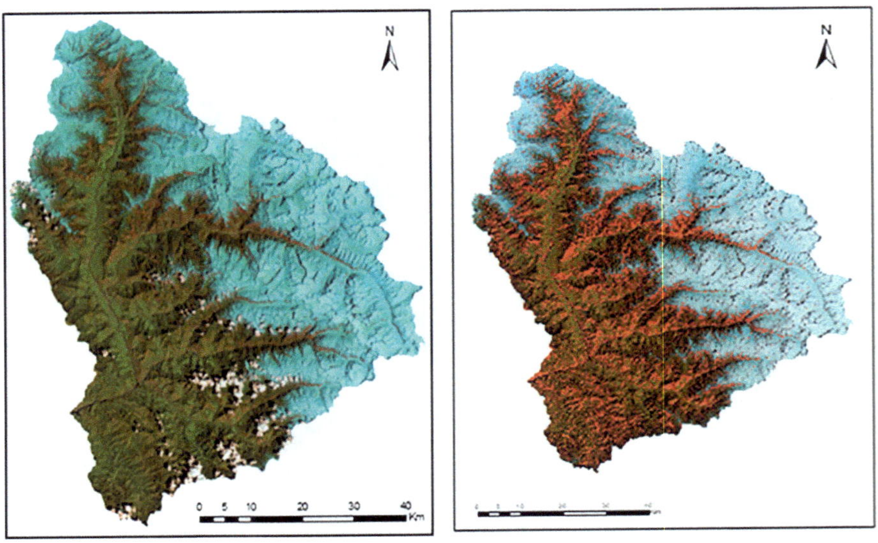

Fig. 5 Cloud removal from Landsat-8 OLI bands composite image (OLI7, OLI5, OLI3, RGB) before (left), after cloud removal (right)

5.3 Comparative Assessment and Discussion

The suggested method exhibited performance that is either equivalent to or greater than those of widely utilized cloud-masking algorithms, including Fmask (Zhu & Woodcock, 2012) and ACCA (Irish, 2000). In contrast to other methodologies, the suggested solution eliminates the need for multi-temporal image pairings or thermal band inputs, hence streamlining operational implementation. The integration of six OLI bands with the cirrus band facilitated the effective differentiation of clouds and their shadows from various land covers. Nonetheless, the method has specific constraints when addressing intricate topography and markedly varied lighting conditions. In steep mountainous regions, shadows produced by the topography may have spectral similarities to cloud shadows, occasionally resulting in misdiagnosis. Furthermore, dense clouds with significant optical depth frequently cover substantial areas of the image, complicating accurate reconstruction due to inadequate contextual information for radiometric correction. A further problem stems from the application of static thresholds, which may not adequately generalize over different seasons or fluctuating atmospheric circumstances. Adaptive or machine-learning-based thresholding may enhance robustness. The suggested method inadequately addresses nearby cloud scattering and haze effects, potentially resulting in persistent spectral pollution at the borders of clouds.

5.4 Practical Implications

Notwithstanding its constraints, this method provides a computationally efficient and data-driven strategy appropriate for regional-scale monitoring in data-scarce mountainous basins. The precision and straightforwardness of cirrus-band-based detection render it especially beneficial for hydrological and climate research, where cloud interference is common. The resultant cloud-free composites can be used into time-series analyses, snowmelt modeling, or land surface temperature estimation procedures.

5.5 Limitations and Future Work

The primary constraints recognized in this study encompass:

1. Sensitivity to Dense Cloud Cover—The method demonstrates efficacy with thin clouds but is ineffective in reconstructing areas obscured by thick clouds.
2. Topographic Effects—Terrain shadows and anisotropic illumination in mountainous areas may lead to the misclassification of cloud shadows.

3. Temporal Limitation—The methodology relies on single-date imagery and is devoid of multi-temporal data necessary for deducing background surface reflectance.
4. Fixed Thresholds—Static threshold values diminish generalizability across varying seasons and meteorological conditions.
5. Lack of Machine Learning—Contemporary deep-learning techniques (e.g., CNNs or U-Nets) could significantly improve the precision of cloud and shadow segmentation.

Future investigations ought to use DEM-based topographic normalization, adaptive thresholding, or AI-driven cloud removal methodologies. Integrating this methodology with multi-temporal Landsat or Sentinel-2 data may provide more reliable cloud-free surface composites and facilitate operational land-surface monitoring in intricate terrains like the western Himalayas.

6 Conclusions

This research effectively established and showcased a proficient method for identifying and eliminating clouds and their shadows from Landsat-8 Operational Land Imager (OLI) data in the Beas River basin of Himachal Pradesh. This approach accurately differentiates cloud-contaminated pixels from clear-sky areas by utilizing the spectral reflectance properties of clouds and incorporating data from six essential OLI spectral bands, especially the cirrus band (Band 9), irrespective of the underlying land cover type. The findings indicated that both thin and dense clouds were accurately identified using the established threshold-based methods, achieving total detection accuracies of 97% for clouds and 96% for cloud shadows. The established single-band (cirrus-based) and multi-band transformation methods facilitated the elimination of thin clouds and the restoration of near-original surface reflectance values, yielding a high-quality, visually coherent cloud-free composite image. The methodology offers a stable solution for creating accurate datasets suited for future geospatial analysis, such as snow-cover mapping, land-use/land-cover categorization, and hydrological modeling in Himalayan contexts where cloud cover is frequent. This method's implementation in the intricate topography of the Beas River basin illustrates its efficacy in generating reliable, high-resolution optical imaging. The study finds specific limitations, such as diminished performance in thick, optically dense clouds, and occasional misinterpretation between landscape shadows and cloud shadows in steep mountainous regions. Moreover, as the methodology relies on single-date imagery and fixed criteria, its applicability may be limited under diverse meteorological and seasonal conditions. This research represents a significant advancement in automated and semi-empirical cloud removal approaches for Landsat datasets, enhancing data usefulness in climate and hydrological studies despite existing limitations. The incorporation of the cirrus band with radiometric correction methodologies guarantees the preservation of critical spectral integrity, improving the interpretability of post-processed images.

To enhance future outcomes, the integration of multi-temporal or multi-sensor data (e.g., Sentinel-2, MODIS) alongside machine-learning adaptive classification approaches may address existing constraints, especially in areas with dense cloud cover and elevated altitudes. These developments will facilitate the near-real-time creation of cloud-free data mosaics, essential for climate change effect assessments, glacier monitoring, and flood forecasting in the Himalayan and other cloud-prone areas. The suggested cloud detection and removal technique offers a computationally efficient, precise, and scalable solution for producing cloud-free Landsat-8 imagery. It connects a crucial divide between conventional threshold-based algorithms and data-driven augmentation techniques, facilitating enhanced monitoring of terrestrial and hydrological systems amid evolving climate circumstances.

Acknowledgements The authors sincerely recognize, the United States Geological Survey (USGS) kindly provided the remote sensing data and Landsat-8 data from the archive utilized in this study work.

Declarations

Author's Contribution All authors contributed to the study conception and design. Material preparation, data collection by (Gopinadh Rongali) and analysis were performed by (Ashok K. Keshari), (Ashvani K. Gosain). The first draft of the manuscript was written by (Gopinadh Rongali) and edited by (Poonam Vishwas). All authors read and approved the final manuscript.

Conflict of Interest No potential conflicts of interest were reported by the authors.

Availability of Data and Materials. Not applicable.

Funding No funding was obtained for this study.

References

Barsi, J. A., Lee, K., Kvaran, G., Markham, B. L., & Pedelty, J. A. (2014). The spectral response of the Landsat-8 operational land imager. *Remote Sensing, 6*(10), 10232–10251.

Berk, A., Bernstein, L. S., & Robertson, D. C. (1989). *MODTRAN: A moderate resolution model for LOWTRAN 7 (GL-TR-89-0122)*. Air Force Geophysics Laboratory.

Berk, A., Bernstein, L. S., Anderson, G. P., Acharya, P. K., Robertson, D. C., Chetwynd, J. H., & Adler-Golden, S. M. (1998). MODTRAN cloud and multiple scattering upgrades with application to AVIRIS. *Remote Sensing of Environment, 65*(3), 367–375.

Chai, D., Newsam, S., Zhang, H. K., Qiu, Y., & Huang, J. (2019). Cloud and cloud shadow detection in Landsat imagery based on deep convolutional neural networks. *Remote Sensing of Environment, 225*, 307–316.

Chen, S., Chen, X., Chen, J., Jia, P., Cao, X., & Liu, C. (2015). An iterative haze optimized transformation for automatic cloud/haze detection of Landsat imagery. *IEEE Transactions on Geoscience and Remote Sensing, 54*(5), 2682–2694.

Foga, S., Scaramuzza, P. L., Guo, S., Zhu, Z., Dilley, R. D., Jr., Beckmann, T., & Laue, B. (2017). Cloud detection algorithm comparison and validation for operational Landsat data products. *Remote Sensing of Environment, 194*, 379–390.

Goodwin, N. R., Collett, L. J., Denham, R. J., Flood, N., & Tindall, D. (2013). Cloud and cloud shadow screening across Queensland, Australia: An automated method for Landsat TM/ETM+ time series. *Remote Sensing of Environment, 134*, 50–65.

Hagolle, O., Huc, M., Pascual, D. V., & Dedieu, G. (2010). A multi-temporal method for cloud detection, applied to FORMOSAT-2, VENUS, LANDSAT and SENTINEL-2 images. *Remote Sensing of Environment, 114*(8), 1747–1755.

Huang, C., Thomas, N., Goward, S. N., Masek, J. G., Zhu, Z., Townshend, J. R., & Vogelmann, J. E. (2010). Automated masking of cloud and cloud shadow for forest change analysis using Landsat images. *International Journal of Remote Sensing, 31*(20), 5449–5464.

Hughes, M. J., & Hayes, D. J. (2014). Automated detection of cloud and cloud shadow in single-date Landsat imagery using neural networks and spatial post-processing. *Remote Sensing, 6*(6), 4907–4926.

Irish, R. R. (2000). Landsat 7 automatic cloud cover assessment. In *Algorithms for multispectral, hyperspectral, and Ultraspectral imagery VI* (Vol. 4049, pp. 348–355). International Society for Optics and Photonics.

Irish, R. R., Barker, J. L., Goward, S. N., & Arvidson, T. (2006). Characterization of the Landsat-7 ETM+ automated cloud-cover assessment (ACCA) algorithm. *Photogrammetric Engineering and Remote Sensing, 72*(10), 1179–1188.

Jin, S., Homer, C., Yang, L., Xian, G., Fry, J., Danielson, P., & Townsend, P. A. (2013). Automated cloud and shadow detection and filling using two-date Landsat imagery in the USA. *International Journal of Remote Sensing, 34*(5), 1540–1560.

Kong, X., Qian, Y., & Zhang, A. (2013). Cloud and shadow detection and removal for Landsat-8 data. In *MIPPR 2013: Remote sensing image processing, geographic information systems, and other applications* (Vol. 8921, p. 89210N). International Society for Optics and Photonics.

Li, Z., Shen, H., Weng, Q., Zhang, Y., Dou, P., & Zhang, L. (2022). Cloud and cloud shadow detection for optical satellite imagery: Features, algorithms, validation, and prospects. *ISPRS Journal of Photogrammetry and Remote Sensing, 188*, 89–108.

Potapov, P., Turubanova, S., & Hansen, M. C. (2011). Regional-scale boreal forest cover and change mapping using Landsat data composites for European Russia. *Remote Sensing of Environment, 115*(2), 548–561.

Qiu, S., He, B., Zhu, Z., Liao, Z., & Quan, X. (2017). Improving Fmask cloud and cloud shadow detection in mountainous area for Landsats 4–8 images. *Remote Sensing of Environment, 199*, 107–119.

Wang, T., Shi, J., Letu, H., Ma, Y., Li, X., & Zheng, Y. (2019). Detection and removal of clouds and associated shadows in satellite imagery based on simulated radiance fields. *Journal of Geophysical Research: Atmospheres, 124*(13), 7207–7225.

Zhang, Y., Guindon, B., & Cihlar, J. (2002). An image transform to characterize and compensate for spatial variations in thin cloud contamination of Landsat images. *Remote Sensing of Environment, 82*(2–3), 173–187.

Zhou, G., Zhou, X., Yue, T., & Liu, Y. (2016). An optional threshold with SVM cloud detection algorithm and DSP implementation. *International Archives of the Photogrammetry, Remote Sensing and Spatial Information Sciences, 41.*

Zhu, Z., & Woodcock, C. E. (2012). Object-based cloud and cloud shadow detection in Landsat imagery. *Remote Sensing of Environment, 118*, 83–94.

Zhu, Z., & Woodcock, C. E. (2014). Automated cloud, cloud shadow, and snow detection in multitemporal Landsat data: An algorithm designed specifically for monitoring land cover change. *Remote Sensing of Environment, 152*, 217–234.

Zhu, Z., Wang, S., & Woodcock, C. E. (2015). Improvement and expansion of the Fmask algorithm: Cloud, cloud shadow, and snow detection for Landsats 4–7, 8, and Sentinel 2 images. *Remote Sensing of Environment, 159*, 269–277.

Human Dimensions, Culture, Education, and Governance

Sacred Landscapes Under Siege: Climate Change, Pilgrimage Tourism, and Cultural Resilience in the Himalayas

Mehraj Din Wani, Shamim Ahmad Shah, and Zubair Ahmad Dada

Abstract Nestled at the intersection of ecological grandeur and metaphysical reverence, the Himalayan sacred landscapes are undergoing an ontological crisis precipitated by the synergistic forces of anthropogenic climate change and the unbridled expansion of pilgrimage tourism. These spiritually charged geographies—long regarded as axial sites of transcendence and custodians of civilizational memory—are now contested terrains where climatic volatility collides with the logics of commodified spiritual mobility. This chapter undertakes a critical interrogation of how climatic disruptions manifested in glacial destabilization, erratic precipitation regimes, and ecosystem fragmentation converge with the infrastructural and socio-cultural burdens of mass pilgrimage to erode the biocultural sanctity of these alpine sanctuaries. Engaging with concepts from political ecology, environmental anthropology, and resilience theory, the chapter theorizes sacred landscapes not merely as physical spaces, but as dynamic ecologies of meaning, practice, and affect. It explores how indigenous ontologies, ritual epistemologies, and localized cosmologies offer counter-hegemonic frameworks for reimagining environmental stewardship in times of planetary crisis. Simultaneously, it interrogates the epistemic limitations and socio-political asymmetries embedded in dominant conservation discourses and regulatory paradigms that often decouple sacred geographies from their living cultural matrices. By foregrounding the dialectics of tradition and transformation, vulnerability and agency, this work advances a multi-scalar vision of cultural resilience one that centers relational ethics, intergenerational continuity, and spiritual ecology as critical pivots for sustainable governance. Ultimately, the chapter issues a call for an ontological recalibration of Himalayan pilgrimage tourism urging a paradigm shift from extractive visitation toward a reverential ethic of coexistence, where sacredness and sustainability are mutually constitutive rather than mutually exclusive.

M. D. Wani (✉) · S. A. Shah
Geography and Disaster Management, University of Kashmir, Srinagar, India

Z. A. Dada
Centre for Distance and Online Education, University of Kashmir, Srinagar, India

R. Ahmed et al. (eds.), *Climate Change, Water Resources, and Natural Hazards in the Himalayas*, Advances in Geographical and Environmental Sciences,
https://doi.org/10.1007/978-981-92-0835-7_16

Keywords Sacred landscapes · Climate change · Pilgrimage tourism · Cultural resilience · Political ecology

1 Introduction

The Himalayas, often celebrated as the "abode of the gods," embody one of the world's most profound intersections of ecology, spirituality, and culture (Gagné, 2020; Howard, 2016). Spanning across eight nations, this formidable mountain system sustains extraordinary biodiversity and functions as a vital water tower for nearly a quarter of humanity (Bernbaum, 2022). Equally significant is its metaphysical resonance, for these landscapes have long been revered as sacred spaces within Hinduism, Buddhism, Sikhism, and diverse indigenous traditions that root their cosmologies in the mountains (Singh & Rana, 2020). Pilgrimage destinations such as Kedarnath, Badrinath, Amarnath, Hemkund Sahib, and Kailash–Mansarovar exemplify the seamless coexistence of ecological grandeur and sacred devotion, illustrating how natural features have historically been woven into spiritual practice and civilizational memory (Eck, 2012; Singh, 2018). The Himalayas are thus not merely physical geographies; they are living landscapes where environmental processes and cultural meanings are inextricably entangled (Wouters, 2024).

In recent decades, however, this delicate balance between sacredness and sustainability has come under profound strain (Kealiikanakaoleohaililani & Giardina, 2016). Climate change has emerged as an existential threat, with accelerated glacial retreat, altered monsoon dynamics, and heightened frequency of extreme weather events destabilizing fragile mountain ecologies (Bolch et al., 2019; IPCC, 2023). These transformations carry material as well as spiritual consequences. For instance, the retreat of the Gangotri Glacier—revered as the mythical origin of the Ganges—has exceeded 1500 m over the past century, not only disrupting hydrological regimes but also challenging deeply held cosmological imaginaries that situate the river as a divine entity (Tiwari, 2017). Similarly, flash floods, landslides, and glacial lake outburst events are reshaping local landscapes in ways that unsettle both community livelihoods and ritual practices tied to seasonal pilgrimage circuits.

Concurrently, the expansion of pilgrimage tourism has intensified ecological and socio-cultural pressures (Wani et al., 2024c). Millions of devotees now traverse the fragile Himalayan terrains each year, supported by an expanding infrastructure of roads, hotels, and commercial facilities designed to accommodate mass visitation (Wani et al., 2023; Biswas & Kumar, 2024). While such pilgrimages serve to reinforce collective faith, cultural belonging, and transregional connectivity, they also leave behind escalating ecological footprints. Deforestation, waste accumulation, traffic congestion, and the overextraction of water resources place severe stress on already vulnerable ecosystems (Rana & Chandra, 2021; Sharma & Rai, 2019). The catastrophic floods of Kedarnath in 2013 stand as a sobering reminder of how the convergence of unregulated development and fragile ecologies can magnify disaster risks, resulting in immense human and material losses (Kaliraj et al., 2025).

This chapter seeks to interrogate the Himalayan sacred landscapes not merely as static geographic spaces but as dynamic socio-ecological systems, where ecological processes, spiritual epistemologies, and cultural practices continually interact and transform one another. It emphasizes that the current conjuncture represents more than an environmental crisis; it is also an ontological one, as the meanings, rituals, and lived experiences that have historically shaped these landscapes are increasingly destabilized by climatic volatility and the commodification of pilgrimage. By engaging with frameworks from political ecology, environmental anthropology, and resilience theory, the analysis foregrounds how climate disruptions and mass tourism reshape these sacred terrains, while also illuminating the capacity of indigenous ontologies, ritual epistemologies, and spiritual ecologies to offer counter-hegemonic pathways toward sustainability (Wani et al., 2025b).

In this sense, the Himalayas are conceptualized simultaneously as a frontline of climate vulnerability and as reservoirs of cultural resilience. The chapter positions the sacred landscapes of the Himalayas as critical sites where struggles over development, spirituality, and environmental stewardship converge, and where possibilities for a paradigm shift toward a relational ethic of reverence, sustainability, and coexistence can be imagined. The central aim is thus not only to reveal the pressures exerted by climate change and pilgrimage tourism but also to explore how local cosmologies and cultural practices might inform alternative governance frameworks that reconcile the imperatives of ecological survival with the sanctity of spiritual life.

2 Literature Review

2.1 Climate Change in the Himalayas

A growing body of scholarship confirms that the Himalayas are warming at a rate significantly faster than the global average, making the region one of the most climate-vulnerable zones on the planet (Ahmed et al., 2025; Bolch et al., 2019; IPCC, 2023; Parrey et al., 2025). The consequences of this warming are evident in rapid glacial retreat, permafrost degradation, and an increased frequency of extreme weather events such as floods, landslides, and avalanches (Haeberli et al., 2017; Saleem et al., 2021). Glaciers across the Himalayan arc have already lost a substantial portion of their mass over the past century, and projections suggest continued acceleration in the coming decades (Bolch et al., 2019; Saleem et al., 2024a). Such changes carry profound implications for downstream hydrology, agricultural productivity, hydroelectric power generation, and ultimately, human settlements that depend on these water systems (Mushtaq et al., 2024; Wani et al., 2024b; Saleem et al., 2024b).

From a cultural perspective, the retreat of glaciers also erodes symbolic geographies of the sacred. The Gangotri Glacier revered as the source of the Ganges has receded by more than 1500 m within the last century, creating dissonance between ecological realities and cosmological imaginaries (Tiwari et al., 2017). Similarly, shrinking snowlines affect pilgrimage routes to Amarnath and Hemkund Sahib, not only disrupting traditional patterns of movement but also amplifying safety risks for pilgrims. In this sense, climate change in the Himalayas constitutes more than an environmental crisis; it is simultaneously an ontological crisis that destabilizes cultural memory and ritual practice (Gergan, 2017; Ranjan, 2024).

2.2 *Pilgrimage Tourism and Environmental Degradation*

Parallel to climatic transformations, the growth of pilgrimage tourism has intensified anthropogenic pressures on Himalayan ecologies (Batool et al., 2024; Gawlik et al., 2022; Wani et al., 2025b). Sacred sites such as Kedarnath, Badrinath, Amarnath, and Kailash–Mansarovar attract millions of pilgrims annually, creating a paradox of devotion and degradation. The construction of roads, guesthouses, and other infrastructure to accommodate large numbers of visitors often involves deforestation, slope destabilization, and disruption of fragile alpine habitats (Ashraf et al., 2024; Semwal & Tripathi, 2025; Sharma & Rai, 2019). The proliferation of plastic waste, inadequate sanitation, and unchecked commercial activity has further degraded local water sources and soils, creating cumulative ecological stress (Rana & Chandra, 2021).

The Kedarnath disaster of 2013 is frequently cited as a tragic reminder of how fragile mountain ecologies can be overwhelmed when climatic volatility converges with unregulated development. Torrential rains, compounded by glacial melt, triggered devastating floods and landslides that claimed thousands of lives and destroyed vast infrastructure (Hait & Sahu, 2024; Spiridonov et al., 2025). Scholars argue that the scale of destruction was not only a consequence of extreme weather but also of unsustainable patterns of pilgrimage expansion that failed to respect ecological thresholds (Ivakhiv, 2016; Shinde & Olsen, 2022). The incident illustrates the urgent need to rethink pilgrimage tourism, shifting away from extractive models toward more ecologically sensitive and culturally respectful practices (Wani et al., 2025c).

2.3 *Sacred Ecology and Cultural Resilience*

While climate change and tourism pose formidable challenges, scholarship also points to the resilience embedded in the cultural and spiritual traditions of Himalayan communities (Kumar et al., 2024; Kumar & Shoshta, 2024). The concept of sacred ecology underscores the reciprocal relationship between spiritual values and environmental management, demonstrating how cultural cosmologies and ritual practices often sustain conservation ethics in practice (Berkes, 2008). In many Himalayan villages, groves, rivers, and mountains are revered as divine entities, and their protection is codified through ritual prohibitions, seasonal taboos, and community-led stewardship systems (Wani et al., 2025d; Singh, 2018). Such practices illustrate that the sacred and the ecological are not distinct domains but mutually reinforcing dimensions of lived experience.

Resilience theory further highlights that communities embedded in these landscapes adapt to ecological changes not only through technical responses but also through cultural strategies. Ritual calendars, oral traditions, and cosmological narratives provide interpretive frameworks that help communities absorb disturbances while maintaining continuity of practice. Scholars such as Adams (2017) argue that these indigenous epistemologies function as counter-hegemonic knowledge systems, challenging modernist conservation frameworks that often exclude cultural meanings. Thus, cultural resilience rooted in sacred ecology becomes an essential dimension of environmental stewardship in the Himalayas (Wani et al., 2024a).

2.4 Integrating Sacredness and Sustainability in Governance

Recent research emphasizes that sustainable governance of Himalayan sacred landscapes requires the integration of traditional ecological knowledge (TEK) with modern conservation and tourism management strategies. Community-led models in regions such as Uttarakhand, Sikkim, and Bhutan demonstrate how sacred values can effectively reinforce ecological stewardship when embedded within governance frameworks (Adams, 2017; Singh, 2018; Wani et al., 2025a). In Bhutan, for instance, Buddhist principles of interdependence inform national policies of Gross National Happiness, linking cultural wellbeing with environmental conservation. Similarly, in Uttarakhand, village councils often regulate pilgrimage flows and resource use by drawing on ritual norms, thereby aligning sustainability with cultural legitimacy.

However, dominant conservation paradigms—often state-centric and technocratic—frequently marginalize local spiritual values, treating sacred landscapes as ecological assets to be managed rather than cultural ecologies to be lived. This disconnect reduces the effectiveness of policy interventions and risks alienating the very communities whose participation is critical for long-term stewardship. The literature thus suggests that policies which recognize sacredness as integral to ecological governance are more likely to succeed in sustaining both environmental integrity and cultural continuity. Effective governance in this context is therefore not merely about balancing ecological protection and tourism revenue; it is about reconfiguring development models to ensure that sacredness and sustainability become mutually constitutive rather than mutually exclusive (Akamani, 2020; Sofield & Li, 2013).

3 Methodology

3.1 Research Design

The study adopts a qualitative research design aimed at exploring the intersection of climate change, pilgrimage tourism, and cultural resilience in Himalayan sacred landscapes. The approach is interpretive and interdisciplinary, informed by political ecology and resilience theory, both of which highlight the dynamic interplay between environmental processes, socio-economic systems, and cultural practices (Adger, 2010; Berkes, 2008). To ensure comprehensiveness and transparency, the review followed a systematic framework modeled on PRISMA-style guidelines (Snyder, 2019), enabling a careful selection, synthesis, and critical evaluation of scholarly and institutional sources published between 2000 and 2025.

3.2 Data Sources

The study draws upon a wide range of materials including peer-reviewed research articles, government reports, international agency assessments, policy documents, ethnographic accounts, and historical records. Climate and environmental change information was obtained from scientific research on glacial retreat, hydrological shifts, and ecosystem degradation (Bolch et al., 2019; Tiwari et al., 2017), as well as from reports by organizations such as the IPCC (2023) and the National Disaster Management Authority. Insights on pilgrimage tourism were gathered from studies on visitor flows, environmental impacts, and infrastructure pressures (Rana & Chandra, 2021; Sharma & Rai, 2019), along with government policies addressing sacred site management. To understand cultural resilience, the analysis considered works on sacred ecology, traditional knowledge systems, and indigenous adaptive practices (Adams, 2017; Berkes, 2008; Singh, 2018), in addition to temple records, ritual documents, and ethnographic studies.

3.3 Data Analysis

Analysis proceeded in three interconnected stages. First, a thematic synthesis was undertaken to identify recurring trends related to ecological stressors, tourism pressures, and adaptive cultural responses. Second, content analysis was applied to policy frameworks and management documents to evaluate how ecological and cultural concerns are integrated into governance structures. Finally, an integrative comparative approach was employed to bring together findings from environmental science, cultural studies, and governance research. This triangulation ensured a holistic understanding of the challenges and adaptive mechanisms shaping Himalayan sacred landscapes, while also identifying gaps in research and policy.

3.4 Limitations

The methodology is not without limitations. Existing studies on Himalayan regions are uneven in their geographic focus, with greater representation from Uttarakhand and Himachal Pradesh compared to other culturally significant sites. Some sources vary in methodological rigor, which may affect comparability across regions. In addition, reliable longitudinal data on tourism volumes and environmental degradation remain scarce for certain pilgrimage centers. While these constraints limit the scope for precise measurement, the interdisciplinary synthesis nevertheless provides a robust basis for analyzing the interlinkages of climate vulnerability, tourism pressures, and cultural resilience.

4 Results and Discussion

The Himalayan sacred landscapes are undergoing transformations of unprecedented scale, shaped by the combined pressures of climate change and pilgrimage tourism. Once celebrated as sanctuaries of ecological stability and spiritual transcendence, these landscapes are increasingly vulnerable to disruptions that threaten not only their biophysical integrity but also their cultural and symbolic vitality. By examining the climate-related shifts, tourism-induced pressures, and the counterbalancing role of cultural resilience, this section illuminates the complex interdependencies that define the contemporary Himalayan experience (Wani et al., 2022).

5 Climate Change Impacts on Sacred Landscapes

The Himalayas, often characterized as the "Third Pole" for their vast cryospheric reserves, are warming at nearly double the global average. This accelerated climatic transformation manifests most visibly in the retreat of glaciers, altered hydrological regimes, and the rising frequency of extreme weather events, each with far-reaching ecological and cultural repercussions, See (Table 1).

The retreat of glaciers is not merely an environmental event but also a cultural rupture. For instance, the Gangotri Glacier, revered as the mythological source of the River Ganga, has receded more than 1500 m in the past century. This retreat threatens ritual practices such as *Gangotri darshan* and the symbolic sanctity of river worship central to Hindu cosmology. The reduction in glacial mass also destabilizes

Table 1 Key climate change indicators in the Himalayas and their ecological and cultural implications for sacred landscapes

Parameter	Observed change	Implications for sacred sites	Source
Glacier retreat	Significant retreat; e.g., Gangotri glacier receded ~1500 m over the last century	Reduced water flow for rivers and rituals; heightened flood risk	Bolch et al. (2019)
Temperature rise	Warming rate ~ 0.06–0.09 °C per year	Alters vegetation zones, affecting sacred groves and ritual practices	Tiwari et al. (2017)
Precipitation variability	Increased rainfall intensity and prolonged dry spells	Flash floods, landslides, and damage to pilgrimage infrastructure	IPCC (2023)
Disaster frequency	Higher incidence of floods and landslides	Disruption of pilgrimages, threats to sacred shrines and pilgrim safety	Kala (2014)

water availability for downstream agricultural communities, intertwining material livelihoods with spiritual anxieties.

Temperature rise alters ecological zones, pushing vegetation belts upward. Such ecological shifts jeopardize the availability of medicinal herbs, sacred groves, and alpine meadows (*bugyals*), many of which hold ritual significance for local populations. These disruptions are not abstract: they reshape the everyday practices of communities who depend on these landscapes for both survival and spiritual continuity.

Equally pressing is the heightened variability in precipitation, which now manifests as erratic monsoons, intense cloudbursts, and prolonged dry spells. These climatic anomalies frequently trigger flash floods and landslides, damaging roads, pilgrimage trails, and sacred shrines. The Kedarnath disaster of 2013 epitomizes this intersection of climate volatility and infrastructural vulnerability, where unseasonal rainfall combined with poorly regulated tourism infrastructure produced catastrophic outcomes. In this sense, climate change in the Himalayas erodes not only physical landscapes but also cosmological imaginaries and ritual geographies, challenging the very sanctity of sacred spaces.

5.1 Pilgrimage Tourism and Environmental Pressures

Parallel to climatic stresses, pilgrimage tourism has emerged as one of the most significant anthropogenic pressures on Himalayan landscapes. Pilgrimage to sites

Table 2 Pilgrimage tourism and associated environmental pressures on selected Himalayan sacred sites

Sacred site	Annual pilgrims	Environmental pressures	Source
Kedarnath	~one million	Solid waste, deforestation, pressure on water resources	Rana and Chandra (2021)
Badrinath	~500,000	Soil erosion, habitat disturbance	Sharma and Rai (2019)
Hemkund Sahib	~50,000	Trail degradation, littering	Sharma and Rai (2019)
Vaishno Devi	8–ten million	Forest disturbance, vehicular emissions, solid waste	Singh (2018)

such as Kedarnath, Badrinath, Hemkund Sahib, Amarnath, Kailash–Mansarovar, and Vaishno Devi represents a profound expression of faith and cultural continuity, but its exponential growth has introduced unprecedented ecological and infrastructural burdens (See Table 2).

The scale of ecological pressure correlates strongly with the volume of pilgrims. At Vaishno Devi, with over eight million visitors annually, vehicular emissions, forest disturbance, and waste accumulation have created a near-permanent state of environmental stress. At Kedarnath and Badrinath, the clearing of forest land for lodging facilities, road expansions, and parking areas has destabilized slopes, increasing susceptibility to landslides. Even smaller sites such as Hemkund Sahib, despite receiving fewer pilgrims, face trail degradation and accumulation of non-biodegradable waste, indicating that even modest levels of tourism can disrupt fragile alpine ecosystems.

Tourism infrastructure, intended to facilitate religious mobility, often exacerbates ecological fragility. The construction of roads, helipads, and hotels not only increases the carbon footprint but also destabilizes mountain slopes, creating conditions ripe for disasters. In some cases, unregulated development directly collides with sacred geographies, replacing spiritual sanctity with commercial commodification. What emerges is a paradox: pilgrimage tourism simultaneously sustains local economies and erodes the ecological and cultural foundations upon which its sacredness rests.

5.2 *Cultural Resilience and Sacred Ecology*

Amid these disruptions, the resilience embedded within Himalayan cultures provides critical adaptive capacity. Local communities have long practiced forms of sacred ecology that intertwine ecological stewardship with religious and ritual obligations. These practices not only sustain ecosystems but also reinforce spiritual worldviews that frame humans as custodians rather than consumers of sacred landscapes (Table 3).

Ritual restrictions often function as ecological safeguards. For example, prohibitions on resource extraction during specific festivals limit overharvesting and allow

Table 3 Linkages between sacred ecological practices and community-based resilience in the Himalayas

Aspect	Observations	Implications for resilience	Source
Ritual stewardship	Restrictions on extraction during festivals, reverence for sacred rivers	Sustains ecological balance, protects water quality	Berkes (2008)
Traditional ecological knowledge	Practices of forest use, soil conservation, disaster preparedness	Enhances adaptive capacity and ecosystem health	Singh (2018)
Community governance	Temple trusts and village councils regulating access, preserving groves	Strengthens cultural continuity and ecological stewardship	Adams (2017)

ecosystems to regenerate. Sacred groves, maintained through taboos and spiritual obligations, act as biodiversity reservoirs and water catchments, simultaneously preserving ecological services and reinforcing cultural identity.

Traditional ecological knowledge remains central to adaptive resilience. Communities employ centuries-old practices of terraced agriculture, soil stabilization, and forest management, which not only buffer against climate-induced risks but also maintain symbolic continuity with ancestral knowledge systems. Temple trusts and local councils provide governance mechanisms that embed ecological stewardship within cultural practice, regulating access to shrines and pilgrimage routes in ways that balance spiritual devotion with environmental care.

Yet these resilience mechanisms are under strain. The commercialization of pilgrimage and the encroachment of consumer-driven tourism have diluted the authority of traditional practices. Ritual obligations are increasingly subordinated to economic imperatives, threatening the longevity of cultural resilience. Without deliberate reinforcement, sacred ecology risks being hollowed out into symbolic performance, stripped of its material capacity to sustain ecological and spiritual integrity.

5.3 Interlinkages Between Climate, Tourism, and Resilience

The interplay of climate change, tourism pressures, and cultural resilience reveals the Himalayan sacred landscapes as complex socio-ecological systems defined by feedback loops and contradictions. Climatic variability amplifies the risks created by unregulated tourism, as seen in the Kedarnath disaster, where extreme rainfall combined with poorly planned infrastructure to produce devastation. Tourism, in turn, often erodes the very cultural practices that historically sustained ecological balance, privileging mass visitation over ritual reverence.

Nevertheless, resilience persists. Ritualized forest protections, seasonal taboos, and community-based governance mitigate some climate and tourism pressures, serving as adaptive strategies that sustain ecological health and cultural continuity. These practices demonstrate that cultural systems are not passive victims of environmental change but active mediators that negotiate between vulnerability and adaptation.

In this way, the Himalayas emerge as contested yet resilient landscapes, where ecological fragility, spiritual significance, and socio-economic pressures converge. The evidence highlights that governance strategies must move beyond narrow conservation frameworks or tourism promotion to embrace an integrated approach—one that aligns ecological sustainability with cultural resilience and spiritual reverence. Only through such a paradigm shift can sacredness and sustainability be mutually reinforcing rather than mutually exclusive.

6 Conclusion

Himalayan sacred landscapes are navigating an unprecedented intersection of climatic upheaval and the pressures of intensified pilgrimage tourism. Glacial retreat, rising temperatures, and erratic precipitation have disrupted the physical and ecological foundations of these regions, threatening rivers, forests, and high-altitude ecosystems that underpin both livelihoods and sacred rituals. Simultaneously, burgeoning pilgrimage flows impose significant environmental stress, from soil erosion and forest disturbance to waste accumulation and overextraction of water resources, while infrastructural expansion designed to accommodate mass visitation amplifies the vulnerability of these fragile terrains to disasters such as floods and landslides.

Amid these converging challenges, the cultural resilience embedded in local practices, ritual stewardship, and indigenous ecological knowledge provides critical adaptive capacity, mediating the impacts of both climate and anthropogenic pressures. Sacred governance, seasonal restrictions, and community-led conservation illustrate how spiritual and cultural frameworks operationalize ecological sustainability. Yet, the commercialization of pilgrimage and policy frameworks oriented toward mass tourism risk undermining these traditional mechanisms, highlighting the tension between economic imperatives and the preservation of sacred ecological and cultural systems. The protection of Himalayan sacred landscapes thus demands an integrative paradigm that unites ecological integrity, cultural continuity, and spiritual reverence. Recognizing these landscapes as living socio-ecological systems, where environmental health, ritual practice, and ethical tourism are inseparably intertwined, is essential for sustainable stewardship. Ultimately, the future of the Himalayas hinges on fostering coexistence—where reverence for nature and culture guides human activity, ensuring that sacredness and sustainability reinforce rather than compromise each other.

7 Implications and Policy Recommendations

The findings of this study underscore the profound and multifaceted implications for the governance and sustainable stewardship of Himalayan sacred landscapes, revealing the intricate entanglement of ecological integrity, cultural continuity, and pilgrimage tourism. The convergence of climate change—manifested through glacial retreat, erratic precipitation, and rising temperatures—with intensifying tourism pressures demonstrates that conventional conservation or infrastructural interventions, in isolation, are insufficient. Sacred landscapes cannot be treated solely as ecological reserves or tourist commodities; rather, their protection necessitates an integrative socio-ecological and spiritual paradigm that recognizes landscapes as dynamic assemblages of natural processes, cultural practices, and ethical obligations.

Rising climatic variability destabilizes the hydrological and ecological foundations of sacred sites, while pilgrimage tourism and associated infrastructure amplify the vulnerability of these fragile systems, threatening the continuity of both ritual practice and ecological function. This duality highlights the need for governance frameworks that are not only ecologically informed but also culturally embedded and adaptive to rapidly shifting environmental and social conditions. Effective stewardship requires the institutionalization of indigenous ecological knowledge, ritualized resource management, and community-based governance within formal policy structures, thereby reinforcing resilience while sustaining intergenerational cultural continuity.

From a policy perspective, the management of pilgrimage tourism must move beyond purely economic or utilitarian approaches toward models rooted in ethical, environmentally sensitive, and culturally responsive principles. Regulating visitor flows, developing infrastructure that aligns with the carrying capacity of sensitive terrains, implementing robust and context-specific waste management systems, and promoting participatory, community-led monitoring initiatives are critical for mitigating the ecological footprint of tourism. Simultaneously, cultivating environmental and spiritual awareness among pilgrims can foster a culture of reverence, transforming tourism from a potential source of degradation into an instrument for sustaining sacred landscapes.

The study further highlights the necessity of multi-scalar, integrative governance approaches, where local communities, religious authorities, state institutions, and civil society organizations collaborate to craft adaptive, context-sensitive strategies. Such frameworks enable decision-making that is responsive to ecological variability, socio-cultural dynamics, and disaster risks, ensuring that interventions are neither top-down impositions nor fragmented initiatives. Embedding respect for both natural and cultural systems at the core of governance reinforces the principle that environmental conservation, cultural vitality, and ethical tourism are mutually constitutive rather than independent objectives.

Ultimately, the implications of this study call for a fundamental paradigm shift—from extractive, growth-driven models of tourism and development to a reverential ethic of coexistence, in which the sacredness of the landscape actively guides human engagement. By foregrounding relational ethics, spiritual ecology, and intergenerational responsibility, Himalayan sacred landscapes can be preserved not only as ecological and cultural reservoirs but also as living laboratories of resilience, demonstrating the potential for integrated stewardship that balances human aspiration with environmental and spiritual imperatives. In embracing this paradigm, the Himalayas may continue to serve as both a sanctuary for biodiversity and a repository of cultural and spiritual knowledge, capable of withstanding the dual pressures of a rapidly changing climate and anthropogenic intervention while nurturing the interlinked human and non-human communities that depend upon them.

8 Limitations

While this study provides a comprehensive synthesis of climate change, pilgrimage tourism, and cultural resilience in Himalayan sacred landscapes, it is subject to several inherent limitations. First, the reliance on secondary data constrains the study to the scope and quality of existing literature, reports, and official records. Variations in data collection methods, temporal coverage, and reporting standards across different sources may affect the consistency and comparability of findings. For example, visitor statistics for pilgrimage sites often differ between government reports and independent studies, potentially influencing assessments of tourism pressures.

Second, the study's reliance on published literature may introduce geographical and topical biases. Certain regions of the Himalayas, such as Uttarakhand and Jammu & Kashmir, are more extensively studied than others, resulting in uneven representation of sacred landscapes. Similarly, specific aspects such as cultural resilience and indigenous ecological knowledge are sometimes underreported, limiting the depth of analysis in these areas. Third, the absence of primary field data means that some site-specific ecological, social, and cultural nuances may not be fully captured. Localized variations in rituals, community governance, and microclimatic conditions could influence resilience outcomes in ways that secondary data alone cannot fully reveal. Furthermore, temporal changes in pilgrimage patterns and climate impacts may not be accurately reflected in older studies, leading to potential gaps in capturing recent dynamics.

Finally, the integration of climate, tourism, and cultural resilience data relies on interpretive synthesis rather than direct empirical measurement. While this allows for a broad understanding of interacting processes, it may limit the ability to quantify precise cause-effect relationships between environmental stressors, human activity, and cultural responses. Despite these limitations, the study provides valuable insights into the complex socio-ecological dynamics of Himalayan sacred landscapes. By triangulating multiple sources and critically evaluating existing literature, it offers a robust foundation for understanding the vulnerabilities and adaptive capacities of these regions, while highlighting areas where further empirical research is needed.

References

Adams, W. M. (2017). Conservation from above: Globalising care for nature. In D. Brockington & R. Duffy (Eds.), *Capitalism and conservation* (pp. 23–40). Routledge.

Adger, W. N. (2010). Climate change, human well-being and insecurity. *New Political Economy, 15*(2), 275–292.

Ahmed, R., Rather, A., Banerjee, P., Wani, G., & Saleem, S. (2025). Impact of climate variability on the feeding glaciers of potentially dangerous glacial lakes in the Jhelum Basin of Kashmir Himalaya, India. *Natural Hazards Review*. https://doi.org/10.1061/NHREFO.NHENG-2140

Akamani, K. (2020). Integrating deep ecology and adaptive governance for sustainable development: Implications for protected areas management. *Sustainability, 12*(14), 5757.

Ashraf, M. V., Misri, S., Pant, S., & Ahmad, S. (2024). Unorganized tourism practices and the allied challenges in the Himalayas. In *The resilience of traditional knowledge systems for a sustainable future: A focus on agriculture and food practices in the Himalayas* (pp. 193–220). Springer Nature Switzerland.

Batool, N., Wani, M. D., Shah, S. A., & Dada, Z. A. (2024). Tourists' attitude and willingness to pay on conservation efforts: Evidence from the west Himalayan eco-tourism sites. *Environment, Development and Sustainability*, 1–19. https://doi.org/10.1007/s10668-024-04679-2

Berkes, F. (2008). *Sacred ecology: Traditional ecological knowledge and resource management* (2nd eddc xx. ed.). Routledge.

Bernbaum, E. (2022). *Sacred mountains of the world*. Cambridge University Press.

Biswas, S. N., & Kumar, P. (Eds.). (2024). *Mountain tourism and ecological impacts: Himalayan region and beyond: Himalayan region and beyond*. IGI Global.

Bolch, T., Kulkarni, A., Kaab, A., Huggel, C., Paul, F., Cogley, J. G., et al. (2019). The state and fate of Himalayan glaciers. *Science, 336*(6079), 310–314. https://doi.org/10.1126/science.1215828

Eck, D. L. (2012). *India: A sacred geography*. Harmony Books.

Gagné, K. (2020). The materiality of ethics: Perspectives on water and reciprocity in a Himalayan Anthropocene. *WIREs Water, 7*(4), e1444.

Gawlik, A., Apollo, M., Andreychouk, V., & Wengel, Y. (2022). Pilgrimage tourism and sacred places in the Himalaya. In *Tourism and development in the Himalaya* (pp. 184–207). Routledge.

Gergan, M. D. (2017). Living with earthquakes and angry deities at the Himalayan borderlands. *Annals of the American Association of Geographers, 107*(2), 490–498.

Haeberli, W., Schaub, Y., & Huggel, C. (2017). Increasing risks related to landslides from degrading permafrost into new lakes in de-glaciating mountain ranges. *Geomorphology, 293*, 405–417.

Hait, M., & Sahu, P. (2024). Natural hazards as a source of environmental pollution in the global south. In *Biomonitoring of pollutants in the global south* (pp. 153–190). Springer Nature Singapore.

Howard, C. A. (2016). Walking with the gods: The Himalayas as (dis) enchanted landscape. In *Transnational frontiers of Asia and Latin America since 1800* (pp. 148–162). Routledge.

IPCC. (2023). *Climate change 2023: Synthesis report of the IPCC sixth assessment report*. Intergovernmental Panel on Climate Change. https://www.ipcc.ch/report/ar6/syr/

Ivakhiv, A. (2016). Green pilgrimage: Problems and prospects for ecology and peace-building. In *Pilgrims and pilgrimages as peacemakers in Christianity, Judaism and Islam* (pp. 85–104). Routledge.

Kala, C. P. (2014). Deluge, disaster and development in Uttarakhand Himalayan region of India: Challenges and lessons for disaster management. *International Journal of Disaster Risk Reduction, 8*, 143–152. https://doi.org/10.1016/j.ijdrr.2014.03.002

Kaliraj, S., Shunmugapriya, S., Lakshumanan, C., Suresh, D., Arun Prasad, K., & Srinivas, R. (2025). Flood risk zone mapping and future projections for the Thamirabarani river basin, southern India: Insights from decadal rainfall trends and GIS-based analytical hierarchy process technique. *Natural Hazards, 121*(5), 5327–5361.

Kealiikanakaoleohaililani, K., & Giardina, C. P. (2016). Embracing the sacred: An indigenous framework for tomorrow's sustainability science. *Sustainability Science, 11*(1), 57–67.
Kumar, P., Gupta, S. K., Korstanje, M. E., & Rout, P. C. (Eds.). (2024). Managing tourism and hospitality sectors for sustainable global transformation. IGI Global.
Kumar, S., & Shoshta, A. (2024). Revitalizing roots: Harnessing traditional socio-cultural institutions for sustainable development of the Himalayas. In *The resilience of traditional knowledge systems for a sustainable future: A focus on agriculture and food practices in the Himalayas* (pp. 265–281). Springer Nature Switzerland.
Mushtaq, S., Saleem, S., Ahmed, R., Tass, M. S., Rather, J. A., & Rather, G. M. (2024). Spatio-temporal analysis of land use land cover changes in South Kashmir region of North-Western Himalayas using Landsat data. *Discover Geoscience, 2*(1), 37.
Parrey, H. A., Shah, S. A., Bhat, M. S., Thoker, I. A., Jeelani, P., & Islam, M. N. (2025). Himalayas: A case study of Kashmir Valley. In *India III: Climate change and landscape issues in India: A cross-disciplinary framework* (p. 355).
Rana, R. S., & Chandra, R. (2021). Environmental impacts of religious tourism in Indian Himalaya: A case study of Kedarnath. *Tourism Recreation Research, 46*(2), 234–246. https://doi.org/10.1080/02508281.2020.1817387
Ranjan, R. (2024). Ecology of grief: Climatic events and disasters in the Himalaya. *Environment and Planning E: Nature and Space, 7*(5), 2058–2077.
Saleem, S., Ahmed Hajam, F., & Rather, J. A. (2021). Spatio-temporal analysis of land use land cover changes in Sind catchment of the Kashmir Valley, India. *Geo Eye, 10*(2), 35–42.
Saleem, S., Rather, J. A., Ahmad, S., Ahmed, R., & Hajam, L. A. (2024b). Assessment of livestock carrying capacity in the alpine grasslands of the Kashmir Himalayas. *Pastoralism: Research, Policy and Practice, 14*, 13541.
Saleem, S., Rather, J. A., Ahmed, S., Mushtaq, S., Ahmed, R., & Malik, I. H. (2024a). Above ground biomass estimation for alpine grasslands of Kashmir Himalayas using remote sensing and field-data. *Rangeland Ecology and Management, 96*, 117–127.
Semwal, R., & Tripathi, N. (2025). Fragile peaks: Understanding the environmental consequences of mountain tourism in the Indian Himalayas. In *Balancing Mountain tourism, cultural heritage, and environmental stability* (pp. 265–280). IGI Global Scientific Publishing.
Sharma, P., & Rai, S. C. (2019). Tourism and conservation: A case study of Sikkim Himalaya. *Current Science, 117*(3), 381–388. https://doi.org/10.18520/cs/v117/i3/381-388
Shinde, K. A., & Olsen, D. H. (2022). Reframing the intersections of pilgrimage, religious tourism, and sustainability. *Sustainability, 15*(1), 461.
Singh, R. B. (2018). *Sacred landscapes of the Himalaya: Conservation and management*. Springer.
Singh, R. P., & Rana, P. S. (2020). Faith and place: Hindu sacred landscapes of India. In *The Routledge handbook of place* (pp. 75–87). Routledge.
Snyder, H. (2019). Literature review as a research methodology: An overview and guidelines. *Journal of Business Research, 104*, 333–339. https://doi.org/10.1016/j.jbusres.2019.07.039
Sofield, T., & Li, S. (2013). Tourism governance and sustainable national development in China: A macro-level synthesis. In *Tourism governance* (pp. 91–124). Routledge.
Spiridonov, V., Ćurić, M., & Novkovski, N. (2025). Exploring natural hazards: From earthquakes, floods, and beyond. In *Atmospheric perspectives: Unveiling Earth's environmental challenges* (pp. 271–306). Springer Nature Switzerland.
Tiwari, P. C. (2017). Climate change and the sustainability of mountain environments in Himalaya. *Journal of Mountain Science, 14*(10), 2261–2278. https://doi.org/10.1007/s11629-016-4235-8
Wani, M. D., Ahmad Shah, S., & Ahmad Dada, Z. (2025d). Eco-conscious tourism in India: Investigating the interplay between green hotel practices and traveler decision-making. *Tourism and Hospitality Research*, 14673584251334238.
Wani, M. D., Akhter, S., Shah, S. A., & Dada, Z. A. (2025c). Does mountain tourism impact the socio-ecological balance of the Indian trans Himalayas, Ladakh? In *Balancing Mountain tourism, cultural heritage, and environmental stability* (pp. 89–108). IGI Global Scientific Publishing.

Wani, M. D., Batool, N., Dada, Z. A., & Shah, S. A. (2024b). Investigating the impact of community-based tourism on the residents' quality of life and their support for tourism. *Community Development*, *55*(1), 138–159.

Wani, M. D., Dada, Z. A., & Shah, S. A. (2023). Behavioural intention and satisfaction levels among pilgrim tourists in South Asia. *Tourism Review*, *78*(4), 1147–1163.

Wani, M. D., Dada, Z. A., & Shah, S. A. (2024a). The impact of community empowerment on sustainable tourism development and the mediation effect of local support: A structural equation Modeling approach. *Community Development*, *55*(1), 50–66.

Wani, M. D., Dada, Z. A., & Shah, S. A. (2024c). Divine experiences: Exploring the impact of tourist satisfaction on religious travel in Jammu and Kashmir, India. In K. Jermsittiparsert, I. Wekke, & P. Suanpang (Eds.), *Sustainable development, humanities, and social sciences for society 5.0* (pp. 387–414). IGI Global Scientific Publishing. https://doi.org/10.4018/979-8-3693-7989-9.ch019

Wani, M. D., Dada, Z. A., Bhat, W. A., & Shah, S. A. (2025b). Community-driven ecotourism in the trans Himalayan region: A sustainable model for cultural and environmental preservation. *Community Development*, 1–23. https://doi.org/10.1080/15575330.2025.2451358

Wani, M. D., Shah, S. A., & Dada, Z. A. (2025a). Fostering sustainable tourism in India: Harnessing cultural and heritage assets for enduring development. In *Integrating architecture and design into sustainable tourism development* (pp. 179–214). IGI Global Scientific Publishing. https://doi.org/10.4018/979-8-3693-7388-0.ch007

Wani, M. U. D., Dada, Z. A., & Shah, S. A. (2022). Building peace through tourism: The analysis of an ongoing Siachen glacier dispute between India and Pakistan. *Asian Journal of Comparative Politics*, *7*(4), 836–848. https://doi.org/10.1177/20578911221118730

Wouters, J. J. (2024). Himalayan climes and multispecies encounters: An introduction. In *Himalayan climes and multispecies encounters* (pp. 1–29). Routledge.

Depleting Waters and Drying Landscapes: Lived Narratives from Pastoralists of Pir Panjal Range of Poonch, Jammu & Kashmir, India

Khalida Bi and Bhat Iqball Majeed

Abstract Climate change has emerged as a serious challenge globally, particularly in ecologically sensitive and fragile mountain regions like the Himalayas. The Pir Panjal range in Jammu and Kashmir, part of the northwestern Himalayas, is experiencing noticeable environmental changes driven by rising temperatures, erratic precipitation, glacial retreat, and landslides. These changes have resulted in the drying up of wetlands/marshlands (locally called Sar Aalo Tallou), reduced snowfall, increased water scarcity, and higher vulnerability of transhumant pastoralist communities who depend heavily on natural resources for their survival. This chapter is based on a pilot study conducted in Jammu Shaheed area of Hari village in Poonch district, used qualitative approach to document the lived experiences, oral narratives and indigenous knowledge of local pastoralist communities. The study focuses on the impact of climate change on landscape drying, particularly marshlands/wetlands and spring, and the growing water crisis in high-altitude grazing areas. Local accounts reveals that past water- rich landscapes have gradually dried due to inconsistent snowfall, tree cutting, landslides and soil erosion. The chapter also highlights how pastoralists are coping with water scarcity by walking long distances, storing snow for later use, or depending on unreliable government pipeline schemes. The findings reveal that change in snowfall patterns, glacial retreat, deforestation, and increased frequency of landslides have severely disrupted local water sources such as springs, wetlands/marshlands and snowmelt streams. These changes not only threatening water scarcity but also disrupting traditional transhumance practices, leading to altered migration routes, reduced milk production, livestock loss, and increased livelihood vulnerability. Drawing from oral histories and ecological observations, the study emphasizes the importance of integrating local knowledge into climate adaptation planning. In conclusion, the chapter emphasizes the need for sustainable and community based solutions, such as spring fed water pipelines, rainwater harvesting, and watershed protection. It also recommends to include pastoralists traditional knowledge in policy planning and improve environmental awareness to help communities to deal with challenges. The chapter contributes to the growing

K. Bi (✉) · B. I. Majeed
Department of Social Work, Central University of Jammu, Suchani, Samba, J&K, India
e-mail: kh23psow01@cujammu.ac.in

R. Ahmed et al. (eds.), *Climate Change, Water Resources, and Natural Hazards in the Himalayas*, Advances in Geographical and Environmental Sciences,
https://doi.org/10.1007/978-981-92-0835-7_17

discourse on climate change and adaptation by highlighting the voices of mountain communities who are on the frontline of climate change.

Keywords Water scarcity · Climate change · Sar aalo tallou · Poonch · Pir Panjal range

1 Introduction

The growing climate crisis is one of the biggest challenges the world is facing today. Every day, we hear or read about climate-related disasters such as cloudbursts, flash floods, irregular rainfall, landslides, melting glaciers, prolonged drought, water shortages, and the drying up of springs, wetlands, and marshlands. Climate change is having a strong impact on the himalayas, fragile and young mountain of himalayas are very sensitive to even small climatic changes. The region is warming faster than the global average which become the cause of glacier melt, leading to formation, growth and disappearance of glacial lakes (Ahmed et al., 2021). A study conducted by researchers of National Institute of Technology Srinagar revealed that around 122 glaciers are shrinking due to climate change in Pir Panjal only (Kashmir Observer, 2024). In 2025, the situation has become alarming in the Indian Himalayan states like Himachal Pradesh, Uttarakhand, Ladakh, and Jammu & Kashmir. In Jammu & Kashmir itself, more than 100 people have lost their lives due to flash floods and cloudbursts. In Chasoti village of Kishtwar district, many precious lives were lost (India Today, 2025). Similarly, in Warwan village of the same district, nearly 190 houses were destroyed and 45 cattles were killed (Kashmir Observer, 2025). Kalaban village was severely affected by the disaster, a village in the Mendhar tehsil of Poonch district. After heavy rainfall in August and early September 2025, the village started sinking. As a result, around 70 houses were completely destroyed, and religious places as well as schools were also damaged (Greater Kashmir, 2025).

The present chapter aims to provide a brief overview on the Pir Panjal mountain range. It then reviews existing research on climate change, mountain ecosystems and water resource vulnerabilities. The chapter also discusses pastoralists and their deep dependency on natural water systems and grazing landscapes, both at the national and international levels. Further, the chapter presents the study area located in the Pir Panjal range of Jammu & Kashmir, documenting local experiences and oral narratives of environmental change. It explores the lived realities of water insecurity and the gradual disappearance of wetlands/marshlands, locally known as *Sar Aalo Tallou*. These insights are drawn from qualitative fieldwork interviews. Moreover, the chapter discussed the main drivers of climate change and ends with conclusion and recommendation.

1.1 Overview of Pir Panjal Mountain Range

The UT of Jammu and Kashmir (J&K) is situated in the northwestern segment of the Himalayan mountain range in India. It lies between latitudes 32.17° and 36.08° North, and longitudes 73.23° and 76.47° East (Pir, 2024). The UT of Jammu & Kashmir is also referred to as Land of Mountain Ranges. These mountain ranges start with the Shiwalik Hills in the southern part of the region, then rise through the Pir Panjal range, followed by peaks of the Greater Himalayas, continued into the Zanskar range, and ultimately reach the high-altitude Karakoram range in the north (Romshoo et al., 2020). Whereas Pir Panjal mountain range acts as a natural barrier between the Kashmir Valley and the outer Himalayas. Banihal Pass is located in the southern east part of the valley, standing at an elevation of 2832 m. During winter, the Pir Panjal range is covered with snow, making it difficult to traverse. At the other end of the range are the Baramulla Pass (1582 m) and the Hajipir Pass (2750 m), the latter providing a connection between Poonch and Uri (Raina, 2002). Pir Panjal region, also known as lesser Himalaya, ranges in elevation from 3000 to 4500 m above mean sea level. It is characterized by steep, mostly barren slopes and forested summits that receive snowfall during certain seasons. The area has a low population density, with minimal human activity (Romshoo et al., 2020). Stretching about 2621 km long and 50 km wide, the Pir Panjal mountain range separates the Kashmir valley from the outer Himalayas (Mishra & Rafiq, 2017). The average highest temperature in the Pir Panjal range ranges from 3.9 °C in February to a peak of 11.6 °C in April. Meanwhile, the mean minimum temperature varies from a low of −4.1 °C in February to a high of 2.5 °C in April (Gusain et al., 2004).

2 Climate Change and Water Insecurity in Mountain Ecosystem

Climate change causes different problems in different places, it create droughts in some areas and floods in others. These extreme weather conditions affect how much food can be produced, which hits poor communities the hardest and makes it harder for people to get enough food. One major effect of climate change is rising temperatures, which increase the need for water for both people and animals. At the same time, heatwaves and droughts reduce the amount and quality of available water (Kushawaha et al., 2021). Climate change poses a serious risk to India's water resources. Based on current patterns of water demand and usage, it is expected that by 2030, only around half of the country's water needs will be met. By 2050, this shortage is expected to rise to approximately 68% (Mir et al., 2021). Climate change affects north-western Himalayas water system by decreasing snowfall, rise in winter rainfall, and rising glaciar melt. Therefore, spring water flow rises, summer shortages become more common, and floods occur more frequently, create challenges for water management and ecological stability (Ahmed et al., 2025). People living in mountainous region face challenging conditions, include physical hardships in daily life, frequent natural hazards, and limited agricultural potential. In fact, only around 3% of mountain land is considered highly suitable for rainfed farming which shows how limited livelihood options are available for many mountain communities (UNEP-WCMC, 2002).

Mountains are among the earliest regions to feel the impacts of climate change. Glaciers in these areas begin to retreat first, triggering a chain reaction that includes rising temperatures, rapid glacier melt, increased flooding, soil erosion, resource shortages, and widespread ecosystem decline. As a result, mountain communities are often the first to be affected (Sharma & Choudhary, 2021). These mountain areas also known as high altitude ranges are used as seasonal grazing lands for livestock. They support the traditional practice of transhumance, followed by more than six lakh transhumant pastoralists of J&K (Saleem et al., 2024). However, climate change, mainly caused by human activities, is creating problems like droughts, floods, and water shortages, which badly affect the lives and animals of pastoralist communities. These people depend on nature and face many other challenges like poverty and weak services (Mir & Batool, 2021). Nomadic pastoralists of Indian Himalayas have borne the brunt of pastureland degradation, the deterioration of these pasturelands also become a cause of water shortages in some villages, as springs and streams that were once water rich are now drying up (Sen, 2025). Climate change is severely affecting Pakistan's pastoral communities by increasing droughts, floods and livestock diseases. Limited access to water and pasture is leading to poor animal health, reduced productivity, and higher livestock losses (Ahmad & Afzal, 2021). A study from Mongolia highlights that climate change has significantly affected ecosystem services in Mongolia's steppes, particularly grassland productivity and water availability. Rising temperatures and altered rainfall patterns have increased evapotranspiration in arid regions, leading to a reduced water resources. This decline has disrupted traditional herding practices by altering migration patterns and increasing pastureland degradation near rivers and streams due to both environmental such as water shortage, pasture degradation and socio-economic pressures (Tugjamba et al., 2021).

Overall, climate change is having a wide ranging and serious impact on water resources, food security, and traditional livelihoods, especially in mountain and pastoral regions. From shrinking glaciers and drying springs in the Himalayas to degraded grasslands in the steppes of Mongolia, the availability of essential resources like water and pasture is rapidly declining. These changes are hitting the poorest and most vulnerable populations the hardest, especially those who depend on livestock for survival.

2.1 Water, Pasture and Livelihoods: A Cross-Regional Perspective

Water sources are either disappearing completely or becoming seasonal. Many springs are drying up, causing water shortages in several parts of the Himalayan region (Sharma & Choudhary, 2021). Approximately 1.2 billion people worldwide currently face water scarcity and this number is projected to rise by an additional 600 million by 2025 (United Nations, 2007). Cooper et al. (2019) conducted a study on the pastoralists in the Afar region of Ethiopia. The study revealed that the livelihoods of these pastoralists are heavily reliant on local environmental resources to meet basic human and animal needs, such as water and food. The Afar region is semi-arid, with temperatures often exceeding 50 °C and some areas receiving less than 200 mm of rainfall annually. Over the years, the livelihoods of the community have become increasingly vulnerable due to the growing pressures of climate change and water scarcity. Waruguru et al. (2021) also showed that during drought prone season pastoral nomads of Kenya, lost their livestock, the only source of their livelihood. Nuvey et al. (2020) conducted study on pastoralist livestock farmers of Ghana revealed that they lost their livestock due to various factors such as climate change events, prolonged droughts, pastures shortages etc. Kimaro (2017) revealed that climate change have negative effects on livestock production, lack of pasture and water shortages. Another study conducted among the pastoralists of Tanzania by Magita et al. (2017) revealed the same results as above which is climate change. Pastoralists lost their livestock due to intense climate change such as shortage of water and pastures. For mitigating their loss the pastoralists community make a hard decision to slaughter the animals that are nearabout death.

Narbat (2023) examined the impact of environmental changes on the wellbeing of the Dhangar community, a group of pastoralists in Maharashtra. The study found that severe drought, water shortages, and lack of pasture and fodder forced the Dhangar community to settle in one place where they noticed better rainfall. They began selling milk from their livestock to support themselves. However, the drought caused them to lose many of their animals, leading to significant financial loss. A year later, Narbat (2024) conducted another study among the Dhangar pastoralists in Satara district, focusing on the vulnerabilities they faced due to drought. The findings were similar to the previous study. Environmental factors, including prolonged drought, water shortages, and the loss of grazing lands, led to the death of livestock and lower production, which negatively affected their livelihoods and health. As a result, they were forced to relocate and adopt a sedentary lifestyle due to these unforeseen circumstances. Similarly, Singh and Bhandari (2022) conducted a study on the problems and challenges faced by the Van-Gujjar community of Yumuna Bhasin during transhumance. The study revealed that decreasing availability of pastureland has led to a reduction in both the quantity and quality of fodder, which, in turn, affects milk productivity. During the transient phase of their movement, the cattle often stop producing milk due to continuous movement, which depletes their energy. Since the livelihood of these transhumant communities is heavily dependent on livestock products, these challenges contribute to a decline in their economic output. The study also notes that changing social, physical, and environmental conditions, such as heat waves, water scarcity, and outbreaks of deadly diseases and epidemics, have further hindered the practice of transhumance.

Ahmed et al. (2023) extensively examined the impact of climate change on the livelihoods of Bakarwal nomads in the western Himalayas. The study highlighted that various environmental and climatic changes have significantly affected the traditional livelihood activities of the Bakarwal community. According to their findings, climate change has forced nomadic herders to migrate earlier than usual, exposing them to a range of difficulties. These challenges include ice cover on mountain passes, lack of pasture along migration routes, and the risk of unexpected avalanches, thunderstorms, hailstorms, and heavy rainfall while traveling to upland areas. Such factors create substantial obstacles for herders to manage livestock. Additionally, the study noted that reduced water availability and an increased risk of environmental diseases are contributing to animal starvation and death, as well as a decline in the quality and quantity of milk, wool, and meat, which in turn lowers the market price of these goods. Moreover, Tufail et al. (2024) have discussed about the climate change effects on the economic and cultural practices of gujjars and bakarwal nomadic community of Jammu & Kashmir. The study revealed that climate change has imposed daily hardships on these communities, compelling many to abandon the traditional practice of transhumance and settle on the lower slopes of the Himalayas. Mir et al. (2022) also discussed about the impact of climate change on patoralists of Jammu and Kashmir. It revealed that 35% participants responded negative effects of climate change on livelihoods of pastoralists.

The evidence from different pastoral communities across South Asia and Africa consistently highlights the severe and growing impact of climate change on their traditional livelihoods. Water scarcity, prolonged droughts, declining pasturelands and extreme weather events have made it difficult for these communities to maintain their traditional ways of life. Many have lost livestock, which is their main source of income and food. As a result, some pastoralists have been forced to settle in one place, stop the practice of transhumance, or find new ways to survive. These environmental challenges have also affected their health, economic condition and cultural traditions. The growing risk caused by climate change show the urgent need for support through sustainable development policies, better access to resources, and climate resilient solutions to help in protecting the future of these vulnerable communities.

3 Material and Method

3.1 Research Approach

While discussing climatic changes and water scarcity at both global and national level. It is important to also consider the growing problem of water scarcity in mountainous regions, which are especially vulnerable to the increasing impacts of climate change. These high-altitude areas, though less populated play a crucial role in sustaining ecosystems and communities. The present chapter is based on the findings of a pilot study carried out as part of a doctoral research project in the Pir Panjal range of Poonch district, Jammu and Kashmir. This region, home to transhumant pastoralist communities, is witnessing a clear signs of ecological stress, including drying springs, shrinking wetlands/marshlands and uncertain weather patterns. Adopting a qualitative approach, the chapter documents the experiences, narratives, and perspectives of pastoralist communities concerning environmental changes and the growing water crisis in the region. A qualitative research approach was adopted for the present study, as it provides a flexible and in-depth way to capture human experiences that are often difficult to quantify. As Bernard (1994) argues, qualitative methods do not depend on numerical measurement but instead focus on rich, narrative understanding. Moreover, Palinkas (2015) highlights that qualitative research enables scholars to explore unaddressed issues, and evaluate the social and emotional issues embedded in people's daily lives. In this context, the voices of the local communities provides not just as data, but as a powerful lens through which the realities of climate change and water insecurity in mountain ecosystem can be better understood.

3.2 Study Site

As a case study, the Jammu Shaheed area of Hari village, located in the Pir Panjal range and known for its unique ecosystem, has been selected. This region serves as a representative example of many mountainous areas, as its strengths and challenges closely reflect those found across similar ecosystems. The traditional practice of transhumance is commonly practiced here. Jammu Shaheed functions both as a village and as meadowland, pastoralists from this village migrate to higher altitudes during the summer, while people from lower areas move into Jammu Shaheed and adjoining areas during the same season. Therefore, it serves as a seasonal settlement for some and grazing ground for others. The area faces various challenges that help in identifying the driving forces behind climate change and seasonal shifts over time. This chapter focuses on key aspects such as water scarcity in mountainous regions, the issue of drying springs and wetlands, public perceptions and attitudes, the resulting impacts, using this case study as a central example.

Hari village located in tehsil Surankote of district Poonch, Jammu & Kashmir (Census, 2011). Poonch district, located in Jammu and Kashmir, India, lies at approximately, between 73° 58′ to 74° 35′ East longitude and 33° 25′ and 34° 10' North latitude (Manzoor, 2025) with an average elevation of 981 meters above mean sea level, the Poonch region exhibits significant topographic variation with elevations ranging from 883. 92 m to 4742.688 above mean sea level. The Poonch region is also homes to Tata Kutti, the highest peak in the Pir Panjal range which stands at 4765 m (Gregwal, 2022). Geographically, it is bordered by the Pir Panjal range to the north and east, Rajouri district to the south, and Pakistan-occupied Kashmir (PoK) to the west. According to National Informatic Centre, Govt of India, Poonch district cover an area of 1674 km^2, divided into six tehsils, 11 blocks and includes a total of 178 villages (NIC, 2025). The Poonch district is chanelled by major rivers such as the Betaar river, Mendhar river, Surran river, and Mandi River, along with various smaller streams. It has numerous permanent springs that provide the primary source of surface water throughout the area. The climate in Poonch is classified as humid subtropical, characterized by warm summers and mild winters. Poonch receives an average annual rainfall of 929.2 mm. The majority of this precipitation occurs during the monsoon season, while only a small amount of rainfall is recorded in March. In the higher elevations, precipitation mainly falls as snow during the winter months (Manzoor & Jazib, 2023) (Fig. 1).

4 Understanding Sar Aalo Tallou, Linguistic and Cultural Meaning of Sar Aalo Tallou

The Pir Panjal Mountain range, home to diverse transhumant pastoralist communities, is undergoing profound ecological transformation. Wetlands and marshlands that were once water-rich-locally known as *Sar Aalo Tallou*, are drying up due to shifts in precipitation patterns, glacial retreat, deforestation, and frequent landslides. According to local ecological knowledge, *Sar* refers to a condition where a person or animal submerged so deeply in the wetlands/marshlands that only the head remains visible, while *Tallou* denotes low-lying terrain prone to water accumulation. *Aalo* functions as a helping verb, collectively describing terrain that is both dangerous and deeply waterlogged. Together, the term illustrates areas where people or animals could sink into the soft, waterlogged ground. However, these landscapes are now drying up. The particularly mentioned *sar aalo tallou* located in selected study site of Jammu Shaheed area of Hari village of Poonch district. Pastoralist communities in this region bearing the consequences of water scarcity across these high-altitude landscapes. They walked across several kilometers of rugged, mountainous terrain to secure water for themselves and their livestock.

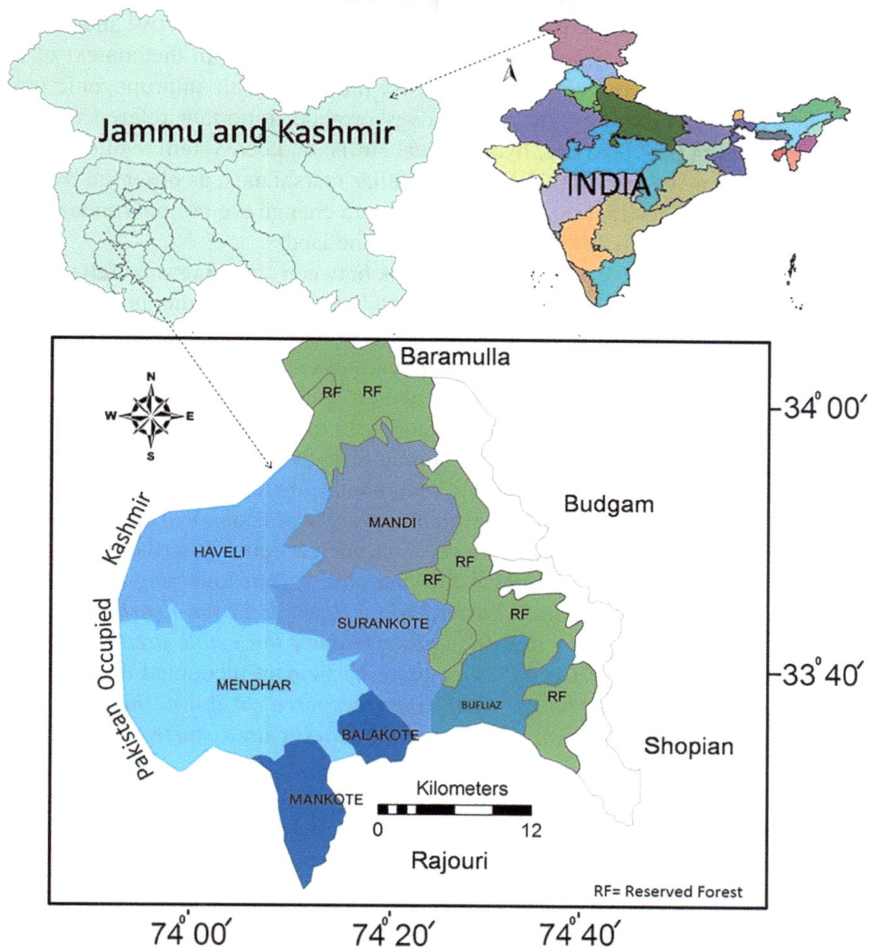

Fig. 1 Location map of study area. *Source* Researchgate (Mughal et al., 2017)

4.1 *Oral Narratives: Understanding Drying Landscape Through Indigenous Perspectives*

Traditional ecological knowledge offers valuable insights into historical environmental conditions and landscape dynamics. As local resident recount, *"Our elders used to say that this low-lying land was once filled with water and mud, like a wetland or marsh. It was so soft and waterlogged that anyone unfamiliar with the area, whether animal or human, could easily get stuck or sink into it"*. Again he replied, "*The low-lying wet/marsh land (locally known as Sar aalo tallou) where people or animals used to sink has slowly dried up over the past 40 to 50 years*".

Such oral histories are not merely folklore; they reflect collective memory of ecosystems that have undergone significant change overtime. In the context of the Himalayan region, where geophysical fragility intersects with anthropogenic pressures and climate variability, these narratives provide important reference points for understanding past land use, hydrological shifts, and ecosystem resilience. The progressive drying of *Sar Aalo Tallou* and similar marshlands, as observed by local communities, confirms that the impacts of climate change are not only measurable but also deeply felt by those who live closest to the land.

Community members have noted a clear link between changing snowfall patterns and the gradual disappearance of local wetlands/marshlands. As another participant replied, *"That low-lying land has dried up because snowfall is no longer consistent, sometimes it's heavy, sometimes light, and it now melts quickly, which causes the area to dry out"*. These observations shows the need to include local climate knowledge when studying how the environment is changing in the area.

Among the elders of the Himalayan communities, vivid recollections of landscape change serve as valuable records of environmental transformation. These personal narratives, passed down through generations, reveal how ecological shifts are experienced and remembered at the local level. One such account describes the gradual degradation of a once water saturated lowland area, *"That low-lying land used to have a lot of water. Slowly, soil started sliding down from the nearby hills. The upper parts of the land broke and fell into it, making the whole area muddy and swampy. And now, it has dried up"*. One more participant recounted the impact of tree cutting on the lands stability, *"When the trees were cut down, the land began to slide. Landslides started happening, and the marshy area started disappearing slowly"*.

Together, these accounts illustrate a cycle of environmental disruption, how deforestation and subsequent landslides caused soil and debris to slide into the wetland areas, disrupt their natural function such as storing water and supporting life, leads to the gradual drying and disappearance of these vital ecosystems. These narratives provide crucial local perspectives on the slow but profound changes in climate and land stability, reflecting broader processes linked to climate change and human activity. When combined with scientific research, such oral narratives deepen our understanding of the complex relation between natural forces and traditional land use in framing the Himalayan landscape.

5 Living with Scarcity: Pastoral Adaptation and Resilience

Access to water resources is closely linked to the health and wellbeing of human communities. In many Himalayan villages, the quest for safe drinking water is a daily ordeal. Access to clean drinking water was not only physically challenging but also a source of social tension within the community. One young woman's account reveals the hardships and struggles of collecting water from a limited natural spring, *"We walked several kilometers to collect drinking water from a natural spring. Since the spring produced very little water, we often had to wait a long time in line. Many other girls would also come to collect water, and sometimes quarrels broke out among us, as everyone wanted to fill their containers quickly and go home"*. Such narratives bring to light the complex realities faced by communities where water scarcity not only strains physical ability but also influences social relationships, underscore the critical need for sustainable water solutions.

The onset of rains, while important for the environment, brought their own set of challenges to water collection, often compromising the quality and availability of spring water. One community member shared how these changes disrupted daily life, *"Whenever it rained, the spring water turned muddy. After that, we couldn't collect water and had to wait for a long time, sometimes even an entire day. It created a lot of problems for us"*. This accounts illustrates the challenges posed by environmental variability on water supply systems and highlights the need for resilient water management strategies.

To combat chronic water shortages, a government program have recently installed pipeline water supply systems to deliver water from distant sources. However, the new infrastructure comes with its own set of challenges, as one resident stated, *"Now this month, water is being supplied to us through pipeline from a far-off meadow, as part of a government scheme. But if the pipe breaks in between, the supply stops, and we have to go a long distance to repair it again"*. While such projects/schemes represent progress in addressing water scarcity, they also require ongoing maintenance and local involvement to ensure reliability.

Access to clean and reliable water is not just a basic need, but a daily struggle for many communities. One resident captured this challenge perfectly when they shared their experience, *"Whenever water arrived, we tried to fill the tanks quickly, and store it in plastic containers. We also managed to make the water last for one or two days. But we want a permanent solution to all these irregularities and challenges. So that we can get water easily both for personal use and for livestock also"*. This account highlights the urgent need for sustainable water solutions. Temporary measures are no longer enough. There is a need of reliable infrastructure that ensures consistent access to water for both households and livelihoods.

Pastoralist communities have long relied on indigenous knowledge and creative coping mechanisms to adapt to harsh environmental conditions. One such strategy shared by a herder, illustrates their resilience in water scarce winters, *"During winter we stored snow in plastic tanks. Once it melted, we used the water for ourselves and for our livestocks"*. While melting snow for water demonstrates remarkable adaptability, it also highlights the vulnerability of these communities to climate variability and long term environmental change. Sustainable water planning must build on such indigenous strategies while ensuring reliability.

Overall, the experiences of Himalayan communities reveal that access to clean and reliable water remains a daily struggle, affects both physical wellbeing and social relations. While government efforts such as pipeline systems offer some relief, challenges like pipeline disruptions, environmental changes, and irregular water supply continue to pose serious problems. Community members, especially pastoralists have

Table 1 Comparative perspectives of pastoralists on landscape changes and water scarcity in the Pir Panjal range of Poonch (before 1980's and 2025)

Theme	Views about past (before 1980's)	Present Situation (2025)
Landscape changes	"Our elders used to say that this low-lying land was once filled with water and mud, like a wetland or marsh, has slowly dried up over the past 40–50 years"	"That low-lying land has dried up because snowfall is no longer consistent, it now melts quickly, which causes the area to dry out"
	"That low-lying land used to have a lot of water. Slowly, soil started sliding down, making the whole area muddy and swampy. And now, it has dried up. When the trees were cut down, the marshy area started disappearing slowly"	
Water scarcity	"We walked several kilometers to collect drinking water from a natural spring, and quarrels broke out"	"The same issue persists today; long treks, dispute, and limited spring water"
	"Whenever it rained, the spring water turned muddy"	"Rain still disrupts water collection"
	"During winter we stored snow in plastic tanks. Once it melted, we used the water"	"We still melt snow for water in winter due to lack of alternatives"
		"Water is being supplied through pipeline from a far-off meadow. But if the pipe breaks, we have to go a long distance to repair it"
		"We tried to fill the tanks quickly and store it, but we want a permanent solution for both personal use and livestock"

Source interview narratives

shown resilience by using traditional methods like storing snow, but these are not long term solutions (Table 1).

6 Disappearing Ecosystem: Drivers of Climate and Landscape Change

Between 1975 and 2005, floods accounted for 35% of all disasters in South Asia. In the Himalayan region, floods takes various forms including glacial lake outburst floods, riverine floods, floods caused by failures of landslide dams, flash floods etc. A significant upward trend in flood related disasters was observed between 1964 and 2018, which has become a major global concern (Shrestha, 2008). Changes in climate, land use and land cover are key global factors contributing to environmental change and they are especially critical in dryland regions as they accelerate the process of desertification (Asner et al., 2004). Wetlands are some of the most threatened ecosystems in the world. In India, they are increasingly being affected by various human activities. Rapid population growth, major changes in how land is used, growing development projects, and poor management of water catchment areas have all led to a major decline in the health and extent of wetlands across the country (Prasad et al., 2002).

Most wetlands are drying up because the air is getting hotter, and there isn't enough rain to replace the water lost through evaporation. Temporary wetlands, especially those without streams or rivers flowing into them, could disappear completely, particularly if there is less rain and more groundwater is used for homes. On the other hand, too much rain create flood situation, making some wetlands bigger and deeper. But heavy rainfall can also wash in more dirt and pollution, which also harm the wetlands (Khan & Patel, 2021). Over half of the worlds wetland are already dried up (Kushawaha, 2021). The State of India's Environment 2020 report highlights a steady decrease in the number of Himalayan springs, mainly due to changes in land use patterns and ongoing environmental degradation. It is also stated that by 2030, the number of precipitation days is likely to increase by 5–10 days, where the eastern part of union territory of Jammu and Kashmir, expected to see a rise of over 15 days (Down to Earth, 2020). According to intergovernmental panel on climate change, the global temperature expected to rise from 1 to 5 °C over a time period of 20–80 years (Bates, 2008). Scientific research has confirmed that climate change driven by human activities is taking place. This is mainly due to the rise in greenhouse gas levels caused by burning fossil fuels and widespread deforestation (Solomon et al., 2007).

Floods and drying wetlands are becoming bigger problems in the Himalayan region and across India because of climate change and human activities. Changes in how land is used, higher temperatures, inadequate rain are a threat to important water sources and natural habitats. Since global temperatures are expected to rise in the future, it is very important to take quick action to protect the environment and manage water and land carefully. Working together with local communities and making good plans can help in reducing the risks of floods, drying wetlands, and other problems caused by climate change.

7 Conclusion and Recommendation

Lack of water is now a serious problem facing people around the world. Around two billion people across the world do not have access to clean and safe drinking water in their daily lives (WHO/UNICEF, 2019). It is estimated that by 2050 around five million will face water shortages for at least one month each year (GCA, 2019). The Pir Panjal region, like many parts of the northwestern Himalayas, is witnessing a profound ecological transformation. The oral narratives and lived experiences of transhumant pastoralists reveal the devastating effects of climate change on water sources such as springs, wetlands, and seasonal snow. These communities, heavily dependent on natural water systems for both domestic and livestock needs, are among the first to feel the pressure of increasing temperatures, erratic precipitation, glacial retreat, and extreme weather events. The disappearance of wetlands/marshlands locally known as *Sar Aalou tallou* and the increasing burden of walking long distances for water are not just environmental challenges but also social and cultural disruptions. Despite this, the community continues to exhibit remarkable resilience through adaptive strategies like snow storage, migration shifts, and participation in government pipeline projects. This chapter shows that pastoralists are not just victims of climate change, they are also key knowledge holders. Their experiences give us important clues about how the land and climate are changing. These voices must be included in future planning and policies.

To support pastoralist communities in the Pir Panjal region, as they are facing growing water scarcity and climate related challenges, there is an urgent need for long term, locally appropriate solutions. Efforts should focus on developing simple and low maintenance water infrastructure, such as spring fed pipelines, rainwater harvesting systems, and protected water tanks that can be managed by the community. It is also important to recognize and integrate traditional ecological knowledge of pastoralists, as their observations and adaptive practices offer valuable insights into local climate change. Futhermore, it is important to strengthen community awareness through environmental education and training programs to help in promoting sustainable practices in water use, grazing and land use. Government and non-government agencies should work closely with local communities to ensure that development schemes are culturally sensitive, easily accessible and effectively maintained.

References

Ahmad, D., & Afzal, M. (2021). Impact of climate change on pastoralists' resilience and sustainable mitigation in Punjab, Pakistan. *Environment, Development and Sustainability, 23*(8), 11406–11426. https://doi.org/10.1007/s10668-020-01119-9

Ahmed R, Saleem S, Shamim T, Javed S, Malik IH et al (2025) Assessing the climate change impacts in the Jhelum Basin of North-Western Himalayas, Nature Environment and Pollution Technology An International Quarterly Scientific Journal, 24 (1):175–185.

Ahmed, R., Wani, G. F., Ahmad, S. T., Sahana, M., Singh, H., & Ahmed, P. (2021). A review of glacial lake expansion and associated glacial lake outburst floods in the Himalayan region. *Earth Systems and Environment., 5*, 695–708.

Ahmed, S., Iqbal, Z., & Prakash, C. A. (2023). Impact of climate change on nomadic herders' livelihoods: Evidence from Bakarwal community in the Western Himalayas in India. *Environmental Development, 48*.

Asner, G. P., Elmore, A. J., Olander, L. P., Martin, R. E., & Harris, A. T. (2004). Grazing systems, ecosystem responses, and global change. *Annual Review of Environment and Resources, 29*, 261–299.

Bates, B., Kundzewicz, Z., & Wu, S. (2008). *Climate change and water*. Intergovernmental Panel on Climate Change.

Bernard, H. R. (1994). *Research methods in anthropology: Quantitative and qualitative approaches*. Sage Publications.

Census. (2011). *Population finder*. Government of India. https://censusindia.gov.in/census.website/data/population-finder

Cooper, S., Hutchings, P., Butterworth, J., et al. (2019). Environmental associated emotional distress and the dangers of climate change pastoralist mental health. *Global Environmental Change*. https://doi.org/10.1016/j.gloenvcha.2019.101994

Down to Earth. (2020). *The state of India's environment 2020 in figures*. Centre for Science and Environment.

Global Commission on Adaptation (GCA). (2019). *Adapt now: A global call for leadership on climate resilience* (p. 3) Global commission on Adaptation. Available from: https://gca.org/assets/2019-09/GlobalCommission.

Greater Kashmir. (2025). *Land sinking in Poonch's Kalaban village damages 63 houses, displaces 70 families*. www.greaterkashmir.com › jammu › land-sinking-in

Gregwal, J. S. (2022). *Glimpses of Poonch, Poonch India's Invincible Citadel*. Lancer Publishers.

Gusain, H. S., Singh, A., Ganju, A., & Singh, D. (2004). Characteristics of the seasonal snow cover of Pir Panjal and great Himalayan ranges in Indian Himalaya. In *Proceedings international symposium on snow monitoring and avalanches (ISSMA)* (pp. 97–102).

India Today. (2025). 45 dead, over 200 missing as cloudburst in J&K's Kishtwar triggers flash flood. Retrieved from www.indiatoday.in

Kashmir Observer. (2024). Floods, water woes on the horizon. 122 glaciers shrinking in Pir Panjal: Report epaper.kashmirobserver.net › uploads › epaper

Kashmir Observer. (2025). Warwan Cloudburst: 190 Houses Damaged, 45 Cattle Killed. kashmirobserver.net › 2025/09/01 › 190-houses

Khan, R., & Patel, V. (2021). The influence of global climate change on freshwater ecosystem. In *Water conservation in the era of global climate change*. Elsevier. https://doi.org/10.1016/B978-0-12-820200-5.00002-6

Kimaro, E. G., Mor, S. M., Gwakisa, P., & Toribio, J. A. (2017). Seasonal occurrence of Theileria Parva infection and management practices amongst Maasai pastoralist communities in Monduli District, northern Tanzania. *Veterinary Parasitology, 246*, 43–52.

Kushawaha, J., Borra, S., Kushawaha, A. B., Singh, G., & Singh, P. (2021). Climate change and its impact on natural resources. In *Water conservation in the era of global climate change*. Elsevier. https://doi.org/10.1016/B978-0-12-820200-5.00002-6

Magita, S. Y., & Anthony, Z. S. (2017). Effects of climate stress to pastoral communities in Tanzania: A case of Mvomero District. *Livestock Research for Rural Development, 29*(8).

Manzoor, J. (2025). Evaluation of prevailing agroforestry systems and their components in Poonch District of Jammu and Kashmir, India. *Environmental Science Archives, IV*(1) ISSN: 2583-5092.

Manzoor, J., & Jazib, M. A. (2023). Local perception on climate change and its impact on livelihood: A case study from Poonch district of J&K, India. *Sustainability, Agri, Food and Environmental Research, 11*(X) (ISSN: 0719-3726). https://doi.org/10.7770/safer-V11N1-art78

Mir, S. A., & Batool, M. (2021). Significance of inherent knowledge to deal with climate changes in India. *International Journal of All Research Writings, 2*(12) https://www.researchgate.net/publication/359512789

Mir, B. H., Kumar, R., Lone, M. A., & Tantray, F. A. (2021). Climate change and water resources of Himalayan region—Review of impacts and implication. *Arabian Journal of Geosciences, 14*, 1088.

Mir, S. A., & Batool, M. (2022). Impact of climate change on Gujjar and Bakarwal communities of Jammu and Kashmir. *Journal of Sustainability and Environmental Management (JOSEM), 1*(2), 1–5.

Mishra, A. K., & Rafiq, M. (2017). Analyzing snowfall variability over two locations in Kashmir, India in the context of warming climate. *Dynamics of Atmospheres and Oceans, 79*, 1–9.

Mughal, R., Malik, A. H., Dar, G. H., & Khuroo, A. A. (2017). Location map of Poonch district. Woody Flora of Poonch District in Pir Panjal Himalaya (Jammu & Kashmir), India. *East Himalayan Society for Spermatophyte Taxonomy, 11*(2), 367–388. https://www.researchgate.net/publication/322153334_Woody_Flora_of_Poonch_District_in_Pir_Panjal_Himalaya_Jammu_Kashmir_India

Narbat, N. (2023). Environmental changes and its impact on the pastoral nomadism: A study of Dhangar community. *International Journal in Management and Social Science, 11*(12) ISSN: 2321-1784.

Narbat, N. (2024). Drought and resilience among pastoral nomads: A case study of Dhangar Community in the State of Maharashtra. *International Journal in Commerce, IT and Social Sciences, 11*(03) ISSN: 2394-5702.

National Informatic Centre. (2025). District Poonch, Government of Jammu and Kashmir. poonch.nic.in

Nuvey, F. S., Kreppel, K., Nortey, P. A., et al. (2020). Poor mental health of livestock farmers in Africa: A mixed methods case study from Ghana. *BMC Public Health.* https://doi.org/10.1186/s12889-020-08949-2

Palinkas, L. A. (2015). Qualitative methods in mental health services research. *Journal of Clinical Child and Adolescent Psychology, 43*(6), 851–861. https://doi.org/10.1080/15374416.2014.910791

Pir, R. A., (2024). Ground Water Monitoring Bulletin May-2024 Jammu & Kashmir, Ministry of Jal Shakti Department of Water Resources River Development and Ganga Rejuvenation, p-4 Government of India. gwdata.cgwb.gov.in

Prasad, S. N., Ramachandra, T. V., Ahalya, N., Sengupta, T., Kumar, A., Tiwari, A. K., et al. (2002). Conservation of wetlands of India-a review. *Tropical Ecology, 43*(1), 173–186.

Raina, A. N. (2002). *Geography of Jammu & Kashmir*. Kashmir News Network.

Romshoo, S. A., Rashid, I., Altaf, S., & Dar, G. H. (2020). Jammu and Kashmir state: An overview. In *Biodiversity of the Himalaya: Jammu and Kashmir state* (Vol. 18). Springer. https://doi.org/10.1007/978-981-32-9174-4_6

Saleem, S., Rather, J. A., Ahmed, S., Mushtaq, S., Ahmed, R., & Malik, I. H. (2024). Above ground biomass estimation for alpine grasslands of Kashmir Himalayas using remote sensing and field-data. *Rangeland Ecology & Management, 96*, 117–127.

Sen, M. (2025). *A new model for ecological restoration of Himalayan pasturelands.* ICLEI South Asia: Local Governments for Sustainability.

Sharma, A., & Choudhary, M. (2021). Changing climate and depleting water resources in the mountains with a case study from the Himalayas. In *Water Conservation in the Era of Global Climate Change.* Elsevier. https://doi.org/10.1016/B978-0-12-820200-5.00002-6

Shrestha, M. S. (2008). Impacts of flood in South Asia. *Journal of South Asia Disaster Studies., 1*(1).

Singh, J., & Bhandari, R. (2022). Problems and challenges in transhumance as a livelihood strategy: A case study of Van-Gujjar in upper Yamuna Basin. *International Journal for Research Trends and Innovation, 7*(7).

Solomon, SD, Qin D, Manning M et al (2007) Climate change 2007: The physical science basis, Working Group I Contribution to the Intergovernmental Panel on Climate Change Fourth Assessment Report.

Tufail, M., Kamble, P. N., Khumukcham, J., et al. (2024). Climate variability and transhumant adaptation among Gujjar and Bakarwal communities, India. *Ecology Environment and Conservation, 30*(1), 222–231.

Tugjamba, N., Walkerden, G., & Miller, F. (2021). Climate change impacts on nomadic herders' livelihoods and pastureland ecosystems: A case study from Northeast Mongolia. *Regional Environmental Change, 21*(105). https://doi.org/10.1007/s10113-021-01829-4

UNEP-World Conservation Monitoring Centre. (2002). *Mountain watch.* UNEP-WCMC.

United Nations. (2007). Water—coping with water scarcity: challenge of the twenty-first century. World Water Day 2007. Available online: http://www.un.org/waterforlifedecade/scarcity.shtml

Waruguru, M., Kingori, J., Lackey, D., Musomi, J., Ngugi, J. G., & Okpaku, S. A. (2021). Mental health Care in Nomadic Populations: The case of Kenya. In S. O. Okpaku (Ed.), *Innovations in global mental health*. https://doi.org/10.1007/978-3-030-57296-9_30

WHO/UNICEF Joint Monitoring Programme. (2019). *Progress on household drinking water, sanitation and hygiene, 2000–2017* (p. 8). UNICEF/WHO. Available from: http://washdata.org/reports

Assessment of Environmental Education: A Strategic Path to Sustainability

Najmah Peerzada, Rooheena Farooq, and Rameez Ahmad Bhat

Abstract The contemporary world is increasingly confronted with severe environmental challenges, most of which are a result of unsustainable human activities, ecological neglect and widespread environmental illiteracy. A critical factor contributing to these issues is the lack of public awareness and understanding of the adverse impacts of human behaviour on the natural environment. In developing regions especially, environmental illiteracy perpetuates unsustainable practices that accelerate ecological degradation. This chapter explores environmental education as a strategic and transformative pathway to sustainability. By integrating environmental literacy with ethical reasoning and practical engagement, environmental education nurtures a culture of sustainability rooted in ecological responsibility and global citizenship. It emphasizes the urgent need to revitalize and strengthen environmental education systems to cultivate a deeper awareness of ecological balance, foster environmental ethics and inspire collective responsibility for environmental stewardship. Environmental education, when applied strategically, has the power to transform an individual's behaviour, influence institutional practices and reform public policy. Thus, environmental education emerges as a foundational strategy for achieving environmental sustainability and securing the well-being of future generations.

Keywords Environmental education · Sustainability · Himalayan region · Ecological balance · Environmental ethics · Environmental literacy

1 Introduction

The current world faces unmatched environmental problems such as climate change, depletion of natural resources, biodiversity, and natural hazards that threaten the environment and human well-being, especially when it comes to environmentally sensitive areas. As a result, achieving sustainability has become a priority that demands not only technological and political measures, but also a paradigm shift concerning

---

N. Peerzada (✉) · R. Farooq · R. A. Bhat
Department of Education, University of Kashmir, Hazratbal, Srinagar, India

R. Ahmed et al. (eds.), *Climate Change, Water Resources, and Natural Hazards in the Himalayas*, Advances in Geographical and Environmental Sciences,
https://doi.org/10.1007/978-981-92-0835-7_18

human attitudes, values, and practices regarding the natural world (United Nations, 2015a, 2015b; IPCC, 2022a, 2022b). Education is one of the most important factors that can bring about this paradigm shift by equipping individuals with the necessary knowledge, skills, and values that enable them to understand and address environmental problems. The United Nations Sustainable Development Goals (SDGs), particularly Goal 4.7, emphasize the urgent need to reorient education systems to foster sustainable lifestyles, global citizenship and appreciation of cultural and environmental diversity (United Nations, 2015a, 2015b). Likewise, India's National Education Policy (NEP) 2020 recognizes environmental education as a core element of the national curriculum, calling for its integration at all stages and forms of learning–formal, non-formal and informal (Government of India, 2020). These policy frameworks reflect the importance of education in empowering individuals and communities to make informed, ethical and sustainable choices. The relevance of environmental education becomes even more pronounced in the context of the Himalayan region, which is characterized by ecological fragility, climate sensitivity and a close interdependence between human livelihoods and natural resources. Rapid environmental changes, glacier retreat, water stress and recurrent natural hazards have heightened the vulnerability of both ecosystems and communities in this region. Addressing these challenges requires context-specific educational interventions that integrate scientific knowledge with indigenous ecological wisdom and local realities (ICIMOD, 2020; UNDRR, 2021). This chapter conceptualizes environmental education as a strategic pathway to sustainability, with particular reference to the Himalayan region. The chapter examines the meaning, scope, goals and objectives of environmental education, its relationship with sustainable development. The chapter also seeks to highlight its critical role in fostering ecological resilience and sustainable futures.

1.1 Environmental Education

Environmental education occupies a critical space in the contemporary discourse on sustainability and human development (Stevenson et al., 2013). Environment, as a terminology under consideration, can, at its broadest meaning, define all natural and man-made factors that affect and support life on earth. These include both biotic components such as plants, animals and micro-organisms and abiotic elements like water, air, soil and climate (Palmer, 2002). In the face of accelerating ecological degradation, climate change and biodiversity loss, there is a growing consensus that education must not merely inform individuals about the environment but must also empower them to engage meaningfully with environmental issues (Reid, 2019). Essentially, environmental education is more than the dissemination of factual information; it is more of a transformative experience that helps in creating critical consciousness, appropriate attitudes, and problem-solving competencies, which, in turn, are highly necessary for living in a sustainable way. It also tries to instill a kind of ecological consciousness within individuals, which enables them to develop a deep understanding of the complexity of nature, the interconnectedness of various forms of life, and the effects of human behavior on nature's delicate ecosystem (McBride et al., 2013). Environmental education emphasizes developing learners' understanding of ecological principles, environmental challenges, and the functioning of natural systems. It adopts an experiential learning model in which the environment itself becomes a dynamic and interactive medium for inquiry, reflection and discovery. This approach encourages students to engage directly with their surroundings, fostering a deeper emotional and intellectual connection with nature. It also focuses on the development of action-oriented capacities that aim to inspire behavioural changes, encourage sustainable practices, and promote collective efforts towards environmental preservation and restoration. Environmental education thus transcends traditional disciplinary boundaries and calls for an integrative interdisciplinary framework. It draws upon insights from the natural sciences, social sciences, ethics and philosophy, so it cultivates ecological literacy and informed citizenship. In doing so, it positions learners not as passive recipients of information but as active participants in the co-creation of a more just, resilient and sustainable world.

1.2 Scope of Environmental Education

The scope of environmental education is broad, comprehensive, and multidimensional, encompassing formal, non-formal, and informal modes of learning across all levels of education. It extends from school education to higher education, teacher education, and lifelong learning, thereby engaging learners, educators, policymakers, and communities in fostering environmental awareness and environmentally responsible behaviours (UNESCO, 2017a, 2017b; Sterling, 2020a, 2020b). Moreover, environmental education does not strictly lie within the domain of curriculum teaching and learning. Instead, it also encompasses experiential learning, community engagement, policy engagement, and institutional efforts that seek to promote sustainable values and environmentally responsible behavior. In the context of the Himalayan region, the scope of environmental education assumes greater significance due to the region's ecological fragility, climate sensitivity, and strong dependence of local livelihoods on natural resources. In the Himalayas, issues like climate change effects, retreat of glaciers, water resources management, biodiversity conservation, and the mounting threat of natural hazards such as floods and landslides are primarily addressed by environmental education initiatives (ICIMOD, 2020; IPCC, 2022a, 2022b). Moreover, environmental education deals with a broad spectrum of modern environmental issues, which include climate change, conservation of natural resources, biodiversity, sustainable water management, and disaster risk reduction. By using a multi-disciplinary and participatory approach, environmental education helps people grasp complex environmental problems and make effective contributions to sustainable development at the local, national, and international levels (United Nations, 2015a, 2015b; Tilbury, 2011). In this way, environmental education acts as a key component in improving ecological resilience and sustainable development, especially in areas that can be vulnerable to the environment, like the Himalayan region.

2 Materials and Methods

2.1 Area Study

The study area for this chapter lies at the intersection of environmental education, sustainability studies, and educational geography. Conceptually, the study focuses on how environmental education contributes to the development of sustainable attitudes, values, and behaviours among learners within formal educational systems. Geographically, the study draws illustrations primarily within the Indian context, with particular reference to ecologically sensitive regions, such as mountainous and Himalayan areas, where environmental degradation, climate vulnerability, and resource constraints intensify the need for sustainability-oriented education (Ahmed et al., 2025; Saleem et al., 2024a). These regions provide a meaningful setting for examining the relevance and effectiveness of environmental education initiatives, particularly in the context of changing land use patterns and ecological stress (Mushtaq et al., 2024; Saleem et al., 2021). Institutionally, the study encompasses secondary schools and teacher education institutions, where environmental education is implemented through curricular subjects, co-curricular activities, and community-based programs. Policy frameworks such as the National Education Policy (NEP 2020) and environmental studies curricula form part of the analytical scope, enabling an examination of how educational policies translate sustainability goals into pedagogical practice (Saleem et al., 2024b).

2.2 Methodology and Data Sources

This chapter is based on a qualitative descriptive research design, employing a systematic review of secondary sources combined with a case study approach. The methodology framework for this study is based on interpretive and analysis approaches to critically assess the role of environmental education as a strategic approach for sustainability. The secondary data for this study were extracted from peer-reviewed journals, edited books, policy files, and reports from national and international organizations like UNESCO and UNEP. Relevant literature was identified through academic databases such as SpringerLink, Scopus-indexed journals, and institutional repositories, using keywords related to environmental education, sustainability, education for sustainable development, and environmental awareness. The gathered data was analyzed using thematic content analysis. The major theme areas, including integration of the curriculum, teaching practices, engagement, sustainable competencies, and participation, were considered., a case study method was employed to contextualize theoretical insights and demonstrate practical implementation at the institutional level. This methodological triangulation strengthens the credibility of the chapter by linking theory, policy, and practice.

3 Goals/Objectives of Environmental Education

The overarching goal of environmental education is to develop environmentally informed, ethically responsible, and socially responsive individuals capable of contributing to sustainable development. It aims to contribute to the holistic understanding of the environment and the interconnectedness between the natural world and humankind, while calling for sustainable values and practices (UNESCO, 2017a, 2017b; Sterling, 2020a, 2020b).

The specific objectives of environmental education include (Fig. 1):

- Cognitive objectives that relate to gaining knowledge and understanding regarding environmental ideas, biological processes, climate change, and sustainability issues.
- Affective objectives, which aim to cultivate environmental values, attitudes, sensitivity, and ethical responsibility towards nature.
- Behavioural aims, which foster active involvement, problem-solving, and ecologically correct behaviour on an individual and a collective basis (Tilbury, 2011; Wals and Benavot, 2017a, 2017b.

In regions such as the Himalayas, the goals and objectives of environmental education further emphasise resilience building, sustainable resource use, disaster preparedness, and community participation. Through the focus of learning goals that cater to regional environmental needs and globally contextualized issues of sustainability, environmental education enables the growth of climate-resilient and environmentally conscious societies (ICIMOD, 2020; IPCC, 2022a, 2022b). It is imperative that environmental education develops from conventional approaches based on content deliveries towards experiential approaches, especially action-based learning that is local and globally responsive (Monroe et al., 2019).

4 Environmental Education and Sustainable Development

Education is the principal driving force that contributes towards the development of a more sustainable planet because it enables people to achieve social aspirations. Education is important because it not only enables students to gain scientific and technical knowledge, but also provides them with the incentive and facilities necessary for its implementation. Education is the core of the collective progress humanity must undertake for a more sustainable and equitable future. It shapes the

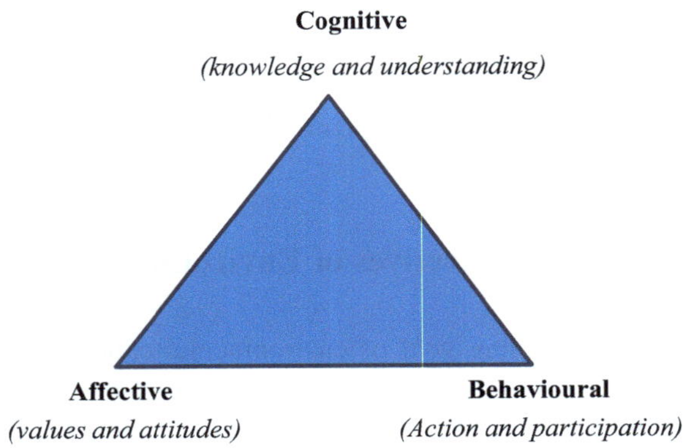

Fig. 1 Show objectives of environmental education

values, behaviours and skills necessary for individuals to respond to environmental, economic and social challenges. By fostering critical thinking, ethical reasoning and a sense of global citizenship, education empowers learners to become proactive agents of change in their communities and beyond. One of society's greatest concerns must be to reorient educational goals to recognize the importance of sustainable development and enhance its scope and quality. Education for sustainability (EFS) refers to a transformative approach that integrates principles, values and practices of sustainable development into all aspects of education and learning (UNESCO, 2017a, 2017b). According to Stephen Sterling, the necessary cultural change can be achieved through sustainable education (SE). Education for Sustainable Development (ESD) enables individuals to take appropriate decisions and undertake committed action on behalf of environmental protection, economic prosperity, and a just society for the present and future generations and respect for diversity (UNESCO, 2018). It encourages lifelong learning and forms a vital component of quality education. ESD is considered to be a goal for education rather than a prescribed body of information and knowledge (Hopkins, 2012). In 2002, the UN General Assembly adopted a resolution that called for a Decade of Education for Sustainable Development (DESD 2005–2014). The prime aim of this decade was to integrate the values inherent in sustainable development into all aspects of learning to encourage changes in behaviour that allow for a more sustainable and just society for all. SDG 4.7 sets the goal that by 2030, all learners should gain the knowledge and abilities essential for advancing sustainable development. This includes, among other aspects, education that promotes sustainable ways of living, respect for human rights, gender equality, a culture of peace and non-violence, global citizenship, and an understanding of cultural diversity and the role of culture in achieving sustainability. The overall idea of education for sustainable development is that it aims at establishing a world in which every individual has access to education that allows them to develop values, attitudes, and practices that help in building a sustainable world (DESD, 2014). To make sustainability successful in education, a number of measures can be taken. These include incorporating concepts of sustainability into curricula across different disciplines, developing experiential learning through projects and community involvement, educating teachers to use sustainability-focused pedagogies and encouraging eco-friendly practices in school buildings. Engaging local communities in activities promoting sustainability also increases the effectiveness of education and fosters social action. Education is the foundation of a sustainable world. Education for sustainability requires a radical, transformative pedagogy, focused on the process of learning to live within ecological limits without human suffering, while also uncovering the ideologies and power relationships that underlie the discourses of sustainable development (Stevenson, 2007). By integrating sustainability into the core of education, we enable individuals and communities to conserve the environment, promote social justice, and foster economic resilience. Ensuring that education systems globally prioritize sustainability at the forefront is not just a requirement but also fundamental to securing a just and sustainable world for current and future generations.

4.1 Environmental Education—The Himalayan Perspective

The Himalayan region occupies a unique ecological and strategic position in the global environmental landscape. As one of the most fragile mountain systems in the world, the Himalayas play a vital role in regulating climate, sustaining biodiversity and supporting water resources through vast networks of glaciers, rivers and watersheds that serve millions of people downstream (ICIMOD, 2019; IPCC, 2022a, 2022b). At the same time, the region is increasingly exposed to the adverse impacts of climate change, including rising temperatures, accelerated glacier retreat, erratic rainfall, declining water availability and an increasing incidence of natural hazards such as floods, landslides and glacial lake outburst floods (IPCC, 2022a, 2022b; UNDRR, 2021). In this sensitive environmental setting, environmental education emerges as a critical pathway for promoting sustainability, resilience and informed environmental decision making. Environmental education in the Himalayan context is particularly significant because of the close and direct dependence of mountain communities on natural resources for their livelihoods. Forests, water sources, agricultural land and mountain ecosystems form the foundation of social and economic life in the region. Unsustainable exploitation of these resources, often driven by lack of awareness, population pressure, and unplanned development, has intensified environmental degradation and human vulnerability (FAO, 2018; ICIMOD, 2020). Environmental education helps learners to develop an understanding of ecological interrelationships, human–environment interactions, and the long-term consequences of environmental mismanagement, thereby encouraging responsible attitudes and sustainable practices (UNESCO, 2017a, 2017b; Sterling, 2020a, 2020b). A distinctive strength of the Himalayan region lies in its rich repository of indigenous and traditional ecological knowledge. Over centuries, local communities have evolved adaptive strategies for water conservation, forest management, agriculture and disaster preparedness that reflect a deep ecological consciousness and respect for nature (Berkes, 2018; Gadgil & Guha, 2019). Environmental education offers an essential framework for recognising, documenting and integrating this indigenous knowledge with contemporary scientific understanding. Such integration not only enriches environmental learning but also ensures that sustainable education remains culturally relevant, locally grounded and socially inclusive (UNESCO, 2017a, 2017b; ICIMOD, 2020). Education that focuses on climate literacy, environmental risk awareness, early warning systems and sustainable water and land-use practices can significantly enhance community preparedness and resilience (UNDRR, 2021; IPCC, 2022a, 2022b). When integrated into school curricula, teacher education programmes and community-based learning initiatives, environmental education empowers individuals to participate actively in conservation efforts, disaster mitigation and climate-resilient development (Tilbury, 2011).

5 Challenges in Implementing Environmental Education

Effective and meaningful environmental education is a challenge we must take seriously if we and future generations want to enjoy the benefits of quality life and natural heritage (Hudson, 2001). There are many challenges that come across while successfully practicing effective environmental education, from resource unavailability to loopholes in policies, social and cultural issues. These issues obstruct an important task of making children aware of environmental issues and of sustainable living. These are some of the most important challenges that come across while practicing effective environmental education:

(a) **Inadequate funding and educational resources**: Another key challenge in this field is the inadequate funding and educational resources. Most educational institutions are faced with funding problems that constrain them from accessing the required educational resources to offer effective environmental education programs. Inadequate funding will directly influence the availability of support resources and may hamper efforts to involve experiential learning in the field of the environment.

(b) **Lack of qualified and trained environmental educators**: In addition to the issue of resources, there is a shortage of qualified and trained environmental educators. Effective environmental education teaching requires teachers who not only possess a good knowledge of environmental issues but also possess skills that improve critical thinking, problem-solving skills, and an affinity with nature. However, environmental teachers in many learning institutions are not trained professionals in environmental education. This culminates in a poor understanding of the subject itself and the relevance it holds. In many cases, environmental education becomes more "add-on" than part of the learning process.

(c) **Challenges in Curriculum Integration**: Curriculum-related challenges are also a major hindrance. The interdisciplinary method for environmental education, which uses sciences, geography, and social sciences, sometimes becomes difficult to integrate into traditional curriculum segments. A lack of time and unclear instructions from curriculum development teams often leave teachers puzzled about how to properly integrate environmental education into their curriculum. A lack of subject knowledge and a feeling that environmental education is all about tough knowledge act as a challenge for teachers.

(d) **Lack of adequate policy and institutional support**: Lack of adequate support in terms of policies to ensure effective implementation is very common. Lack of attention to the processes involved in the monitoring workshop and the inability to construct the whole-school model for learning for the environment can render initiatives ineffectual. This is because the institutional support provided by the school administration is very limited, and the overall public policies are inadequate, with a low priority for the integration of environmental education.

(e) **Socio-economic and cultural factors**: When less developed nations face such challenges, they are accompanied by a unique level of hindrances that are socio-economic and infrastructural in nature. The poverty rate is consistently at higher levels, and the pressing needs of the population get priority over the concern of the environment. The cultural and traditional beliefs of the population often generate resistance to new forms of learning like the incorporation of environmental education.

(f) **Operational challenges in implementing environmental education**: Real-world classroom management concerns, including class size, can compound the challenges of implementing environmental education, which generally functions best in smaller, more engaging learning settings. Large class sizes can hinder the interactive and hands-on approach that environmental education often requires. Additionally, concerns related to safety and liability during outdoor activities create further logistical difficulties for educators. Overcoming these multifaceted issues demands a collaborative approach, where policymakers, educators, and communities work together to prioritize and effectively support environmental education as a pathway to a sustainable future.

6 Effective Strategies for Strengthening Environmental Education

(a) **Instilling environmental responsibility in learners**: One of the foundational aims of environmental education is to develop a sense of responsibility towards nature among students. This can be achieved by nurturing their emotional connection with the natural world and raising awareness about the consequences of human activities on the environment. Teachers can incorporate sustainable practices into daily routines, initiate waste reduction and recycling programs, and promote eco-friendly behaviours. Interactive class discussions, environmentally themed projects, and real-life case studies of environmental leaders can help foster ecological consciousness. Highlighting inspiring stories of individuals who have contributed meaningfully to environmental protection encourages students to take similar initiatives in their communities.

(b) **Promoting critical thinking and problem-solving skills**: Environmental education serves as a platform where students' critical thinking and problem-solving skills can be developed effectively. Educators can help students collect and analyze environmental data and come up with possible solutions for environmental issues. Models of project-based learning work well in the process. When students are working together on environmental issues surrounding local communities, they achieve a higher level of knowledge as well as develop a sense of ownership, thus increasing motivation and engagement.

(c) **Integrating experiential learning methods**: Experiential learning can provide students with hands-on experience with their surroundings. Projects like tree plantings, studying the purity of water, and biodiversity studies can be taken up that would give the students substantial experience. This would allow the students to relate whatever they are studying in their classrooms to real-world issues and make them think and act in a proactive way about environmental issues. By getting actual first-hand experience with environmental projects, students can inculcate environmental values in themselves.

(d) **Building community partnerships**: Community partnerships add relevance to environment education. Community partnerships with NGOs, government agencies, and communities give students insight into what is going on in their environment and give them an opportunity to get involved in projects that address issues in their environment. These involve projects that give students an opportunity to participate in service-learning projects, enhance teamwork and leadership skills, and give them experience in environment conservation projects.

(e) **Leveraging technology for environmental learning**: Technology has an extremely important role to play in making learning for the environment an engaging one. Technology-based learning resources, applications, and portals open up access to quality material, including simulations, documentaries, and live data on the environment. Virtual reality-based environments enable students to discover the environment, spot wildlife, and study the effects of the environment engagingly. In addition, technology-based applications facilitate students to interact with one another for environment-based assignments, connect despite geographical constraints, and conduct global campaigns.

7 Case Study (*School-Based Environmental Education as a Strategy for Sustainability*)

A school-based case study is used to explain and provide the practical application of environmental education. The case focuses on integrating environmental education in a secondary school through curricular and co-curricular interventions for sustainable practices. In this case, environmental education was integrated in subjects such as geography, science, and social sciences with an emphasis on interdisciplinary learning. Students were involved in structured activities like tree plantation, waste management and segregation at source programs, water conservation practices, and eco-club activities. Experiential learning strategies like project-based learning, field observations, and community awareness campaigns were used to develop ecological understanding and its application in real life. The results of the intervention showed a significant improvement in the environmental awareness, responsible behaviour, and participation of students in the sustainability-related activities. The teachers observed higher learner engagement and critical reflections concerning local environmental issues. The case study illustrates that by the systematic implementation of environmental education; it proves to become an effective tool for developing competencies of sustainability and the promotion of long-term environmental behaviour among the learners. The case study above supports the claim that environmental education is more than just an academic discipline, as it is the key that enhances the connectivity between education and development within sustainable objectives.

8 Conclusion

Environmental Education appears as a strategic and essential component in the quest for environmental sustainability in the sensitive Himalayan region. Environmental Education promotes awareness, responsibility, and appropriate actions through individuals and society in a comprehensive way that can address issues relating to environmental challenges arising out of Climate Change, Water Resource Utilization, and Disasters (UNESCO, 2017a, 2017b; IPCC, 2022a, 2022b). Coupling Scientific Knowledge with Indigenous Ecological Knowledge through participatory learning processes builds resilience in individuals and societal groups for appropriate interactions between humans and the environment supports long-term ecological balance (Berkes, 2018; Sterling, 2020a, 2020b). For the Himalayas and its environment, the improvement of environmental education and training is not only an educational goal; rather, it has become a major urgent demand for ensuring the sustainability of the environment (United Nations, 2015a, 2015b; ICIMOD, 2020).

References

Abbass, K., Qasim, M. Z., Song, H., Murshed, M., Mahmood, H., & Younis, I. (2022). A review of the global climate change impacts, adaptation, and sustainable mitigation measures. *Environmental Science and Pollution Research, 29*(28), 42539–42559.

Agbedahin, A. V. (2019). Sustainable development, education for sustainable development, and the 2030 agenda for sustainable development: Emergence, efficacy, eminence, and future. *Sustainable Development, 27*(4), 669–680. https://doi.org/10.1002/sd.1931

Ahmed, R., Rather, A., Banerjee, P., Wani, G., & Saleem, S. (2025). *Impact of climate variability on the feeding glaciers of potentially dangerous glacial lakes in the Jhelum Basin of Kashmir Himalaya.* Natural Hazards Review. https://doi.org/10.1061/NHREFO.NHENG-2140

Berkes, F. (2018). *Sacred ecology.* Routledge.

Boca, G. D., & Saraçlı, S. (2019). Environmental education and student's perception, for sustainability. *Sustainability, 11*(6). https://doi.org/10.3390/su11061553

Bonnett, M. (2002). Education for sustainability as a frame of mind. *Environmental Education Research, 8*(1), 9–20. https://doi.org/10.1080/13504620120109619

Brody, M., Dillon, J., & Wals, A. E. (2013). In R. B. Stevenson (Ed.), *International handbook of research on environmental education* (pp. 1–6). Routledge.

Brundtland, G. H. (1987) Our common future: Report of the world commission on environment and development, oxford university press.

Centre for Environment Education (CEE) (2021) Paryavaran mitra program report. Ministry of environment, forest and climate change.

Chavula, Y., Abdi, E., Uwimbabazi, A., Habowa, C., Mensah, G., Amanzi, L., Lungu, G., & Kayusi, F. (2024). Bridging environmental education and sustainable development: An integrated approach for a greener future.

DESD (Decade of Education for Sustainable Development). 2014.

Dhara, C., & Singh, V. (2021). The elephant in the room: Why transformative education must address the problem of endless exponential economic growth. arXiv preprint https://arxiv.org/abs/2101.07467.

Eulefeld, G. (1979). The UNESCO-UNEP programme in environmental education. *European Journal of Science Education, 1*(1), 113–118. https://doi.org/10.1080/0140528790010114

FAO (2018). Mountain agriculture and sustainability.

Gadgil, M. & Guha, R. (2019). Ecology and equity.

Government of India. (2020). *National Education Policy 2020.* Ministry of Human Resource Development.

Hajj-Hassan, M., Chaker, R., & Cederqvist, A.-M. (2024). Environmental education: A systematic review on the use of digital tools for fostering sustainability awareness. *sustainability, 16*(9), 3733.

Hopkins, C. (2012). Reflections on 20+ years of ESD. ESD high. Educ. *Prof. home, 6*, 21–35. https://doi.org/10.1177/097340821100600108

Hudson, S. J. (2001). Challenges for environmental education: Issues and ideas for the 21st century: Environmental education, a vital component of efforts to solve environmental problems, must stay relevant to the needs and interests of the community and yet constantly adapt to the rapidly changing social and technological landscape. *Bio Science, 51*(4), 283–288. https://doi.org/10.1641/0006-3568(2001)051[0283:CFEEIA]2.0.CO;2

ICIMOD. (2019, 2020). Himalayan environment assessments.

ICIMOD. (2020). *Hindu Kush Himalaya assessment: Mountains.* Climate Change.

IPCC. (2022a). Sixth assessment report.

IPCC. (2022b). Sixth assessment report: Impacts, adaptation and vulnerability.

Jensen, B. B., & Schnack, K. (1997). The action competence approach in environmental education. *Environmental Education Research, 3*(2), 163–178. https://doi.org/10.1080/1350462970030205

Kollmuss, A., & Agyeman, J. (2002). Mind the gap: Why do people act environmentally and what are the barriers to pro-environmental behavior? *Environmental Education Research, 8*(3), 239–260.

Kopnina, H. (2018). Education for sustainable development (ESD): The turn away from 'environment'in environmental education? In *Environmental and sustainability education policy* (pp. 135–153). Routledge.

Koul, M. A., Tandon, B., & Dua, B. (2024). Nep 2020's vision for environmental sustainability in education. *Integrating environmental sustainability into education insights from nep, 2020*, 1. https://doi.org/10.25215/9392917465

Le Roux, C., & Maila, W. (2004). *Issues and challenges regarding environmental education policy implementation.* https://doi.org/10.1080/18146620408566282

Lukman, R., Lozano, R., Vamberger, T., & Krajnc, M. (2013). Addressing the attitudinal gap towards improving the environment: A case study from a primary school in Slovenia. *Journal of Cleaner Production, 48*, 93–100.

Martian, I. (2015). The environmental education dimensions for sustainable development. *The Journal of Legal Studies, 16*(30), 18–30. https://doi.org/10.1515/jles-2015-0007

McBride, B. B., Brewer, C. A., Berkowitz, A. R., & Borrie, W. T. (2013). Environmental literacy, ecological literacy, ecoliteracy: What do we mean and how did we get here? *Ecosphere, 4*(5), 1–20.

Ministry of Education. (2020). Government of India (2020) National Education. *Policy*. Retrieved from.

Monroe, M. C., Andrews, E., & Biedenweg, K. (2008). A framework for environmental education strategies. *Applied Environmental Education & Communication, 6*(3–4), 205–216. https://doi.org/10.1080/15330150801944416

Monroe, M. C., Plate, R. R., Oxarart, A., Bowers, A., & Chaves, W. A. (2019). Identifying effective climate change education strategies: A systematic review of the research. *Environmental Education Research, 25*(6), 791–812.

Musah, E., & Wangila, C. (2024). Inquiry-based ethnoecological conservation approach to improve students' critical thinking skills. *International journal of Ethnoscience and technology. Education, 1*(2), 97–115.

Mushtaq, S., Saleem, S., Ahmed, R., Tass, M. S., Rather, J. A., & Rather, G. M. (2024). Spatio-temporal analysis of land use land cover changes in South Kashmir region of North-Western Himalayas using Landsat data. *Discover Geoscience, 2*(1), 37.

Nibedita, P. (2020). *Sustainable development and education ISBN*, 97888854528.

Palmer, J. (2002). *Environmental education in the 21st century: Theory, practice, progress and promise*. Routledge.

Pavlova, M. (2013). Towards using transformative education as a benchmark for clarifying differences and similarities between environmental education and education for sustainable development. *Environmental Education Research, 19*(5), 656–672.

Reid, A. (2019). Climate change education and research: Possibilities and potentials versus problems and perils? *Environmental Education Research, 25*(6), 767–790.

Saleem, S., Ahmed Hajam, F., & Rather, J. A. (2021). Spatio-temporal analysis of land use land cover changes in Sind catchment of the Kashmir Valley. *India. Geo Eye, 10*(2), 35–42.

Saleem, S., Rather, J. A., Ahmad, S., Ahmed, R., & Hajam, L. A. (2024b). Assessment of livestock carrying capacity in the alpine grasslands of the Kashmir Himalayas. *Pastoralism: Research, Policy and Practice, 14*, 13541.

Saleem, S., Rather, J. A., Ahmed, S., Mushtaq, S., Ahmed, R., & Malik, I. H. (2024a). Above ground biomass estimation for alpine grasslands of Kashmir Himalayas using remote sensing and field-data. *Rangeland Ecology and Management, 96*, 117–127.

Saluja, A. (2024). Fostering environmental literacy through the curriculum at school level: NEP 2020's approach and its implications. *Journal of research in social science and humanities, 4*(1), 15–20. https://doi.org/10.47679/jrssh.v4i1.58

Sterling, S. (2008). Sustainable education—Towards a deep learning response to unsustainability. *policy & practice. A development education review, 6*(Spring), 63–68.

Sterling, S. (2010). Transformative learning and sustainability: Sketching the conceptual ground. *Learning and teaching in higher education, 5*, 17–33.

Sterling, S. (2020a). Sustainable education.

Sterling, S. (2020b). Sustainable education: Re-visioning learning and change.

Stevenson, R. B. (2007). Tensions and transitions in policy discourse: Recontextualizing a decontextualized EE/ESD debate. *In Researching education and the environment* (pp. 49–62). Routledge.

Tilbury, D. (2011). Education for sustainable development: An expert review.

Tomar, A. (2007). Environmental education. ISBN 8178355485.

Transforming Our World (2019) The 2030 Agenda for sustainable development. Available online: (accessed on 14 May 2019).
UNDRR (2021). Disaster risk reduction framework.
UNESCO (2017a). Education for sustainable development goals: Learning objectives.
UNESCO (2017b). Education for sustainable development.
UNESCO. (2005). *The DESD at a glance*. UNESCO.
United Nations (2015a). Sustainable Development Goals.
United Nations (2015b). Transforming our world: The 2030 Agenda for sustainable development.
Leicht, A., & Heiss, J. (2018). Issues and trends in education for sustainable development (Vol. 5). UNESCO publishing.
Van De Wetering, J., Leijten, P., Spitzer, J., & Thomaes, S. (2022). Does environmental education benefit environmental outcomes in children and adolescents? *A meta-analysis. Journal of Environmental Psychology, 81*, 101782.
Wals, A. E. J. & Benavot, A. (2017a). Can we meet the sustainability challenges?. The role of education and learning.
Wals, A. E. J. & Benavot, A. (2017b). ESD and transformative learning.

Bridging Policy and Practice: Strengthening Governance for Sustainable Development in the Indian Himalayan Region

Shazia Farooq, Aasia Maqbool, and Nighat Manzoor

Abstract The Indian Himalayan Region (IHR), a landscape of immense ecological richness and cultural diversity, faces the dual pressures of climate change and unsustainable development. Despite ambitious global frameworks such as the Sustainable Development Goals (SDGs) and national missions like the National Action Plan on Climate Change (NAPCC), governance in the Himalayas often struggles to bridge policy intent with ground realities. Fragmented institutions, inadequate stakeholder participation, and short-term development priorities undermine resilience in this fragile ecosystem. Yet, examples such as Van Panchayats and the Chipko Movement illustrate the transformative potential of inclusive and participatory governance. This chapter examines how governance can serve as a bridge between policy and practice, analyzing barriers to sustainability, highlighting community-based innovations, and proposing governance models that integrate local knowledge with institutional accountability. By emphasizing collaboration, equity, and long-term resilience, it positions governance not merely as regulation but as a dynamic driver of sustainable futures in the Himalayas.

Keywords Governance and sustainability · Indian Himalayan region (IHR) · Policy practice gap · Participatory governance · Sustainable development goals (SDGs)

1 Introduction

In the twenty-first century, the world stands at a critical juncture marked by accelerating social, economic, and environmental transformations. Climate change, biodiversity loss, persistent poverty, and widening social inequalities have collectively challenged traditional models of governance and development. The complexity and interconnectedness of these crises have revealed that governance is no longer a matter

S. Farooq (✉) · A. Maqbool · N. Manzoor
Department of Education, University of Kashmir, Srinagar, India

R. Ahmed et al. (eds.), *Climate Change, Water Resources, and Natural Hazards in the Himalayas*, Advances in Geographical and Environmental Sciences,
https://doi.org/10.1007/978-981-92-0835-7_19

of mere administrative control, but a multidimensional process involving diverse actors and institutions working across scales. According to Tiwari (2015), governance has evolved from the limited scope of bureaucratic regulation to a dynamic platform that unites communities, stakeholders, and institutions in addressing complex socio-ecological concerns. The growing realization that unsustainable governance systems intensify vulnerability has prompted a global rethinking of policy priorities. In this context, the United Nations' adoption of the Sustainable Development Goals (SDGs) in 2015 marked a paradigm shift toward integrating environmental protection, social equity, and economic progress into a single development agenda. Yet, as Dasgupta et al. (2023) emphasize, translating these goals into meaningful action remains an immense challenge, particularly in developing regions where institutional capacity, socioeconomic inequalities, and ecological fragility intersect. Effective governance thus emerges as the critical link between visionary global frameworks and the lived realities of communities struggling with environmental and developmental pressures. India exemplifies this global tension between growth and sustainability. As one of the world's most ecologically diverse nations, it faces the twin challenge of promoting economic advancement while safeguarding natural and cultural resources. The government's commitment to sustainability is articulated through policy frameworks such as the National Action Plan on Climate Change (NAPCC) and its sub-missions, which seek to integrate environmental concerns into mainstream development. However, whether these frameworks can bridge the gap between policy intent and grassroots action remains uncertain (Rana et al., 2022). This concern becomes particularly pronounced in the Indian Himalayan Region (IHR)—A region of immense ecological, cultural, and geopolitical significance. Comprising eleven states and two Union Territories, the IHR functions as the country's vital ecological lifeline, supplying water, energy, and biodiversity services to millions (IHCAP-DST-SDC, 2018). Yet, it remains highly vulnerable to the pressures of unplanned urbanization, livelihood insecurity, and climate variability. National missions such as the National Mission for Sustaining the Himalayan Ecosystem (NMSHE) were established to address these concerns, but their effectiveness is limited unless supported by inclusive, participatory, and adaptive governance frameworks (Badola et al., 2015). Traditional community-based institutions, such as Van Panchayats in Uttarakhand, illustrate how decentralized and participatory governance can contribute to long-term sustainability. These local mechanisms, rooted in indigenous knowledge and collective decision-making, demonstrate how governance at the grassroots can complement formal state-led initiatives. However, the growing centralization of authority and inadequate policy integration have gradually eroded their autonomy and effectiveness (Naaz & Sahu, 2025).

Bridging these divides between institutional design and local knowledge, and between top-down governance and bottom-up participation is fundamental for achieving resilience and sustainable development in the Himalayas. Therefore, this chapter seeks to explore how governance serves as a bridge between policy and practice within the Indian Himalayan Region. It examines the evolution of governance mechanisms, evaluates policy frameworks and their implementation, and highlights innovative participatory approaches that can inform a more sustainable and inclusive governance paradigm for the Himalayas.

1.1 Contextualizing Governance and Policy Frameworks in the Indian Himalayas

Building upon the broader discussion of sustainability and governance challenges introduced earlier, the Himalayan region offers a compelling case for examining how policy and governance mechanisms operate in ecologically fragile and socially diverse landscapes. Despite numerous governmental and non-governmental initiatives, the region continues to face critical governance and policy challenges. Environmental degradation has intensified socio-economic issues such as water scarcity, fuelwood shortages, and reduced agricultural productivity. These persistent problems raise fundamental questions about whether inappropriate policy design or the weakening of traditional local institutions is to blame. Scholars argue that most development initiatives fail to incorporate the "mountain perspective"—that is, the unique ecological fragility, remoteness, and adaptive lifestyles specific to mountain environments (Ives, 2013; Price, 2015; Jodha, 2000, 2005a, 2005b). Historically, decentralized local institutions in the Hindu Kush Himalayan (HKH) region played an essential role in managing natural resources sustainably. However, with the gradual centralization of administrative structures, these traditional systems were marginalized, disrupting the equilibrium between communities and ecosystems (Foggin, 2018; Jodha, 2005a, 2005b). The resulting fragmentation in governance, weak coordination between formal and informal institutions, and persistent financial and trade-related conflicts have constrained effective environmental management. Furthermore, local governance structures often lack the institutional capacity and gender-inclusive participation necessary for implementing sustainability-oriented policies (Agrawal et al., 2016; UNEP, 2002, 2008). Environmental challenges in the Himalayan region are inherently complex and long-term, requiring adaptive and integrative strategies. Yet, governments have often failed to prioritize regional coordination or adaptive governance frameworks (Underdal, 2010). Although certain policy instruments now attempt to address ecosystem services, they remain inadequate due to limited understanding of ecological interlinkages and interdependencies (Pagiola et al., 2004; Barbier, 2007; TEEB, 2010; CBD & UNEP-WCMC, 2012). Ecosystem-based governance approaches, which promote multi-level and cross-sectoral collaboration, have been suggested to enhance resilience and reduce community vulnerability (Andrade et al., 2011).

The Indian Himalayas, which comprise about 18% of India's landmass, hold immense ecological and cultural significance. The region contributes nearly 50% of India's forest cover and harbors 40% of its endemic species (Semwal et al., 1999). Serving as the nation's "water tower," it provides essential ecosystem services to millions downstream (Government of India, 2010). However, mounting population pressures, resource overexploitation, and climate change have led to glacial melt, landslides, and habitat degradation (Liu & Chen, 2000; Dyurgerov & Meier, 2005; INCCA, 2010; Singh et al., 2011; Wulf et al., 2012). Consequently, the weakening of both natural and social systems has undermined local economies, tourism, and food and water security (WHO, 2012; Badola et al., 2014; Smith, 2014). Despite the region's significance, the Himalayas remain marginalized in policy frameworks (http://himlayaforum.org). This neglect stems largely from a failure to recognize the economic and ecological value of Himalayan ecosystem services (Singh, 2002; Blaikie & Muldavin, 2004). Their provisioning, regulating, and cultural functions are rapidly declining, endangering both local and downstream populations. Inclusive policymaking that respects the region's diversity, fragility, and indigenous knowledge systems is therefore crucial (Gautam et al., 2013).

For governance in the Himalayas to be effective, there must be alignment between policies, institutions, and local practices. As Ostrom (1991) notes, institutions define authority, rules, incentives, and accountability mechanisms. Hence, formal governance structures—such as laws and administrative frameworks—must operate in synergy with informal systems like traditional leadership and community-based decision-making. Strengthening these linkages can bridge policy and practice, fostering governance that is inclusive, adaptive, and grounded in local realities. Ultimately, promoting sustainable development in the Himalayas requires reimagining governance as a collaborative, multi-level process that integrates traditional wisdom, modern science, gender equity, and cross-border coordination (UNEP, 2008; Foggin, 2018). Recognizing the Himalayas not merely as an ecological zone but as a living social–ecological system is essential for developing resilient governance capable of sustaining livelihoods and ecosystem services for generations to come (Blaikie & Muldavin, 2004; Gautam et al., 2013).

1.2 Materials and Methods

1.2.1 Study Area

Sustainable towns in the context of policy-oriented disaster risk governance and environmental resilience have become a widely desirable target for urban planners and policymakers, especially in disaster-prone regions of South Asia. Cities such as Srinagar, Shimla, and Guwahati face increasing risks due to rapid urbanization, climate change, and unsustainable land use practices. Hence, the integration of disaster risk reduction (DRR) strategies with sustainable urban development frameworks is of utmost importance. The study area focuses on regions that are ecologically fragile yet socioeconomically vital, where urban growth intersects with environmental vulnerability. These towns have been selected based on their exposure to hydro-meteorological hazards, such as floods and landslides, and their inclusion in national-level sustainable urban development policies. In this context, the concept of "Sustainable Towns" aligns with the broader framework of the National Disaster

Management Policy (2009) and the National Mission on Sustainable Habitat (2010), which emphasize adaptive, risk-informed governance. The study area thus represents a dynamic interface between urban sustainability, institutional governance, and environmental security in the face of climate-induced disasters.

1.2.2 Data Source

This study is based exclusively on secondary data sources, ensuring a comprehensive and policy-informed understanding of disaster risk governance and sustainability in urban contexts. Data were systematically collected from official policy documents, government reports, and international frameworks that address sustainability and disaster resilience. The primary sources included reports from the National Disaster Management Authority (NDMA), Ministry of Environment, Forest and Climate Change (MOEFCC), NITI Aayog, and the United Nations Office for Disaster Risk Reduction (UNDRR).

Additionally, peer-reviewed journal articles, books, and conference proceedings were reviewed to synthesize theoretical and empirical insights on urban sustainability and disaster governance. The Sendai Framework for Disaster Risk Reduction (2015–2030), the Sustainable Development Goals (SDGs), particularly Goal 11 (Sustainable Cities and Communities) and Goal 13 (Climate Action), provided the global policy context for this research. Policy analyses and case studies from similar developing regions were also examined to compare adaptive strategies, institutional mechanisms, and governance challenges.

All data were evaluated based on credibility, relevance, and recency, with special attention given to documents published between 2015 and 2024, ensuring alignment with contemporary sustainable development and climate governance discourse.

1.2.3 Methodology

The study adopted a qualitative and descriptive analytical approach, utilizing secondary data to examine the intersection between disaster risk governance and sustainable urban development. A policy analysis framework was employed to evaluate the content, objectives, and implementation of relevant national and international strategies. The data were organized thematically under key domains such as institutional capacity, policy coherence, resilience planning, and sustainability integration.

A content analysis method was used to interpret and synthesize information from policy reports and academic literature. Each document was reviewed to identify recurring themes, such as community participation, inter-agency coordination, and policy gaps in urban resilience. Comparative analysis was conducted to highlight best practices and challenges in implementing disaster-resilient policies within urban areas of developing economies.

This methodology allows for a critical understanding of how policy instruments and governance frameworks contribute to sustainable and resilient town development. It also provides a conceptual foundation for evaluating existing gaps and proposing policy-level improvements in disaster management and sustainability governance.

2 Result and Discussion

Governance and policy processes are still mostly at the national level, despite the HKH's growing recognition as a region with interwoven transboundary landscapes. Even though the nations in the region have different political, administrative, and legal structures and varying degrees of economic development, they have all developed laws and policies to protect the environment and promote the sustainable use of natural resources (Chettri et al. 2008). Since the 1950s, as central governments' power and responsibility in formulating policies and planning for development have grown, environmental governance in particular has accelerated. From then on, and particularly following the 1992 United Nations Conference on Environment and Development, nations began developing policies pertaining to environmental protection and sustainable natural resource management. There have been more and more regulatory changes in the area to control the environment and common pool resources (CPRs), especially water, forests, and rangelands. In general, the paradigm for conservation policies and practices has gradually changed from one that is centered on species and habitat to one that is centered on participatory livelihood-based landscapes. (Sharma et al., 2010).

3 Multi-Level and Participatory Governance Approaches for Sustainable Development in the Himalayas

According to Larson and Soto (2008), stakeholder participation is widely recognized as a crucial component of policymaking, development planning, and the implementation of sustainable natural resource management in global environmental governance reforms. The HKH's environmental governance reflects this tendency, which may be characterized as the emergence of hybrid modes of governance spanning the state, market, and community. According to Khan and Hossain (2006), there are three different levels with two interfaces: 1. international and regional level, where the main actors are foreign governments, donor agencies, international NGOs, and international governing bodies; 2. national level, where the players are the government and national-level NGOs along with the private sector; and 3. grass-roots level, where the primary actors are local NGOs, community-based organizations (CBOs), local governments, and development agencies. The state, business sector, and civil society are the three categories of social institutions, and each of these levels has unique roles and ways of collaborating with them. There are two interfaces where these operational levels coordinate and collaborate: national–international and national–grass roots (Khan and Hossain 2006). Although not on an equal footing, the public, private, and community sectors are becoming participants in resource governance in the HKH. In Bhutan, for instance, sectoral policies are put through a rigorous stakeholder consultation process and a policy screening mechanism to make sure they adhere to the Gross National Happiness (GNH) ideals and principles. The foundational development theory that serves as the foundation for sector-specific policies and programs is Gross National Happiness. Fundamentally, it advocates for a well-rounded and coordinated strategy for socioeconomic growth, environmental and cultural heritage conservation, and the advancement of good governance. In India, the judiciary, civil society, and media are all becoming more outspoken in their support of mountain environments. Additionally, any development project that has an impact on natural resources must be approved by the local panchayat institutions, which are responsible for natural resource governance, rural development, and conflict resolution in many South Asian nations, such as Bangladesh, India, Nepal, and Pakistan (Ghosh 2008). Public consultations regarding new government policies are also held by communities and advocacy organizations. With the growing acknowledgment of the private sector, stakeholder participation in Nepal has been growing (Timsina and Gotame 2014).

4 Policy Integration and Governance Challenges in Climate Change Adaptation: Evidence from the HKH

Every nation in the HKH has complied with the UN's request to mainstream environmental considerations into development planning and decision-making in order to fulfill the twin goals of reducing poverty and ensuring environmental sustainability. Orienting governance, institutions, and political systems to improve the environment in ways that both benefit the poor and lessen susceptibility to the effects of climate change is a crucial component of mainstreaming the environment (UNDP-UNEP 2009). With the ratification of the 2015 Paris Agreement, the adoption of the Sustainable Development Goals, and assistance from international funders, the area has started to mainstream environmental and climate change initiatives (UNFCCC 2015). Unfortunately, the policies frequently follow a linear approach with distinct processes for policy formation and execution, failing to take into account the governance issues that are present (Friend et al., 2013). As documented in the instance of climate policy development in Nepal (Nightingale, 2016; Ojha et al., 2016), this means that the process is frequently influenced by global imperatives rather than through venues for national politics. Although adaptation is still a top concern, the HKH nations are also working to lower carbon emissions through mitigation measures. India, China, Bangladesh, and Nepal were some of the first countries in the world to ratify the Paris Agreement. The mismatch between different levels of governance makes it difficult to adopt climate change into development strategies, according to experience. Due to the highly centralized design and implementation process, the involvement of primarily government players, the short timelines, and the scattered activities, initiatives in Nepal—which are frequently funded by donors—have had little impact on the ground. Regmi and Star (2015) examined two case studies to examine how climate change is mainstreamed in Nepal. The results imply that the disparity in implementation strategies is one of the biggest obstacles to the primary streaming of community-based adaptation into the development process. The results of Regmi and Star's study demonstrate that bottom-up strategies, including the Local Adaptation Plan of Action (LAPA), were effective in raising awareness and organizing neighbourhood organizations. They were unable to impact government institutions, though, and as a result, the center did not provide support or continuity, underscoring the crucial gap between policy and reality.

5 Policy and Governance Challenges in Achieving Sustainability across the Himalayas

The experience of the Hindu Kush Himalaya (HKH) countries reveals that while policy-level mainstreaming of environmental and sustainability goals has shown progress, significant gaps persist between policy design and implementation. Numerous studies have identified barriers such as policy ambiguity, inadequacy of institutional frameworks, and weak regulatory enforcement mechanisms. For instance, although China's *Environmental Protection Law of 2014* is considered one of the most progressive, it remains insufficient due to its fragmented and overlapping governance structures, limited avenues for civic legal recourse, and conflicts of interest within administrative systems (Zhang and Cao, 2015). Similarly, policies in Pakistan have been described as rhetorical and politically driven, often fluctuating with changes in government, while Bangladesh's forestry sector policies have been criticized for being dominated by rhetoric rather than effective implementation (Khan and Harriss-White, 2012). In China and India, where mountain regions constitute only a small proportion of total land area, national policies frequently overlook mountain-specific needs, thereby constraining community options and increasing vulnerability to economic and environmental stress (Sharma and Chettri, 2005; Jodha 2005a, 2005b). Nepal's experience with pro-poor community forestry shows that compartmentalization within government line agencies has further restricted effective policy execution (Pokharel et al., 2007). Moreover, persistent donor dependence in countries such as Afghanistan, Bangladesh, Bhutan, Myanmar, Nepal, and Pakistan has hindered sustainable, locally driven policy implementation due to limited institutional, human, and financial capacities.

Beyond these execution gaps, the Himalayas continue to face broader governance challenges that impede the realization of sustainability goals. The region's fragile ecosystems and complex socio-cultural diversity demand long-term and integrated governance approaches, yet most institutional arrangements remain short-term and fragmented. Despite widespread recognition of the Himalayas' ecological significance, sustainability efforts are constrained by weak institutional commitment, insufficient accountability, and inadequate resource allocation (ICIMOD, 2019). Developmental priorities in many mountain states still emphasize immediate economic gains over ecological resilience, as seen in the recurring floods and landslides across Uttarakhand and Himachal Pradesh, which expose institutional unpreparedness for environmental management and disaster risk reduction (Satendra, 2025). Fragmented policies across hydropower, tourism, and forestry sectors have produced conflicting outcomes; for instance, while hydropower projects are promoted as clean-energy alternatives, they frequently disrupt local livelihoods and forest ecosystems (Gulati & Gupta n.d.), reflecting a lack of cross-sectoral integration in planning (Ojha et al., 2019). Equally critical is the limited participation of local stakeholders. Mountain communities possess deep ecological knowledge, yet their exclusion—particularly of women and indigenous groups—has reduced the legitimacy and effectiveness of sustainability policies (Singh and Pandey, 2020). True participatory governance that integrates local and scientific knowledge remains essential for equitable resource management (Ojha et al., 2019). Political and economic constraints further complicate sustainable policy implementation. The Himalayan economies' dependence on agriculture, tourism, and natural resource extraction limits the financial investment in sustainability initiatives, while short-term political agendas often prioritize infrastructure expansion, even in ecologically sensitive zones (Times of India, 2025). Cross-border tensions over shared rivers and biodiversity add another layer of complexity to collaborative governance (ICIMOD, 2019).

Ultimately, the persistent tension between short-term development ambitions and long-term sustainability continues to weaken the region's environmental resilience. Expanding road networks and tourism infrastructure bring immediate economic benefits but simultaneously exacerbate deforestation, waste generation, and vulnerability to natural disasters (Satendra, 2025). Without an integrated, forward-looking, and resilience-based policy framework, the Himalayas will remain increasingly susceptible to the combined pressures of climate change, institutional fragmentation, and unsustainable growth (ICIMOD, 2019).

6 Collaborative Governance and Institutional Accountability for Sustainable Development in the Himalayas

The Himalayan region's transboundary significance, sociocultural diversity, and delicate ecosystem make it a special governance issue. Governmental, business, civil society, and community institutions all have specific roles to play in ensuring the sustainability of the area. Governmental organizations are tasked with creating and carrying out policies that strike a balance between development objectives and ecological protection. History demonstrates that local concerns were frequently overlooked by top-down efforts. A grassroots response to state-sponsored commercial logging that endangered forests and livelihoods, for example, led to the Chipko Movement in the 1970s, which ultimately forced the government to prohibit green felling in the Himalayan region for 15 years (Guha, 1989; Frontline, 2022). By elevating the voices of the oppressed and holding institutions responsible, civil society organizations play a crucial mediating role. Establishing itself in Dehradun in 1996, the Society for the Promotion of Himalayan Indigenous Activities (SOPHIA) has continuously assisted the Van Gujjar pastoralists in obtaining forest rights under the 2006 Forest Rights Act, demonstrating how NGOs can help close the gap between people and policy (SOPHIA, 2023). Similar to this, the Centre for People's Forestry has promoted community-based natural resource management through participatory forest governance (CPF, 2023). Local communities provide traditional knowledge and adaptive techniques, making them essential stewards of Himalayan ecosystems. In the Chipko Movement, women were particularly important because of their leadership in embracing trees to oppose their felling, which showed both gendered agency in governance and ecological dedication (Shiva & Bandyopadhyay, 1986; FAO, 1997). Such instances emphasize the significance of diversity and show how grassroots activity changes institutional agendas.

In the Himalayas, multi-stakeholder partnerships are becoming more and more important for sustainable governance. This dynamic is demonstrated by the Mid-Himalayan Watershed Development Project in Himachal Pradesh, which brought together the World Wildlife Fund, foreign organizations, Gram Panchayats, and state officials for an impartial review. In addition to increasing openness, this cooperative design improved community involvement and responsibility (World Bank, 2012).

Despite this, the Himalayan governance still faces long-standing injustices and power imbalances. Even though marginalized groups are disproportionately at risk from environmental degradation, they are frequently left out of decision-making processes, particularly women, indigenous communities, and the rural poor. In order to achieve sustainability goals, policymakers are reminded by Chipko's legacy that fair representation and institutional accountability must be ingrained in governance systems (Guha, 1989; PolSci Institute, 2023).

7 Integrated and Adaptive Governance Strategies for Sustainable Development in the Himalayas

The Himalayan region offers a crucial perspective for comprehending the necessity of integrated and adaptive governance because of its delicate ecosystems, cultural variety, and socioeconomic vulnerabilities. Economic, social, and environmental policies frequently function in isolation, leading to disjointed strategies that fall short of tackling the region's interrelated problems. For example, economic growth projects like hydroelectric dams might threaten social cohesion and ecological stability, while environmental regulations focused at forest conservation may clash with the demands of the local population. Therefore, a conscious effort to combine these sectors is necessary for effective administration in the Himalayas, guaranteeing that social justice, environmental preservation, and economic development are pursued in harmony rather than at odds.

A promising approach to handling this complexity is adaptive governance. Adaptive governance places more emphasis on adaptability, ongoing learning, and adaptation to shifting social and ecological circumstances than strict, top-down policy structures. Policies in the Himalayan region must be developed to adapt to new information and changing conditions, as climate change is accelerating glacial melt, biodiversity loss, and disaster risks. This entails combining scientific knowledge with indigenous knowledge systems and strengthening institutional capacity for observation, feedback, and modification.

Furthermore, the Himalayan governance paradigm needs to change from one of regulatory control to one of cooperation and co-creation. Although official agencies, scholars, local communities, and civil society organizations all have an interest in the region's sustainable future, their contributions are sometimes overlooked during formal policy processes. Participatory watershed management, community-led forest protection, and cross-border collaboration on river basin management are examples of collaborative governance platforms that show how multi-stakeholder engagement can provide more resilient and equitable results. Governance in the Himalayas may become a transformative instrument for sustainability, resilience, and justice by overcoming institutional barriers and giving local voices more clout.

8 Toward Adaptive and Collaborative Governance for Sustainable Development in the Himalayas

Strengthening governance for sustainable development in the Himalayan region requires a comprehensive and integrated approach that aligns institutional, technological, and community-based strategies. To begin with, institutional accountability must be reinforced through the establishment of transparent systems for monitoring and assessment, the implementation of clear accountability structures, and the introduction of community-based auditing and social accountability mechanisms. Such measures would ensure effective policy execution and promote public trust in governance processes. Equally important is the need to enhance inter-agency coordination among national, state, and local institutions. Clearly defining institutional roles and responsibilities can minimize duplication of efforts and foster collaborative mechanisms for policy implementation. Inclusive decision-making systems should also be prioritized by involving youth, women, indigenous groups, and local communities in governance structures. Incorporating traditional ecological knowledge into policy frameworks will not only strengthen local participation but also enrich contemporary sustainability practices with indigenous wisdom.

Furthermore, efforts should be made to integrate traditional knowledge with modern scientific practices to develop context-specific, sustainable solutions. Recognizing and preserving traditional ecological wisdom while integrating it with scientific advancements can generate innovative approaches that are both culturally rooted and environmentally adaptive. At the policy level, it is crucial to mainstream sustainability across all sectors, including tourism, agriculture, health, education, and infrastructure. District and state development plans must align with the Sustainable Development Goals (SDGs), emphasizing long-term ecological resilience over short-term economic gains. The role of technology in promoting transparent and efficient governance cannot be overstated. Digital platforms for policy feedback, participatory monitoring, and data sharing can significantly enhance accountability and citizen engagement. Similarly, the use of remote sensing, GIS mapping, and early warning systems can strengthen environmental management and disaster preparedness, particularly in addressing the region's vulnerability to landslides, floods, and glacial hazards. Robust disaster preparedness mechanisms should focus on community-based risk reduction, resilient infrastructure development, and capacity-building programs for local stakeholders. To advance sustainable governance further, multi-stakeholder partnerships must be encouraged. Collaboration among government agencies, academia, the private sector, and civil society can facilitate knowledge exchange, joint research initiatives, and the mobilization of technical and financial resources. Aligning regional policies with global sustainability agendas is also essential, ensuring that the SDGs are contextualized to the specific socio-ecological realities of the Himalayan states. Establishing reporting frameworks consistent with international sustainability metrics will help track progress effectively.

Lastly, governance structures in the region should evolve toward adaptive and collaborative models. Moving beyond rigid, top-down systems toward flexible, learning-oriented frameworks will enable institutions to respond dynamically to changing environmental and social conditions. Cross-border and interstate cooperation is equally vital for managing shared ecosystems and ensuring collective responsibility for sustainable development in the Himalayas.

9 Conclusion

In the fragile yet vital ecosystems of the Himalayas, governance emerges not merely as a regulatory apparatus but as a transformative driver of sustainable development. The region's unique ecological sensitivities, coupled with its socio-cultural and economic complexities, demand governance approaches that move beyond traditional policy enforcement. Instead, they must foster collaborative frameworks that integrate the voices of local communities, academic institutions, policymakers, and civil society. By embracing inclusivity, governance in the Himalayas can ensure that development pathways respect cultural traditions while addressing pressing needs such as livelihood security, climate adaptation, and biodiversity conservation. Equally, resilience must be embedded within governance systems, enabling them to adapt to unpredictable environmental shifts, economic volatility, and social disparities that characterize the region. Ultimately, the Himalayan experience underscores that sustainable futures cannot be dictated top-down. They must be co-created through partnerships, knowledge-sharing, and adaptive institutions that are responsive to both global imperatives and local realities. Governance, therefore, becomes a platform for collective responsibility—anchoring equity, collaboration, and resilience as the essential pillars of sustainable development in one of the world's most critical ecological landscapes.

References

Agrawal, N. K., Krishnan, A., Leikanger, I. C. P., Sherpa, J. D., Choudhury, D., & Pantha, R. H. (2016). *Action for mountain adaptation: Solutions beyond the boundaries of science, policy, and practice.* ICIMOD.

Andrade, A., Córdoba, R., Dave, R., Girot, P., Herrera-F., B., et al. (2011). Draft principles and guidelines for integrating ecosystem-based approaches to adaptation in project and policy design: A discussion document. CEM/IUCN, CATIE.: Commission on ecosystem management, IUCN.

Badola, R., Barthwal, S., & Hussain, S. A. (2015). Assessing the effectiveness of policies in sustaining ecosystem services in the Indian Himalayas. *Journal of Mountain Science, 12*(2), 338–353.

Badola, R., Hussain, S. A., Dobriyal, P., & Barthwal, S. (2014). *An integrated approach to reduce the vulnerability of local communities to environmental degradation in Western Himalayas.* Study report.

Barbier, E. B. (2007). Valuing ecosystem services as productive inputs. *Economic Policy, 22*, 178–229. https://doi.org/10.1111/j.1468-0327.2007.00174.x

Birch, J. C., Thapa, I., Balmford, A., Bradbury, R. B., Brown, C., Butchart, S. H., et al. (2014). What benefits do community forests provide, and to whom? A rapid assessment of ecosystem services from a Himalayan forest, Nepal. *Ecosystem Services, 8*, 118–127. https://doi.org/10.1016/j.ecoser.2014.03.005

Blaikie, P. M., & Muldavin, J. S. S. (2004). Upstream, downstream, China, India: The politics of environment in the Himalayan region. *Annals of the Association of American Geographers, 94*(3), 520–548. https://doi.org/10.1111/j.1467-8306.2004.00412.x

CBD, & UNEP-WCMC. (2012). *Best policy guidance for the integration of biodiversity and ecosystem services in standards.* Secretariat of the Convention on Biological Diversity and the United Nations Environment Programme-World Conservation Monitoring Centre.

Chettri, N., Shakya, B., Thapa, R., & Sharma, E. (2008). Status of a protected area system in the Hindu Kush-Himalayas: An analysis of PA coverage. *International Journal of Biodiversity Science and Management, 4*, 164–178.

CPF. (2023). *Centre for People's forestry: Mission and activities.* CPF.

Dale, V. H., Joyce, L. A., McNulty, S., Neilson, R. P., Ayres, M. P., Flannigan, M. D., Hanson, P. J., Irland, L. C., Lugo, A. E., Peterson, C. J., et al. (2001). Climate change and forest disturbances. Bioscience, 51(9), 723–734. doi: https://doi.org/10.1641/0006-3568(2001)051[0723:CCAFD]2.0.CO;2.

Dasgupta, R., Hashimoto, S., & Basu, M. (2023). Ecosystem services as systemic enablers for sustainable development. *Ecosystem Services, 61*, 101520.

Dyurgerov, M. B., & Meier, M. F. (2005). *Glaciers and the changing earth system: A 2004 snapshot.* Institute of Arctic and Alpine Research, University of Colorado.

FAO. (1997). Women in forestry: The Chipko movement in India. Food and Agriculture Organization of the United Nations.

Foggin, J. M. (2008). Depopulating the Tibetan grasslands: National policies and perspectives for the future of Tibetan herders in Qinghai Province, China. *Mountain Research and Development, 28*(1), 26–31.

Foggin, J. M. (2018). Environmental conservation in the Tibetan plateau region: Lessons for China's Belt & Road Initiative in the mountains of Central Asia. *Land, 7*(2), 52. http://www.mdpi.com/2073-445X/7/2/52

Friend, R., Jarvie, J., Reed, S. O., Ratri, S., Pakamas, T., & Canh, T. V. (2013). Mainstreaming urban climate resilience into policy and planning: Reflections from Asia. *Urban Climate, 7*, 6–19.

Frontline. (2022). 1973: Chipko Movement starts. Frontline, The Hindu.

Gautam, M. R., Timilsina, G. R., & Acharya, K. (2013). *Climate change in the Himalayas: Current state of knowledge. Policy research working paper WPS6516*. The World Bank Development Research Group, Environment and Energy Team.

Ghosh, D. K. (2008). Governance at local level: The case of panchayats in West Bengal. *The Indian Journal of Political Science, 69*(1), 71–88.

Government of India. (2010). National Mission for sustaining the Himalayan ecosystem under National Action Plan on climate change. http://dst.gov.in/scientificprogramme/NMSHE_June_2010.pdf.

Guha, R. (1989). *The unquiet woods: Ecological change and peasant resistance in the Himalaya.* University of California Press.

Gulati, A. K., & Gupta, H. K. (n.d.). An analysis of policy framework for mountain development in the Northwest Himalayas,. FAO. https://www.fao.org/4/XII/0759-C1.htm

ICIMOD. (2012). *A strategy and results framework for ICIMOD.* ICIMOD.

ICIMOD. (2019). *The Hindu Kush Himalaya assessment: Mountains, climate change, sustainability and people.* Springer & ICIMOD.

IHCAP-DST-SDC. (2018). Climate vulnerability assessment for the Indian Himalayan region. Department of Science & Technology, Government of India.

INCCA. (2010). *Climate change and India: A 4x4 assessment*. Ministry of Environment and Forests, Government of India. http://www.moef.nic.in/downloads/public-information/fin-rpt-incca.pdf

Ives, J. (2013). *Sustainable mountain development—Getting the facts right*. Himalayan Association for the Advancement of Science.

Jodha, N. S. (2000). Poverty alleviation and sustainable development in mountain areas: Role of highland–lowland links in the context of rapid globalisation. In M. Banskota, T. S. Papola, & J. Richter (Eds.), *Growth, poverty alleviation, and sustainable resource management in the mountain areas of South Asia* (pp. 541–570). Deutsche Stiftung fur internationale Entwicklung.

Jodha, N. S. (2005a). Adaptation strategies against growing environmental and social vulnerabilities in mountain areas. *Himalayan Journal of Sciences, 3*(5), 33–42.

Jodha, N. S. (2005b). Economic globalisation and its repercussions for fragile mountains and communities in the Himalayas. In U. M. Huber, H. K. M. Bugmann, & M. A. Reasoner (Eds.), *Global change and mountain regions: An overview of current knowledge* (pp. 583–591). Springer Netherlands.

Khan, M. R., & Hossain, A. (2006). *Regulatory vs participatory governance and environmental sustainability in Asia.* North South University.

Khan, N. A., & Harriss-White, B. (2012). The ghost in the machine: Deconstructing forest policy discourse in Bangladesh. *Economic and Political Weekly, 47*(17), 100–110.

Larson, A. M., & Soto, F. (2008). Decentralization of natural resource governance regimes. *Annual Review of Environment and Resources, 33*, 213–239.

Liu, X., & Chen, B. (2000). Climatic warming in the Tibetan plateau during recent decades. *International Journal of Climatology, 20*(14), 1729–1742. https://doi.org/10.1002/1097-0088(20001130)20:14<1729::AID-JOC556>3.0.CO;2-Y

Naaz, S., & Sahu, A. (2025). Revisiting governance of Van panchayats: Experiences from Pauri Garhwal, Uttarakhand. *The Indian Journal of Social Work, 80, 525.*

Nightingale, A. J. (2016). Adaptive scholarship and situated knowledges? Hybrid methodologies and plural epistemologies in climate change adaptation research. *Area, 48*(1), 41–47.
Ojha, H., et al. (2019). Governance for sustainability in the Hindu Kush Himalaya. In *The Hindu Kush Himalaya assessment* (pp. 567–601). Springer.
Ojha, H. R., Ghimire, S., Pain, A., Nightingale, A., Khatri, D. B., & Dhungana, H. (2016). Policy without politics: Technocratic control of climate change adaptation policy making in Nepal. *Climate Policy, 16*(4), 415–433.
Ostrom, E., Burger, J., Field, C. B., Norgaard, R. B., & Policansky, D. (1999). Revisiting the commons: Local lessons, global challenges. *Science, 284*(5412), 278–282. https://doi.org/10.1126/science.284.5412.278
Pagiola, S., von Ritter, K., & Bishop, J. (2004). Assessing the economic value of ecosystem conservation. Environment department paper no. 101.: World Bank environment department.
Pokharel, B. K., Branney, P., Nurse, M., & Malla, Y. B. (2007). Community forestry: Conserving forests, sustaining livelihoods and strengthening democracy. *Journal of Forest and Livelihood, 6*(2), 8–19.
PolSci Institute. (2023). Chipko movement: Grassroots environmentalism in the Himalayas
Price, M. F. (2015). *Mountains: A very short introduction.* Oxford University Press.
Rana, S., Joshi, M., & Sharma, V. (2022). Scientific research in the Himalaya: Current state and policy needs. *International Journal of Environmental Research and Public Health, 19*(8), 4562.
Regmi, B. R., & Star, C. (2015). Identifying operational mechanisms for the mainstreaming of climate change adaptation in Nepal. *Climate and Development, 6*(4), 306–317.
Satendra, S. (2025, August 19).: Need for strong monitoring, response, and governance. The Pioneer. https://www.dailypioneer.com/2025/columnists/himalayas--need-for-strong-monitoring--response--and-governance.html
Saxena, K. G., Rao, K. S., Sen, K. K., Maikhuri, R. K., & Semwal, R. L. (2001). Integrated natural resource management: Approaches and lessons from the Himalaya. *Conservation Ecology, 5*(2), 14.
Semwal, R. L., Nautiyal, S., Rao, K. S., Maikhuri, R. K., & Bhandari, B. S. (1999). Structure of forests under community conservation: A preliminary study of Jardhar village initiative in Garhwal Himalaya. *ENVIS Bulletin: Himalayan Ecology & Development, 7*, 16–27.
Sharma, E., & Chettri, N. (2005). ICIMOD's transboundary biodiversity management initiative in the Hindu Kush-Himalayas. *Mountain Research and Development, 25*(3), 278–281.
Sharma, E., Chettri, N., & Oli, K. P. (2010). Mountain biodiversity conservation and management: A paradigm shift in policies and practices in the Hindu Kush-Himalayas. *Ecological Research, 25*(5), 909–923.
Shiva, V., & Bandyopadhyay, J. (1986). *Chipko: India's civilizational response to the forest crisis.* Gandhi Peace Foundation.
Singh, P., & Pandey, A. (2020). Impact of institutional overlapping on water governance of Himalayan city, Nainital, India. *Water Policy, 22*(4), 601–618.
Singh, S. P. (2002). Balancing the approaches of environmental conservation by considering ecosystem services as well as biodiversity. *Current Science, 82*(11), 1331–1335.
Singh, S. P., Bassignana-Khadka, I., Karky, B. S., & Sharma, E. (2011). *Climate change in the Hindu Kush-Himalayas: The state of current knowledge.* International Centre for Integrated Mountain Development.
Smith, T. (2014). *Climate vulnerability in Asia's high mountains: How climate change affects communities and ecosystems in Asia's water towers* (pp. 1–104). World Wildlife Fund.
SOPHIA. (2023). *Society for the promotion of Himalayan indigenous activities (SOPHIA) annual report*. SOPHIA.
TEEB (The Economics of Ecosystems and Biodiversity). (2010). The economics of ecosystems and biodiversity for local and regional policy makers. http://www.teebweb.org/wp-content/uploads/StudyReport/TEEB

Times of India. (2025, August 9). Experts: Himalayas on slippery slope, fault lines in master plan. The Times of India. https://timesofindia.indiatimes.com/city/chandigarh/experts-himalayas-on-slippery-slope-fault-lines-in-master-plan/articleshow/123196273.cms

Timsina, N. P., & Gotame, B. (2014). *Political economy of forestry sector of Nepal: Analysis of actors' engagement and policy processes*. ForestAction.

Tiwari, P. C. (2015). Local and regional institutions and environmental governance in Hindu Kush Himalaya. *Environmental Science & Policy, 49*, 66–76.

Underdal, A. (2010). Complexity and challenges of long term environmental governance. *Global Environmental Change, 20*(3), 386–393. https://doi.org/10.1016/j.gloenvcha.2010.02.005

UNEP. (2002). *Civil society statement on international environmental governance* (p. 2002). In Seventh special session of the UNEP Governing Council/GMEF.

UNEP. (2008). International environmental governance and the reform of the United Nations. In *XVI Meeting of the Forum of Environment Ministers of Latin America and the Caribbean.*

WHO (World Health Organization). (2012). *Atlas of health and climate*. WHO.

World Bank. (2012). *Himachal Pradesh mid-Himalayan watershed development project: Project signing press release*. The World Bank.

Wulf, H., Bookhagen, B., & Scherler, D. (2012). Climatic and geologic controls on suspended sediment flux in the Sutlej River valley, western Himalaya. *Hydrology and Earth System Sciences, 16*, 2193–2217.

Zhang, B., & Cao, C. (2015). Four gaps in China's new environmental law: Implementation and accountability will remain challenging, especially at the local level. *Nature, 517*(7535), 433–435.

MIX
Papier aus verantwortungsvollen Quellen
Paper from responsible sources
FSC® C105338

If you have any concerns about our products,
you can contact us on
ProductSafety@springernature.com

In case Publisher is established outside the EU,
the EU authorized representative is:
Springer Nature Customer Service Center GmbH
Europaplatz 3, 69115 Heidelberg, Germany

Printed by Libri Plureos GmbH
in Hamburg, Germany